AWS Certified Advanced Networking – Specialty (ANS-C01) Certification Guide

A pragmatic guide to acing the AWS ANS-C01 exam

Tim McConnaughy

Steve McNutt

Christopher Miles

AWS Certified Advanced Networking – Specialty (ANS-C01) Certification Guide

Authors: Tim McConnaughy, Steve McNutt, and Christopher Miles

Reviewer: Vivek Velso

Publishing Product Manager: Sneha Shinde

Development Editor: Akanksha Gupta

Digital Editor: M Keerthi Nair

Presentation Designer: Shantanu Zagade

Editorial Board: Vijin Boricha, Megan Carlisle, Ketan Giri, Saurabh Kadave, Aaron Nash, Alex Mazonowicz, Gandhali Raut, and Ankita Thakur

First Published: January 2025

Production Reference: 1300125

Published by Packt Publishing Ltd.

Grosvenor House

11 St Paul's Square

Birmingham

B3 1RB

ISBN: 978-1-83508-083-2

www.packtpub.com

Contributors

About the Authors

Tim McConnaughy is a Technical Marketing Engineer with 15 years of experience working in networking. He has worked on some of the world's largest networks as a contractor, employee, or consultant. At Aviatrix, he creates technical whitepapers, demonstrations, and other collateral showcasing the power of cloud networking and blending traditional and cloud networking expertise. Prior to Aviatrix, Tim worked at Cisco on Cisco's SD-WAN solution in the customer proof of concept (CPOC) labs and as a dCloud developer. He has spoken at Cisco Live and written several books. He is also a co-host of the Cables2Clouds Podcast, focusing on hybrid cloud networking.

Tim currently lives in Raleigh, North Carolina with his wife and two daughters, two cats, and a dog. When not studying or helping others with networking, he spends time with his family, studies Japanese, and plays video games.

LinkedIn profile: `https://www.linkedin.com/in/tmcconnaughy/`

Steven McNutt is an educator, speaker, and cybersecurity solutions engineer. He holds over 20 industry certifications and has a master's degree in cybersecurity and information assurance. He is a multiple time distinguished speaker at Cisco Live and co-author of the Cisco XDR training course.

Steven lives with his wife Caroline and their two dogs in Western North Carolina and is an avid mountain biker and snowboarder.

LinkedIn profile: `https://www.linkedin.com/in/smcnutt/`

Christopher Miles brings over 13 years of deep networking expertise to this certification guide. As a Cisco Certified Internetwork Expert (CCIE) in Enterprise Infrastructure and a Senior Solutions Engineer at Aviatrix, Chris specializes in multicloud networking and security architectures. With a robust background spanning traditional enterprise networking technologies like SD-WAN, MPLS, and L3VPN, he has successfully navigated the complex transition to cloud networking.

Passionate about knowledge sharing, Chris is a respected content creator in the networking community. He regularly contributes insights through his blog, The Control Plane, and co-hosts the Cables2Clouds Podcast, which explores hybrid cloud networking trends and technologies.

Currently based in Sydney, Australia, Chris brings a global perspective to networking challenges, balancing his professional expertise with an active lifestyle that includes team sports and skateboarding.

LinkedIn profile: `https://www.linkedin.com/in/christopherleemiles/`

About the Reviewer

Vivek Velso is an AWS Ambassador with over 25 years of industry experience. He helps startups and enterprises build scalable cloud infrastructure, migrate applications, and adopt IaC and DevOps practices. He holds all 15 AWS certifications and is a subject matter expert (SME) for AWS certification exams and an active contributor to AWS forums. Vivek shares his expertise through insightful blogs and innovative cloud solutions.

Table of Contents

3

Networking Across Multiple AWS Accounts   75

4

AWS Direct Connect   91

5

Hybrid Networking with AWS Transit Gateway   125

6

Connecting Third-Party Networks to AWS   157

9

AWS Elastic Load Balancing   285

10

AWS CDN and Global Traffic Management   303

11

Security Framework   339

14

Data Analytics and Optimization   411

15

Conclusion   437

16

Accessing the Online Practice Resources   445

Appendix 1

Network Fundamentals   451

Preface

The cloud has been made famous by the apps that run there. Application developers and IT infrastructure professionals have embraced public clouds such as **Amazon Web Services** (**AWS**) to build highly available and resilient resources. However, just like how IT has functioned for the last few decades, when these services are not functioning, who gets blamed first? The network.

The network engineers of today have been building scalable and resilient networks for decades and the networks running within AWS cloud environments are no different. AWS cloud computing teams need engineers and architects with strong networking capabilities to support the design and deployment of networks just as they do on-premises. However, networking in the public cloud differs from the way things are done on-premises. Network operators need to be able to effectively build and operate complex networks within AWS to function as an extension of the enterprise networks they are responsible for.

This is the purpose of the **AWS Certified Advanced Networking – Specialty** (**ANS-C01**) exam: to demonstrate the skills and knowledge to build and maintain AWS cloud networks. As defined by the blueprint for this exam, individuals looking to take this exam should be able to perform the following tasks:

- Design and deploy core networking functionality within and across AWS accounts
- Implement hybrid networking solutions to connect with traditional networks and third-party networks
- Integrate other core AWS services with networking solutions for functionality with DNS, load balancing, and content distribution
- Implement security within AWS networks utilizing native services and constructs
- Deploy AWS network solutions using **infrastructure as code** (**IaC**) for optimization and automation
- Operate AWS networks using native analysis tools and services

The goal of this book is to prepare candidates specifically for the *ANS-C01* version of the AWS Certified Advanced Networking – Specialty exam. This book will not only cover the topics outlined in the exam blueprint but also provide real-world examples about operating AWS cloud networks from the experience of the authors.

Who This Book Is For

This guide is meticulously crafted for experienced cloud networking professionals, network architects, and solution designers who are committed to mastering advanced networking concepts within the **Amazon Web Services (AWS)** ecosystem. It is specifically tailored to support candidates preparing for the *ANS-C01* examination, providing comprehensive coverage of complex networking technologies, design patterns, and implementation strategies unique to AWS infrastructure.

Networking and AWS Constructs You Should Already Know

This certification guide is going to make some assumptions about what you should already know with respect to computer networking and AWS services.

When going by the AWS blueprint for the *ANS-C01* exam, there is an expectation that candidates need to be very comfortable working with computer networking. The following description is taken directly from the *AWS Certified Advanced Networking – Specialty (ANS-C01)* exam guide:

The target candidate possesses the knowledge, skills, experience, and competence to design, implement, and operate complex AWS and hybrid networking architectures. The target candidate is expected to have 5 or more years of networking experience with 2 or more years of cloud and hybrid networking experience.

> **Note**
> The details for the AWS Certified Advanced Networking – Specialty exam, including the blueprint, can be found here: `https://aws.amazon.com/certification/certified-advanced-networking-specialty/`.

Here are the key network fundamentals that you should have a strong level of comfort with before proceeding with this chapter. If any of the topics below are unfamiliar to you, please refer to the appendices to develop a baseline on these concepts:

- The OSI layer - (*Appendix 1, Network Fundamentals*)
- IP Addressing fundamentals (v4 and v6) - (*Appendix 1, Network Fundamentals*)
- IP Subnetting & CIDR notation - (*Appendix 1, Network Fundamentals*)
- Public vs Private IP Addressing - (*Appendix 1, Network Fundamentals*)
- IP Routing - (*Appendix 2, IP Routing Fundamentals*)
- Dynamic Routing (for later chapters) - (*Appendix 2, IP Routing Fundamentals*)
- BGP Basics (for later chapters) - (*Appendix 2, IP Routing Fundamentals*)

In addition to the above network fundamentals, there are a few AWS constructs that you will need to be familiar with to proceed. These items are typically covered within another certification track, like the AWS Certified Cloud Practitioner. Additional reading for each of these components will be linked below.

Here are the AWS constructs you should already be familiar with before proceeding:

- AWS Regions and Zones

- Amazon Simple storage service (S3)

- Amazon Elastic cloud compute (EC2)

- Amazon Elastic block store (EBS)

- AWS Service Quotas

> **Note**
>
> Further detail on the above mentioned AWS topics can be found at the links below
>
> https://docs.aws.amazon.com/AWSEC2/latest/UserGuide/using-regions-availability-zones.html
>
> https://docs.aws.amazon.com/AmazonS3/latest/userguide/Welcome.html
>
> https://docs.aws.amazon.com/AWSEC2/latest/UserGuide/concepts.html
>
> https://docs.aws.amazon.com/AWSEC2/latest/UserGuide/storage_ebs.html
>
> https://docs.aws.amazon.com/servicequotas/latest/userguide/intro.html

AWS Certified Advanced Networking – Specialty Exam Outline

The *ANS-C01* exam is divided into separate domains. Each domain includes specific statements that outline what candidates should be able to demonstrate. This outline briefly covers each domain and how these statements should be interpreted by candidates for the exam.

Test Domains

The *ANS-C01* exam is divided into four test domains. These domains identify, at a high level, areas of expertise that the candidate will be tested on. You can see in the following table that each domain holds a significant amount of weight on the exam; therefore, the candidate needs to make sure they can demonstrate their knowledge and skills adequately in each domain. These weights only represent the scored content on the exam.

Domain	Weight (%)
Domain 1: Network Design	30%
Domain 2: Network Implementation	26%
Domain 3: Network Management and Operation	20%
Domain 4: Network Security, Compliance, and Governance	24%
Total	**100%**

Table 0.1: Test domain weighting

Task Statements

Each domain contains a set of task statements relating to the content covered on the exam. While not a comprehensive list of the content of the exam, each task statement on the blueprint provides additional context to help candidates prepare for the exam. The additional context per task statement identifies items that the candidate will need **knowledge of** versus **skills in**. As the reader, you can logically separate these two statements as theory versus hands-on experience.

For areas that require **knowledge of**, you will be expected to demonstrate your knowledge of how that item fits into an overall solution. This may require detailing the components of the technology/service being asked about and/or what use case it can address. For areas that require **skills in**, you will be expected to demonstrate your knowledge of how to configure, validate, or troubleshoot a selected technology or AWS service.

For example, a **knowledge of** example for Amazon VPC would involve understanding its components, such as subnets, route tables, and NAT gateways, and how they fit together in the network architecture. A **skills in** example would involve hands-on experience configuring a NAT gateway to allow private subnet instances to access the internet while ensuring proper route table and security group settings.

> Note
>
> For a comprehensive analysis of the chapters that correspond to specific domains, please refer to the *List of Task Statements per Test Domain* section in the *Appendix 1*.

What This Book Covers

In this comprehensive guide, you'll learn everything you need to know to become an expert in AWS advanced networking. The content has been divided into strategic technology domains, covering all critical aspects of AWS networking.

Chapter 1, Advanced VPC Networking, is part of *Domain 3* and explores sophisticated VPC connectivity design patterns, including VPC peering and AWS Transit Gateway, demonstrating how to interconnect multiple VPCs effectively.

Chapter 2, VPC Traffic and Performance Monitoring, is part of *Domain 3* and showcases various traffic and performance monitoring tools available to AWS network operators, ensuring comprehensive network transparency.

Chapter 3, Networking Across Multiple AWS Accounts, is part of *Domain 2* and walks through strategies for providing network connectivity across multiple AWS accounts, addressing enterprise needs for resource separation and integrated networking.

Chapter 4, AWS Direct Connect, is part of *Domain 3* and examines AWS Direct Connect as a primary method for establishing direct connectivity between on-premises data centers and AWS cloud networks.

Chapter 5, Hybrid Networking with AWS Transit Gateway, is part of *Domain 3* and explores advanced integration techniques for connecting Direct Connect with AWS Transit Gateway architectures.

Chapter 6, Connecting Third-Party Networks to AWS, is part of *Domain 2* and investigates connectivity solutions for third-party locations, including temporary and edge networking scenarios like SDWAN integration.

Chapter 7, AWS Route 53: Basics, is part of *Domain 2* and introduces DNS management within AWS, covering domain registration and hosted zone fundamentals.

Chapter 8, AWS Route 53: Advanced, is part of *Domain 1* and delves into complex routing policies, health checks, and resolver endpoints for sophisticated DNS management.

Chapter 9, AWS Elastic Load Balancing, is part of *Domain 1* and covers the comprehensive suite of load balancing options, including Application, Network, and Gateway Load Balancers.

Chapter 10, AWS CDN and Global Traffic Management, is part of *Domain 1* and explores services like Amazon CloudFront and AWS Global Accelerator for optimizing content delivery and reducing latency.

Chapter 11, Security Framework, is part of *Domain 4* and provides a comprehensive overview of network security principles and best practices in cloud environments.

Chapter 12, AWS Security Services, is part of *Domain 4* and examines the robust portfolio of AWS security products for protecting network traffic and preventing vulnerabilities.

Chapter 13, Infrastructure as Code, is part of *Domain 1* and demonstrates the power of **Infrastructure as code (IaC)** principles in AWS networking, showcasing automation and API-driven network management.

Chapter 14, Data Analytics and Optimization, is part of *Domain 3* and presents strategies for comprehensive network traffic logging, performance optimization, and operational insights.

Chapter 15, Conclusion, outlines how you can get the most out of this book and offers guidance on building a strong learning strategy to pass the *ANS-C01* exam. It also includes instructions on how to register for the exam.

Appendix 1, Network Fundamentals, provide foundational networking concepts for professionals with less than 5 years of networking experience or those seeking a comprehensive review.

Appendix 2, IP Routing Fundamentals, provide foundational routing and **Border Gateway Protocol (BGP)** concepts and will aid you in learning how AWS networking works.

Appendix 3, VPC Networking Basics, introduce the core constructs of **Virtual Private Cloud (VPC)** and the fundamental networking component in AWS.

Appendix 4, VPC Security and External Connectivity, provides the best practices for securing VPCs against unauthorized access and configuring them for optimal security.

Authors' Assumptions

This book is meant to prepare you with the skills and knowledge to pass the AWS ANS-C01 exam. This book is *not* meant to be a be-all-and-end-all guide to the world of computer networking. The authors will be making some assumptions about the skills you'll need to be comfortable with before diving into this book.

You should be familiar with the OSI model. The OSI model is the industry standard framework for describing how applications communicate over a network. It is often the first concept taught when conducting network training to those who are brand-new to the field. This book will contain many references to different layers of the OSI model. A brief explanation of the OSI model can be found in *Appendix 1, Network Fundamentals.*

You should be comfortable with and have knowledge of layer 2 concepts such as MAC addresses and VLANs. While the cloud is often branded as "a world without layer 2" because of the lack of these concepts being present for consumers of AWS VPC, there will still be usage of these concepts. Specifically, when discussing hybrid networking with AWS, layer 2 concepts such as VLANs and MAC addresses play a large role in building connectivity to on-premises.

You should have a general understanding of network appliances and their roles. There are specific appliances that make up the proverbial stool that computer networking sits upon, such as routers, switches, and firewalls. This book will contain many references to these types of appliances and will expect you to have knowledge of what these appliances are used for. A brief explanation of these appliances can be found in *Appendix 2, IP Routing Fundamentals.*

You should have a general understanding of layer 4 protocols such as TCP and UDP. The flow and behavior of network traffic heavily depend on the transport protocol that is used. While you do not need to be an expert on TCP and UDP, this book will contain references to the general operation of these protocols and the considerations that come with them.

Use of Example Organization: Trailcats

In order to facilitate understanding of how AWS services solve real-world problems, this book makes use of a fictious business called Trailcats. Trailcats is an outfitter for people who want to take their cats hiking. Trailcats uses AWS services to scale and expand its global supply chain footprint as it grows in popularity.

Online Practice Resources

With this book, you will unlock unlimited access to our online exam-prep platform (Figure 0.1). This is your place to practice everything you learn in the book.

> **How to Access These Materials**
>
> To learn how to access the online resources, refer to *Chapter 16, Accessing the Online Practice Resources*, at the end of this book.

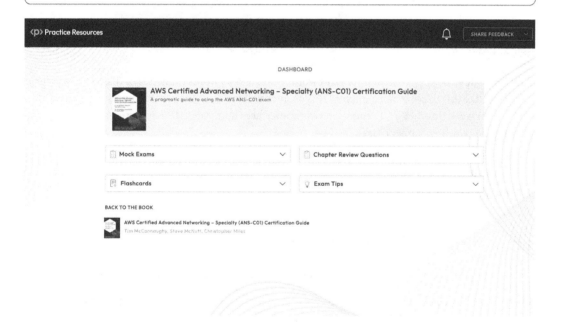

Figure 0.1: Online exam-prep platform on a desktop device

Sharpen your knowledge of ANS-C01 concepts with multiple sets of mock exams, interactive flashcards, hands-on activities, and exam tips, accessible from all modern web browsers.

Conventions Used

Several text conventions are used throughout this book.

Code Words

`Code in text`: Indicates code words in text, database table names, folder names, filenames, file extensions, pathnames, dummy URLs, and user input. Here is an example: "An ENI can be created using the AWS CLI `create-network-interface` command."

A block of code is set as follows:

```
AWSTemplateFormatVersion: "2010-09-09"
Resources:
  myvpc:
    Type: AWS::EC2::VPC
    Properties:
      CidrBlock: 10.16.0.0/16
  mysubnet:
    Type: AWS::EC2::Subnet
    DependsOn: myvpc
```

Bold: Indicates a new term or an important word. For example, words in menus or dialog boxes appear in the text like this. Here is an example: "This chapter will focus on the critical building blocks within a **virtual private cloud** (**VPC**) around network interfaces and public IP addresses, and then shift to focusing on building connectivity."

> **Tips or important notes**
> Appear like this.

Accessing the Code Files and Colored Images

You can find the complete code files of this book at `https://packt.link/9seca`.

The high-quality color images used in this book can be found at `https://packt.link/vgd38`.

Get in Touch

Feedback from our readers is always welcome.

General feedback: If you have any questions about this book, please mention the book title in the subject of your message and email us at customercare@packt.com.

Errata: Although we have taken every care to ensure the accuracy of our content, mistakes do happen. If you have found a mistake in this book, we would be grateful if you could report this to us. Please visit www.packtpub.com/support/errata and complete the form. We ensure that all valid errata are promptly updated in the GitHub repository at https://packt.link/9seca.

Piracy: If you come across any illegal copies of our works in any form on the internet, we would be grateful if you could provide us with the location address or website name. Please contact us at copyright@packt.com with a link to the material.

If you are interested in becoming an author: If there is a topic that you have expertise in and you are interested in either writing or contributing to a book, please visit authors.packtpub.com.

Share Your Thoughts

Once you've read *AWS Certified Advanced Networking – Specialty (ANS-C01) Certification Guide*, we'd love to hear your thoughts! Scan the QR code below to go straight to the Amazon review page for this book and share your feedback:

https://packt.link/r/1835080839

Your review is important to us and the tech community and will help us make sure we're delivering excellent-quality content.

Download a Free PDF Copy of This Book

Thanks for purchasing this book!

Do you like to read on the go but are unable to carry your print books everywhere?

Is your eBook purchase not compatible with the device of your choice?

Don't worry – now, with every Packt book, you get a DRM-free PDF version of that book at no cost.

Read anywhere, any place, on any device. Search, copy, and paste code from your favorite technical books directly into your application.

The perks don't stop there: you can get exclusive access to discounts, newsletters, and great free content in your inbox daily.

Follow these simple steps to get the benefits:

1. Scan the QR code or visit the link below it:

https://packt.link/free-ebook/9781835080832

2. Submit your proof of purchase.
3. That's it! We'll send your free PDF and other benefits to your inbox directly.

1

Advanced VPC Networking

The *AWS Certified Advanced Networking – Specialty (ANS-C01)* exam focuses on advanced networking. This chapter will focus on the critical building blocks within a **virtual private cloud** (**VPC**) around network interfaces and public IP addresses, and then shift to focusing on building connectivity.

Within your AWS cloud environment, there may be several scenarios where you need to build interconnectivity between your VPCs. This could be as simple as two servers in separate VPCs needing to communicate directly, or hundreds of VPCs needing access to shared services hosted in another VPC.

In addition, your VPCs could need access to other AWS services such as S3 for storage. Lastly, testing whether this connectivity is correctly configured is a crucial point of verification. Understanding how VPCs communicate is important for the exam and real-world applications, as most cloud deployments involve multiple VPCs and services.

Making the Most of This Book – Your Certification and Beyond

This book and its accompanying online resources are designed to be a complete preparation tool for your **ANS-C01 exam**.

The book is written in a way that means you can apply everything you've learned here even after your certification. The online practice resources that come with this book (*Figure 1.1*) are designed to improve your test-taking skills. They are loaded with timed mock exams, chapter review questions, interactive flashcards, case studies, and exam tips to help you work on your exam readiness from now till your test day.

> **Before You Proceed**
>
> To learn how to access these resources, head over to *Chapter 16, Accessing the Online Practice Resources*, at the end of the book.

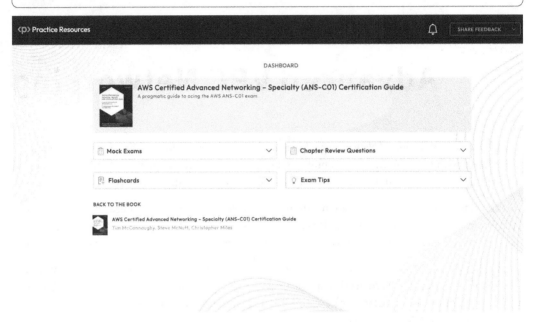

Figure 1.1: Dashboard interface of the online practice resources

Here are some tips on how to make the most of this book so that you can clear your certification and retain your knowledge beyond your exam:

1. Read each section thoroughly.

2. **Make ample notes**: You can use your favorite online note-taking tool or use a physical notebook. The free online resources also give you access to an online version of this book. Click the BACK TO THE BOOK link from the dashboard to access the book in **Packt Reader**. You can highlight specific sections of the book there.

3. **Chapter review questions**: At the end of this chapter, you'll find a link to review questions for this chapter. These are designed to test your knowledge of the chapter. Aim to score at least **75%** before moving on to the next chapter. You'll find detailed instructions on how to make the most of these questions at the end of this chapter in the *Exam Readiness Drill – Chapter Review Questions* section. That way, you're improving your exam-taking skills after each chapter, rather than at the end of the book.

4. **Flashcards**: After you've gone through the book and scored **75%** or more in each of the chapter review questions, start reviewing the online flashcards. They will help you memorize key concepts.

5.   **Mock exams**: Revise by solving the mock exams that come with the book till your exam day. If you get some answers wrong, go back to the book and revisit the concepts you're weak in.

6.   **Exam tips**: Review these from time to time to improve your exam readiness even further

This chapter will cover the following topics:

- Elastic network interfaces
- Elastic IP addresses
- Subnet configuration and optimization
- Prefix lists
- Connectivity between AWS VPCs
- IP address overlap management

By the end of the chapter, you will understand how to configure **elastic network interfaces** (ENIs) and **Elastic IP addresses** (EIPs) and manage advanced IP address configurations, including handling IP address overlaps and preventing IP address depletion. You will have learned how to optimize subnet configurations to support auto-scaling and avoid IP address depletion using secondary CIDR blocks. This chapter will also help you gain an understanding of additional VPC networking services and features, and how to utilize VPC Reachability Analyzer and other tools to test and analyze network connectivity and troubleshoot issues.

Elastic Network Interfaces (ENIs)

All Amazon EC2 instances are connected to a specific VPC. This is done using **ENIs**. Within a VPC, an ENI is a construct that represents a virtual network interface card. Just as with any computer, server, or even mobile device, a network interface card is responsible for processing network traffic in and out of that instance. **Network interface cards**, often referred to as **NICs**, are also responsible for owning both the IP address (layer 3) and MAC address (layer 2) for the infrastructure they are attached to. ENIs connect AWS virtual machines and services to a VPC at the network layer and are commonly just referred to as **network interfaces**. For the sake of this certification guide, any time you hear a reference to a network interface, you can understand that to be an ENI.

ENIs are created and attached to instances. In turn, ENIs are bound to a **single Availability Zone** (**AZ**) and belong to a single subnet. When creating an ENI, you are required to specify the VPC and subnet where the ENI will reside. In addition, you can configure any IP setting, such as dynamic/static addressing, as well as settings such as TCP/UDP idle timeout tracking.

> **Note**
> Idle timeout is the amount of time it takes for a tracked TCP or UDP session on a specific ENI to time out after the last packet was received.

Primary Network Interfaces

All EC2 instances deployed require a primary network interface. An EC2 instance must always have a primary network interface attached, and the interface cannot be detached or deleted. You can attach additional ENIs to each EC2 instance. The maximum number of permitted network interfaces is defined per EC2 instance type.

Figure 1.2 shows EC2 instances with multiple ENIs attached to separate subnets. Keep in mind that since EC2 instances cannot span AZs, this means that **all attached ENIs must also belong to a single AZ**.

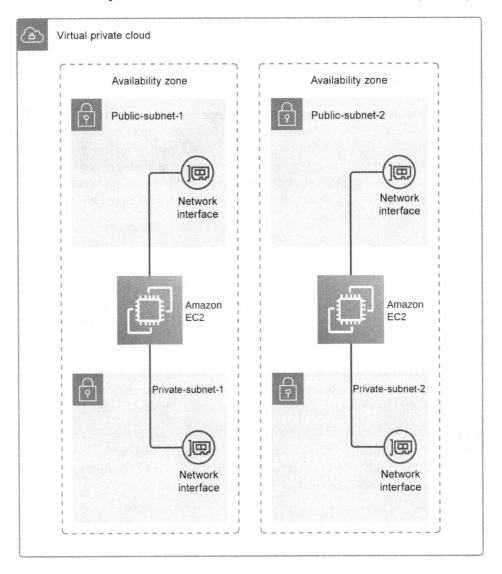

Figure 1.2: EC2 with dual ENI

As shown in the preceding diagram, Amazon EC2 instances can have multiple ENIs attached to them and these interfaces can be in different VPC subnets.

> **Note**
>
> An exhaustive list of EC2 instance types and their maximum number of supported network interfaces can be found here: `https://docs.aws.amazon.com/ec2/latest/instancetypes/ec2-instance-type-specifications.html`

Configuration of Network Interfaces

The configuration of an ENI can support several ways of assigning an IP address for the instance, but the configuration also controls certain behaviors for that network interface as well. The following sections will cover the assignment of IPv4 and IPv6 addresses and configurations for ENIs.

Auto-Assigned Public IPv4 Addresses

ENIs are attached to subnets within VPCs. Therefore, IP address on the ENI will belong to one of the CIDR blocks assigned to the VPC – more specifically, the subnet in which it resides. The IP address can be dynamically or statically assigned.

VPCs are commonly deployed with private, RFC 1918-based IPv4 address. However, subnets allow for the auto-assignment of public IPv4 addresses, as well. If an instance is deployed into a subnet that has this enabled, then a public IPv4 address is automatically associated with that instance. This public IPv4 address is automatically pulled from Amazon's pool of public IP addresses. While this public IPv4 address is associated with the instance, it is technically set up with a one-to-one **network address translation (NAT)** to the address on the primary ENI.

It is important to understand that these auto-assigned addresses are not specifically allocated to your AWS account. They are somewhat ephemeral in nature and purely active based on the state of the instance. These addresses will automatically be released if the instance is **stopped**, **hibernated**, or **terminated**. Once an auto-assigned public IPv4 address is released, you will not be able to retrieve it.

For use cases that require persistent usage of public IPv4 addresses, it is recommended to use *EIPs*, which will be covered later in this chapter – for example, if you have an app server that requires a consistent public IP address to be used so that you can associate a DNS entry or have a single IP to whitelist on your own network for this application.

Auto-Assigned Private IPv4 and IPv6 Addresses

When instances are deployed, the residing subnet may also have enabled auto-assignment of IPv4 and IPv6 addresses. This means that when the primary ENI is launched into that subnet, it will automatically be assigned an IPv4 and/or an IPv6 address from the associated CIDR block. The subnet configuration settings that automatically provision new public IPv4 or IPv6 addresses when an ENI is created in the subnet are shown in *Figure 1.3*:

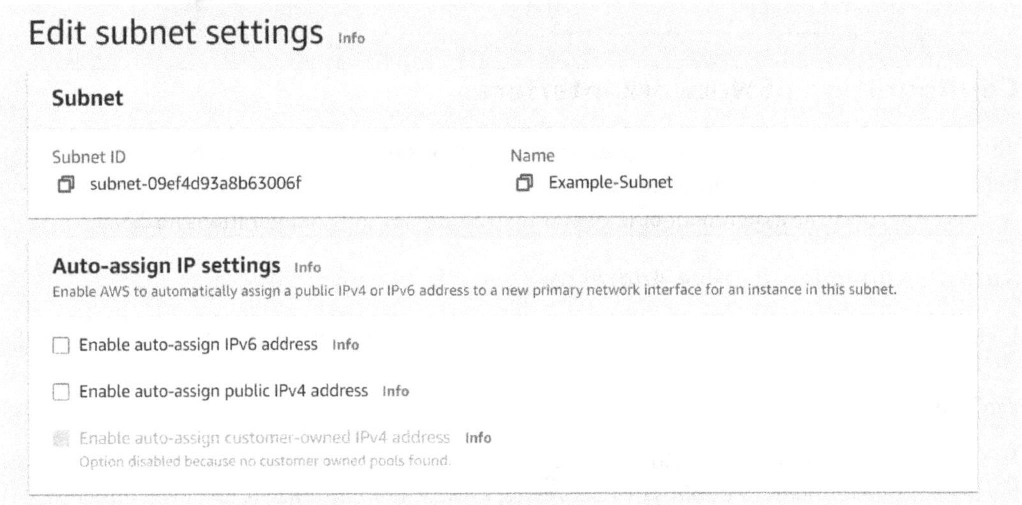

Edit subnet settings Info

Subnet

Subnet ID
📋 subnet-09ef4d93a8b63006f

Name
📋 Example-Subnet

Auto-assign IP settings Info
Enable AWS to automatically assign a public IPv4 or IPv6 address to a new primary network interface for an instance in this subnet.

☐ Enable auto-assign IPv6 address   Info

☐ Enable auto-assign public IPv4 address   Info

☐ Enable auto-assign customer-owned IPv4 address   **Info**
   Option disabled because no customer owned pools found.

Figure 1.3: Subnet auto-assign IP settings

If the boxes in *Figure 1.3* remain unchecked, no resources created within this subnet will receive automatic assignment of IPv6 addresses or IPv4 public addresses. Conversely, if every resource in this subnet should have one or both of these addresses, this setting can automate that process.

> **Note**
>
> The auto-assign IP address settings can be overridden when deploying an EC2 instance into the subnet if desired. During the launch, you can specify any specific IPv4 and IPv6 addresses you would prefer to use on the instance, assuming it is an address that belongs to the assigned CIDR blocks. You are also able to configure secondary IPs at the time of launch.

Termination Behavior

Once an ENI has been attached to an EC2 instance, you can change its **termination behavior**. This setting allows you to adjust whether the ENI will be terminated once the attached instance has been terminated. This can sometimes assist with ensuring resources are properly cleaned up if they are not needed after the parent instance is decommissioned. Refer to *Figure 1.4* for a screenshot of this setting in the AWS Management console where this setting can be configured under the VPC | Network Interfaces | Actions menu:

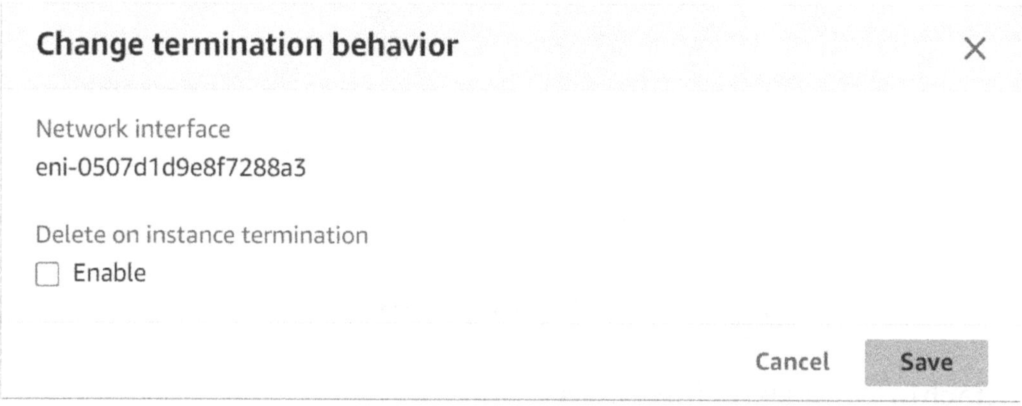

Change termination behavior   ✕

Network interface
eni-0507d1d9e8f7288a3

Delete on instance termination
☐ Enable

Cancel   Save

Figure 1.4: ENI termination behavior

The termination behavior affects whether the ENI will be deleted if its attached EC2 instance is deleted.

Source/Destination Check

By default, all ENIs have a setting enabled that performs a **source/destination check**. When this setting is enabled, the ENI will ensure that any packets processed by the ENI have the ENI's IP address in either the **source or destination field** of the IP header. This applies to both IPv4 and IPv6 traffic.

This setting must be disabled if the instance in question is performing any kind of process such as IP routing, NAT, or even firewall functions. This is common for things such as **network virtual appliances (NVAs)** because when performing a task such as IP routing or NAT, the traffic is rarely destined to go *to* that device, but *through* that device. This means that NVA devices are primarily focused on moving packets from endpoint to endpoint and are rarely the target device for data traffic. Refer to *Figure 1.5* for a look at this setting in the AWS console:

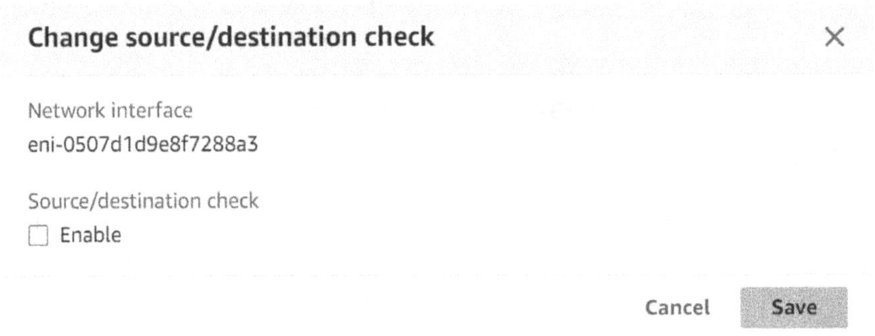

Change source/destination check   ✕

Network interface
eni-0507d1d9e8f7288a3

Source/destination check
☐ Enable

Cancel   Save

Figure 1.5: ENI source/destination check

Attaching and Detaching ENIs

As the name suggests, ENIs are elastic and hence are able to scale up or down as demand requires, allowing them to exist outside the EC2 instances to which they are assigned, and able to move between them with ease. ENIs support attachment in three different scenarios, which are classified as **hot**, **warm**, and **cold** attachments:

- **Hot attachment**: Attachment while the instance is in the *running* state
- **Warm attachment**: Attachment while the instance is *stopped*
- **Cold attachment**: Attachment when the instance is initially being *launched*

> **Note**
> Hot and warm attachments are automatically recognized by instances running Amazon Linux or Windows Server. However, other operating systems may require secondary ENIs to be configured manually.

While you can attach multiple ENIs to an instance within the same subnet, it is important to note that this **does not increase the network bandwidth** to or from the instance. These limits are still bound to those of the EC2 instance type. As an example, a **t2.micro** EC2 instance can only have two total ENIs, while an **m5.4xlarge** instance can support up to eight.

Secondary ENIs can be detached in the same fashion, regardless of whether the instance is running or stopped. As noted before, the primary ENI cannot be detached.

Configuring ENIs

To configure a network interface, you can simply navigate to the EC2 dashboard of the AWS console, select Network Interfaces, and choose the Create network interface option. As shown in *Figure 1.6*, you must give the ENI a name and assign the subnet in which to create the interface:

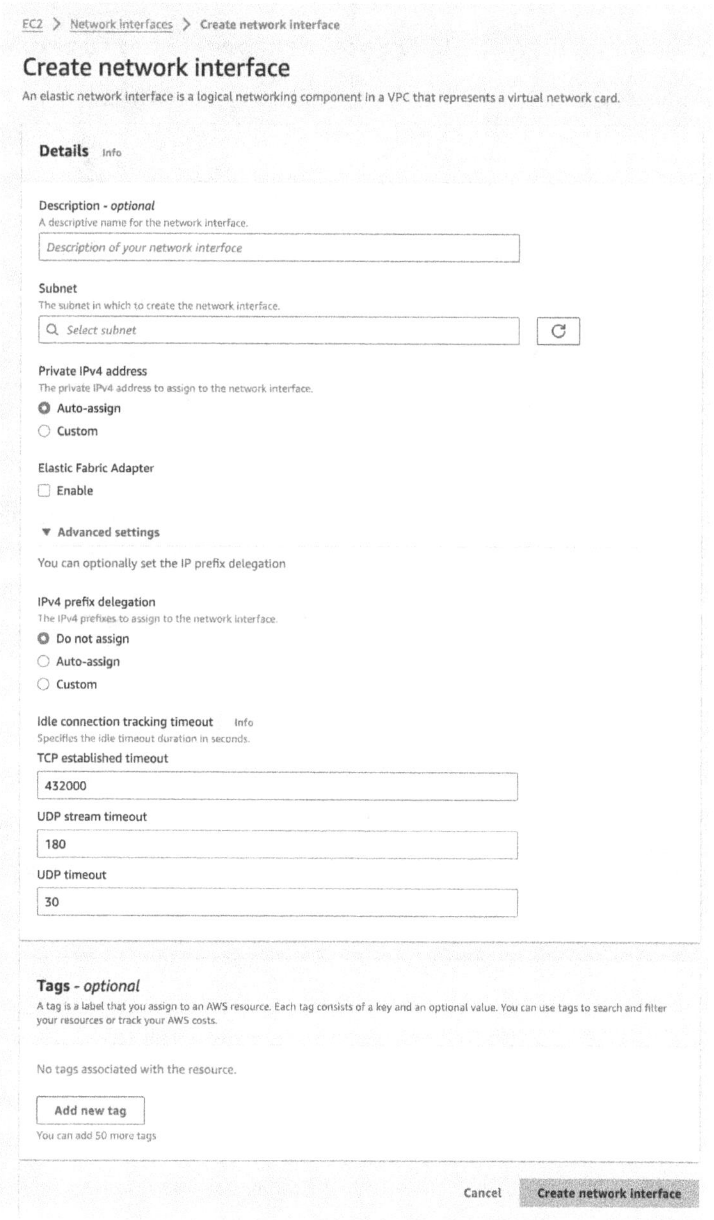

Figure 1.6: Create network interface details

Creating an ENI is independent of an EC2 instance, but the specific capabilities of an ENI must be configured. These settings will persist with the ENI regardless of what instance or service it is attached to from the time of creation.

The ENI will be detached from any EC2 instance until it is attached and can only be attached to EC2 instances that can be deployed in that subnet.

An ENI can be created using the AWS CLI `create-network-interface` command.

For example, the following command will create an ENI using the AWS CLI:

```
aws ec2 create-network-interface \
  --subnet-id subnet-12345678 \
  --description "My ENI" \
  --groups sg-12345678 sg-87654321 \
  --private-ip-address 10.0.1.10
```

AWS gives you multiple ways to create and configure resources. The AWS console is fully click-based, while the AWS CLI allows resource creation via commands or automation. The result is the same.

Elastic IP Addresses

The use of auto-assigned public IPv4 addresses can be useful, but they are somewhat ephemeral in nature. Often, you may need to allocate static IPv4 addresses for your AWS resources that are more controlled in nature and, again, elastic. You will use **EIPs** for these use cases. A perfect example of an EIP is when replacing a workload that has externally facing services that must be reachable via the same IP address; when the workload is replaced, the EIP can be migrated and associated with the new one.

An **EIP** is a public IPv4 address that is allocated and associated with your AWS account. Much like ENIs, these EIPs can be moved between resources as needed. EIPs are allocated first and then associated with specific resources. From an EC2 perspective, you can associate an EIP with either an EC2 instance or a network interface. When associating an EIP to an EC2 instance, the EIP will be associated with the IP address assigned to the primary network interface. Additionally, any EIPs assigned to secondary ENIs attached to an instance will also show up on the EC2 dashboard as being associated with the instance.

> **Note**
> EIPs can also be assigned to resources like elastic load balancers and NAT gateways.

When an EIP is reassigned from one instance to another, the public IPv4 address is reassigned and associated with the private IP of the interface on the new instance. If you recall the one-to-one NAT association that is built on the **internet gateway (IGW)**, this NAT entry is what gets updated. This process is represented in *Figure 1.7*:

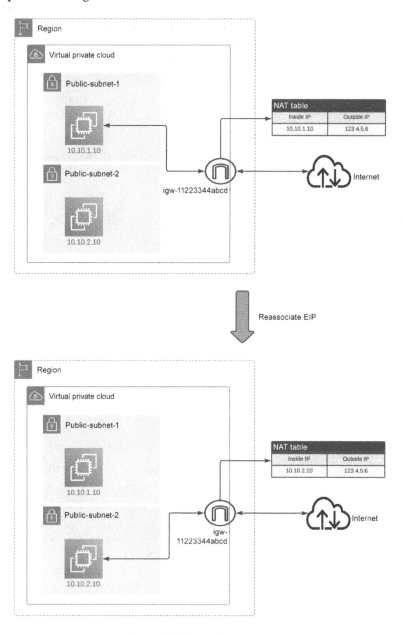

Figure 1.7: Reassociate an EIP

Reassociating an EIP is just configuring the one-to-one NAT entry on the internet gateway of the VPC.

Configuring Elastic IP Addresses

This section details the creation of an EIP both from the AWS console and using the AWS CLI. To configure an EIP, navigate to the `EC2` dashboard of the AWS console, select `Elastic IPs`, and choose the `Allocate Elastic IP address` option. As shown in *Figure 1.8*, you must give the EIP a name and define what AWS network border group to allocate the EIP from. This is a geographic representation of the AWS border and governs from which public IP address pool the IP should come.

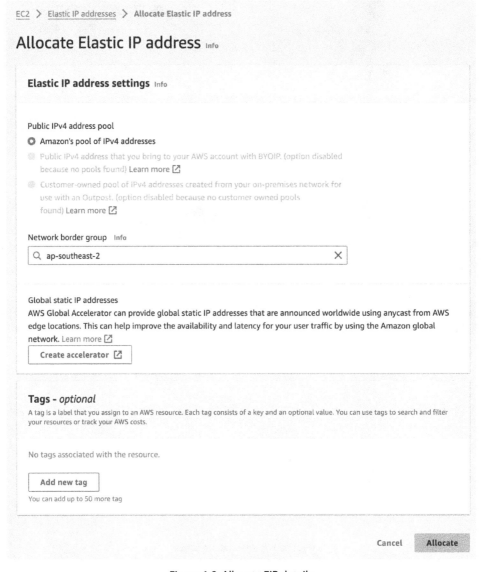

Figure 1.8: Allocate EIP details

The main choice to make with an EIP is to which regional area the IP should be allocated; this should match the geographic area in which you expect to use the IP.

Next, you will need to associate the EIP with a resource. The process of associating the EIP with either an EC2 instance or a network interface is shown in *Figure 1.9* and continued in *Figure 1.10*. *Figure 1.9* shows the AWS console menu where the association action can be selected.

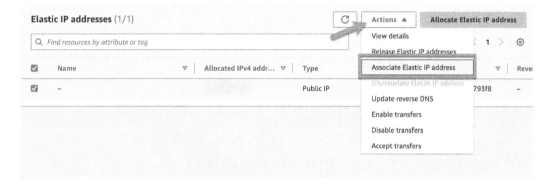

Figure 1.9: Associate Elastic IP address

This begins the process of choosing the ENI or EC2 instance with which to associate the Elastic IP.

Figure 1.10 offers two options: the EC2 instance (and its default network interface), or an ENI that is not the default network interface of an EC2 instance.

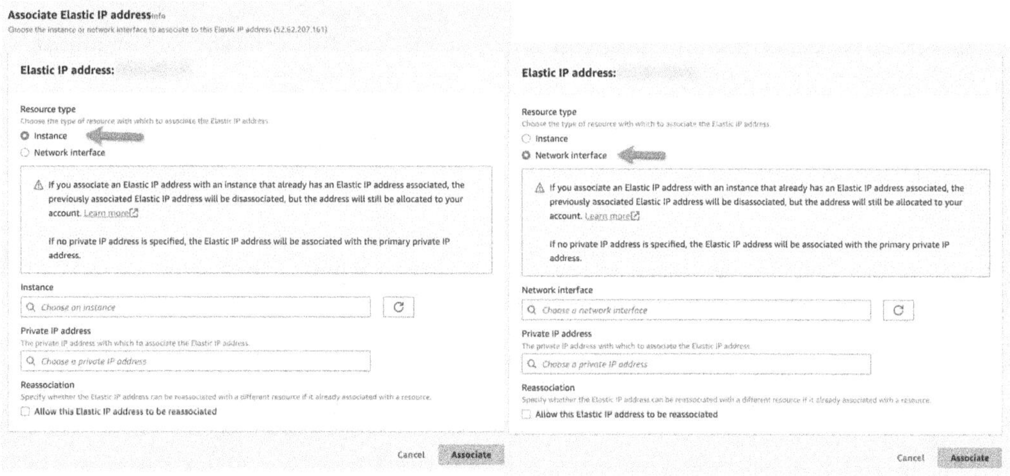

Figure 1.10: Associate EIP details

Selection of one or the other is a matter of whether the ENI will be associated with an EC2 instance that will use the interface in a dedicated fashion with the Elastic IP, or whether the EC2 instance will use its default network interface for traffic related to the associated Elastic IP.

An EIP can be created using the AWS CLI `aws ec2 allocate-address` command.

For example, the following code will allocate an EIP using the AWS CLI:

```
aws ec2 allocate-address --domain vpc
```

An EIP can be associated using the AWS CLI `aws ec2 associate-address` command.

For example, to associate an EIP using the AWS CLI, use this code:

```
aws ec2 associate-address --instance-id i-12345678
--allocation-id eipalloc-12345678
```

Subnet Configuration and Optimization

When deploying elastic and scalable solutions in the AWS cloud, you need to ensure that your VPC networks are properly configured to allow for that behavior. AWS services such as EC2 and **Elastic Load Balancing** (**ELB**) enable solutions to automatically horizontally scale dynamically based on capacity needs. This short section will detail some considerations that should be made when deploying subnets to support ELB and EC2 Auto Scaling.

> **Note**
> There will be a later chapter that dives deeper into Elastic Load Balancing and the details that come with ELB configuration.

Subnet Considerations for ELB

When deploying various types of elastic load balancers in AWS, the ELB will need to be mapped to specific subnets within the VPC that it belongs to. When this mapping is done, the ELB will have an ENI deployed into each of those subnets. These ENIs are fully managed by the ELB and cannot be moved to other resources such as an EC2 instance. However, it is important to consider that these ENIs will need to consume IP addresses out of each of the subnets as well.

This sort of mapping is common when using an ELB as the frontend connection of a website. The ELB becomes the internet-facing website DNS entry that customers connect to, and when the connections come into the ELB, the routing rules of the load balancer determine into which subnet the traffic is delivered. Delivering traffic to these subnets requires the ENI deployment.

This may seem like a small thing to consider when you need to only account for one IP address in each subnet for the ELB. While only a single IP address, it is important to consider this when deploying your subnets within a VPC, especially if multiple ELBs may reside there. If you are using something small such as /28 for your subnets, there are only 11 usable addresses (14 minus the 3 AWS addresses) to assign to resources. The addresses must be used sparingly.

Refer to *Figure 1.11* for a simplistic view of multiple application load balancers mapped to the same subnets and consuming multiple addresses:

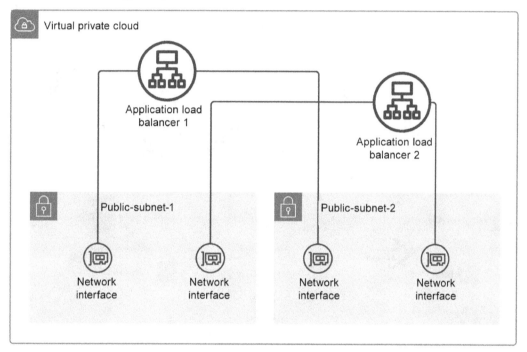

Figure 1.11: ALB with ENIs

ALBs are usually deployed in such a fashion for resiliency and redundancy in case an AZ is impacted by a failure.

Subnet Considerations for Auto-Scaling

When configuring EC2 Auto Scaling, you need to make similar considerations for the number of IP addresses that will be consumed out of your VPC subnets. Since EC2 Auto Scaling allows you to set minimum and maximum desired capacity units for EC2 instances, you will need to ensure your subnets have adequate IP addresses, not just for normal operation, but also to account for failure.

Take the following example. Trailcats has an **Auto Scaling group** (**ASG**) spread across three AZs within a VPC. During peak times, the maximum capacity on the ASG is configured for nine instances, so they are spread evenly across the AZs. If there is an AZ failure, then those instances will need to be redeployed in other subnets. In this scenario, all VPC subnets need to have enough IP addresses to account for all the EC2 instances that the ASG could deploy into the subnet during an AZ failure. Refer to *Figure 1.12* to see an example of how the ASG could react to an AZ failure:

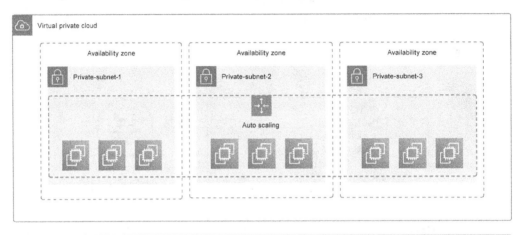

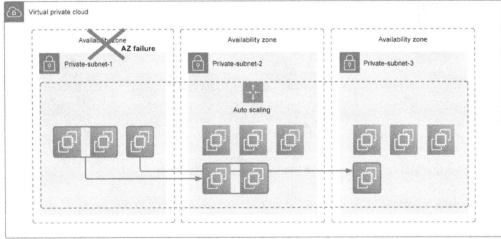

Figure 1.12: Subnets with auto-scaling configuration

Auto-scaling is both based on demand and resiliency; in the preceding figure, auto-scaling is being used for resiliency to ensure adequate resources are available.

Prefix Lists

Configuring rules for security groups or routes for route tables can sometimes be difficult to manage. When an organization has several CIDR blocks they need to allow access from or route to, maintaining all those entries can come with a management overhead to keep up with them. This can become quite cumbersome when you are responsible for maintaining these entries in multiple VPCs and potentially across multiple regions. Using **managed prefix lists** to maintain an up-to-date list of these CIDR blocks can make this process easier.

Prefix lists are lists that contain multiple IP CIDR blocks, which can be IPv4- or IPv6-based. These lists can then be referenced in rules belonging to security groups and route tables. In turn, if you have a prefix list containing 10 prefixes, a single security group rule or route table route entry will apply for all 10 of those prefixes. Prefix lists can be a good way to maintain consistency with security groups and/or route tables across all your resources and even between AWS accounts.

There are two types of prefix lists within AWS: **customer-managed** and **AWS-managed**.

Customer-Managed Prefix Lists

Customer-managed prefix lists are created and maintained by you within your AWS account(s). You are responsible for adding/removing IP prefixes from these lists as necessary. As prefixes are added/removed, any references to them in a security group rule or route table entry will automatically be updated in place.

A customer-managed prefix list is a **regional** construct, meaning it only exists within a single AWS Region. Here are a few other characteristics of customer-managed prefix lists to consider. A prefix list supports either IPv4 or IPv6 addressing, but not both. If you require the use of both, this task will require two separate prefix lists. A prefix list also requires a limit of the **maximum** number of prefix entries to be set. By default, the maximum number this value can be set to is 1,000. To help manage the life cycle of a prefix list, it also supports **versioning**. When entries are added/removed, a new version of the prefix list is automatically created and promoted. This allows for simple restoration to previous versions.

Be careful when referencing a prefix list in another resource. The number of prefix entries applies to the service quota for that resource. For example, if a prefix list with 25 entries is referenced in a VPC route table, then that is equivalent to 25 separate route entries. This can quickly consume entries and is inefficient.

Refer to *Figure 1.13* for an example of a customer-managed prefix list within the AWS dashboard. Make note of the prefix list version, max entries, and address family.

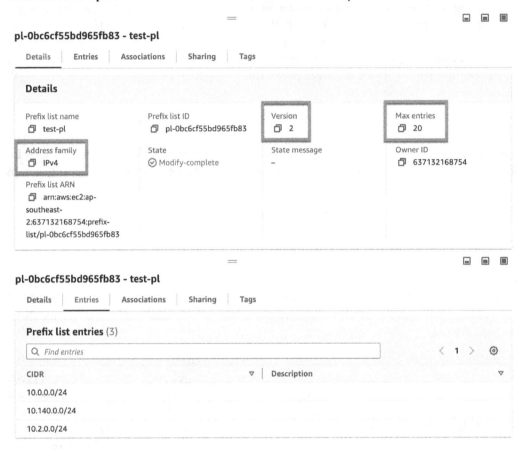

Figure 1.13: Customer-managed prefix list

Customer-managed prefix lists offer granular logical grouping based on specific needs, but as with all custom features in the cloud, this granularity comes with strict limitations such as low maximum entries.

> **Note**
>
> Customer-managed prefix lists are also supported within AWS **Transit Gateway** (**TGW**) route tables, which will be covered later in this chapter.

AWS-Managed Prefix Lists

In addition to customer-managed prefix lists, there are also AWS-managed prefix lists. These can be referenced in a very similar fashion to customer-managed ones, but all the entries are owned and maintained by AWS. These prefix lists are automatically created within each AWS region and can be referenced as needed.

These prefix lists are created for several AWS services and are populated with all the IP prefixes associated with those services. These lists can be referenced by security groups or route tables to ensure resources can interact with these services in a secure manner.

Refer to *Table 1.1* for a list of AWS-managed prefix lists and their corresponding names:

AWS Service	Prefix List Name
Amazon CloudFront	com.amazonaws.global.cloudfront.origin-facing
Amazon DynamoDB	com.amazonaws.region.dynamodb
AWS Ground Station	com.amazonaws.global.groundstation
Amazon Route 53	com.amazonaws.region.ipv6.route53-healthchecks
	com.amazonaws.region.route53-healthchecks
Amazon S3	com.amazonaws.region.s3
Amazon S3 Express One Zone	com.amazonaws.region.s3express
Amazon VPC Lattice	com.amazonaws.region.vpc-lattice
	com.amazonaws.region.ipv6.vpc-lattice

Table 1.1: AWS-managed prefix lists

These AWS-managed prefix lists make it simpler for customers to reference them when creating a policy that needs to reference AWS services.

> **Note**
> You can find more details about AWS-managed prefix lists here: `https://docs.aws.amazon.com/vpc/latest/userguide/working-with-aws-managed-prefix-lists.html`

Configuring Customer-Managed Prefix Lists

This section details the creation of a customer-managed prefix list from either the AWS console or AWS CLI.

To create a customer-managed prefix list, navigate to the VPC console of the AWS dashboard and select `Managed prefix lists`. This section contains both customer-managed and AWS-managed prefix lists. Select the `Create prefix list` option to create a new customer-managed prefix list. Next, you will define the name of the prefix list, the maximum number of entries, and any specific prefix entries. This process is shown in *Figure 1.14*:

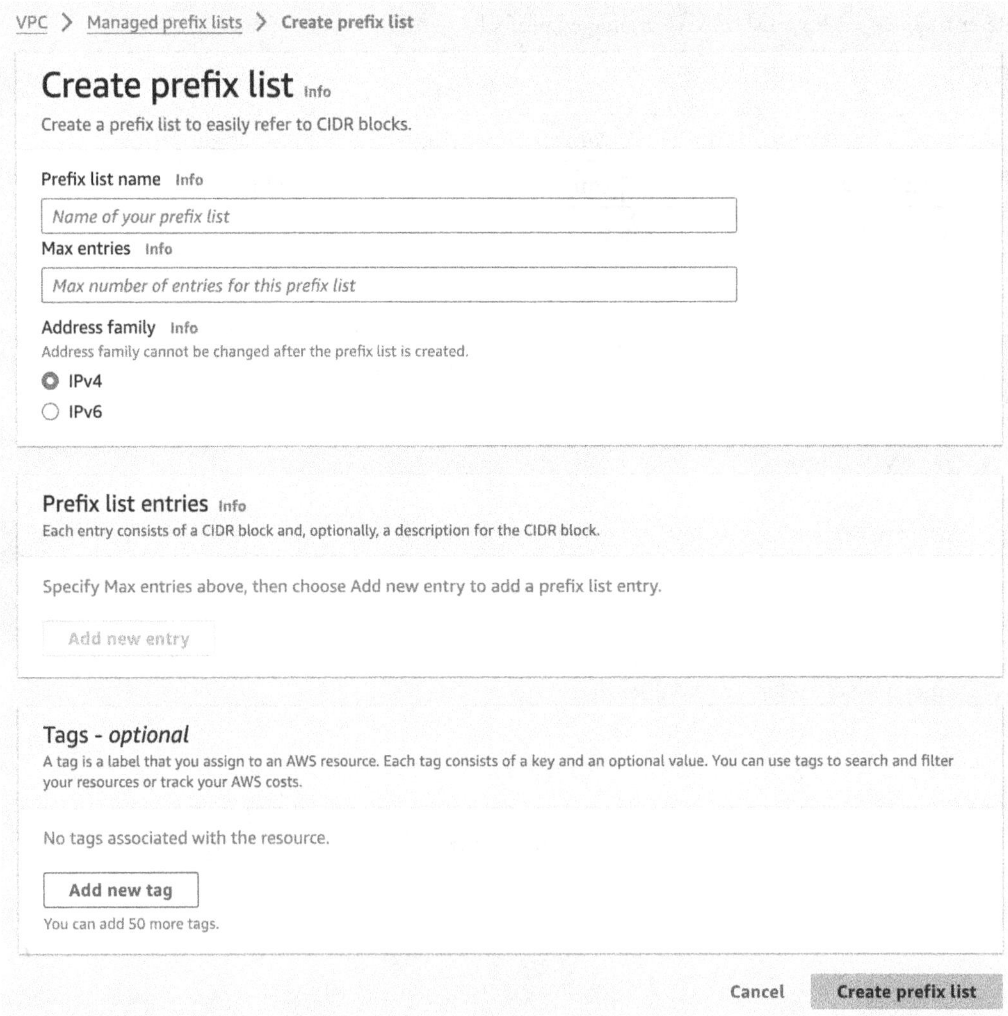

VPC > Managed prefix lists > **Create prefix list**

Create prefix list Info
Create a prefix list to easily refer to CIDR blocks.

Prefix list name Info

Name of your prefix list

Max entries Info

Max number of entries for this prefix list

Address family Info
Address family cannot be changed after the prefix list is created.

◉ IPv4

○ IPv6

Prefix list entries Info
Each entry consists of a CIDR block and, optionally, a description for the CIDR block.

Specify Max entries above, then choose Add new entry to add a prefix list entry.

Add new entry

Tags - *optional*
A tag is a label that you assign to an AWS resource. Each tag consists of a key and an optional value. You can use tags to search and filter your resources or track your AWS costs.

No tags associated with the resource.

Add new tag

You can add 50 more tags.

Cancel   **Create prefix list**

Figure 1.14: Create prefix list details

The prefix list is a custom way to maintain a list of interesting prefixes that should be referenced in the same way – for example, you may wish to create a prefix list with the prefixes of all development resources across your VPCs; a prefix list can do this.

A customer-managed prefix list can be created using the AWS CLI `aws ec2 create-managed-prefix-list` command.

For example, to create a custom prefix list using the AWS CLI, follow the given code:

```
aws ec2 create-managed-prefix-list --prefix-list-name
MyPrefixList --max-entries 10 --address-family IPv4 --entries
Cidr=10.0.0.0/16,Description="My CIDR block"
```

The next section will cover how to connect VPCs together.

Connectivity between AWS VPCs

Amazon VPCs are great for housing applications within your AWS account, but it is rare that you will need only a single VPC. Many AWS customers use multiple VPCs. This can range from five VPCs in a single account or hundreds of VPCs across several AWS accounts. This is due to a number of factors; different teams may own different AWS accounts and VPCs, or there may be security concerns that require decentralizing workloads. The next sections will focus on building connectivity between your Amazon VPCs.

VPC Peering

One of the simplest ways to build connectivity between VPCs is using a **VPC peering** connection. VPC peering is a simple connection that is built between two VPCs. This peering does not use a specific type of gateway or external connection, so it does not possess any limits on throughput for connectivity across the peering.

Refer to *Figure 1.15* for a visual of VPC peering:

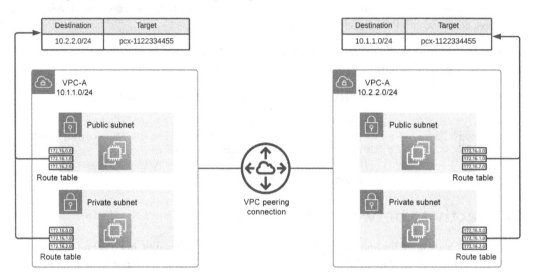

Figure 1.15: VPC peering connection

In the preceding figure, you will see that there is a VPC peering connection built between **VPC-A** and **VPC-B**. Every VPC peering connection will have a **VPC peering connection ID** associated with it. That peering connection ID can then be used within the route tables of the VPC as a target for any route entries that are to use the VPC peering. In this example, **VPC-A** has an IPv4 CIDR of 10.1.1.0/24 and **VPC-B** has a CIDR of 10.2.2.0/24. Each VPC has local route tables configured with routes to the peered VPC CIDR. The target of the routes is the VPC peering connection ID, which is formatted as pcx-xxxxxxxxxx.

There are a few limitations and constraints you need to consider about VPC peering. This may influence your decision of whether to implement VPC peering for connectivity or choose something such as AWS TGW, which will be discussed in *Chapter 5, Hybrid Networking with AWS Transit Gateway*.

You may have occasions where VPCs have multiple VPC peering connections to other VPCs. As the number of VPCs within your environment grows, your number of VPC peering connections may also grow. It may be easy to assume that you could just "daisy-chain" VPCs together and essentially route through a central VPC to get to others. This is commonly referred to as a **hub-and-spoke network**.

However, you need to be aware that VPC peering connections are **non-transitive**. This means that traffic cannot transit through a VPC and across another VPC peering connection. Refer to *Figure 1.16* as an example:

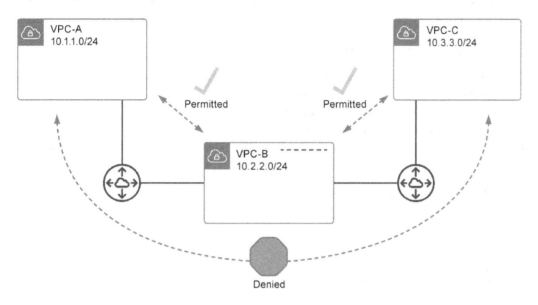

Figure 1.16: Non-transitive VPC peering

There are three VPCs in this example, with **VPC-B** peered to both **VPC-A** and **VPC-C**. Any traffic from **VPC-A** or **VPC-C** destined for **VPC-B** will be permitted and vice versa. However, any traffic destined from **VPC-A** to **VPC-C** will be blocked due to the non-transitive nature of VPC peering. For a transitive solution, it is often recommended to use a service such as AWS TGW. Alternatively, you could opt for a **full-mesh** VPC peering solution. This means that every VPC is peered to every other VPC. As you may imagine, once the number of VPCs reaches a certain scale, managing this number of VPC peering connections can turn out to be a difficult task.

> **Note**
>
> This non-transitive nature also applies to services such as internet gateways, NAT gateways, Direct Connect, and gateway endpoints. In other words, VPC peering cannot be used for resources in a VPC to access these services within another VPC.

Overlapping CIDR Blocks

When creating a VPC peering connection, the two VPCs cannot have **overlapping IP CIDR blocks**. This applies to both IPv4 and IPv6, with both primary and secondary IP blocks considered. If you attempt to create a VPC peering between VPCs with overlapping CIDRs, the peering connection will fail.

Inter-Region Maximum Transmission Unit (MTU)

Creating VPC peering connections between VPCs that reside in different AWS regions is a supported configuration. However, it is important to consider that the **MTU** across inter-region peering connections is **1,500 bytes**. This differs from intra-region peering connections, which support a jumbo MTU, or 9,001 bytes. This can be an important consideration with traffic that may have issues with the fragmentation of payloads, that is, a packet larger than 1,500 bytes needing to be fragmented into multiple packets and sent across VPC peering.

Refer to *Figure 1.17* for an example:

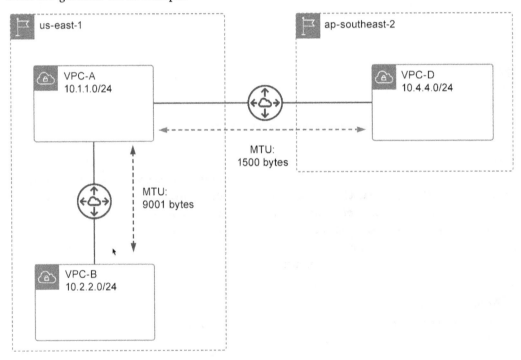

Figure 1.17: Inter-region VPC peering

Within a region, the MTU is much higher, and applications could use jumbo frames. However, traffic between AWS regions is restricted to a standard 1,500 bytes and the applications must be able to detect this and adjust the packet payload size accordingly.

The VPC peering connection between **VPC-A** and **VPC-B** in region **us-east-1** has an MTU of 9,001 bytes. The inter-region peering between **VPC-A** in **us-east-1** and **VPC-C** in **ap-southeast-2** has an MTU of 1,500 bytes.

Other Considerations for VPC Peering Connections

There are several other characteristics of VPC peering connections that need to be considered when using this solution for providing inter-VPC connectivity. A few of them are listed here:

- Routing across VPC peering connections is all dependent on **static routes**. This means all route tables must be updated within both VPCs to ensure bidirectional communications.

- The number of VPC peerings per VPC is limited by a service quota. The default amount is **50** peering connections, but this is adjustable up to **125**.

- You cannot create multiple VPC peering connections between two VPCs.

- Unicast reverse path forwarding is not supported for VPC peering. Therefore, if multiple VPCs with the same CIDR block are peered to the same VPC, you will need to implement **longest match route table entries** to achieve symmetric traffic.

> **Note**
>
> For a full list of VPC peering limitations, refer to the AWS documentation page here: `https://docs.aws.amazon.com/vpc/latest/peering/vpc-peering-basics.html#vpc-peering-limitations`

Provisioning Process for VPC Peering Connections

Provisioning a VPC peering connection is a multi-step process. The process includes a VPC requesting to peer with another VPC, then that request being approved before the VPC peering connection is established. The VPC initiating the peering connection request is known as the **requester VPC** and the receiving VPC is known as the **accepter VPC**.

This process is quite simple when peering VPCs within the same AWS account, as the request and the approval can all be done within the account. However, this process is allowed across AWS accounts, so the requester VPC could be in one AWS account, while the accepter VPC is in another AWS account. The accepter VPC could be within an account that is under the same AWS Organization or a separate one.

A VPC peering connection will go through a series of stages within both the provisioning and deprovisioning processes. These stages represent the current state of the peering connection and whether it is ready for use. The stages are outlined here:

- **Initiating request**: The request to form a VPC peering connection has been made and is in the initiation state. From here, the process will move to pending acceptance, unless there is a failure.

- **Failed**: The VPC peering connection has failed after initiation. The peering connection cannot be recovered from this state – it must be re-initiated.

- **Pending acceptance**: The VPC peering connection has been initiated and is awaiting approval from the accepter VPC.

- **Expired**: The peering connection has expired.

- **Rejected**: The accepter VPC has rejected the request for a VPC peering connection to be created.

- **Provisioning**: The accepter VPC has accepted the peering request and it is in the process of being provisioned.

- **Active**: The VPC peering connection is active and ready for use. VPC route tables can be updated with route entries to use the VPC peering connection ID as the target.

- **Deleting**: The current VPC peering connection has been requested for deletion and is in the process of being removed.

- **Deleted**: The VPC peering connection has been removed and is no longer available for use.

Understanding the VPC peering connection status will give insight into whether the connection is active and usable or whether there is a problem.

Configuring VPC Peering Connections

This section details the creation of a VPC peering connection both from the AWS console and using the AWS CLI.

To configure an EIP, you can simply navigate to the VCP dashboard of the AWS console, select Peering connections, and choose the Create peering connection option. The process and required VPC peering details are shown in *Figure 1.18*.

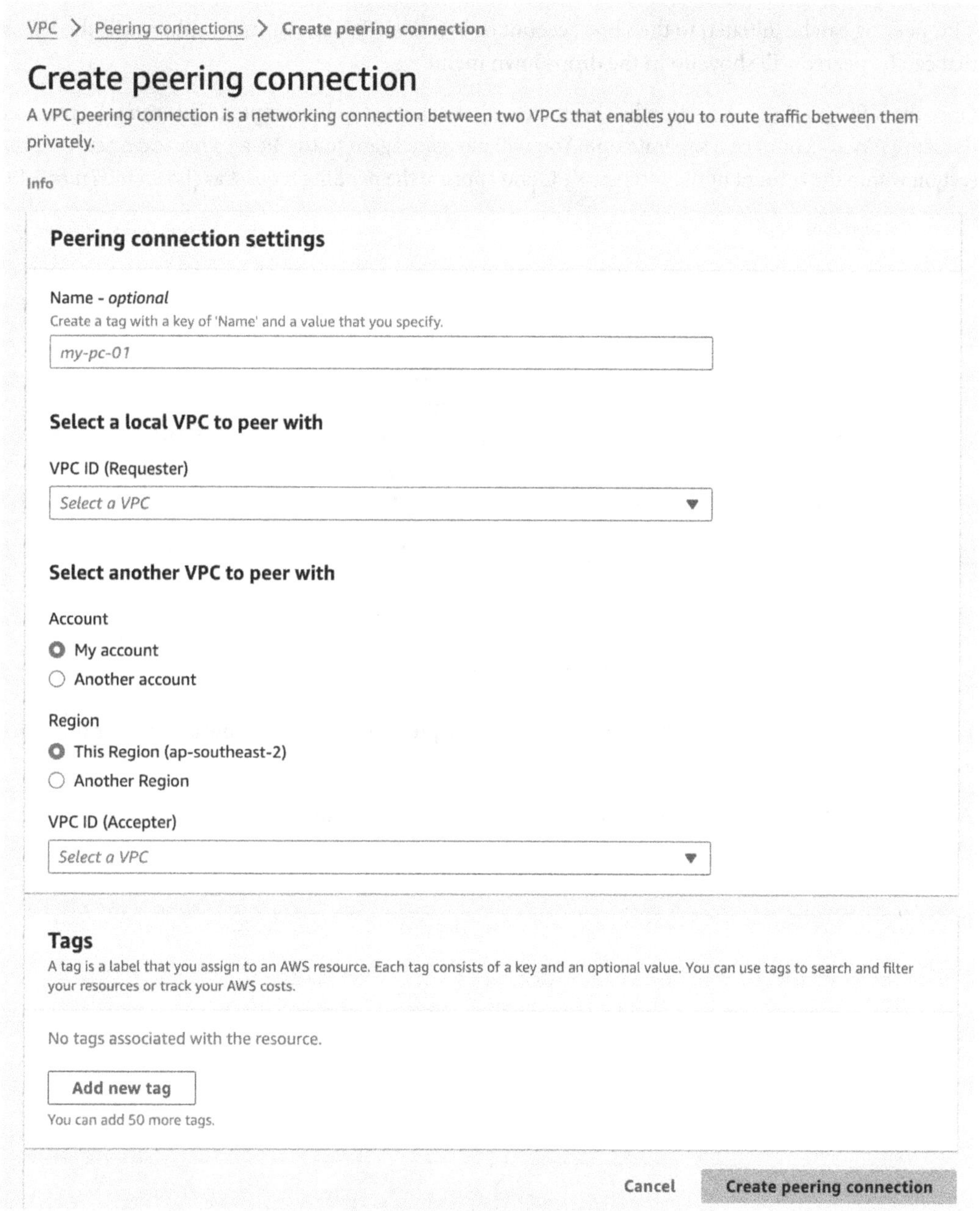

VPC > Peering connections > Create peering connection

Create peering connection

A VPC peering connection is a networking connection between two VPCs that enables you to route traffic between them privately.

Info

Peering connection settings

Name - *optional*
Create a tag with a key of 'Name' and a value that you specify.

```
my-pc-01
```

Select a local VPC to peer with

VPC ID (Requester)

```
Select a VPC                                          ▼
```

Select another VPC to peer with

Account
● My account
○ Another account

Region
● This Region (ap-southeast-2)
○ Another Region

VPC ID (Accepter)

```
Select a VPC                                          ▼
```

Tags

A tag is a label that you assign to an AWS resource. Each tag consists of a key and an optional value. You can use tags to search and filter your resources or track your AWS costs.

No tags associated with the resource.

```
Add new tag
```

You can add 50 more tags.

Cancel   **Create peering connection**

Figure 1.18: Create VPC peering details

VPC peering can be initiated to the same account or between AWS accounts; in either case, the VPCs that can be peered will show up in the drop-down menu.

Once the VPC peering connection has been created, it will need to be accepted. This can take place in the same AWS account or a separate one. You will navigate again to the `Peering connections` section within the account of the accepter VPC and approve the pending request as shown in *Figure 1.19*.

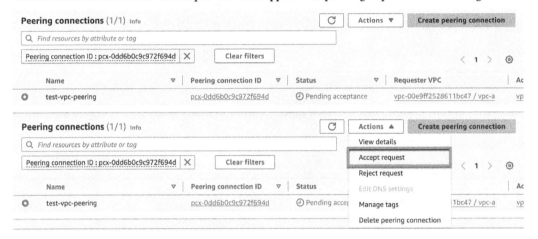

Figure 1.19: Accept VPC peering

Before the VPC peer is active, the request must be accepted, either by the same account or the other account to which the request was made.

A VPC peering connection can be created using the AWS CLI command:

```
aws ec2 create-vpc-peering-connection
```

For example, to create a VPC peering request using the AWS CLI, use the following command:

```
aws ec2 create-vpc-peering-connection --vpc-id vpc-12345678
--peer-vpc-id vpc-87654321 --peer-region us-west-2
```

A VPC peering connection can be accepted using the following AWS CLI command:

```
aws ec2 accept-vpc-peering-connection
```

As an example, to accept a VPC peer request, use the following AWS CLI command:

```
aws ec2 accept-vpc-peering-connection --vpc-peering-
connection-id pcx-12345678
```

Hub-and-Spoke VPC Architectures

As mentioned before, one common network topology that is used for connecting multiple IP networks together is **hub-and-spoke**. This topology is composed of a central "hub" network that then has connections with separate spokes. Typically, all traffic is backhauled through these hubs when attempting to talk to another network outside of their own. Given that VPCs are non-transitive, it is not possible to build this topology within AWS simply using VPC peering connections. You have the option to use a concept called a **transit VPC** to achieve this topology. In addition, you need to properly evaluate when to use this transit VPC or a service such as AWS TGW. This section will cover both topics.

Since hub-and-spoke networks cannot be directly achieved with VPC peering connections, an alternative solution is the transit VPC. The concept of a transit VPC is based on the use of IPsec VPN connectivity **over the top** of the standard VPC connectivity. This can be achieved by deploying **NVAs** into a central VPC and then configuring a series of IPsec VPN tunnels from these appliances to other VPCs. Additionally, these appliances could build IPsec connectivity to on-premises or other third-party networks. For building connectivity to other Amazon VPCs, you have a couple of options when using a transit VPC:

- **Option 1**: Build IPsec connectivity from the NVAs in the transit VPC to **virtual private gateways** (**VGWs**) residing in the spoke VPCs

- **Option 2**: Build IPsec connectivity from the NVAs in the transit VPC to additional NVAs within the spoke VPCs

Both options are outlined in *Figure 1.20*:

Option 1:

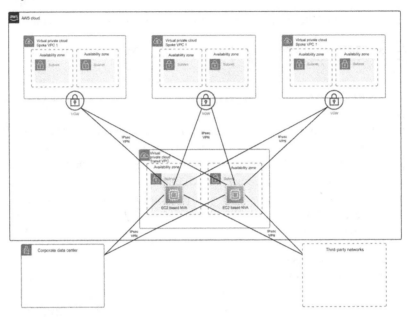

Option 2:

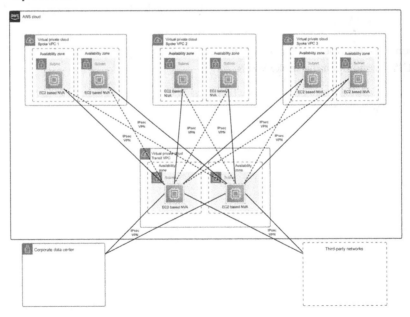

Figure 1.20: Transit VPC options

In *Figure 1.20*, both the preceding options have the option to utilize either static or dynamic routing for overlay connectivity via IPsec tunnels. The dynamic routing option is fully dependent on the capabilities of the vendor used for the NVA. The VGW will only support **Border Gateway Protocol (BGP)** as a routing protocol.

Consider a scenario where Trailcats had a need for expanded connectivity due to some quota limitation, such as a maximum number of routes in the VPC routing table, or if connecting to a third-party vendor that required a different connectivity solution than that offered by native cloud networking. This is common when working with vendors and a major driver of enterprises adopting NVA-based solutions.

The strength of **Option 1** in this situation is that a minimal deployment of NVAs might meet the needs of vendor connectivity while allowing workload VPCs to be connected to the vendor and each other. This option works so long as there is no need to propagate large amounts of routes to the workload VPCs, as the VPC route tables have a low maximum number of routes.

The weakness of this option is evident in that a large portion of connectivity becomes manual – namely, the VPC route tables and the VPN connections to the VPC-attached VGWs.

Option 2 requires more up-front work to deploy NVAs to workload VPCs, but often these solutions are software-defined, which facilitates route exchange and connectivity between the NVAs.

The weakness of this option is it requires far more life cycle management of NVA instances and potential software problems if the NVA vendor releases software with bugs or other problems.

> **Note**
>
> More details on AWS transit VPCs can be found at the following URL: `https://docs.aws.amazon.com/whitepapers/latest/aws-vpc-connectivity-options/transit-vpc-option.html`

IP Address Overlap Management

During AWS cloud deployment, you may need to provide connectivity between resources that have overlapping IP address ranges. This could be for several reasons, such as a developer configuring a VPC without checking whether the IP address range was available within their **IP address manager (IPAM)** solution or even a merger/acquisition between two companies. Nonetheless, from a networking perspective, you may be tasked with making connectivity happen. This short section will touch on how to accomplish this with AWS NAT gateways, but there is another method using AWS PrivateLink covered later, in *Chapter 3, Networking Across Multiple AWS Accounts*.

Using Private NAT Gateways for IP Overlaps

As mentioned in the previous chapters, AWS NAT gateways can be used to allow private subnets to talk to resources on the public internet or even other private resources within the AWS cloud or on-premises. Allowing communication between resources with overlapping IP address space can be one of those particular use cases. Refer to the following example.

Trailcats has acquired another company called **Mountain Felines** (**MF**), which also uses the AWS cloud for its applications. An MF VPC needs to communicate with some Trailcats resources, but it is using an IPv4 CIDR that is already in use by a Trailcats VPC, 10.100.0.0/16. Trailcats has attached the acquired MF VPC to their existing AWS TGW but cannot have two routes to the same destination. To allow the two VPCs to communicate with other resources, a secondary CIDR is used within the VPC to house an AWS NAT gateway. The NAT gateway can then be used as the next hop for traffic from the overlapping subnets to any other resources. This (in addition to some DNS adjustments) would allow for the VPC to initiate and establish connectivity. This is shown in *Figure 1.21*:

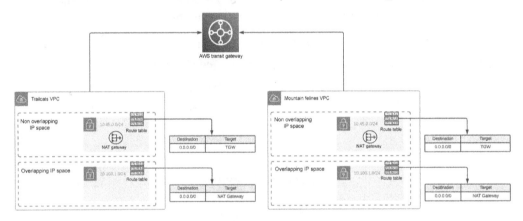

Figure 1.21: Overlapping IP space with NAT gateways

The use of a private NAT gateway allows workloads with overlapping IPs to communicate; the nature of a NAT gateway requires that the workload using the NAT gateway is the one to initiate the communication. This is why if both sides initiate communication, a NAT gateway is needed for each subnet.

> **Note**
> This same behavior could not be achieved with VPC peering because two VPCs cannot be peered together if they have overlapping CIDR ranges.

Service Quotas Quick Reference

The following table includes some AWS service quotas you should be aware of for the AWS ANS-C01 exam. Service quotas are artificial limitations put in place by AWS to ensure that no single customer can impact other customers through resource denial.

Name	Default Limit	Adjustable?	Maximum
Network interfaces per instance	Varies by instance type	No	
Network interfaces per region	5,000	Yes	AWS' discretion
Elastic IP addresses per region	5	Yes	AWS' discretion
Elastic IP addresses per NAT gateway	2	Yes	8
Prefix lists per region	100	Yes	AWS' discretion
Versions per prefix list	1,000	Yes	AWS' discretion
Maximum number of entries per prefix list	1,000	Yes	AWS' discretion
Active VPC peering connections per VPC	50	Yes	Up to 125

Table 1.2: VPC service quotas

> **Note**
> These service quota details were pulled directly from the following AWS documentation pages:
>
> `https://docs.aws.amazon.com/vpc/latest/userguide/amazon-vpc-limits.html`
>
> `https://docs.aws.amazon.com/vpc/latest/peering/vpc-peering-connection-quotas.html`

Summary

In this chapter, you looked deeper into AWS networking components that can be used to build upon the fundamental networking services. This chapter focused on building connectivity and on how to control connectivity in certain scenarios. Advanced VPC networking concepts such as ENI, EIP, VPC peering, and prefix lists are a necessary foundation upon which the rest of this guide will be built. The next chapter focuses on monitoring the traffic and performance of VPCs.

Exam Readiness Drill – Chapter Review Questions

Apart from mastering key concepts, strong test-taking skills under time pressure are essential for acing your certification exam. That's why developing these abilities early in your learning journey is critical.

Exam readiness drills, using the free online practice resources provided with this book, help you progressively improve your time management and test-taking skills while reinforcing the key concepts you've learned.

HOW TO GET STARTED

- Open the link or scan the QR code at the bottom of this page

- If you have unlocked the practice resources already, log in to your registered account. If you haven't, follow the instructions in *Chapter 16* and come back to this page.

- Once you log in, click the START button to start a quiz

- We recommend attempting a quiz multiple times till you're able to answer most of the questions correctly and well within the time limit.

- You can use the following practice template to help you plan your attempts:

Working On Accuracy		
Attempt	Target	Time Limit
Attempt 1	40% or more	Till the timer runs out
Attempt 2	60% or more	Till the timer runs out
Attempt 3	75% or more	Till the timer runs out
Working On Timing		
Attempt 4	75% or more	1 minute before time limit
Attempt 5	75% or more	2 minutes before time limit
Attempt 6	75% or more	3 minutes before time limit

The above drill is just an example. Design your drills based on your own goals and make the most out of the online quizzes accompanying this book.

First time accessing the online resources? 🔒

You'll need to unlock them through a one-time process. **Head to** *Chapter 16* **for instructions**.

Open Quiz	
https://packt.link/ansc01ch1	
OR scan this QR code →	

2

VPC Traffic and Performance Monitoring

When administrating a cloud network, it can be important to understand the various problems that can arise and how to effectively diagnose and troubleshoot them to restore services quickly. Business recovery and continuity are among the most critical needs of any company, and neither is possible without understanding what is happening within the technical resources underpinning the applications that drive business revenue and outcomes. AWS has come a long way from the first-generation monitoring options, and their inclusion of smarter networking has brought with it smarter monitoring tools. Understanding how to monitor traffic and performance within a VPC will help minimize any business impact, which is a crucial skill for a cloud network engineer.

In this chapter, you will cover the following topics:

- Potential cloud network problems
- Metrics and logging
- AWS performance monitoring services
- Monitoring and troubleshooting
- Troubleshooting packet size issues

By the end of this chapter, you should be able to understand the potential problems that can arise, how to diagnose them, and how to troubleshoot them using AWS tools.

Potential Cloud Network Problems

Depending on the application and design, not all things you can monitor in the cloud are associated with problems. AWS gives its users an impressive number of metrics associated with its various services, and the only guarantee is that you will never need all of them at once. In the following sections, you will learn about the various issues that impact cloud networking and how to diagnose and ultimately resolve many of them.

IP Address Allocation

Situations where multiple resources have been misconfigured with the same address can happen in the cloud, and discovery of this problem can be challenging. Other common problems include exhausting the IP addresses in a subnet or VPC or allocating IP addresses to multiple VPCs, then having to implement a solution to deal with overlapping IP addresses because the resources in those VPCs must communicate with each other. This is very common in scenarios where different teams are responsible for different VPCs, and later there is a need to connect them together for applications to communicate.

When diagnosing problems with IP addresses, the main problem is often simply understanding where the IP addresses are in use and whether the **IP Address Management (IPAM)** solution is appropriate to what is planned and deployed. IPAM can help manage the allocation and planning of IP addresses and should be the first place you look when suspecting IP address allocation problems.

Routing Problems

AWS uses a hyperplane to manage its impressive network. The hyperplane is a large, automated network of data centers managed by an equally large and automated control plane. This hyperplane is what provisions and runs the environment for customers. To ensure that no single customer (or multiple customers) can negatively impact the health of the entire hyperplane, AWS implements limits known as quotas on certain resources available to customers.

Some of these limitations are enforced by default but can be extended or ignored with the help of the AWS support organization by opening a support ticket. These limitations are soft quotas. One example of a soft quota is the number of static routes that can be added to a VPC route table, as in *Table 2.1*. In this case, AWS will not allow you to add more than 50 such routes in a single route table until you open a ticket with AWS support and they approve the increase. Even then, the increase has a maximum limit, a hard quota, of 1000 routes.

Other features have hard quota limits and cannot be extended or increased without going to the top levels of AWS. Functionally, these limits should be considered set in stone, though if you work for a large enough company, AWS may consider increasing even those quotas. Route scale is one of the hardest quotas to work with in AWS and will influence the network design and operation of every cloud network. *Table 2.1* shows some common AWS network resources and their soft/hard route quota limits as of mid-2024:

Network Resource	Route Option	Soft Quota Limit	Hard Quota Limit
VPC route table	Non-propagated routes	50	1,000
	Propagated routes	100	100
Transit gateway (TGW)	Static routes	10,000	Approved on a case-by-case basis
	Routes advertised from TGW to **Network virtual appliance (NVA)/ Site-to-Site (S2S) VPN**	5,000	5,000
	Dynamic routes from **Customer gateway (CGW)** to S2S VPN on TGW	1,000	1,000
Virtual private gateway (VGW)	Dynamic routes from CGW to S2S VPN	100	100
	Routes advertised from S2S VPN on VGW to CGW	1,000	1,000
	Static routes from CGW to S2S VPN on VGW	100	100

Table 2.1: Common route quotas in AWS

Being aware of these limitations is important to monitoring network operation and performance, especially when these quotas are exceeded. Often, when a route quota is exceeded, for example, one of the only warnings will be missing routes that were expected to be advertised or received. In some extreme cases, routing neighborships between two network devices can be torn down and rebuilt, only to be torn down again until the limitation is addressed. Monitoring quota limits before they are exceeded can help to avoid these impactful events.

Besides route scale limitations, other common concerns with cloud routing include ensuring that routes are statically or dynamically propagated where they are needed.

Many native cloud constructs only support the creation of static routes, and the route tables for these must be maintained manually. Other cloud constructs support dynamic routing (with limits) but only support the **Border Gateway Protocol (BGP)** routing protocol, and only certain features (such as some static BGP communities) of the protocol. AWS also has tools such as the VPC Reachability Analyzer (with an optional Amazon Q generative AI assistant for network troubleshooting) to assist in understanding the end-to-end reachability of a source and destination, including the routing used to determine the path, which will be examined in more detail later in this chapter when discussing troubleshooting tools.

Mismatched Packet Sizes

The AWS cloud network, like all hyperplanes, is a large network fabric built on top of an underlying network infrastructure. This infrastructure is heavily automated and abstracted from its consumers. Part of that hyperplane includes a fair amount of tunneling technologies to overcome legacy limitations found in Layer 2 of the OSI model, as well as extensions that make multi-tenancy possible. In short, the average packet size allowable within a cloud tends to vary based on the networking service. This can wreak havoc on applications that do not have transmission-size-discovery mechanisms.

Just like an on-premises network, there can be a severe impact on the data payloads of packets if the packets are too large to transmit without fragmentation, a process by which a single payload is disassembled, transmitted as multiple packets with payload fragments, and reassembled on the far end.

Diagnosing issues with application data due to packet size mismatches and **maximum transmission unit (MTU)** problems can be a complicated affair, even when you have full control of the end-to-end network. In the cloud, it can be almost impossible due to the limited visibility of the infrastructure. Luckily, AWS has several options, which will be covered later in the chapter.

Bandwidth and Throughput

No examination of potential network problems could be complete without talking about the speed of the network.

In AWS, the underlying network utilizes speeds that most on-premises networks could not hope to match. However, an individual customer cannot be allowed to utilize the full speed and power under the hood, lest they commit a **denial of service (DoS)** attack on the other customers utilizing the same hardware.

For this reason, much as the quotas seen for other services, AWS restricts how much bandwidth an individual network construct can consume. Unlike other service quotas, the bandwidth/throughput limitations cannot generally be increased by a support ticket. *Table 2.2* outlines some of the common bandwidth quotas as of mid-2024:

Network Resource	Transit Option	Soft Quota Limit	Hard Quota Limit
Transit gateway (TGW)	VPC Attachment	Up to 50 Gbps	N/A
	VPN Tunnel	Up to 1.25Gbps (per tunnel)	N/A
	TGW Connect (GRE Tunnel)	Up to 5 Gbps	N/A
VPC Peering	Peered VPCs	No published limit	N/A
AWS Direct Connect (DX)	DX or DX Gateway	Up to 10GBps per DX	N/A

Table 2.2: Common bandwidth quotas as of mid-2024

When planning a deployment, it is important to be aware of the limitations of throughput to ensure that data will not be bottlenecked and impact performance. For example, if the application requires a high volume of data throughput to function correctly, care must be taken to ensure there is sufficient network bandwidth for the application when planning network connectivity between all parts of the application itself. Failure to do so may impact the responsiveness or functionality of such an application.

Packets per Second Limits

Of all the limitations placed on you by the hyperplane, this one might be the most difficult to diagnose. As stated earlier, AWS runs a hyperplane, and part of good stewardship of the hyperplane involves ensuring that no single customer can use all the available resources. This includes the network processing power itself. Imagine a scenario where an application uses very small packets—just a few hundred bytes—but that application (or group of applications, multiple servers, etc.) is sending an immense amount of those packets. From a bandwidth/throughput perspective, all should be well. Unfortunately, as many on-premises network engineers know, the amount of processing power needed to send that many packets can exceed what is available on a network device.

In the case of AWS, these same laws of network physics still apply, but an artificially low ceiling is placed by AWS to keep this from happening. What this often means is that based on the service, infrastructure, or virtual machines in play, it's possible (and likely) that you will hit this limit long before exceeding others. The symptoms of hitting this limit tend to make correlation with the problem difficult. However, **packets per second (PPS)** limits won't manifest as anything but dropped packets in most network devices, for instance, when the real culprit is the hyperplane discarding them rather than a problem with the device itself. To illustrate how the PPS limit is used, here is a short table of common network services and the PPS limits as of mid-2024:

Network Resource	Network Service	Soft Quota Limit	Hard Quota Limit
Transit Gateway (TGW)	TGW Attachment (any)	5,000,000	N/A
	VPN Tunnel	140,000	N/A
	TGW Connect Peer	300,000	N/A
Site-to-Site (S2S) **VPN**	VPN Tunnel	140,000	N/A
Elastic Compute Cloud (EC2) Virtual Machines	Non-Network Optimized	Unpublished, but larger instances have higher limits	N/A
	Network-Optimized	Unpublished, but this class has higher PPS limits than non-optimized EC2	N/A

Table 2.3: Common network services and the PPS limits as of mid-2024

In general, when dealing with EC2 Instances, the PPS limit tends to be ranked in the following order (largest to smallest):

- Network-optimized (larger sizing)
- Network-optimized (smaller sizing)
- Non-network-optimized (larger sizing)
- Non-network-optimized (smaller sizing)

Note that the limitation is per instance (ingress and egress), not per interface. This means that the entire instance has this limit, regardless of what direction traffic is flowing or how many interfaces.

Packet Loss

Applications run on data, and missing data can often severely impair or prevent applications from functioning. As a cloud network engineer, your role is to ensure the health of application communication. Packet loss can often negatively impact applications by failing to deliver vital data. Worse, the effects of packet loss are often different based on the type of application experiencing the loss. TCP-based applications have a built-in data recovery mechanism through the stateful delivery mechanism in TCP itself. Other applications may be built with their own error recovery mechanisms, and still others may have neither the stateful delivery mechanism of TCP nor any innate error recovery.

Because it is very unlikely that only a single application will suffer when packets are lost, the effort of diagnosing the root cause based on application performance alone can be maddening. The underlying transit may also be unreliable and offer no obvious indication of failure, such as in the case of an S2S **virtual private network** (**VPN**) over the internet, which is a medium that has few, if any, service-level agreements for reliability. Altogether, the detection, root cause analysis, and remediation of this failure can test the limits of fault detection mechanisms due to its multilayer impact and potentially inconsistent manifestation.

When internet service providers and WAN providers have intermittent problems themselves, sometimes just getting them to admit that the problem lies with them is difficult. Troubleshooting methods such as ping and traceroute can surface issues but not necessarily what they are and under which organization's jurisdiction they lie. This makes packet loss one of the most difficult problems to solve administratively, if not technically.

Misconfigured Security

That security is a requirement in networks is a non-negotiable fact. However, it's well known that as more security is added, the complexity and possibility of unintended consequences increase. Examples of these consequences may include firewalls that unintentionally block traffic and threat detection that returns false positives, and the diagnosis of problems now involves even more work because there is an added layer of complexity and another potential fault to consider.

On-premises, this would often manifest in access control lists, firewall rules, and security false positives. In the cloud, many of these causes still exist but are often harder to track down. **Security Groups (SGs)** attached to network interfaces often end up being a cause of unintended denial of traffic because they do not exist on-premises. **Network access control lists (NACLs)** are the AWS access control lists applied to subnets and work in a similar way to how **access control lists (ACLs)** work on-premises but are different enough to trip up engineers who expect them to work the same. AWS offers us several ways to diagnose security issues between application flows, which will be shown in the *Monitoring and Troubleshooting* section of this chapter.

Multiple Administrators/Changes

There's nothing quite like configuring something, testing it, and closing out the work request only to be woken up at 2 A.M. by the help desk because something has failed. There's also nothing quite like investigating to find out that someone else changed the routing to execute a different work request, resulting in your changes being broken. The joy of having multiple administrators is a tale as old as time, but luckily, AWS gives cloud network engineers a tool to verify changes in AWS CloudTrail. Knowing which administrator made the change will not prevent the change from being made (or breaking things), but it will provide a clear audit trail so the process can be improved and future problems avoided.

Metrics and Logging

What sort of information would be needed to diagnose some of these problems? How would a system effectively provide that information to a cloud network engineer? Most importantly, how could that information be consumable and actionable? In this section, you will learn about the kind of metrics and logging that would make the diagnosis of problems possible. Metrics are the measurements of the performance of a cloud network, while logging focuses on the recording of problems and events that can aid in the discovery of problems in a cloud network.

IP Address Management

One of the most fundamental metrics affecting cloud networking is simply knowing how the IP addresses available to use are allocated, where they are allocated, and what remains that can be used to expand the network. AWS provides an **IPAM** solution that will keep track of all the IP addresses in use and available across your deployment. With this information centrally managed, AWS IPAM allows you to do the following:

- Create IP address pools for certain regions and domains
- Allocate those IP addresses at scale across multiple VPCs and AWS accounts
- Monitor the usage of currently allocated IP addresses

- Identify IP address conflicts among resources due to misallocation or misconfiguration

- Audit IP address allocation and usage over time for capacity planning and policy

IPAM can be administrated by a single account or multiple accounts using **AWS Organizations**. In *Chapter 3, Networking Across Multiple AWS Accounts*, AWS Organizations will be discussed further.

Routing Problems

Diagnosing routing problems often requires the use of multiple tools because the complete picture is rarely found in one source of information. If it becomes necessary to understand why certain routes are present, missing, or directing traffic to the wrong destination, what data would be needed to diagnose that problem? The following are the biggest questions to ask when troubleshooting:

- What route or routes are needed?

- Where should these routes be present to support the expected traffic flow?

- Are the routes needed already installed where they are needed?

- If they are installed, do they direct traffic to the correct next hop?

- If they are not installed, should the route be updated manually or dynamically?

- If updated dynamically, is the route being learned, or is a limit being exceeded?

When troubleshooting cloud network problems associated with routing, the main culprits are often tied to the absence or presence of routes and the interoperability with other routing solutions such as NVAs or other cloud providers.

Packet Size Mismatches

Packet size mismatches refer to the tendency of applications to offload the discovery of the size of how large a data packet can be to the network instead of managing it themselves. This means that the network itself must be capable of understanding the maximum transmissible unit of size of data. The MTU is affected by packet headers such as those introduced by network tunnels, and this means that the smallest MTU along any given path is the maximum MTU across that path. However, unlike some network problems, packet size mismatches do not give a clear indication that something is wrong; that is, it will almost never be the network itself that suffers, but applications.

Diagnosing packet size mismatches and MTU problems often requires looking at the application data itself. This means capturing packets in flight along the path to see whether the packet size changes or fragmentation occurs. AWS supports a form of port mirroring and packet capture to further dig into the details of the application traffic. VPC port mirroring and packet analysis will be covered further in the section on VPC port mirroring.

AWS also captures certain network metrics to use in the diagnosis of packet size issues, such as the number of packets being reordered (a common symptom of fragmentation). If AWS captures a metric, often an alert can be set to trigger in **Amazon CloudWatch**, the monitoring service provided by AWS.

Another valid test to diagnose packet size problems is to send ICMP ping tests between the source and destination having the problem, specifying a certain packet size and with the **Don't Fragment (DF)** bit set. This is a useful test because if the packet needs to be fragmented to be passed along but cannot be (due to the bit), it will send an error back to the originator, which identifies where the packet sizing problems are occurring.

Bandwidth/Throughput Limitations

Exceeding the line rate is a very common problem with oversubscribed networks, and the cloud is no exception. It can be harder to remember what the line rates are in the cloud because the physical limitations are hidden from customers and, instead, are based on the hyperplane's artificial limits. The best metrics possible to watch bandwidth utilization and limit exceeding involve the peak usage, the sustained usage over time, as well as the number of packets dropped, all of which can be viewed in Amazon CloudWatch.

Importantly, the metric covering packet drops does not include the precise reason so another correlation is needed. Here are some examples of relevant metrics and details to help correlate packets being dropped due to the bandwidth limit being exceeded:

- The `PacketDropCount` metric can alert if it passes a threshold.
- `NetworkPacketsIn` and `NetworkPacketsOut` can be used to understand the peak and sustained traffic. Compared with the service or infrastructure sizing, this metric can offer insight into the limitations if not explicitly printed by AWS.
- For VPN connections, `TunnelDataIn` and `TunnelDataOut` can help see whether the maximum tunnel bandwidth is being exceeded, the maximum being 1.25 Gbps per tunnel.

Some Amazon EC2 network adapters include enhanced monitoring. For example, the **Elastic Network Adapter (ENA)** offers metrics that include a reason for packet drops due to exceeding bandwidth on ingress or egress. This is a premium feature tied to the usage of an enhanced networking suite with the ENA.

PPS Limitations

In the same vein as bandwidth and throughput limitations, PPS limits can be tough to diagnose by normal metrics alone. While there are plenty of AWS services that publish PPS limits, this information is lacking when it comes to EC2 instances. What this means is that, often, a PPS limitation could result in packet drops and not be clearly attributable to that limit. AWS offers no explicit details for the PPS limits on each EC2 instance size but suggests running a test using the `iperf` network benchmarking utility to find the approximate PPS limits on an instance.

The simplest way to correlate suspected drops with a PPS limit being exceeded is to look at the number of packets being sent and received and compare it to a good baseline. If there are significantly more packets moving through an instance (ingress and egress), which is causing packet drop counters to increment, but bandwidth is well within the normal acceptable rate, suspect that the drops are from the artificial PPS limit imposed by the hyperplane.

Alternatively, as with the bandwidth/throughput metrics, a specific metric does exist pertaining to packet drops due to PPS limits but is only available when using the ENA adapter with the instance. This adapter is intended for more detailed and granular metrics around networking.

Packet Loss

The bane of every network engineer in or out of the cloud, packet loss often defies explanation while severely impacting applications. What metrics or logs could help explain packet loss in the cloud, and how can they be captured?

Often, packet loss can be a symptom of a different problem entirely. An example of this would be a third-party router or firewall from a vendor running as an EC2 instance. Relevant metrics might include the CPU and memory usage of these appliances. If they are undersized, they may be dropping packets due to the inability of that instance to keep up with demand. Alerting on NVA performance is an excellent way to get proactive notification of impending problems.

If there is no virtualized network infrastructure, the same can be true of cloud workloads running the applications themselves. Packet loss can be observed from an overwhelmed end host, though the alerting and diagnosis of this often fall into the responsibility of the system administrators.

If there is no virtualized instance at all (as in the case of serverless architectures), the possibility of troubleshooting the problem yourself is almost nil. Your cloud network runs on a completely managed infrastructure that cannot be monitored by you as the cloud consumer. However, AWS offers a comprehensive status page that covers the health of its regions and services to help you understand the state of the underlying infrastructure so that you can understand whether the AWS hyperplane is experiencing issues that may impact your cloud deployment.

Additionally, **AWS Network Manager** offers an infrastructure performance view that allows you to subscribe to real-time monitoring of AWS Regions and Availability Zones inside of CloudWatch. Of course, other reasons exist for packet loss, as a cloud consumer, you have no control over the physical layer issues or other underlying infrastructure problems that can occur.

Security Misconfiguration

Unintended denial of connectivity is often attributable to incorrect security posture or application of policy. Determining that security is to blame and not any number of other causes of connectivity loss can be a real challenge. Security rules can prevent connectivity as completely as if it were complete packet loss or missing routes, causing you to waste time verifying basic connectivity. The best way to combat a misconfigured security policy involves tools that can test reachability end to end and incorporate the routing and security policies that come into play on a tested flow.

Depending on where the security policy or rule at fault lies, testing can be done with tools that AWS provides or with a third-party security policy tool. The location of the fault determines what tool will correctly indicate a problem. If the misconfigured firewall rule is on a third-party NVA firewall, tools such as **AWS Reachability Analyzer** will not be able to see the problem because native cloud tools lack insight into the operations of a third-party network appliance, and vice versa. Likewise, if an NACL tied to a VPC subnet is to blame, the third-party firewall tool is unlikely to see the problem. The best way to diagnose misconfigured security rules is to understand the impacted flow end to end and determine where security may be applied in-line using AWS Reachability Analyzer in conjunction with any third-party vendor tools that govern the security solution in place. Once this is done, the security policies and rules in play can be inspected to ensure they are applied correctly.

Multiple Administrators Making Changes

When trying to deal with problems associated with changes, the best way to track down the issue is with an audit tool. A good audit tool should be able to correlate what exactly was changed, when that change occurred, and what account implemented the change.

AWS has a service called **CloudTrail**, which can correlate all these data points, providing the answer to any questions concerning changes through AWS. CloudTrail delivers logging to an Amazon **Simple Storage Service (S3)** bucket for review and can be retained indefinitely based on the retention policy.

Of course, an audit tool is only as good as the organizational structure it supports. If accounts are sharing credentials among multiple users, for example, accountability could be lost. CloudTrail can help best when it is enabled where needed and when accounts are traceable to a single user.

AWS Performance Monitoring Services

In this section, you will explore the AWS services supporting monitoring and performance in detail using the AWS tools provided for this purpose, such as the following:

- CloudWatch
- CloudTrail

- VPC Flow Logs
- VPC Traffic Mirroring
- Network Manager

The purpose of monitoring is to aid in capacity planning through the interpretation of trends, to diagnose and troubleshoot issues occurring across your network, and to gain a better understanding of the behavior of applications across your organization. By the end of this section, you should understand the tooling that AWS provides, and, at a high level, which tool is appropriate to use when diagnosing or troubleshooting problems that occur.

Amazon CloudWatch

CloudWatch is the AWS metric logging and alerting service. This book does not focus on the hundreds of AWS services that can use CloudWatch for logging, alerting, and visualization. Instead, the focus of CloudWatch in terms of the *Advanced Networking Specialty* exam is on network services, specifically, how to log, alert, and visualize them effectively. Before getting specific, however, you need to spend some time exploring CloudWatch itself.

CloudWatch is a collector of metrics and logs from AWS services that can also take basic actions on the data it receives. It is a service that can be configured to capture one, some, or all metrics from an AWS service such as S3 object storage, EC2 instances, RDS databases, Transit Gateway, and others. CloudWatch can also be configured to visualize that data in a dashboard, send alerts based on preconfigured thresholds, and even take actions based on the alerts.

The previous sections of this chapter covered the problems that can occur on a cloud network as well as the information needed to diagnose and alert on them. CloudWatch provides the answers to a lot of these problems tied to metrics, logging, and alerting. The following are some of the primary concepts used in CloudWatch:

- **Alarm**: CloudWatch can be set up with certain rules to act based on the metrics it consumes. The metric must be collected over a configured time and include a threshold at which action should be taken. A popular use for CloudWatch alarms is to trigger an Auto-Scaling policy that will build and add more servers to an application group if the servers come under heavy load.

- **Namespace**: This is a container for CloudWatch metrics used to separate the metrics into isolated environments to avoid mixing data.

- **Metrics**: This refers to a set of data that is published to CloudWatch from a source. Each metric is tied to a particular set of data, ordered into slices of time, and consumable by users. Metrics can be preconfigured, such as EC2 CPU metrics, or a custom metric from a custom application that can publish to CloudWatch. Metrics are tied to the region in which they are collected and consist of a name, a namespace, a time stamp, and data that can be measured.

- **Dimension**: This is a category for the characteristics of a metric. These dimensions can be used to filter or order the data from the metric. For example, if a metric is published that includes the tags for the resource, the tag is a dimension that can be used to filter or order the metric data.

- **Period**: This refers to the length of time associated with a statistic, representing an aggregate time period in which metric data was collected. As an example, a period of 15 minutes would include all data captured within a 15-minute period.

- **Percentile**: Very useful in tracking down anomalies, the percentile is used to mark the value of data within a set. If looking for evidence of an unexpected spike in traffic, a graph of traffic sent within that time would illustrate the effect by standing out, perhaps showing that within a short period of time, traffic spiked to within 90% of the baseline measurement that will have been established by steady state usage. Percentile is a measurement that is very reliant upon a baseline to be effective.

AWS CloudTrail

In terms of the *Advanced Networking Specialty* exam, CloudTrail is mostly useful for understanding how to audit network changes for correlation. While CloudTrail has a rich ecosystem dealing with compliance, governance, and risk auditing, these responsibilities do not typically fall within the responsibility of a cloud network engineer. Now, you will explore CloudTrail further in the context of the exam.

CloudTrail is to change control and audit what CloudWatch is to metrics and logging. Most activities that can be taken in AWS are logged and auditable by CloudTrail, whether through the AWS console, **application programming interface (API)**, **Command Line Interface (CLI)**, or **Software Development kit (SDK)**. The source of the changes is not important, just the identity and permissions of the user or role implementing the changes. Importantly, not all AWS resources can be monitored by CloudTrail. The following are the primary concepts used with CloudTrail:

- **Events**: Events are a record of activity within an AWS account taken by a user, role, or service that can be monitored by CloudTrail. Events are logged in a JSON format and are broken into event types.

- **Management event**: These events deal with operations related to the administration of AWS accounts and resources, such as logging in to the AWS console or administrating a user access policy with **Identity and Access Management (IAM)**.

- **Data event**: These events deal with operations related to data-plane-impacting actions, such as executing an AWS Lambda function or deleting an object in an S3 storage bucket.

- **Insight event**: These events only exist if the CloudTrail Insights service is turned on. The service looks for unusual API call rates or error rates in the AWS account that might indicate anomalous behavior such as an attack. For example, if an API call spikes to 10 times its normal call rate within a short period of time, or if there are a very high number of `AccessDeniedException` errors logged on resources, CloudTrail Insights will log this activity to an S3 bucket for that purpose. CloudTrail Insights logging can also be watched by CloudWatch, and alerts sent based on this unusual behavior.

VPC Flow Logs

As a cloud network engineer, it is very likely that you will make use of the VPC Flow Logs service. Cloud networking is often an opaque window into which customers have little visibility. VPC Flow Logs attempt to give some visibility beneath the covers into the decisions that the AWS network is taking for traffic flows and what is impacting them. It is not the same as a packet capture, instead offering basic metadata about a flow and the forwarding decision made by the VPC router and AWS network. VPC Flow Logs excels in the following tasks:

- Showing `ACCEPT` or `REJECT` forwarding decisions associated with cloud-native security rules
- Validating the traffic that is passing through a VPC to an instance
- Understanding what traffic is being sent and received to an instance or within a VPC/subnet

At the most basic level, VPC flow logs are logs of the forwarding decisions made for a network flow with a consumable frontend for customers. The logs can be filtered, certain metadata included or excluded, and formatted in a certain way. In that respect, they are somewhat like CloudWatch, though VPC Flow Logs are focused entirely on the traffic within a VPC. This tool is meant for both real-time analysis as well as archival, though archival could become expensive based on how widely the logging criteria are set. These logs can be archived to S3, published to CloudWatch, or even sent to an Amazon Kinesis Firehose service, which is an AWS service meant to ingest a lot of real-time data. Think of VPC Flow Logs as a basic NetFlow record that can be exported or viewed directly.

The following is the default format of a basic VPC flow log:

```
[version, accountid, interfaceid, srcaddr, dstaddr, srcport, dstport,
 protocol, packets, bytes, start, end, action, logstatus]
```

Here is a VPC flow log example using that format. This traffic was between two EC2 instances, one connecting to the web socket of another:

```
2 123456789010 eni-1334c3ca123453186 10.30.15.220 10.30.15.211 21622
443 6 32 14302 2318522010 2318528010 ACCEPT OK
```

VPC flow logs are a powerful way to troubleshoot and verify traffic within a VPC, even if the source or destination is outside the VPC. They can also be custom formatted to troubleshoot specific problems, such as TCP handshake connectivity issues related to missing a SYN or ACK. Here is an example of a VPC flow log format that captures the TCP flag sequence for troubleshooting:

```
version vpc-id subnet-id instance-id interface-id account-id type
srcaddr dstaddr srcport dstport pkt-srcaddr pkt-dstaddr protocol bytes
packets start end action tcp-flags log-status
```

Here is the example log in practice:

```
3 vpc-0140149df3f335532 subnet-0e0e2ad681ddcb029 i-01e9e43234abbc72a
eni-0676fc5401bc7193e 795297372191 IPv4 172.31.39.209 172.31.35.65
22 38836 172.31.39.209 172.31.35.65 6 2081 10 1690898079 1690898136
ACCEPT 19 OK
```

Importantly, for this test, the **Secure Shell** (**SSH**) request was refused by the other server due to mismatched SSH keys at the application level to illustrate an important point about flow logs: the flow logs do not have any awareness of the application-level decisions, only the network level.

VPC flow logs can be captured at three different layers within a VPC based on what resolution of traffic is needed. Logs can be captured at the **elastic network interface** (**ENI**) level, the subnet level, or the VPC level. The ENI level is the least noisy and most specific but will only log traffic involving that network interface. Creating flow logs at the VPC level will give the largest amount of context but also create the most noise and cost significantly more.

The correct place to create VPC flow logs depends on a series of qualifiers:

- How much is already known about the traffic needing to be logged?
- How much other traffic is needed to understand the context of what is being investigated?
- Will the VPC flow log criteria metadata match what needs to be captured for investigation?

Often, VPC Flow Logs are created to investigate a reported problem. Depending on how much data can be ascertained about the problem, the resolution level of VPC flow logs needed might be clear. For example, if a problem is reported for a single server, the log can probably be created at the ENI level of the server to further investigate. If the problem is very broad or without many details, it makes more sense to capture it at the VPC level and inspect the traffic in CloudWatch or Amazon Kinesis Firehose. If the issue is reported between several application servers in the same VPC, it would make the most sense to capture flow logs at the subnet level.

VPC Traffic Mirroring

Traffic mirroring is a fancy term for packet captures. Given the fabric nature of the cloud and the virtual nature of networking within it, things such as capturing packets "on the wire" have become a complicated affair. AWS offers VPC Traffic Mirroring as a service to aid cloud network engineers in capturing packets directly from ENIs for the purpose of troubleshooting and visibility.

Put simply, traffic mirroring copies network traffic from an interface to a destination target interface attached to an instance created to receive the traffic for analysis. This is commonly a security appliance for intrusion detection purposes or content inspection associated with troubleshooting problems. However, the target could also be a load balancer with listeners behind it.

For any on-premises network engineers, this behavior will seem familiar. Essentially, AWS is allowing the creation of a **SPAN** port as you would find on a network switch. The following concepts are associated with VPC Traffic Mirroring:

- **Source**: The source interface to copy traffic from
- **Target**: The destination interface to copy the traffic to
- **Filter**: Any criteria that limits what traffic is mirrored between the source and target
- **Session**: The relationship between source, target, and filter

You will explore these in more detail in the following sections.

Source

The source of a VPC Traffic Mirroring session must be an ENI. Specifically, it must be an ENI attached to an EC2 instance. Unfortunately, the ENIs associated with other AWS services such as **NAT Gateway** or **Transit Gateway** are not eligible for VPC Traffic Mirroring. This means that traffic mirroring will not be a good method of troubleshooting if packet-level analysis is needed on native AWS services outside of EC2.

Target

The target of a traffic mirroring session must be one of the following:

- **ENI of the interface type**: The simplest option for the target of a traffic mirror is another interface. This interface will be attached to some form of EC2 instance ready to receive the copied traffic and do something with it, most likely inspecting or capturing it. Because the target is an ENI, traffic between the source and target will be routed in a one-to-one manner using normal routing techniques such as TGW or VPC peering.

- **Network Load Balancer:** AWS Network Load Balancer is a cloud-native network construct that consists of a frontend IP that uses flow-based five-tuple load balancing to backend listeners, usually EC2 instances. The load balancer will only deliver traffic to backend listeners marked as operational by accepting a health check from the **Network Load Balancer** (**NLB**). In the case of VPC Traffic Mirroring, these listeners need to accept traffic on UDP port 4789 for traffic delivery. VPC Traffic Mirroring does not have fault detection or recovery, so it will continue to deliver to the target even if the load balancer is discarding traffic due to the absence of backend listeners.

  For high availability, it is recommended to deploy the NLB in multiple **Availability Zones** (**AZs**). The backend instances that will consume the mirrored traffic should be deployed in a different AZ as well. To ensure that load balancing can continue in the case of failure, cross-zone load balancing should be enabled.

  Perhaps the most important thing that sets the NLB apart from the **Gateway Load Balancer** (**GWLB**) is that the NLB will route traffic to the backend listeners using standard routing options such as VPC peering or TGW.

- **Gateway Load Balancer:** The AWS GWLB is an entire topic all on its own, but for the purposes of VPC Traffic Mirroring, the important thing to understand is that the load balancer can be the target of a traffic mirror session, and the backend listeners can be in an entirely different VPC with no connectivity between. This is due to how the GWLB encapsulates traffic and, using the AWS underlay, delivers it between endpoints.

  Unlike the NLB, the GWLB endpoint listens for all IP packets and forwards to the backend listener. Because the GWLB uses the GENEVE tunnel protocol, the maximum MTU of mirrored packets is reduced to 8,446 (IPv4) or 8,426 (IPv6).

  For high availability, it is recommended to deploy the GWLB in multiple AZs. The backend instances that will consume the mirrored traffic should be deployed in different AZs as well. To ensure that load balancing can continue in the case of failure, cross-zone load balancing should be enabled.

Filter

This is a rule set that determines what traffic is mirrored from the source to the target. As with VPC Flow Logs and the choice to log at the VPC, subnet, or interface level, the information at hand will be important in deciding how to build a traffic filter. For example, in the case of a server suffering from problems related to DNS, it may be most helpful to limit the traffic mirror to DNS traffic to and from that instance for investigation. Alternatively, if the problem is broad and could have multiple root causes, the filter should be wider to ensure the issue is not missed.

Like firewall rules, VPC Traffic Mirroring filter rules are built and evaluated from a top-down perspective. The first rule that is evaluated where traffic matches the filter will result in that traffic being mirrored. Here is an example of a set of filter rules for the inbound direction and the resulting action:

Number	Action	Protocol	Source Port Range	Dest Port Range	Source CIDR	Dest CIDR
10	Reject	TCP (6)		443	Management Network	VPC Network
20	Accept	TCP (6)			0.0.0.0/0	0.0.0.0/0

Table 2.4: Sample Inbound Traffic Mirroring filter example 1

Table 2.4 ensures that SSL traffic (which will be encrypted) from the management network to any host in the VPC is not selected for the traffic mirroring, while all other TCP traffic will be mirrored. What if the rules were reversed? *Table 2.5* shows the table when the rules are reversed:

Number	Action	Protocol	Source Port Range	Dest Port Range	Source CIDR	Dest CIDR
10	Accept	TCP (6)			0.0.0.0/0	0.0.0.0/0
20	Reject	TCP (6)		443	Management Network	VPC Network

Table 2.5: Sample Outbound Traffic Mirroring filter example 2

Traffic mirror rules are evaluated from the top down, and the first rule that matches the traffic will select that traffic for mirroring. In this case, because the first rule matches all TCP traffic, the SSL management traffic would be selected and mirrored, ignoring the second rule completely. If you are familiar with how NACLs, route maps, and firewall rules work, this behavior should be familiar.

Session

A **session** is like a container for the previous concepts, the table of a relational database that intersects how these concepts relate to each other. By combining a source, target, and filter into a session, the traffic mirror has all the criteria needed to capture relevant traffic.

When a VPC Traffic Mirroring session is set up, AWS creates and allocates a tunnel encapsulated with the **Virtual eXtensible LAN** (**VXLAN**) protocol for the express purpose of mirroring traffic between a source and target interface. Multiple sessions can be sent to the same target, but it is up to the capturing application to sort out the different streams. Importantly, multiple sessions can also pull from the same source, meaning you can have different filters applied and different targets as well.

Consider a situation where you may want one application's traffic to go to a dedicated capture device but want other traffic to go to a more generalized intrusion detection system, such as with applications that require the traffic to be captured for later analysis. It is possible to create two sessions using different filters and targets but the same source.

The source and target of a VPC Traffic Mirroring service do not have to exist within the same VPC. Commonly, the target resides in a shared services VPC or other dedicated appliance VPC for this purpose while the source could be anywhere. Based on where the source and target are in relation, and whether the target is an ENI or a load balancer, the connectivity between the two may have different requirements. Traffic mirroring can even be done between AWS accounts, with one account owning the source and another the target. *Table 2.6* explains how this works:

Source Owner	Source VPC	Target Owner	Target VPC	Connectivity
Dev Account	Dev VPC	Dev Account	Dev VPC	No additional configuration needed
Dev Account	Dev VPC	Dev Account	Dev VPC 2	Intra-region VPC peer, TGW, or GWLB endpoint
Dev Account	Dev VPC	Database Account	DB VPC2	Cross-account intra-region VPC peer, TGW, or GWLB endpoint
Dev Account	Dev VPC	Database Account	DB VPC	VPC sharing (explained in *Chapter 3*)

Table 2.6: VPC Traffic Mirroring source/target options

The main detail to take away from this table is that VPC Traffic Mirroring will not work with a VPC peering across regions. A TGW can route the session across regions through TGW peering (since TGW is a regional service). A GWLB endpoint can only load balance to VPCs in the same region, though no connectivity between the source and target VPC needs to exist outside the endpoint and load balancer itself.

> **Note**
>
> You can read more about traffic mirroring and target connectivity options here: `https://docs.aws.amazon.com/vpc/latest/mirroring/traffic-mirroring-connection.html`

Transit Gateway Network Manager (part of AWS Network Manager)

AWS has provided a unified dashboard to understand and manage cloud and hybrid cloud connectivity with the **Transit Gateway Network Manager** (**TGNM**). This system is part of AWS Network Manager and allows customers to build a logical and graphical network topology using a logical grouping of containers to visualize an entire network. Resources can be added to the containers in a way that helps to organize the network deployment logically using relevant criteria.

Network Manager (**NM**) is part organizational tool, part troubleshooting tool, and part visualization tool. AWS allows customers to register their current AWS cloud infrastructure easily with NM, offering full integration with TGW. Customers can also create containers for network objects that have no network data plane impact, such as sites and countries, and then register on-premises routers, firewalls, and other resources within these containers for organizational purposes. While AWS does not use the information registered about on-premises equipment to make routing and forwarding decisions, by registering the information with NM, the data is available and searchable in context from within the dashboard.

NM straddles two AWS products, including CloudWAN. This version of the AWS *Advanced Network Specialty* exam does not cover CloudWAN, so there will be no extra space devoted to it in this book. It is mentioned now because during preparation and research for the exam, you will find that AWS tends to use TGNM (in scope for this exam), NM (the overall solution for management of both TGNM and CloudWAN), and CloudWAN itself (not in scope for this exam) interchangeably. In fact, the documentation for NM is found under the documentation for CloudWAN, despite the Transit Gateway Network Manager not being part of the CloudWAN solution. This can get confusing, so you can organize the hierarchy as follows:

- NM:
- TGNM
- CloudWAN

Of these three, only TGNM and NM (in the context of TGNM) are in scope for this exam. Even focusing on only TGNM, NM allows us to provide those logical containers in a hierarchy, as follows:

- Global network:
  - Sites
  - Devices
  - Transit Gateway Network

An example of NM for the sample organization Trailcats is shown in *Figure 2.1*, including several sites:

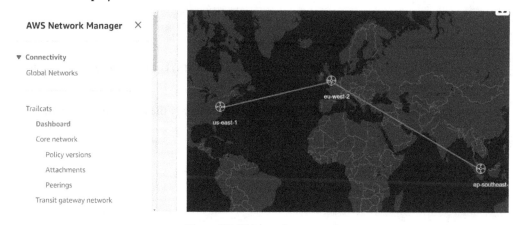

Figure 2.1: NM – Sites

In the case of TGNM, the only data plane, cloud-specific resources are the Transit Gateways themselves. They are registered with NM and form the basis of the visualization of the global network that a customer builds. *Figure 2.2* shows an example of the Trailcats global network, currently only showing three TGWs deployed:

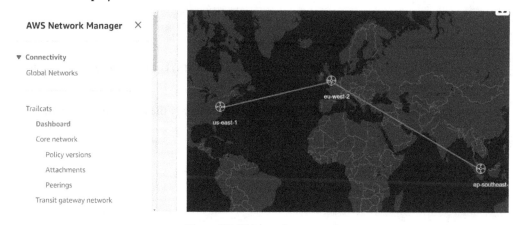

Figure 2.2: NM sample geography

Along with the containers, there are other attributes that are important but do not necessarily affect the data plane. This will be further explored in *Chapter 14, Data Analytics and Optimization*.

Some SD-WAN vendors have integrations with NM, and these vendors have features that will automatically register with NM for monitoring and visualization. For companies with large SD-WAN deployments, this lessens the amount of manual effort needed to account for the devices and sites in NM and accelerates the monitoring of the global network from within the solution. This will be explored more in *Chapter 6, Connecting Third-Party Networks to AWS*.

Monitoring and Troubleshooting

This section focuses on how to create logging and alerts for the various services in AWS, followed by some common troubleshooting methods you may be called on to use as a cloud network engineer.

Log Creation for CloudWatch and S3

Logs must be exported somewhere for analysis. No matter what service is being logged, CloudWatch requires a logging group to be created to serve as a container and target for those logs. Commonly, log groups are segregated by function or service.

Creating a log group in CloudWatch is very simple. It can be done via the AWS CLI, SDK, or API, but in this case, it will be done using the AWS console. If the Trailcats administrators found themselves in a situation where a particular server or AWS service needed to be logged for diagnosis of an intermittent, recurring issue, for example, they would need to create a CloudWatch log group to serve as the target to collect these logs. Here is how they could do that:

1. From the CloudWatch dashboard, select `Log Groups`.

2. Select `Create Log Group`.

3. For `Log group details`, enter the following:

   - `Log group name`: Name for the log group

   - `Retention setting`: How long to retain the logs

   - `KMS key ARN`: Which **Key Management Service (KMS)** key to use for encryption, if the data should be encrypted (optional)

   - `Tags`: Standard key/value tags to help classify and identify the log group (optional)

Figure 2.3 shows the log group details filled in:

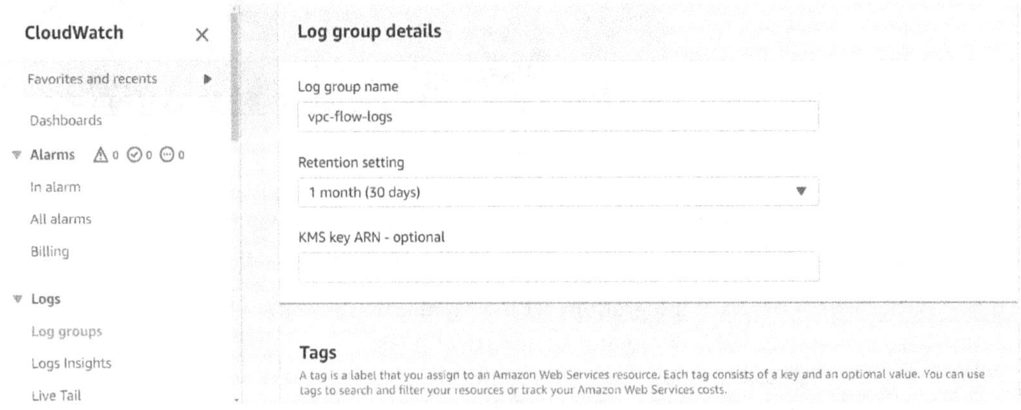

Figure 2.3: CloudWatch log group creation

> **Note**
> AWS often changes the AWS console so this may look different on your system.

Alternatively, many AWS services allow the logs to be exported to an S3 bucket for retention. In the case of Trailcats, suppose that the administrators were interested in logging a high-volume application for later trend analysis and these logs would take multiple gigabytes, or need a longer retention than CloudWatch provides. It would make more sense to use an S3 bucket and then import the data from the bucket into CloudWatch as needed. Set up an S3 bucket for this purpose by going through the following steps:

1. Open the Amazon S3 dashboard.

2. Select `Create Bucket`.

3. Now, enter the following information and make selections (not exhaustive):

    *   `Bucket name`: This must be unique across AWS

    *   `AWS Region`: The region in which the S3 bucket is to be created

    *   `Object ownership`: Use a bucket policy or bucket ACL to determine who can read/write

    *   `Block public access`: This blocks any attempt to access the bucket from the internet

    *   `Enable or disable bucket versioning`: With versioning enabled, updates to a file result in new file creation instead of overwriting existing files

    *   `Encryption method`: By default, AWS uses AWS-managed keys with **Server-Side Encryption S3 (SSE-S3)**, but you can also use **AWS Key Management System (AWS SSE-KMS)** if you have keys in KMS

    *   `Tags`: Specify cloud-based object tags using the standard key/value system (optional)

Figure 2.4: S3 bucket creation for logging

Alternatively, many of these settings can be copied from another S3 bucket, if one exists. In *Figure 2.4*, this option is highlighted. This is a helpful feature if the S3 buckets will be capturing logs for different services or different parts of your organization, and logical segregation is needed.

When the S3 bucket is created, it can be targeted by AWS service logging options as a destination. In most cases, configuring the logging under the service will automatically create a resource policy that allows the service to write to the S3 bucket, but in rare cases, it may be necessary to create a resource policy that allows writing to S3 buckets using IAM and allow the service to assume that role for write permissions.

VPC Flow Log Creation

Now that the most common log repositories are covered, it's time to move on to specific services. VPC flow log creation allows you to target the VPC, subnet, or ENI level for flow logging. The log destination can be a CloudWatch log group, an S3 bucket, or Amazon Kinesis Firehose. Firehose is beyond the scope of this exam, but for completeness, it is a data stream service that accepts real-time logging, can optionally use AWS Lambda to transform the data, and then forwards that result to another AWS service for consumption. VPC flow logs can be created using the AWS CLI, SDK, API, or console.

If Trailcats was experiencing an intermittent network problem, VPC Flows Logs could help isolate that by analyzing the flow and behavior of traffic for a particular VPC, ENI, or service.

Follow these steps to set up a VPC flow log at the VPC level using the AWS console:

1. From the VPC dashboard in the region where the VPC is located, select the VPC.

2. From the `Actions` menu, select `Create Flow Log`. Alternatively, select `Create Flow Log` from the details of the VPC below the list.

3. Enter the following information:

   - `Flow log name`: This is optional

   - `Filter`: Whether to capture only accepted or rejected traffic, or everything

   - `Maximum aggregation interval`: The maximum interval of aggregation for the flow log

   - `Destination`: CloudWatch, S3, or Kinesis Firehose in the same or different account

   - Details of the destination based on selection (CloudWatch log group, S3 bucket, etc.)

   - `Log record format`: The format of the flow log record (AWS default or custom)

   - `Tags`: Cloud-based object tagging in standard key/value format (optional)

The process is the same for creating flow logs at the subnet or ENI level; the flow log must be created from that resource and supply the same information to begin logging.

VPC flow logs are often created to diagnose a specific problem. For example, suppose that two application servers in a VPC are unable to communicate, and it is not clear why. VPC flow logs can be used to diagnose the problem. Assuming the following topology, think of all the possible causes why a server in the App1 VPC would be unable to connect to the server in the App2 VPC:

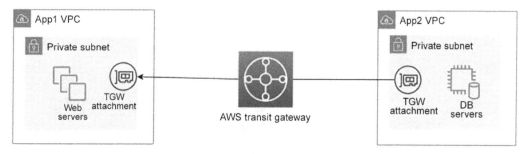

Figure 2.5: Sample VPC connectivity diagram showing direction of traffic

Suppose the users of Trailcats reported a problem with the Trailcats website, and some initial diagnosis determines that the problem is due to a lack of connectivity between the two servers. Although the problem is isolated to an issue between the servers, the issue could involve one of the servers or anything in between. In this case, as the problem relates to a lack of connectivity, the flow log should focus on rejected traffic first. The flow log can be created using the AWS console by performing the following steps:

1. From the EC2 console, select `Network Interfaces`.

2. Locate the ENI of the server to be logged. Select it.

3. Select `Actions` > `Create Flow Log`. Alternatively, from the pop-up detail menu, select `Flow Logs` > `Create Flow Log`.

4. Enter the information. Set the filter to `REJECT`.

5. Set the aggregation period to 1 minute.

6. Set the destination to CloudWatch logs to a previously created logging group.

Now that the flow logging is set up, it's time to inspect the flow logs to see what the problem is. In this example, you will just use **Internet Control Message Protocol** (**ICMP**) to simulate the server connection.

There's just one problem. There are zero flow logs coming in, which could mean the issue is not with rejected traffic, or the traffic might be getting rejected before it reaches the server. Troubleshooting often involves multiple tests to gather data points. The next step might be to widen the net and see whether the issue is somewhere else within the VPC. Creating a VPC flow log at the subnet level yields a different result, which can be determined by setting up the flow log at the subnet level and then going to CloudWatch to inspect the result, as shown in *Figure 2.6*:

Figure 2.6: CloudWatch inspecting VPC flow log events

Now it makes sense why the server ENI was seeing no rejected traffic. This was because it was being rejected before it got to the server. Inspecting the VPC flow logs further, you can see the following:

- The version of the flow log is the default (version 2).

- The AWS account ID (obscured for anonymity).

- The ENI that sent the flow. This is tracked back to the expected server interface.

- The packet was seen leaving the source server 10.200.1.36 with a destination IP of 10.200.2.12, as expected.

- The source port and the destination port look fine (this is ICMP after all).

- There were several bytes sent, and there are Unix timestamps for the beginning and the end of the flow.

- The flow was rejected by something within the VPC, likely a network ACL since this is happening at the subnet level and not at the ENI level.

All the evidence points to an NACL associated with the App1 server subnet. Upon inspecting the subnet and seeing an NACL attached, the culprit is found, as shown in *Figure 2.7*:

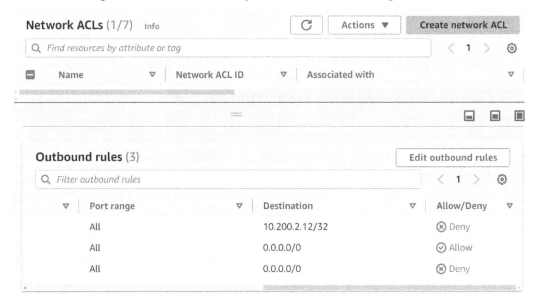

Figure 2.7: NACL outbound rules

A deny rule in the outbound NACL is blocking traffic. VPC Flow Logs can also include custom formats for the logs depending on what is needed. Because AWS often changes the platform, it would be pointless to write an exhaustive list here.

> **Note**
>
> You can find the up-to-date VPC Flow Logs fields available for use here: `https://docs.aws.amazon.com/vpc/latest/userguide/flow-logs.html#flow-logs-fields`.

Load Balancer Access Logs

Troubleshooting load balancer issues did not become simpler upon moving to the cloud. Luckily, AWS provides the ability to generate logs for multiple load balancer types, as well as set up CloudWatch metrics to look for problems. This section will focus on what details the load balancer can give you and how to use them effectively. For the purposes of this chapter, you will focus solely on how to set up logging, and in *Chapter 14, Data Analytics and Optimization*, you will focus on interpreting what's logged.

Network Load Balancer

Access logs for the NLB is an optional feature and is disabled by default. The unfortunate thing about NLB access logging is how little can be logged. Though the NLB itself supports a wide range of connection types, only **Transport Layer Security (TLS)** connections can be logged.

> Note
>
> As stated, AWS NLB can only log TLS requests in access logs, and only if there is a TLS listener configured. The access logs will only contain information about TLS requests. If an exam question refers to an NLB and uses access logs to log any other type of connection, it's false.

NLB access logs can only be exported to an S3 bucket for further analysis. The NLB access logs are sent every five minutes, not in real time, and they are treated as eventually consistent, which means the logs themselves may not all show up at once. The S3 bucket must be in the same region as the NLB sending logs, and the bucket policy must allow the AWS log delivery service to write to the bucket. The data can, of course, be encrypted with AWS KMS. The log files themselves are a free service with NBL. Their storage with S3, however, as with all data, does carry a change.

> Note
>
> You can find further lists of the fields and file format output here: `https://docs.aws.amazon.com/elasticloadbalancing/latest/network/load-balancer-access-logs.html`.

Application Load Balancer

All the things that are true for NLB are also true for the **Application Load Balancer (ALB)**. The biggest difference is that the ALB is an application-layer load balancer that terminates incoming requests. This means the requests expected are web-layer by default, and the app-layer information is used to load balance the requests.

Like the NLB, the access logs are disabled by default and must be enabled to use. The logs can be sent to an S3 bucket with the same requirements as the NLB logging capabilities.

Exam Tip

Unlike the NLB, the ALB access logs are the best effort, meaning there is no guarantee of delivery. They are intended to be used as a request behavior tool, not to search for specific requests. If there is a question on the exam about analyzing ALB logs to find a specific request, it is false.

Unlike the NLB, the ALB access logs are extremely detailed in the nature of requests and how that request was treated. Reprinting the entire list would take up a lot of space, but here are a few examples:

- Actions taken:

    - `Authenticate`: The request was validated, the user authenticated, and the user information was added to the header of the request

    - `Redirect`: The incoming request was redirected to another URL as specified in the rules

    - `WAF`: The load balancer forwarded the request to AWS **Web Application Firewall (WAF)**

- Classification (if the request was not a valid request, as stated in RFC7230):

    - `Bad Header`: The header of the request contains null characters or carriage returns

    - `Empty Header`: The header is empty or contains only spaces

    - `Bad Version`: The version request is malformed

- Error codes:

    - `Auth Invalid Cookie`: The authentication cookie is invalid

    - `Auth Invalid Token`: The authentication token is invalid

    - `Auth Unhandled Exception`: The load balancer hit an unhandled exception

Note

You can find an exhaustive list of other codes and responses in the access logs at `https://docs.aws.amazon.com/elasticloadbalancing/latest/application/load-balancer-access-logs.html`.

CloudFront Logs

CloudFront is the AWS **content delivery network** (**CDN**) and allows users to access content for websites and applications from the closest geographical location. The details of CloudFront will be explained further in *Chapter 10, AWS CDN and Global Traffic Management*. In this chapter, the focus is simply on how to set up logging to monitor and troubleshoot. CloudFront has multiple logging options available based on what part of the service is being monitored.

Access Logs

Just like the ALB/NLB, CloudFront access logs involve exporting the data to an S3 bucket. With CloudFront, the log files are based on the CloudFront distribution itself, which is associated with the URLs that users are accessing. Different edge locations associated with the same distribution will still end up logging to the same S3 bucket and log files. The entries cover each request; if there is no request, there is no log. This helps constrain the amount of log files to focus on where the users are using the CloudFront CDN.

It is important that the S3 buckets that serve the content associated with a distribution are not used for logging as well to simplify lifecycle policies and maintenance. Not every AWS region supports CloudFront access logging. Make sure to check for supported regions before deciding on an access logging strategy and where the data will be kept. As with most S3 resources, the logs can be encrypted using AWS KMS, but with some caveats. KMS must use a customer-managed key for this, and the policy must be configured to allow CloudFront to deliver logs using that key.

> **Exam Tip**
> S3 buckets that will be used for CloudFront logging must use a bucket ACL and not a bucket policy, and permissions must be granted to a special account called `awslogsdelivery`. If an exam question refers to a bucket policy for CloudFront logging, it is false.

As with the ALB access logs, the logs themselves are eventually consistent and not in real time and should be used to understand the nature of the requests, not used to troubleshoot a real-time problem.

Real-Time Logs

To help monitor and troubleshoot real-time issues, CloudFront supports real-time logging. AWS still advises that even real-time logging is not a full account of every request and will not show every transaction. Real-time logs are delivered on a best-effort basis as well. Unlike standard access logs, real-time logs support a configurable sampling rate, fields, and patterns for which to receive logs.

Real-time logs can be delivered only to Amazon Kinesis Data Streams or Kinesis Firehose, two AWS services associated with log ingestion and analysis (which can then be delivered to an S3 bucket). This is consistent with the AWS service offerings in general, where streaming telemetry is best directed to a stream-capable service.

There is a charge for real-time logs and Kinesis, and depending on the volume of requests, the requests may have to be broken into multiple streams for consumption. The AWS Pricing Calculator can help estimate the data throughput to Kinesis stream requirements.

There are 40 (as of mid-2024) fields that can be configured for real-time logs. This list is not exhaustive but gives a sample of some of the configurable fields:

- **X-Edge-Location**: The AWS Edge location that served the request
- **X-Host-Header**: The domain name of the CloudFront distribution (for example, `xyz123.cloudfront.net`)
- **X-Forwarded-For**: If a proxy was used (or ALB), this field includes the IP of the original requester
- **C-Country**: The country of request origin

Because real-time delivery is to an Amazon Kinesis stream, CloudFront requires an IAM role with sufficient rights to send logs to those services. The permissions also need to include permissions to use KMS if the stream should be encrypted.

> **Note**
> You can find the exhaustive list of all 40 configurable log fields and the IAM policy details here: `https://docs.aws.amazon.com/AmazonCloudFront/latest/DeveloperGuide/logging.html`.

Troubleshooting Packet Size Issues

This section will be dedicated to the sort of problems that happen when packets and frames are mismatched along the path of a network flow, as well as how to capture, diagnose, and correct the problem.

Trailcats recently migrated an application to the cloud from the on-premises data center as part of a migration project. After migrating the application to the cloud and setting up connectivity to an AWS TGW over an S2S VPN, users have complained about slowness in application performance. It's time to investigate the problem. *Figure 2.8* shows the relevant traffic flow:

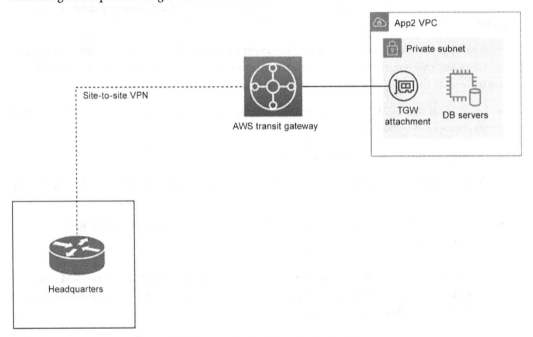

Figure 2.8: On-premises traffic going to cloud resource

In this case, the application is hosted in the AWS cloud, and the users are connecting to the application over the new WAN connection. To further investigate the slowness problem, consider what changed when migrating the connectivity between a locally hosted data center to one connected over the internet with a VPN.

Assuming the application and users haven't changed, the underlying transport is the only current culprit to be investigated. The application is working but slow, which should rule out issues such as security and routing reachability. The first thing to look at is the application itself, and for that, a VPC traffic mirror is needed to capture traffic for analysis.

Capturing packets in the cloud is not as simple as connecting a laptop to a switch port and spanning the server port in question. Because the underlying infrastructure is opaque, it's necessary to use the services AWS provides for troubleshooting.

The order of operations looks something like the following:

1. Deploy an EC2 instance to capture the mirrored traffic. Make sure the instance has an ENI dedicated just to the traffic mirror so you can still manage the device during an active mirroring session.

2. Set up the VPC traffic mirror:

    - Set up the source as the ENI of the server to be mirrored.

    - Set up the destination as the ENI of the dedicated interface on the instance you deployed.

    - A filter is required but can be wide open (0.0.0.0/0). Don't forget to set an inbound and outbound rule to capture traffic entering and leaving the server.

3. On the EC2 instance to be used for the traffic mirror, start whatever packet capture application you plan to use.

4. Start the application test.

5. On the EC2 instance being used for the mirror, stop the capture after sufficient data has been captured and inspect it.

For the purposes of this example, the **application traffic** is a crafted ICMP packet meant to emulate a realistic application payload. For the test, an on-premises **user** at 10.123.1.1 sends traffic to the application server at 10.200.1.7. The following code is captured, which will be analyzed in the following lines (some of the output is removed for brevity):

```
Internet Protocol Version 4, Src: 10.123.1.1, Dst: 10.200.1.7
    Total Length: 1480
    Time to Live: 253
    Protocol: ICMP (1)
    Source Address: 10.123.1.1
    Destination Address: 10.200.1.7
Internet Control Message Protocol
    Type: 8 (Echo (ping) request)
    Data (1452 bytes)
```

This seems normal, and no immediate culprit jumps out here. Continuing to look through the capture, it becomes apparent there was unexpected behavior, and this could be the cause of the problem:

```
Internet Protocol Version 4, Src: 10.123.1.1, Dst: 10.200.1.7
    Total Length: 1480
     Time to Live: 253
     Protocol: ICMP (1)
     Source Address: 10.123.1.1
     Destination Address: 10.200.1.7
Internet Control Message Protocol
     Type: 8 (Echo (ping) request)
     Data (470 bytes)
```

That was unexpected. This application should have only sent a single packet for its payload, and yet there are two packets showing on the capture. Inspecting the details further, you will see that though the total length of the payload was 1480 bytes, the payload of this second packet is 470 bytes. What's more, the first packet was 1452 bytes. Clearly, the application payload is being changed.

MTU, MSS, and the Impact of Tunnels in Networking

All Ethernet-based networking has rules about how the frames are built that must comply with the IEEE standard. Among those rules are the maximum size of any standard-sized Ethernet frame. The Ethernet standard specifies that a standard frame cannot exceed 1,500 bytes, including the payload (the actual data) and all the headers, such as VXLAN encapsulation, **Generic Routing Encapsulation** (**GRE**) tunnel overhead, IPsec encryption headers, and so on.

This means that to comply with the standard, the maximum payload available for an application sent in a single packet/frame must fit within the total headers equaling at most 1,500 bytes. Of course, there are jumbo frame standards that expand the MTU to 9,000 bytes, but this is reserved for private networks and between devices that support it such as within a data center or CSP underlay. Over the internet, the 1,500-byte standard is strictly enforced.

Why does this matter? It matters because if a payload plus headers exceeds the maximum MTU of 1,500 bytes, the payload must be fragmented into sizes that fit and reassembled, usually by the receiving host. This requires more processing overhead as it must both transmit the device and receive it (or whatever device is responsible for reassembly). Also, the application itself may slow down or mangle requests because of having its data mangled and reassembled, especially ones that require quick processing times. Here's how tunneling can start impacting the payload:

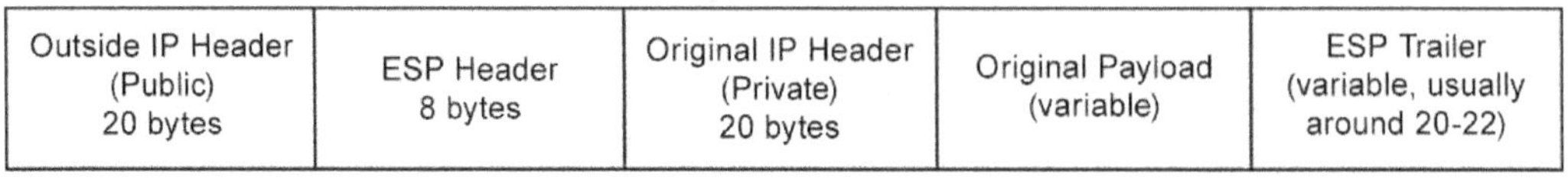

Outside IP Header (Public) 20 bytes	ESP Header 8 bytes	Original IP Header (Private) 20 bytes	Original Payload (variable)	ESP Trailer (variable, usually around 20-22)

Figure 2.9: IPSec tunnel overhead

In this scenario, the only overhead being added comes from the site-to-site IPsec tunnel between headquarters and AWS TGW. However, there are many ways to tunnel, and in fact, it's becoming common to do tunneling inside of tunneling for some use cases. Many new solutions simply add a VXLAN header and a GRE header and it is repackaged as some form of new tunnel technology. Tunneling lowers the payload amount a single packet/frame can hold, and fragmentation has a negative effect on the performance of the network. Because of this, there are ways to combat it at different layers.

At a pure network layer, network interfaces can be set with a configured lower MTU rather than the default of 1,500 bytes. This allows network operators to force packets that will exceed 1,500 bytes to be fragmented prior to adding the tunnel encapsulation. This setting is not possible on the AWS side of a connection as it is managed by AWS, but on the on-premises hardware, the MTU setting can be configured per the vendor's instructions. Note that this does not actually prevent fragmentation, it just artificially lowers the MTU so the device knows fragmentation will be needed prior to encapsulation with most but not all vendors.

Another way to combat fragmentation is at the TCP layer using the **maximum segment size (MSS)** setting. Without exhaustively explaining TCP, the protocol is negotiated between two hosts and deals with reliable traffic delivery. This means that some settings for the connection are negotiated as well. MSS is not strictly negotiated, but the sender is made aware of the maximum size available by the receiver. The MSS strictly determines how much data can be sent in a single TCP segment and has the effect of modifying the payload, which can avoid fragmentation altogether.

The catch is that this is only supported in applications using TCP for delivery, and the receiving host must be aware of the packet size limitations or be artificially configured with a lower limit. Alternatively, most network vendors do allow you to set an artificial TCP MSS adjustment value on the network interface, which will artificially lower the TCP MSS in the SYN packet of a TCP negotiation, effectively accomplishing the same thing. Again, this is only configurable on the on-premises hardware, not in AWS.

The most effective way to solve the packet size mismatch problem associated with tunneling is at the application layer. Remember that the payload is by far the bulk of a frame/packet, and it is up to the application to decide how much data to send at a time. An application that supports a feature called **Path MTU Discovery (PMTUD)** can send probes with the DF bit set that helps understand the maximum allowable MTU along a connection. This works because the behavior of ICMP is to send a specific type of error message back to the originator if fragmentation is required but not allowed. The application can use this information to adjust its payloads, ensuring fragmentation is not needed.

So, returning to the scenario with Trailcats, the problem and possible solutions should be clear. The **application traffic** is being fragmented, causing unwanted processing and data reassembly. This has the effect of slowing down the application. Among the proper solutions, the most effective one is usually to update the application to use PMTUD. However, in practice, this is usually the toughest to implement because the application may be critical and unable to be refactored to use PMTUD, it may be an off-the-shelf solution that cannot be modified, or it may take too long to fix. It's more likely that if the application uses TCP, then TCP MSS adjustment can be used at the network layer, or if not using TCP, the MTU can be artificially lowered. The last one isn't really a solution as fragmentation will still be needed.

One last solution that would not help in this case but may help in others where the full transit is private and not using the internet involves setting the connection between AWS and the on-premises device to use jumbo frames. Note that this must be supported along the entire path between AWS and your equipment to be useful; if leasing from a colo provider or traveling across a provider WAN, it will be important to verify that there is jumbo frame support there as well.

AWS Direct Connect does support jumbo frame settings at 9,001 bytes for private VIF, and AWS TGW does support it at 8,500 bytes for Transit VIF. The setting is changed in AWS under the Direct Connect console, and changing this setting will briefly impact connectivity as the settings are changed in AWS. Of course, you will also need to update the on-premises devices as well.

Different Network Interfaces in AWS

To finish this chapter, it's time to cover in detail what the different network interfaces in AWS are, how they differ, and with what AWS resources they can be used.

Elastic Network Interface

What makes the ENI truly elastic is that it is a logical construct that can be moved between resources. There are some limitations to this, however. For example, the ENI is tied to a VPC and AZ, but within that, an ENI can be attached, detached, and reattached as needed between instances. The operating system running on the EC2 instance may react in unpredictable ways, but at the virtual infrastructure level, this is a simple operation.

Some features of the ENI are as follows:

- An IPv4 address from the address range of the VPC is allocated to the primary ENI. This ENI cannot be detached from the EC2 instance.

- A MAC address is allocated to each ENI.

- SGs are tied to an ENI and not an instance! An ENI can also have multiple SGs.

- A source/destination check flag, which is enabled by default, is used with each ENI. This check ensures that traffic destined to or from the ENI is allowed, but not traffic through it. This should be disabled for network infrastructure to allow the forwarding of packets through the ENI.

- ENIs can have one primary IPv4 and multiple secondary IPV4 addresses allocated. The actual limit of this is determined by the EC2 instance size and type (compute-optimized, network-optimized, general use, etc.).

- A single public elastic IP can be allocated to an ENI, but if you have your own customer pool of elastic IPs in AWS, you can associate one per private IPv4 IP address attached to an ENI.

> **Note**
>
> There can be only one AWS-supplied elastic IP attached to the primary IPv4 address of the ENI. If an exam question refers to attaching more than one AWS public elastic IP to an ENI or suggests attaching it to a secondary IPv4 address, it is false.

- One or more IPv6 addresses can be assigned to an ENI.

- ENIs can also have a prefix delegation, whereby you specify an IPv4 or IPv6 CIDR range to allocate to the ENI. This is mostly used for container networking where a container network interface needs more IPs to allocate for worker nodes.

Enhanced Networking with Elastic Network Adapter

The **Elastic Network Adapter** (**ENA**) is a specialized networking option within AWS that is supported on certain instance types. It allows enhanced networking metrics to be collected and uses a specific network technology called **Single-Root I/O Virtualization** (**SR-IOV**) to dedicate increased capacity to virtual network interfaces. SR-IOV is a virtualization method by which the hypervisor of physical infrastructure can be partially bypassed by dedicating slices of physical hardware. The result of SR-IOV is increased throughput, lower latency, and faster PPS performance. In addition, the enhanced metrics made available can be exported to CloudWatch for more granular control of alerting. If the instance supports it, it is recommended that you use enhanced networking as it comes with no added cost.

The requirements for using enhanced networking vary by the EC2 instance type and operating system of the virtual machine. Generally, the instance should be launched using a supported instance type, which will enable this automatically.

Tests can be performed to verify that the EC2 instance supports enhanced networking, and if not, some instances can download a package to enable it. For the exam, you will most likely not need to know the details of exact implementation but should understand the general steps on what enhanced networking is (and what ENA means) as well as that it is not supported in all cases.

The main reason to use the ENA with enhanced networking is to get better performance and to expose a lot of extra granular metrics in CloudWatch, covered earlier.

Elastic Fabric Adapter

The highest-performing network interface is the **Elastic Fabric Adapter** (**EFA**). The EFA is a highly specialized interface in AWS that is intended to be used with **high-performance computing** (**HPC**) and machine learning applications. Because it is highly specific to these applications, there is a tightly controlled list of supported instance types, libraries, and operating systems. The EFA uses **Message Passing Interface** software to reach its highest performance.

> **Note**
>
> You can find the exhaustive listing of all the currently supported resources here: `https://docs.aws.amazon.com/AWSEC2/latest/UserGuide/efa.html`.

The EFA is only supported on non-Windows instances and a subset of optimized instance types. The EFA is a specialized ENA and includes the features of that interface type. The EFA supports extra OS-bypass capabilities for increased performance. Like the ENA, the EFA is a free network interface upgrade for supported workloads. The EFA requires installing specific software packages on the EC2 instance to use the EFA capabilities based on the type of MPI software being used.

The reason to use an EFA instead of an ENI or ENA deals with the need to support HPC or machine learning applications. The MPI stack is used to support this special application, so the EFA is needed.

Summary

In this chapter, we examined the definition and purpose of monitoring and troubleshooting within the AWS cloud by listing the potential problems, pitfalls, and issues that can cause problems with a cloud network deployment, providing details on each of these potential issues, and the most effective tooling available in AWS to diagnose and troubleshoot them. We then walked through the details of creating some of the most useful (and used) diagnosis and troubleshooting tools in AWS. Understanding the problems that can plague a cloud network, how they can be monitored, and how the problems can be discovered using AWS tools is an important part of administrating a cloud network.

In the next chapter, you will read about networking across multiple AWS accounts.

Exam Readiness Drill – Chapter Review Questions

Apart from mastering key concepts, strong test-taking skills under time pressure are essential for acing your certification exam. That's why developing these abilities early in your learning journey is critical.

Exam readiness drills, using the free online practice resources provided with this book, help you progressively improve your time management and test-taking skills while reinforcing the key concepts you've learned.

HOW TO GET STARTED

- Open the link or scan the QR code at the bottom of this page

- If you have unlocked the practice resources already, log in to your registered account. If you haven't, follow the instructions in *Chapter 16* and come back to this page.

- Once you log in, click the START button to start a quiz

- We recommend attempting a quiz multiple times till you're able to answer most of the questions correctly and well within the time limit.

- You can use the following practice template to help you plan your attempts:

Working On Accuracy		
Attempt	Target	Time Limit
Attempt 1	40% or more	Till the timer runs out
Attempt 2	60% or more	Till the timer runs out
Attempt 3	75% or more	Till the timer runs out
Working On Timing		
Attempt 4	75% or more	1 minute before time limit
Attempt 5	75% or more	2 minutes before time limit
Attempt 6	75% or more	3 minutes before time limit

The above drill is just an example. Design your drills based on your own goals and make the most out of the online quizzes accompanying this book.

First time accessing the online resources? 🔒

You'll need to unlock them through a one-time process. **Head to** *Chapter 16* **for instructions**.

Open Quiz	
https://packt.link/ansc01ch2 OR scan this QR code →	

3
Networking Across Multiple AWS Accounts

Given a cloud consumption model where the entirety of goods and services belong to the cloud service provider, some conflicts may arise within a business about how to build, manage, and pay for IT infrastructure. Cloud infrastructure uses the **operational expenses (OpEx)** model, which is also known as the recurring charge model, and not a **capital expenses (CapEx)** model (buy once and depreciate over time, more common in traditional on-premises deployments). This is one of the main value propositions of cloud computing, after all. However, shifting from purchasing a large amount of equipment once every few years to a more consumption-based model can bring a corporate finance system to its knees as it struggles to pivot the budgetary requirements with the change in mind.

It's not uncommon in an enterprise for the IT networking budget to be centralized, but nor is it uncommon for it to be decentralized. A decentralized IT budget model, where each department or business unit funds its own IT services spending, tends to be less valuable in a world where assets are purchased and depreciated. By contrast, the central model, wherein each business unit funds a central organization charged with the acquisition and execution of business objectives, makes the most sense.

In the cloud world, however, where everything is charged based on consumption, it becomes a simpler and potentially more valuable exercise to attribute usage more closely with individual business units. For this reason, AWS supports either model through the use (or lack thereof) of services such as AWS Organizations, AWS Control Tower, and AWS Resource Access Manager.

In this chapter, you'll focus on how AWS allows different networking options across different accounts, and what that provides. Here's a brief rundown of the topics covered:

- AWS Organizations
- AWS Resource Access Manager
- Sharing VPC resources between accounts
- AWS PrivateLink
- Third-party network appliance connectivity

The focus of the AWS Advanced Networking Specialty exam is not on becoming an expert in account governance and policy, but it is important for a cloud network engineer to understand how to use network services across different AWS accounts. For this reason, you will focus less on the specifics of setting up account governance/policy and more on how it impacts network design. The blueprint for the ANS-C01 exam devotes very little space to multi-account networking, but to understand the purpose and value of multi-account setups to a business, you will learn a little more about the context.

AWS Organizations

Understanding the purpose of a service is a shortcut to understanding how to use it. Imagine a scenario where a business dips its toes into the cloud, and perhaps developers in a business unit start using it as a development playground. Over time, it gains traction and adoption. Other parts of the business begin leveraging cloud services to some degree, each swiping a credit card and consuming as needed. Before long, each department is separately in the cloud, with varying levels of governance and policy, with all charges being billed to the same company, potentially through different budgets. Now pretend to be that business' finance, infosec, and IT departments, charged with IT governance, security posture and policy, and financial auditing.

This is an unfortunate but common occurrence. AWS Organizations is a free service that allows a business to safely manage its business units in the cloud with a unified policy and governance framework, as well as consolidated billing; though, of course, the services used by your members still cost money. The Organizations service allows the implementation of categorization and hierarchy like Microsoft Entra ID. It allows you to create new user accounts under the organization, import existing ones, and consolidate their management.

For the purposes of the Advanced Networking exam, the most important part of AWS Organizations is simply understanding the framework and what it allows you to do with network resource sharing. Here is a high-level diagram of a typical AWS Organizations structure:

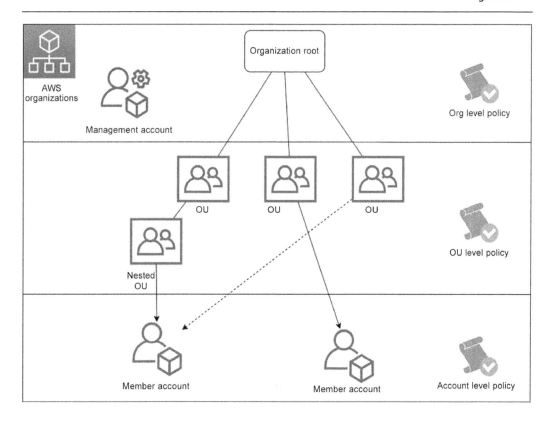

Figure 3.1: Typical AWS Organizations structure

This sort of structure allows individual member accounts to own certain services and share those services with the rest of the organization, such as VPC subnets and **transit gateways** (**TGWs**).

Since you haven't yet learned about the structure itself, this diagram may or may not appear simple, depending on your expertise. Therefore, to make certain that you have a strong foundation, the following subsections will review each component of this structure from the top down.

Organizational Root

The parent container for the entire AWS organization is the root. You may only have one parent container per AWS organization, and it is automatically created and cannot be modified. The root level has only a management account that's responsible for administrating the organization. Since the structure is top-down, any policy applied at the root level is inherited by all other objects below it.

Organizational Unit (OU)

An **organizational unit** (OU) is a logical container for accounts. While accounts are the main focus of organizations, OUs allow for the logical grouping and administration of accounts to align with business needs. For example, an OU containing AWS accounts associated with a product shipping business unit should contain only those AWS accounts in that business unit. An AWS policy associated with accounts that need to use AWS services for shipping products should be applied at this level, not the root level. The policy could be applied at the member account level, but that would require administrating each account and applying each policy separately. An OU allows bulk administration and prevents duplication of effort.

Since most businesses are not completely flat in their organizational structure, OUs can be nested within others. A good example of this might be a top-level IT OU that contains sub-OUs for networking, system administration, database administration, and others. Policies applied at the top level of an OU are inherited by sub-OUs as expected.

Member Accounts

This is the standard AWS account itself, in the context of the larger AWS Organizations structure. Member accounts will interact with the services and resources of the AWS cloud based on the configured policies of the organization. It is very common to deputize member accounts to be the administrator of specific services, including sharing those services with other accounts. Member accounts can only belong to one OU.

Remember that an AWS account is not a user account. The entire network team might have a single AWS network account but use individual user accounts with their own permissions and audit trails. AWS Identity and Access Management would be used to create a user account and grant access to the AWS account for whatever purpose is required. This is true even for the root-level management account. It should have a Delegated Administrator role for user accounts that will use the management account to administrate AWS Organizations. It is important to understand that an AWS account is a different entity from the users that will use those accounts.

Policy

There are two main policy types within AWS Organizations: authorization policy and management policy. Authorization policies are the most straightforward and cover the **service control policy** (SCP) that governs permissions. Management policies, by contrast, are more complex and involve ways to leverage AWS services to centrally manage all the accounts in your organization. As an example, a tag policy applied will not allow resources to be created under the level at which the policy is applied without including the requisite tags.

The SCP is a guardrail permissions policy that does not grant permissions by itself but instead works to limit the maximum permissions that can be granted or used by resources created under the policy. The IAM users and roles still must be granted the permissions themselves, but the SCP serves to limit the maximum effective permission to the policy level. Importantly, the SCP can only be used to restrict IAM users and roles created within the organization itself.

These SCP permissions also follow a hierarchical model, where permissions that explicitly deny access override those that grant permissions. This means that an SCP applied at the OU level granting access to certain resources may be effectively nullified by another SCP with an explicit deny applied at a level above.

Management policies also work on a hierarchical model, focusing more on enforcing compliance than on denying access. There are only three management policies, as of mid-2024:

- AI service opt-out policies (to avoid having content used to improve AWS AI services)
- Backup policies (create and enforce backup plans for your organization)
- Tag policies (enforce tag styles, names, and frameworks across the organization)

The next section will cover the tools used to create and manage account sharing.

AWS Resource Access Manager

While AWS Organizations is the framework for managing different AWS accounts, AWS **Resource Access Manager** (**RAM**) is the tool that allows the sharing of AWS resources between accounts inside or outside of an organization. Not every resource in AWS is shareable between accounts, even if they belong to the same organization, and this chapter will focus on the network resources that can be shared in this manner.

> Note
>
> AWS RAM does not work with AWS Organizations by default. The setting must be enabled in the AWS RAM console under `Settings`.

Resource Sharing

The act of sharing resources using AWS RAM involves three objects: the resource to be shared, the principal to which the resource share will be granted, and the permissions applied to the share, which determines what the principal can do within the share.

Note that RAM works with or without AWS Organizations! RAM allows any AWS account to share resources with other AWS accounts. The main benefit of using AWS Organizations is that it extends RAM functionality by allowing RAM to use Organizations objects for sharing, such as OUs. The second benefit of using Organizations is that the invite/acceptance step is omitted. When inviting an AWS account to share resources, normally, if the two accounts are not in the same Organization, an invitation is generated and sent to the consuming account for acceptance. With Organizations, this formality is skipped for accounts under the same organization root.

> **Note**
>
> Any test question that involves steps to accept a resource share across accounts should detail whether the two accounts are in the same organization. If this is not mentioned, assume they are not, and make sure the step to accept the invitation is included in any answer.

The list of resources that RAM can share is inclusive of many AWS network objects. Rather than exhaustively listing every one of them, we will cover the ones that are most impactful to network design and architecture across accounts:

- **AWS Route 53**: Route 53 Resolver rules can be shared with other accounts, which allows DNS queries from a VPC to be forwarded to a Resolver endpoint in another account. This can be helpful to avoid having to deploy multiple DNS resolvers in each account. Centrally managed DNS resolver rules can prevent the duplication of effort.

- **AWS Network Firewall**: AWS Network Firewall rules and policies can be created by one AWS account and then shared with other accounts. This allows a consistent experience across accounts and prevents more duplication of effort.

- **There are specific AWS VPC components that can be shared between accounts as well**:

  - **Subnets**: Subnets can be shared from a centrally managed VPC with other AWS accounts only within an AWS organization and cannot be used to share between accounts that are not a member of the same AWS organization. This resource sharing allows AWS accounts to create resources within the shared subnets without needing to own the VPC. Multiple AWS accounts can share subnets, allowing then to share resources and communicate without needing to use VPC peering or a TGW.

  - **Transit Gateway**: Sharing a TGW between accounts allows those other accounts to build connectivity to the shared TGW, facilitating connectivity between resources within accounts. Unlike subnet sharing, TGW sharing can be done with any AWS account, not only ones within the same AWS organization.

- **VPC Traffic Mirror**: By sharing VPC Traffic Mirror targets, AWS can centralize the collection of VPC Traffic Mirror sessions. This makes sense as VPC Traffic Mirroring targets are virtual machines or load balancers to send session data to backend collectors for analysis. A lot of architecture would be duplicated if this could not be shared because each AWS account would need its own infrastructure. For example, if the VPC Traffic Mirror targets were restricted to a single account, each account would have to set up its own monitoring infrastructure to capture VPC traffic, driving up costs significantly. For this reason, most organizations opt to create one or two sets of traffic mirror infrastructure and share the targets with other accounts.

Outside of AWS RAM, other AWS network services support multi-account sharing within their own workflows. As an example, a **Direct Connect gateway** (**DXGW**) in one account can be shared with a **virtual private gateway** (**VGW**) or TGW belonging to any AWS account. The association proposal mechanism is used to send and accept an invitation; as part of the invitation, all or a subset of prefixes can be shared with the DXGW.

AWS RAM provides a simple way to enable resource sharing between AWS accounts. Sometimes, the resource itself becomes part of the share, as shown in the next section.

Sharing VPC Resources between Accounts

Each account can create its own VPCs and subnets. First, in a central IT model, it makes sense for a network services team to manage the allocation of resources, especially IP addresses. This team would be responsible for IP addressing on-premises, so it makes sense that the same team is responsible in the cloud.

VPCs can have overlapping IP addresses, but this model only works if all the resources within a VPC can stay entirely within that VPC. Otherwise, two VPCs using the same IP address space is a serious problem for communication due to the confusion caused when trying to route traffic to the correct resources. Consider the deployment shown in *Figure 3.2*, where each team selects its own IP addressing for the VPCs it owns and the resulting problem it presents when each team selects the same IP address schema:

App VPC
10.25.0.0/16

Web VPC
10.25.0.0/16

Database VPC
10.25.0.0/16

Figure 3.2: Overlapping VPC IPs

In *Figure 3.2*, there is no way to facilitate communication between the three tiers of this application. Each VPC overlaps the other, and while that's an extreme case, the problem should be evident. Without governance and management of the IP address schema, different AWS teams and AWS accounts could easily end up overlapping with each other and then have no simple way to communicate, even with cross-account VPC peering or TGWs.

The cloud has traditionally been the domain of the developers, and developers often created network resources using a bare-minimum-effort approach. In the case of a completely self-contained application playground, there was no problem, but when constructing enterprise-grade applications using AWS principles such as loose coupling, the poor network foundation translated to high complexity and difficulty once interoperation was needed. A perfect example is when developers built a sprawl of VPCs without concern for IP address conservation, summarization, and overlap between other VPCs. Connecting these VPCs together to support applications adds extra complexity.

Another issue with non-shared cross-account connectivity has to do with the data charges incurred by connecting different resources using things such as TGWs and VPC peering, as well as cross-**Availability Zone (AZ)** charges. Consider this diagram, which shows VPC peering across different AWS accounts:

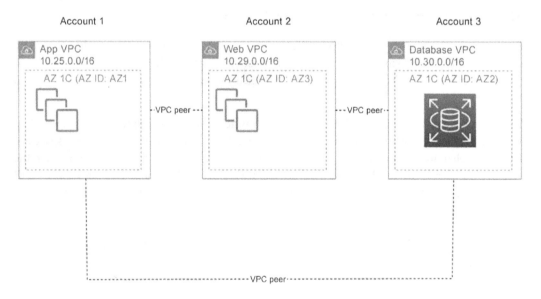

Figure 3.3: Cross-account VPC peering

In this case, different AWS accounts still own their network infrastructure, and each has established cross-account VPC peering to facilitate app communication. Notice that in this scenario, the VPC IP CIDRs do not overlap, as VPC peering does not support peering VPCs with overlapping CIDRs.

This setup will work, and the three-tier application should be able to communicate. However, this is a potentially inefficient use of resources that will cost the business more money. While VPC peering within the same AZ is free, cross-AZ peering is charged at standard data rates. Where this really comes into play is in understanding that each AWS account has its AZs mixed up to load balance demand across AWS data centers. In *Figure 3.3*, each AWS account deployed its resources into AZ 1C from its perspective, but the AZ IDs are the real indicator of what data center is being used. This app deployment will undoubtedly pay cross-AZ charges throughout its normal operation, which could quickly get expensive.

How does cross-account VPC sharing help with this? *Figure 3.4* illustrates the same three-tier architecture but built in a shared VPC model:

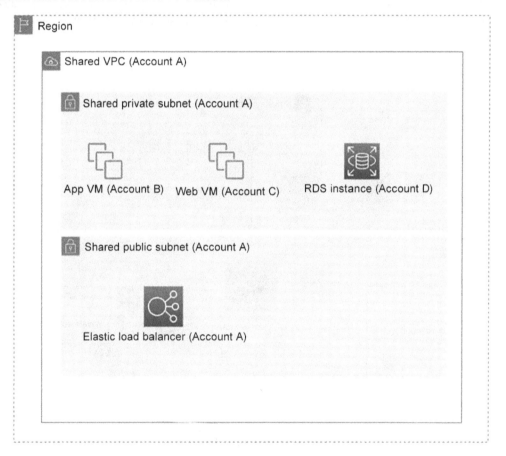

Figure 3.4: Shared VPC architecture

This deployment solves both the problems brought up earlier. Because the subnet of the VPC is shared between accounts, there is no need to concern yourself with whether the AZ IDs match. When sharing the VPC subnet, account A will determine which AZ to use, and when the other accounts deploy within the shared subnet, the AZ ID will automatically match. This allows efficient communication between resources without paying data transfer charges.

The other problem solved by this type of deployment is that the consumer accounts (B, C, and D) do not have to manage their own IP address allocation or networking. Account A handles all the management, and the other accounts are free to focus on what matters to them.

You can now explore some other design patterns that cross-account sharing enables. If there is a need to split workloads across different VPCs but connect them for communication, this is easily handled without requiring each account to maintain its own network infrastructure. Here is an example TGW shared network architecture:

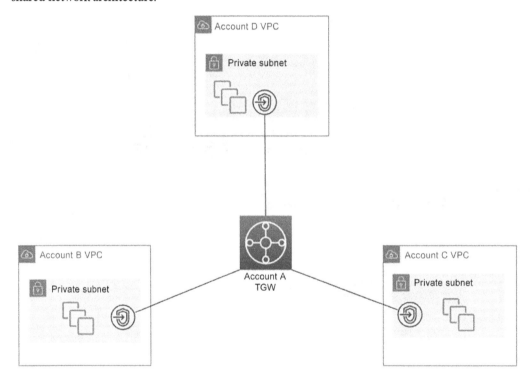

Figure 3.5: Shared TGW architecture

In this deployment, each account continues to manage its own basic network infrastructure. This is common in enterprises where the cloud has been in use for some time, each business unit has already deployed infrastructure, and a shared VPC model is not possible for logistical or historical reasons.

Account A uses AWS RAM to share the TGW with other accounts, either within or without the AWS organization. The TGW appears in each consumer account's TGW list once the share is accepted (automatically if within the same AWS organization, as mentioned before). Each consumer account is then responsible for creating a TGW attachment using the shared TGW to attach to its own VPC. The TGW route table is managed by account A as is the propagation of routes for attached VPCs, but within each account VPC, the subnet route table is managed locally by the individual consumer accounts.

AWS PrivateLink

Unlike other connectivity options, like VPC peering or TGWs, AWS PrivateLink is an endpoint connectivity service that privately connects VPCs to services as if those services were directly connected to the VPC. This service uses a managed network resource called a VPC endpoint. The use of a VPC endpoint allows you to connect private VPC resources to a service that exists outside the VPC without using a NAT gateway or internet gateway to reach it. This VPC endpoint service can be deployed into a VPC and made the target of traffic destined for the service at the other end of the endpoint by adding routes to the VPC subnet route table.

The VPC endpoint uses DNS for reachability, which means that DNS resolution within the VPC where PrivateLink resides is needed to provide the connectivity. This is orchestrated by AWS as part of the PrivateLink service, so no special steps are needed to utilize the VPC endpoint. However, it's important to know that while the VPC endpoint deployed as part of the PrivateLink service uses a private address from the subnet in which it is deployed, workloads connecting to the service will do so using DNS and not the private IP address.

PrivateLink/VPC endpoints are used to connect to AWS services but can also be used to connect to third-party services that are exposing a service endpoint, or even other custom endpoints within your own account. The endpoint is attached to a VPC and an AZ within that VPC (and hence, a subnet). Each VPC endpoint requires a security group to be attached to restrict communication. Commonly, the security group will include the expected traffic to cross that endpoint and nothing else.

In *Figure 3.6*, you will see a diagram of how a PrivateLink VPC endpoint works when connecting to a third-party service in AWS:

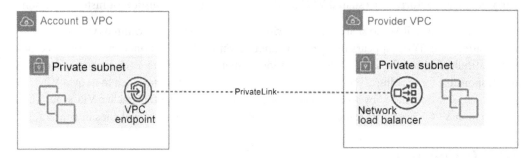

Figure 3.6: Basic PrivateLink connectivity to third-party VPC service

This is a common deployment when a third-party provider is exposing a service in its own environment to be used by customers, such as some form of ETL function that consumes data and transforms it into something else. The customer creates a VPC endpoint in a private subnet with a target of an exposed third-party service. On the provider side, the service exposed is a network load balancer tied to the AWS PrivateLink service, which will accept customer connections and pass along the data to its own backend pool of resources.

In *Figure 3.7*, a VPC endpoint deployment connects to a public AWS service, which can be accomplished without a NAT gateway or internet gateway:

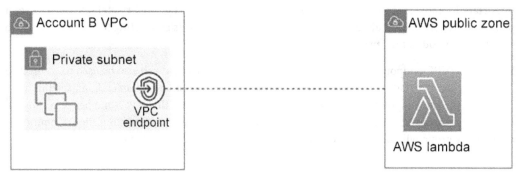

Figure 3.7: Basic VPC interface endpoint connectivity to public AWS services

This is a common connectivity pattern when connecting a private VPC to public AWS services. Unlike PrivateLink services, this kind of connection is not available to a third-party provider and is reserved specifically for connecting to AWS public services.

The next section focuses on network virtual appliances and providing cross-account connectivity.

Third-Party Network Appliance Connectivity

In this final section of this chapter, you will review how the use of other third-party connectivity can be impacted using cross-account connectivity. Commonly, the network team owns the on-premises and extra cloud connectivity into the AWS cloud environment. When using network virtual appliance-based connectivity, normally the network team owns the VPCs where the **Network Virtual appliances (NVAs)** are deployed. These NVAs exist primarily to connect other environments to the cloud, as is the case with **software-defined WAN (SD-WAN)** connectivity. SD-WAN is an industry term describing the use of software-based control planes to deliver network connectivity in a scalable way. The advantage over traditional network solutions is that SD-WAN scales efficiently and offers a single method of pushing network configuration and policy changes to a fleet of network devices. Multiple vendors offer different versions of their own SD-WAN solutions, and while knowledge of each is not in the scope of this exam, it's important to understand how an SD-WAN solution will generally integrate with an AWS cloud deployment. Here is a diagram of a typical SD-WAN deployment:

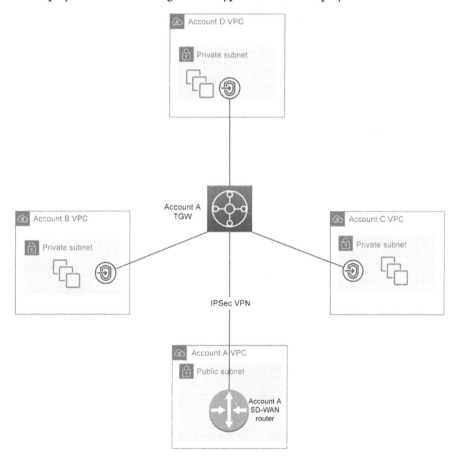

Figure 3.8: Basic SD-WAN integration with a TGW

Not pictured in *Figure 3.8* are the devices and locations connected to the SD-WAN network fabric outside of the cloud. This sort of integration allows the TGW and SD-WAN to exchange routes via BGP and for users attached to the SD-WAN fabric to access the other account's VPC resources. For example, in *Figure 3.7*, users connected to the SD-WAN on account A would have access to account B and account C VPC resources. Without cross-account connectivity, each AWS account would need its own connectivity to the SD-WAN for communication. That would be an inefficient use of resources.

Summary

In this chapter, you have covered several key components and services for managing and connecting multiple AWS accounts within an organization. You learned about AWS Organizations, AWS RAM, and sharing VPCs in a multi-account environment using multi-account connectivity options and resources such as TGWs, PrivateLink, and VPNs. You should now be able to understand how to provide network connectivity across multiple AWS accounts using various methods.

In the next chapter, you will learn about AWS Direct Connect.

Exam Readiness Drill – Chapter Review Questions

Apart from mastering key concepts, strong test-taking skills under time pressure are essential for acing your certification exam. That's why developing these abilities early in your learning journey is critical.

Exam readiness drills, using the free online practice resources provided with this book, help you progressively improve your time management and test-taking skills while reinforcing the key concepts you've learned.

HOW TO GET STARTED

- Open the link or scan the QR code at the bottom of this page

- If you have unlocked the practice resources already, log in to your registered account. If you haven't, follow the instructions in *Chapter 16* and come back to this page.

- Once you log in, click the START button to start a quiz

- We recommend attempting a quiz multiple times till you're able to answer most of the questions correctly and well within the time limit.

- You can use the following practice template to help you plan your attempts:

Working On Accuracy		
Attempt	Target	Time Limit
Attempt 1	40% or more	Till the timer runs out
Attempt 2	60% or more	Till the timer runs out
Attempt 3	75% or more	Till the timer runs out
Working On Timing		
Attempt 4	75% or more	1 minute before time limit
Attempt 5	75% or more	2 minutes before time limit
Attempt 6	75% or more	3 minutes before time limit

The above drill is just an example. Design your drills based on your own goals and make the most out of the online quizzes accompanying this book.

First time accessing the online resources? 🔒
You'll need to unlock them through a one-time process. **Head to** *Chapter 16* **for instructions**.

Open Quiz	
https://packt.link/ansc01ch3	
OR scan this QR code →	

4

AWS Direct Connect

In this chapter, you will learn about AWS **Direct Connect** (**DX**), how it is connected to the cloud physically and logically, and what options exist to bridge the connectivity gap between a customer and the cloud. Hybrid connectivity is an important part of any real-world deployment, especially those where cloud migration is a factor or where certain workloads will remain in a data center for reasons such as cost or legacy support models. Because AWS DX is the preferred method of establishing high-speed, private connectivity directly to AWS, you can expect it to feature in the exam. This chapter covers the different ways to deploy AWS DX and the design considerations of each.

Here's a brief rundown of the topics covered:

- Direct Connect Overview
- Creating a Direct Connect Connection
- Layer 2 and Direct Connect
- Direct Connect Gateways
- Border Gateway Protocol

This chapter will be very heavy on technical detail and process and requires an understanding of some concepts you have covered already, especially around routing. If needed, go back and recap the section on **Border Gateway Protocol** (**BGP**) and routing in *Appendix 1, Networking Fundamentals,* to catch up. By the end of this chapter, you should be able to understand how Direct Connect is physically and logically connected, what the different types of Direct Connect are, and how to use BGP to exchange routes with the different Direct Connect types.

Direct Connect Overview

Having access to build, borrow, or buy solutions in the cloud is a major benefit to organizations. The cloud's value propositions include the ability to quickly adapt to changing market conditions through rapid application change, as well as the elasticity of workload management that allows businesses to pay only for what they are using. This level of business agility is difficult to replicate in a traditional data center and is a large reason that the cloud has exploded in recent years.

But the data center won't disappear anytime soon – if ever. Some workloads are static and belong in a static environment where costs are well-known, understood, and planned. Users still consume applications in data centers that have not moved, or will not be moved, to the cloud. In short, the future of the cloud is hybrid connectivity, and that means that the cloud needs to support multiple extra cloud connectivity options to meet the needs of customers. One such connectivity option is a secure, private circuit between a customer and AWS, or a third-party collocated service such as Equinix or Megaport. This private circuit is managed by AWS and by the collocated service provider to extend cloud connectivity to customers, and it's called a **DX** onnection.

You are unlikely to get any specific questions in the exam about why anyone needs a DX connection, but before getting into the technical details, some context will be helpful. Consider the possible reasons why a business may want private connectivity across a traditional WAN:

- Connectivity is private and cannot be snooped on or captured by an attacker
- The use of private connectivity could obviate the need for a VPN or encrypting the packets between endpoints using a private connection
- It is easier to share routes between two locations to establish reachability
- **Service-level agreements** (**SLAs**) of private circuits are more reliable than those of a public best-effort connection such as the internet, though this comes with added cost
- Higher bandwidth throughput is possible using dedicated provider hardware and cables

The same business drivers that lead customers to choose a private WAN circuit also lead them to purchase private cloud connectivity. Before a solution such as DX existed, the only way to connect the cloud to an extra cloud location was via a VPN over the internet. That offered connectivity with lower bandwidth (limited by the maximum bandwidth of an IPSec VPN) and without any sort of business SLA. This worked fine when the cloud was a playground for development, but it is wholly unacceptable to a business that needs fast, reliable data transfers between the cloud and an extra cloud location, be it an on-premises data center or a collocated one.

In the case of Trailcats, our fictional enterprise, suppose that they had an initiative to move applications from their on-premises data center into the cloud. They could do so with IPSec VPN, but this would be a slow connection over the internet with no SLA to guarantee performance and reliability. Trailcats would want to leverage a private, dedicated, high-bandwidth circuit with an SLA to support its critical application migration activities from the data center to the cloud.

In short, the main reasons to select DX are privacy, reliability, and bandwidth. Let's move on to discussing how to acquire a DX connection.

Creating a DX Connection

Before any DX circuit can be implemented, there's a process that must be followed to order and provision the DX circuit. In this section, let's go over the process by which a DX connection is created.

Direct Connect Location

The very first decision to make when ordering a DX circuit is the physical location to which the circuit will be delivered. A DX is a physical infrastructure component, so it is tied to a physical location. The data centers where AWS offers Direct Connect are usually partner collocated data centers in which AWS has its own infrastructure. The Direct Connect connection can be delivered as a cross-connection directly from AWS to customer equipment, or through an AWS Direct Connect partner. The reasons for using a partner are usually twofold:

- The partner offers sub-rate connectivity if the customer does not require (or cannot afford) the fully committed DX rates

- The partner will take over delivery of the DX connection to the customer at another location (such as over a WAN or its own private circuit)

This diagram shows a common AWS-delivered DX connection to a third-party service provider that resells the DX service:

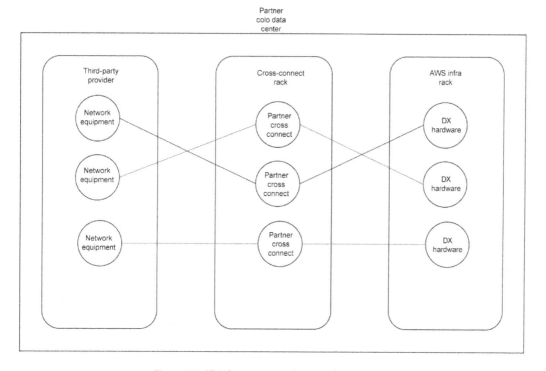

Figure 4.1: Third-party provider DX infrastructure

In this case, an AWS partner facility has racks of its own dedicated equipment for connectivity between AWS and a third-party provider. The DX hardware is cabled from the AWS rack to a partner-managed rack to facilitate cross-connectivity to the third-party provider, who can then subdivide the DX circuit to offer to end customers. The end customers may have equipment in this data center as well, or the provider may offer connectivity to the end customer over a WAN or their own dedicated private circuits.

Often, a customer has their own equipment and connectivity to a partner's collocated data centers, and in those cases, AWS can offer the DX connection delivered directly to their own equipment. The total bandwidth of the DX connection is governed by what's available at the data center but usually ranges between 1 Gbps, 10 Gbps, and 100 Gbps. If a customer wants less bandwidth, they must use a third-party AWS DX partner to sub-rate the DX, meaning the partner can provide a smaller, more cost-effective portion of the total bandwidth, and if the customers desire more than 10 Gbps but less than 100 Gbps, they can order multiple DX circuits.

Here is a diagram showing how **partner colo facilities** usually deliver DX circuits directly to customers:

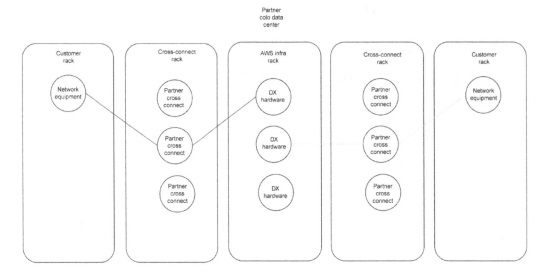

Figure 4.2: DX infrastructure delivered to customer

Note that facilities are almost always a mix of these connectivity options in the same racks and the diagram is merely to illustrate the difference in connectivity.

To get the connection to a DX port, the DX must be ordered from AWS at the partner's data center. AWS provisions a physical port on whose equipment, and then, with your information, provides a **Letter of Authorization (LOA)** to cross-connect the DX port to the customer's or provider's gear. Note that this DX port is a single port and has no resiliency, so commonly, an enterprise looking for redundancy will order multiple DX ports and bind them together using a **link aggregation group (LAG)**, which will be covered shortly.

> **Note**
>
> While a LAG provides a small measure of redundancy, such as a single port failure, it cannot protect against hardware failure where the members of a LAG share the same physical infrastructure such as switches, cross-connect hardware, or entire data centers. For the exam, the features of a LAG are not related to redundancy but to increase DX speeds and manage them as a single connection.

The LOA is a concept that should be familiar to anyone who has ever rented rack space in a collocated data center. It authorizes the data center technicians to perform work in the data center on customer equipment. In the context of AWS Direct Connect, the LOA form is created by AWS when a DX connection is ordered and includes information authorizing the partner's data center technicians to physically cable from the AWS DX rack to the cross-connect or customer equipment to provide the DX service. The official name for this document is a **Letter of Authorization and Connecting Facility Assignment (LOA-CFA)** and it is given to the customer requesting the DX for use. If the DX is to be connected directly to the customer equipment, the LOA-CFA should be given to the partner facility by the customer for cabling work, but if a third-party provider will be connecting the DX because the customer has no equipment or requires sub-rate speeds, the form should be given to them instead.

> **Note**
>
> To understand more about downloading the LOA, visit `https://docs.aws.amazon.com/directconnect/latest/UserGuide/dedicated_connection.html#create-connection-loa-cfa`.
>
> It should be noted that AWS has a Direct Connect presence in many partner data centers run by different companies and some specific processes may differ. AWS provides a Direct Connect cross connect resource for specifics, at `https://docs.aws.amazon.com/directconnect/latest/UserGuide/Colocation.html`.

Direct Connect and Physical Layer Requirements

When ordering a DX connection from AWS directly, there are specific hardware requirements for the cross connect. Here are the specifics:

- The DX speed can be 1 Gbps, 10 Gbps, or 100 Gbps
- The following single-mode fiber optics are required, specifically these types:
  - 1000BASE-LX for 1 Gbps
  - 10GBASE-LR for 10 Gbps
  - 100GBASE-LR4 for 100 Gbps
- Port speed and duplex must be hardcoded and not negotiated

Of course, if using a third-party provider to extend a DX to the customer, the physical requirements are dictated by the provider.

Figure 4.3 presents a diagram of a DX extended from AWS to a third-party provider and then to an end customer as an example:

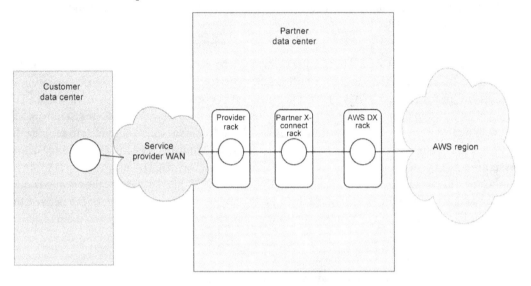

Figure 4.3: Sample connectivity using a third-party provider

This diagram shows the physical cabling path that connects the customer to AWS through the collocated data center. In this case, the WAN connectivity would most likely be a virtual private LAN service or some other emulated layer 2 connection to extend the layer 2 connectivity through the DX and into AWS. You will cover the layer 2 considerations of a DX in the next section.

The last physical layer concept to cover is one we briefly touched on already: the **LAG**. A LAG is a grouping of multiple physical connections into one logical aggregated link. In the traditional networking world, a LAG is often used to solve several problems, some of which would require their own book to really get into. For context, however, the benefits of a LAG include the following:

- A LAG can protect against cabling faults
- A LAG can protect against physical port failures
- A LAG can allow redundant links to the same network equipment to be used without causing a layer 2 loop with a spanning tree

Luckily, in the context of Direct Connect, the last one is managed by AWS. Here's a diagram of what a LAG looks like:

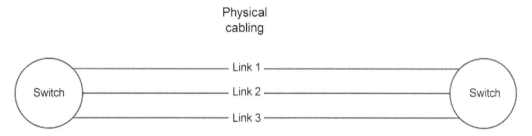

Figure 4.4: Physical LAG cabling diagram

Figure 4.4 shows three individual physical cables connecting two switches. It is physically the same structure, but the cables have been logically organized so they act as one connection, known as a LAG. When a LAG is created, the network devices treat the physical cables as one **logical cable** presented for use:

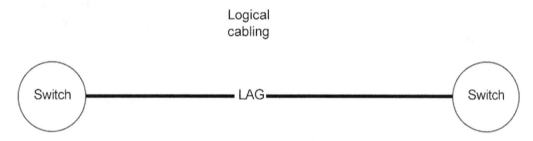

Figure 4.5: Logical LAG cabling diagram

The primary value of a LAG in the context of Direct Connect, at least from the AWS perspective, is the ability to get higher speeds and throughput in a single managed DX. For example, by binding multiple 1G DX cables together in a LAG, the speed can be extended without having to go to a full 10G port.

The LAG does come with some restrictions. The restrictions to be aware of are as follows:

- Members of a LAG must reside in the same DX location

- All members of the LAG must be at the same speed

- If jumbo frames are enabled, they must be enabled across all links in a LAG

- There are (as of now) a maximum of four members in a LAG available if less than 100G, or two if 100G, for a maximum of 200G speed on Direct Connect

- Customers must determine how many members in a LAG can fail before the LAG is considered failed, to avoid the DX becoming overutilized and effectively unusable for applications

The largest restriction on a LAG is that a LAG must be created on the same AWS hardware, and port capacity will determine whether a LAG can be created. This, of course, extends to the cross-connect hardware and, ultimately, either the customer or provider hardware as well. For instance, if a customer wants to create a LAG with four 10 Gbps links, all four links must terminate on the same AWS hardware device to form a valid LAG, and the port must be able to handle 10 Gbps. For example, in *Figure 4.3*, if a LAG were needed, there would need to be available ports on the same AWS equipment in the path.

Lastly, it should be noted that a LAG can be created from existing links (assuming they meet restrictions), as a greenfield connection, or a mix of the two.

Layer 2 and Direct Connect

AWS uses the layer 2 VLAN concept to extend the Direct Connect **virtual interface** (**VIF**), for connectivity to AWS resources across the DX. While the physical cabling of Direct Connect can be provisioned using as little as a single cable, or it can be as complicated as extending across an entire WAN, the VIF architecture is the same irrespective of the physical connectivity. But before we start talking about VIFs, let's spend a little time understanding the concept of a VLAN and how it ties back to VIFs.

At OSI layer 2, an Ethernet frame includes encapsulation for the payload, or the application data, and a few other fields. Without getting into the deep details of the OSI model and Ethernet technology, the most important fields include the source and destination MAC address (the addressing mechanism at layer 2) and a VLAN tag. **VLAN** stands for **virtual local area network** and is a logical way to separate networks utilizing the same physical infrastructure. Imagine the cabling diagrams (*Figure 4.4* and *Figure 4.5*), and within each of those cables are potentially hundreds or a thousand different networks, differentiated by a tag applied to each Ethernet frame.

Here is an Ethernet frame at layer 2, which is not exhaustive but illustrative:

Ethernet frame
headers

Destination MAC address	Source MAC address	802.1q VLAN tag	Data/payload	Frame check sequence

Figure 4.6: Simplified Ethernet frame headers

As referenced in *Figure 4.6*, the 802.1q VLAN tag is a differentiator that indicates to which virtual network the frame belongs. For example, a VLAN of 10 would be on a different network from VLAN 20, and there would not be any communication between devices in VLAN 10 and VLAN 20 without a device that is configured to route traffic between them. This VLAN tagging model comes into play as we start to discuss the concept of VIFs, as VIFs are also separated using the 802.1q VLAN tag as the mechanism by which to indicate to which VIF, and ultimately, to what destination in AWS, traffic belongs.

Remember that from a physical perspective, one or more cables are ultimately delivering the DX to equipment and there is not a VIF per cable or any physical separation in play. The VIF is a completely logical concept, but the VIFs created are tied to the DX itself. Having set that expectation, let's discuss what a VIF is and how it relates to AWS.

Virtual Interfaces and Direct Connect

There are three types of VIF in AWS:

- Public VIF
- Private VIF
- Transit VIF

These three VIFs cover all the possible ways to connect to AWS over Direct Connect. On a single Direct Connect, only 51 VIFs are allowed, and of those, only four can be transit VIFs. It's time to examine each in detail to understand what each VIF provides.

Public VIF

The purpose of the public VIF is to provide dedicated Direct Connect connectivity to AWS public services. Remember that AWS public services are normally accessible via the internet, which limits the reliability and bandwidth available to consume these services. By using an AWS public VIF over Direct Connect, customers have high throughput and dedicated, reliable access to AWS public services such as S3. AWS provides this by advertising the AWS IP addresses for public services over Direct Connect to the customer's or provider's equipment. This advertisement tends to be more specific than the general default route from an internet connection, making it preferable to use DX to reach those AWS public services.

This diagram helps visualize how on-premises DX connected to an AWS public VIF may be preferable to using the internet:

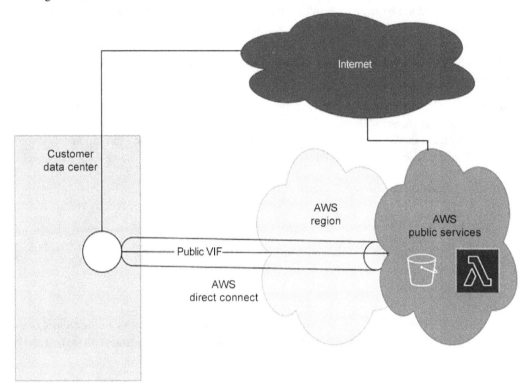

Figure 4.7: AWS Direct Connect with public VIF

In this example, the customer data center is connected directly to an AWS DX connection, and to the internet. DX has a higher throughput and a service-level agreement for reliability, so it is the preferred method to connect with public AWS services. By preferring the path over DX due to more specific prefix matching for AWS services, the customer can use the best path available without complicated traffic engineering.

Using a public VIF does require having public IP addresses available to advertise to AWS over the VIF. These prefixes will only be used to route AWS public services to you over DX; AWS will not advertise your public IP addresses to any third parties. Another important thing to know about a public VIF is that a DX in one region can access AWS public services in any region using DX; it is not region-locked like the other DX VIF types.

You will cover more details about using BGP to prefer paths for public services later in the chapter, in the *Border Gateway Protocol* section.

Private VIF

The purpose of a private VIF is to connect over DX to private AWS resources. These resources live within a VPC or VPCs and should be reachable via a reliable high-speed link without requiring a VPN for privacy. Commonly, the private VIF connects in-cloud EC2 resources to extra cloud resources (such as those in a data center or for users). Here is a diagram showing how a typical private VIF looks when connecting a customer data center to an AWS region:

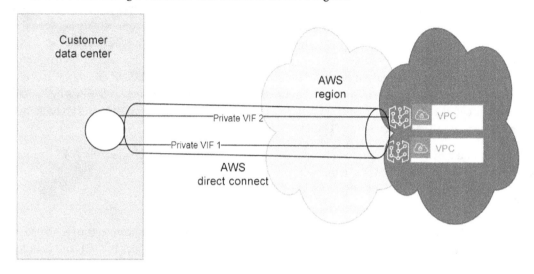

Figure 4.8: AWS Direct Connect with private VIFs

> **Note**
>
> A private VIF connecting to a VPC via a **virtual private gateway** (**VGW**) can only do so within the same AWS region. VPCs in other AWS regions are not accessible via a private VIF terminating on a VGW. If you see an exam question that involves connecting using a private VIF to a VGW in another AWS region, it is not the correct option choice.

Notice that, under normal circumstances, with a private VIF, each VIF is tied to one VPC via a VGW. There is another construct that can be used, which we will cover shortly, but the standard deployment uses a VGW. The customer equipment requires a different layer 3 IP address interface for each VIF, a separate 802.1q VLAN tag, and a separate BGP neighborship as well to learn and advertise the routes for each VPC via the VGW.

Because this is a private VIF, for BGP neighbors the BGP **Autonomous System Number** (**ASN**) needs to be between 64512 and 65535 in the reserved private BGP ASN range. The ASN should not overlap with other private VIFs. For private VIFs, the BGP route limit is 100 prefixes advertised to the AWS side over a single VIF. The VGW will, in turn, advertise the VPC CIDR and BGP peer IPs across DX to the customer's or provider's end.

This presents some interesting (and expensive) architectures when trying to connect between VPCs in the cloud. Imagine the setup in *Figure 4.8*, where the on-premises customer router is connected to both VPCs over Direct Connect. In this case, the customer router offers connectivity between the VPCs, especially if that router is advertising the BGP learned routes of each VPC to the other VPC. This may or may not be a desirable outcome (probably not). Direct Connect has an hourly charge and a data charge, but you also pay an egress fee for traffic leaving the cloud via DX. In this way, you could end up paying an astronomical sum, providing unintended connectivity between VPCs by having the traffic egress the cloud over a private VIF, hairpin through the customer router, and go back into another private VIF to a different VPC via the same DX connection.

You will find a better way to provide inter-VPC connectivity in the next chapter, via AWS **Transit Gateway** (**TGW**), which leads us to the next VIF option.

Transit VIF

The transit VIF is a special VIF that is neither public nor private but instead connects to AWS TGW via an AWS Direct Connect gateway. The Direct Connect gateway terminates the VIF and handles routing between DX and the customer equipment (or provider, if hosted). AWS Transit Gateway is a network connectivity construct within AWS that allows many other network constructs to connect to it, offering a hub and spoke style of communication. By using a transit VIF, a single VIF could gain access to many AWS services and VPCs based on what is connected to Transit Gateway. Here is a diagram of a typical DX connection using a transit VIF:

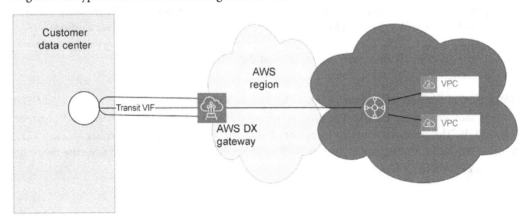

Figure 4.9: AWS Direct Connect with transit VIF

In this case, a single VIF is in use, and through Transit Gateway, many other VPCs are accessible. This moves the burden of routing from the on-premises router into the cloud, where it should be. It also prevents the hairpinning of traffic across DX and back into the cloud as traffic will use Transit Gateway instead. In many ways, the TGW is a managed router and provides all the benefits of a router inside the cloud. As an example, if a DX connection is terminated on TGW, any VPC attached to TGW should be able to use TGW for traffic to other attached VPCs, or to the on-premises networks being advertised to TGW. This is the common implementation when connecting DX to TGW.

From a route exchange perspective, commonly, the on-premises router will share the extra cloud routes with TGW, and TGW will either propagate the routes to the attached VPCs or not, as configured. The reverse is also true. TGW can advertise routes for the **Classless Inter-Domain Routing (CIDR)** of the attached VPCs over Direct Connect to the on-premises router or not, as configured. This will be discussed when covering hybrid networking with Transit Gateway in more detail in the next chapter. When cross-region DX connectivity is needed, TGW is no longer useful; the next section covers what to use in that case.

Direct Connect Gateways

AWS created **Direct Connect gateways (DXGWs)** to solve a limitation in private VIF architecture. Specifically, a DXGW allows you to overcome two very specific limits:

- A private VIF can only be used to connect to VPCs in the same AWS region as the DX location
- Only one VPC can be connected to a single private VIF (limited to 50 per DX)

The DXGW allows a private VIF in one AWS region to attach to VPCs in another AWS region and extends the number of VPCs that can be attached to a single private VIF to 10 instead of just one. The DXGW is a global construct not tied to any region, which is what allows this cross-region DX connectivity. A DXGW in this type of deployment works like a network switch, extending DX and VIF outside of their normal limitations.

Here is a diagram of a basic DXGW deployment with private VIFs extended to other regions and VPCs:

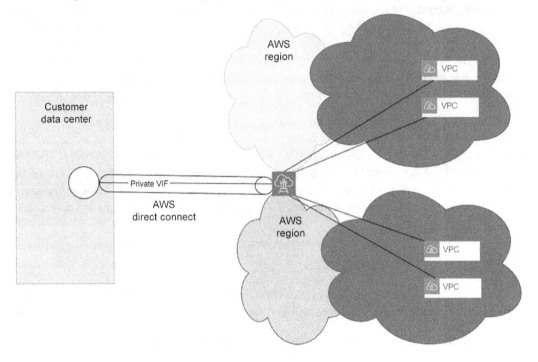

Figure 4.10: Direct Connect gateway and private VIF

What's interesting is that each private VIF can support a DXGW with 10 VPCs, and each DX connection can support up to 50 private VIFs, which scales the connectivity options up to 500 VPCs in the same or multiple regions.

In this deployment, the DXGW can extend the connectivity of the private VIF, but it cannot route traffic between VPCs connected to the DXGW. It works more like a network switch, just connecting two endpoints. As stated before, a customer or provider router with multiple connections can route between the VPCs, but this is rarely an intended behavior. This is unlike Transit Gateway, which can route traffic between attached VPCs without going across Direct Connect to the customer's or provider's router first.

The DXGW can also integrate with AWS Transit Gateway using the transit VIF on Direct Connect.

Remember that each DX connection can have up to four transit VIFs, and a DXGW supporting these VIFs can be connected to a DX connection. A DX gateway allows a single transit VIF to extend to up to three Transit Gateways instances in any region. A DX gateway can be connected to private VIFs or a transit VIF, but not both. This means that a separate DX gateway would be needed for each VIF type.

> **Note**
>
> If you see any questions in the exam that involve connecting a private VIF and transit VIF to the same DX gateway, it is not the correct option choice.

As with private VIFs, a transit VIF attached to a DX gateway can't route between Transit Gateway instances attached to the DX gateway. TGW has its own peering connection, which will be needed if TGW should be connected. A DX gateway only provides connectivity between the endpoints connected to the DX gateway, but it won't route traffic between them. Consider the following diagram:

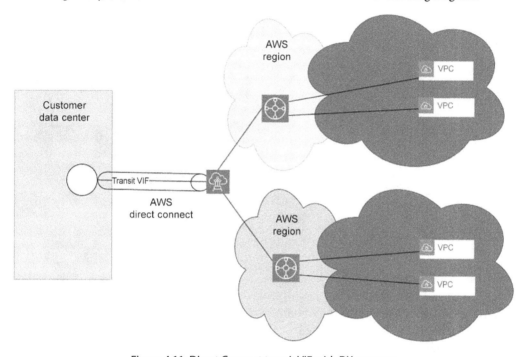

Figure 4.11: Direct Connect transit VIF with DX gateway

In this case, the VPCs in one AWS region would not be able to communicate with VPCs in another AWS region without using the customer router, as the AWS TGW instances are not peered. From a routing perspective, TGW would send traffic between regions down the DX connection to the customer router, which incurs charges and is highly inefficient, as has been discussed previously.

It's important to understand the implications of DX gateway connectivity; it is not transitive in any way. While a DX gateway extends connectivity between objects at either end of the DX gateway, it offers no connectivity between other endpoints using the same DX gateway.

That is what Transit Gateway is for. Consider this architecture:

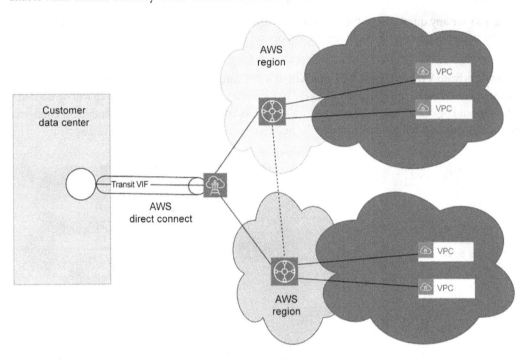

Figure 4.12: Transit VIF with DX gateway and peered TGW

In *Figure 4.12*, the TGW instances are peered across regions and the VPCs can communicate without leaving the cloud. The limitation here is that TGW peering is not dynamic; the route tables must be updated manually for the prefixes that should be reachable over the TGW. Therefore, each TGW instance must have its route tables updated to include the CIDRs of the VPCs reachable in the other region to communicate.

What's been shown so far are isolated deployments meant to highlight individual concepts, but real cloud networks are never that simple. For example, a TGW instance can connect to 20 different DX gateways, though a single DX gateway can only connect to three TGW instances. When having to split up DX gateways to focus on a private or transit VIF, dealing with these different limitations, and then peering TGW instances together, it can quickly become a nightmare.

The following architecture is illustrative, but not a best practice:

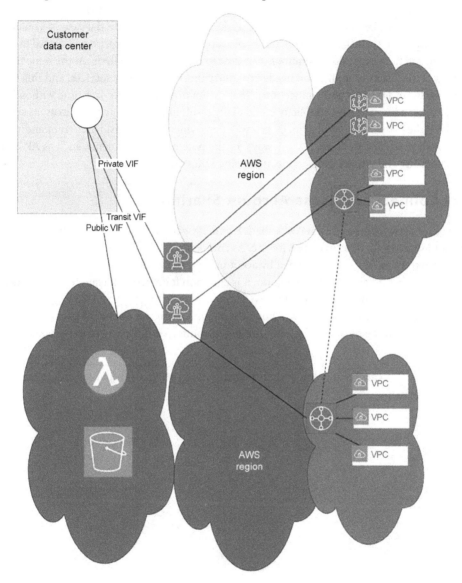

Figure 4.13: Complex DX gateway architecture

Figure 4.13 shows a poorly planned DX gateway architecture – often the result of growth over time without proper planning.

This is not even moderately complex; it just starts to show the horrors you can unleash on a poorly planned cloud deployment. Being measured and deliberate with a Direct Connect deployment is crucial to effectively plan a resilient but uncomplicated cloud network. In the case of Trailcats, the fictional example enterprise being used in this book, suppose that the enterprise had started in a single AWS region, and then expanded operations over time to include several others. At the same time, the developers of the Trailcats application needed to utilize more AWS public services, and this resulted in a slow sprawl of DX VIFs and connections. That sort of organic growth over time without regard to end state and administrative simplicity could easily result in DX gateway complexity, as shown in *Figure 4.13*. The proper way to plan a DX gateway deployment is to consider current and possible future growth needs and dedicate DX gateways for the potential requirements, such as AWS public zone connectivity and cross-region connection using DX.

Direct Connect and Cross-Account Sharing

Just like other network resources in AWS, the DX and DX gateway constructs can be owned by one account and shared with other accounts for use. As covered in *Chapter 5*, *AWS Account Sharing*, most network resources in AWS can be owned by accounts and shared with others. Direct Connect and DX gateways are no different. In this scenario, it is common for a network team to own the network constructs and allow association with other AWS accounts using them. With AWS Organizations, the association proposal and acceptance are automatic, but otherwise, the DX or DX gateway association must be shared, and the owner of the endpoint being connected must accept the proposal, as with other network services.

In *Figure 4.14*, you can see a color-coded diagram illustrating a sample cross-account Direct Connect deployment:

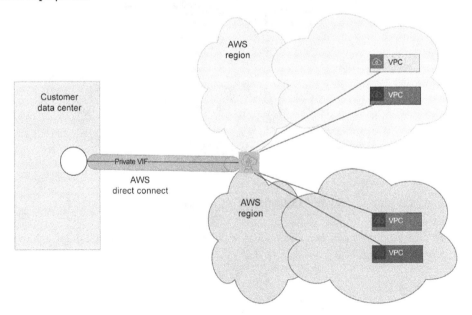

Figure 4.14: Shared Direct Connect deployment

Direct Connect and Encapsulation/Encryption

The Direct Connect architecture itself provides basic connectivity between customer or provider equipment and one of the AWS network constructs, such as a VGW or TGW instance. But if you need a higher route scale and more routing options than either of those constructs provides, you will need to use a **network virtual appliance (NVA)**. These virtual appliances are usually released by third-party vendors in the AWS Marketplace and serve to perform much more granular networking than those offered by native constructs. These NVAs are often released by popular network vendors as an EC2 instance of their own product and include the network stack and OS of that vendor, allowing far greater scale and granularity.

In this scenario, Trailcats still needs to make connections over DX to AWS native constructs. They are adding a layer of connectivity on top of that to facilitate two appliances communicating. The following diagrams depict the process of making this over-the-top connectivity using some form of tunneling.

Step 1: Create Connectivity

The first connectivity needed is between the customer or provider router and AWS Direct Connect over a private VIF. The other side of the VIF terminates into either a VGW (shown here) or a DX gateway, which then terminates into the VGW. This establishes the pseudo-layer 2 adjacency needed to build a BGP peering for route exchange, as shown in *Figure 4.15*.

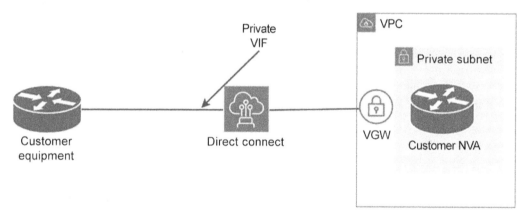

Figure 4.15: Encapsulation/encryption to NVA: Step 1

Step 2: BGP Peering

BGP peering is set up between the equipment and the AWS VGW (or DX gateway) and routes are exchanged. The AWS side will share the CIDR of the VPC it's connected to on the VIF. The customer equipment can share any prefixes configured up to the limit (100 prefixes), but in this scenario, the goal is not to share the needed prefixes over Direct Connect at all, so it's more likely the equipment will only share the address of the interface that will be used to build a tunnel over Direct Connect. This is shown in *Figure 4.16*.

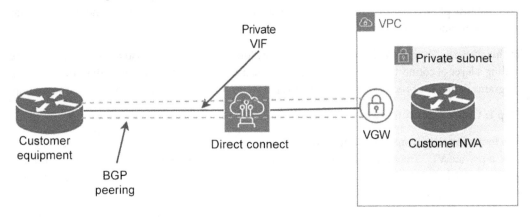

Figure 4.16: Encapsulation/encryption to NVA: Step 2

Step 3: Route Propagation

The VGW or DX gateway shares the VPC CIDR with the on-premises equipment. This is important so that the NVA interface IP can be reached for tunnel establishment, as shown in *Figure 4.17*.

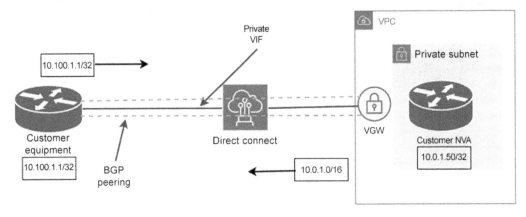

Figure 4.17: Encapsulation/encryption to NVA: Step 3

Step 4: Overlay Tunnel Establishment

A tunnel can be built between the equipment and the cloud NVA appliance. When this tunnel is built, a new BGP (or, candidly, any routing protocol supported by both endpoints) is established, and the network prefixes can be shared irrespective of native cloud limitations. *Figure 4.18* shows a completed tunnel overlay deployed using DX as an underlay transport:

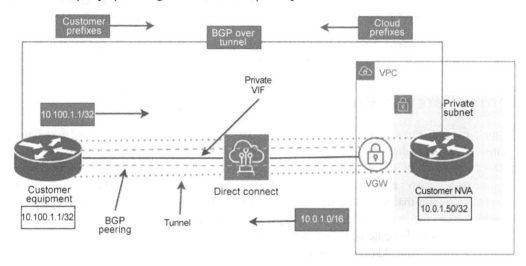

Figure 4.18: Encapsulation/encryption to NVA: Step 4

Encryption is often included but not necessarily required. For the purposes of setting up an overlay using NVA in which the DX is an underlay, any tunnel technology, such as **Generic Routing Encapsulation** (**GRE**) will work with this setup. If IPSec is included, it will be at the tunnel layer, encrypting the traffic between the equipment and appliance. Direct Connect will only see encrypted traffic with the source and destination IP address of the appliances and have no visibility of what is inside the tunnel.

An NVA is not required to establish a GRE or IPsec VPN, however. In the same way that a site-to-site VPN can be established over the internet, Direct Connect can also carry a VPN. In that case, a dedicated VPN is created between the equipment and the AWS VGW attached to a VPC over the Direct Connect private VIF in the same manner as if it were using the internet. AWS also supports doing this with a transit VIF to TGW, landing a VPN to TGW instead of a VGW. The process is the same as other site-to-site VPN options in AWS, creating a **customer gateway** (**CGW**) (the equipment connected to Direct Connect), and associating the VPN connection using that CGW with a TGW.

All encryption options come with a cost. The scale limitations of a single VPN max out at between 1.25 Gbps and roughly 2 Gbps, depending on the encryption algorithms and equipment, which could be a problem if higher bandwidth is needed on Direct Connect. Using GRE or no tunneling does not have the same impact. Understanding the trade-offs is important for the exam.

There is a way to encrypt traffic at layer 2 across Direct Connect without running into these bandwidth limitations, but it requires the equipment used by the customer or provider to support it, using a technology called **MACsec**. Without delving too far into MACsec directly in this chapter, it is a layer 2 standard security technology that provides frame encryption instead of using layer 3 IPSec-based encryption. It is currently only supported on 10G or 100G DX connections and uses a pre-shared key method for creating the encryption. This encryption is only over Direct Connect itself, between the equipment and the AWS DX location. The MACsec encryption does not extend into the AWS region.

Now that we have covered the different ways to encapsulate and encrypt data over Direct Connect, it's time to delve into BGP options with Direct Connect.

Border Gateway Protocol

Before launching into an explanation of the requirements and limitations of routing over Direct Connect, it would be helpful to give a short recap on BGP in the context of how Direct Connect utilizes the protocol. **BGP** is a route exchange application that allows network devices to advertise and consume route prefixes known as **Network Layer Reachability Information (NLRI)**, which build a map of the networks that are owned and can reach other networks for the purpose of communication. Network devices require an understanding of where to send traffic, and that means having a route table of destinations. It would be tedious to have to manually maintain such a list, especially as new networks are added and old ones are removed, so BGP does this dynamically.

The most important concepts to understand with BGP and Direct Connect are the ASN, the NLRI, the best path selection, and, to a lesser extent, BGP communities. NLRI was just explained, and *Appendix 1, Networking Fundamentals,* covers BGP ASN and path selection. So, what are BGP communities?

BGP Communities

BGP communities are a tagging system that can be attached to NLRI and can be utilized by the network devices that receive the NLRI. As an example, the BGP community 7224:9100 is used when advertising public prefixes to AWS to indicate to the AWS network that the prefixes being advertised should not be sent to other AWS regions. This works because AWS has configured in its own BGP route policy that this BGP community tag will result in that treatment.

AWS publishes the BGP community list for its service so that customers know what BGP communities to use to get certain outcomes for its prefixes advertised to AWS. AWS also directs customers on which BGP communities it will use when advertising prefixes to customers and their significance so customers can configure their own route policy based on the desired outcome.

Think of a BGP community as a tag attached to the prefix NLRI in BGP between the customer (or provider) and AWS that indicates certain actions or treatments of the NLRI should take place. This is most used on a public VIF where public prefixes are advertised to and from AWS. When covering public VIFs and BGP, this will be explained in greater detail.

BGP and Private VIFs

Using BGP with a private VIF is very simple. By default, when using a private VIF, the on-premises equipment forms a BGP neighborship with AWS Direct Connect, and that connects to the VGW attached to a VPC. AWS advertises the CIDR of the VPC to which it is attached to the other side of the DX connection using BGP. The on-premises equipment advertises whatever NLRI is configured in AWS, but this has a very low limit, so care must be taken. The limit of prefixes that can be propagated to a VPC via a VGW is very low; as of mid-2024, it is 100 only. This means that, often, the prefixes advertised need to be summarized or filtered. It is common to advertise only a superset of routes in this way, such as 10.0.0.0/8 or even 0.0.0.0/0 to attract all non-VPC traffic.

When using a DX gateway, the BGP neighbors are still formed with DX and the DX gateway will propagate the routes advertised from on-premises to any VGW that is connected to the DX gateway. AWS will also advertise the CIDRs of any VPC with a VGW attached to the DX gateway over Direct Connect to the on-premises equipment, whether the premises is a customer-owned data center or a customer-rented rack in a partner colocation data center. Here is a diagram of what connectivity between a customer-owned router at a colocation data center and AWS Direct Connect would look like:

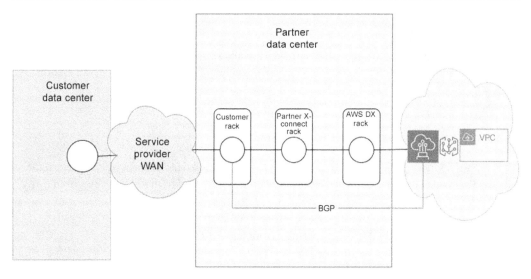

Figure 4.19: Private VIF with BGP from colo data center

In this case, Direct Connect is extended to a customer-owned router in the partner data center. The private VIF is built through Direct Connect to a VGW attached to a VPC. BGP is established between the customer router and the Direct Connect service, not the VGW directly. The Direct Connect service (almost certainly just an AWS-managed router) automatically propagates any BGP routes learned from the customer router (up to the limit of 100 prefixes) to the attached VGW and VPC route table. Direct Connect automatically advertises the CIDR of the VPC attached to the VGW to the customer router.

BGP and Private VIF Considerations

If more than 100 pieces of NLRI are advertised from the customer or provider side, the BGP neighborship will be torn down automatically by AWS and will flap, that is, it will continually try to re-establish and fail, until the advertisement is within limits.

When provisioning a private VIF to a VGW or a DX gateway, the BGP ASN used on the AWS side of the VIF must be a private ASN (between the range of 64512 and 65534 or 4,200,00,00 and 4,294,967,294). This is set within AWS when creating the DX gateway or VGW.

BGP neighbors across a private VIF support BGP attributes such as AS_PATH, local preference, and the well-known NO_EXPORT BGP community, which indicates to the BGP peer not to further advertise the routes tagged with this community.

The most common use case in which this becomes an issue is when peering with on-premises routers that carry massive route tables. Without proper route filtering on the remote end, this limitation will be violated very easily. The best way to mitigate it is to filter all but a summary or supernet route from the on-premises side as much as possible to stay under the limit.

> **Note**
> Keep these limitations in mind when considering any scenarios involving a Direct Connect private VIF connection. If an exam question violates either of these limitations, it is an incorrect option choice.

AWS Private VIF and BGP Communities

AWS supports sending BGP communities across DX to indicate which routes will be preferred over which VIF. This is done by attaching certain BGP community tags to the routes advertised to AWS over the VIF and determining what path AWS will use to send traffic to those destinations. As an example, these BGP communities are used to advise AWS about which path to send traffic via to these destinations, when advertised from on-premises routers over DX:

- 7224:7100 – Low preference
- 7224:7200 – Medium preference
- 7224:7300 – High preference

These communities cannot coexist on the same routes across the same BGP peer because it would be counterintuitive to specify multiple priorities. AWS will use the tag to evaluate the priority of the path used to send traffic to the destination. If the same NLRI prefix is advertised to AWS over two different Direct Connect VIFs with the same BGP community tag priority, the two paths will be candidates for multipathing.

Other attributes are evaluated as well, such as AS_PATH, so it is only a candidate. To ensure that traffic is load balanced over multiple paths, it is important that all attributes match across all paths. This is true of VIFs on the same DX connection or VIFs on two different DX connections.

> **Note**
>
> You can revise BGP attributes in *Appendix 1, Networking Fundamentals.*

BGP and Public VIFs

Of the ways to utilize BGP with AWS, using BGP with a public VIF is one of the more complex. Because a public VIF allows DX to be used to access AWS public services, the customer's or provider's equipment needs a public IP for this VIF, and that public prefix must be owned by the customer or provider advertising the prefix to AWS over the public VIF.

Unlike a private VIF, the BGP requirements for route exchange are very strict, with tight rules on authentication and session parameters, as well as rigorous validation and filtering of routes that can be advertised or received. This is needed to ensure proper routing of traffic between the AWS backbone for public services and Direct Connect. In addition, care is needed to ensure that neither AWS nor the end user of Direct Connect becomes a transitive routing destination for other internet networks. In normal circumstances, and with no filtering or special BGP communities, something like this could happen:

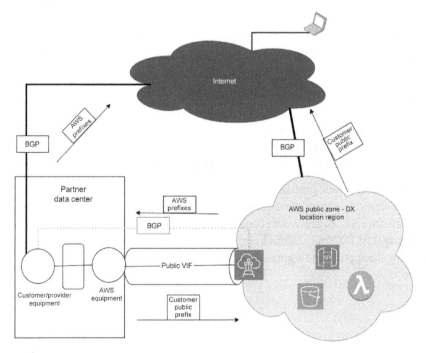

Figure 4.20: Transitive BGP advertisement without filtering

Figure 4.20 illustrates the sloppy and dangerous routing situation in this (thankfully not real) scenario, in which a customer has an active public VIF connected to the AWS public zone, but also has a connection to the internet, and is advertising AWS prefixes to the internet that it is learning from AWS. AWS would, in turn, advertise the customer's public prefixes to the internet too. If this sort of deployment were allowed, AWS would advertise its public zone internet prefixes to the internet at large and across the Direct Connect public VIF to the customer or provider. The equipment owner on the other side of the public VIF would advertise its public prefix to the AWS public zone for traffic to return over Direct Connect.

So far, this is not a problem, but without proper filtering and route management, AWS would advertise the received prefix to the internet, and the equipment owner would advertise AWS prefixes to its own internet provider. This makes each transitive for the other. The internet is a very big place and, based on BGP routing path preference, it's possible that other users may end up using the customer/provider as the best path to reach AWS, or vice versa. This is highly undesirable, and that's why AWS enforces strict BGP peering and advertisement requirements.

In the case of Trailcats, our fictional enterprise, consider a scenario in which the business needs to advertise specific prefixes to AWS to ensure traffic will route over DX to and from the public zone. If Trailcats has multiple Direct Connect instances across the world, care must be taken when sending and receiving routes to the AWS public zone over each DX to ensure that there will not be suboptimal traffic flow, such as routing traffic intended for one region over another entirely.

Here is a rundown of the BGP limitations and peering requirements in AWS for the AWS public VIF and why they exist:

- **Public IP on customer/provider equipment**: The DX connection itself is a private, managed connection, but connecting to the AWS public zone over a public VIF is not connecting with a private endpoint, so both sides of the connection need to be using publicly routable addressing.

- **Globally unique, customer/provider-owned BGP ASN**: For the same reason, a public IP is required, a real, non-private BGP ASN is required by the organization connecting to the public VIF. This ASN must not be private (between the range of 64512 and 65534 or 4,200,00,00 and 4,294,967,294 in 2-byte and 4-byte notation, respectively). Technically, you can use a private BGP ASN, but you will lose access to most of the traffic-shaping features available on the public VIF.

- **Advertise only public prefixes owned by the organization**: This one may seem obvious, but AWS validates the source of traffic to the AWS public zone and will drop traffic that does not originate from the organization connected to the public VIF. This means that AWS strictly guards against DX connections being used in a transitive fashion. This validation includes checking the appropriate regional internet registry for ownership.

- **AWS DX prefix attributes**: AWS advertises prefixes to the organization using DX and will include at least three BGP ASNs in `ASPATH`. These prefixes will also come with a well-known `NO_EXPORT` BGP community so that the prefixes advertised do not end up being advertised by the organization to any other BGP neighbors. This is to prevent the customer from becoming transitive for AWS public zone traffic, as discussed previously. Other than AWS China, expect that, under normal circumstances, the AWS public zone in the DX region will include advertisements for ALL AWS public regions. This can be filtered by the customer directly using the AWS BGP communities attached to the NLRI advertised to the customer/provider. AWS will use the inbound advertisement, along with any communities attached by the advertising organization, to decide how far within the AWS public zone to advertise the NLRI being advertised from that organization.

AWS Public VIF and BGP Communities

When advertising NLRI prefixes to AWS over a public VIF, an important consideration to think about is how far you want the advertisement to go within the AWS public zone. Here are the options and BGP communities that should be attached to the prefix advertisement to AWS, remembering that AWS China is a walled garden and is not included in this, as AWS China is essentially its own AWS cloud:

- 7224:9100 – Local AWS region only
- 7224:9200 – All AWS regions for a continent, based on the DX location of the advertiser:
  - North America
  - Asia Pacific
  - Europe, the Middle East, and Africa
- 7224:9300 – Global (all public regions except China)

The reason to restrict the advertisement of your prefixes to certain AWS regions is that whatever you advertise to AWS will likely be the return path as well. For latency reasons, you may not wish to advertise a DX location in the US to the AWS Asia Pacific public zone, you may have a different DX location in **Asia-Pacific (APAC)** that is closer, or perhaps you have no need to reach services in APAC at all. These BGP communities are attached to the prefixes you advertise to AWS and set the expected return path for public services.

What about filtering how much AWS advertises to you? This is also done with BGP communities on the advertisements you receive from AWS. Commonly, organizations will have a BGP route map, which will look for BGP communities and take some action based on the community being present, or not. In this way, if you only want to use DX to reach certain public regions, AWS sets a BGP community so that you can selectively discard or use those regional prefixes for route preference.

The BGP communities used for this purpose by AWS are as follows:

- 7224:8100 – Local AWS Region only

- 7224:8200 – All AWS Regions for a continent, based on the DX location of the advertiser:

  - North America

  - Asia Pacific

  - Europe, the Middle East, and Africa

- 7224:8300 – Global (all public regions except China)

Here is an example of these BGP communities in action, and a customer-side route map to modify the advertisement based on preference:

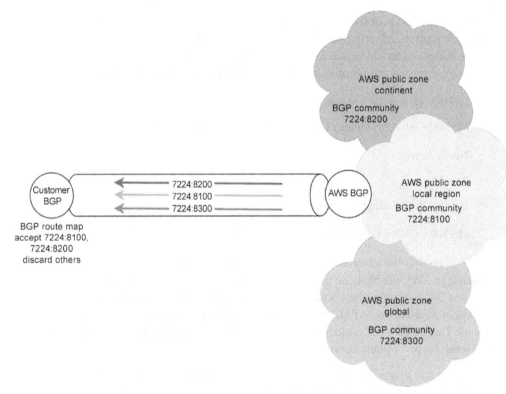

Figure 4.21: AWS public VIF advertisement with BGP communities

In this case, Direct Connect is sending all AWS public prefixes to the customer, but the customer is only interested in the NLRI prefixes from the local region and other public zones on the same continent. The customer configures a BGP route map to match the NLRI with the BGP community that is desired and discards the other prefixes entirely.

This allows traffic to AWS public services within a reasonable geographical limit to be reached over DX. If services in another geography are needed, the internet can be used to get there, or – more likely – that other geography will have its own DX with access to an AWS public zone that will be local.

Consider a scenario where there are multiple DX connections belonging to an organization across the globe, and how it may be advantageous to prefer both ingress and egress paths, like this:

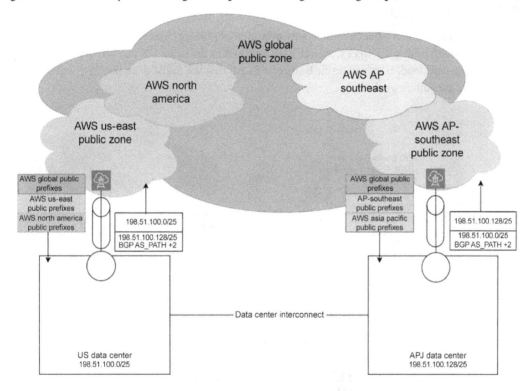

Figure 4.22: Using BGP to prefer paths over multiple DX connections

Given this scenario, the data center in the US has a connection to the data center in the Asia Pacific region and has a DX connection to AWS in a partner facility (not pictured for brevity). The data center in the US advertises both public prefixes to AWS over DX for redundancy, but to keep AWS routing traffic as locally as possible, it uses extra BGP ASNs in the AS_PATH attributes for the APJ data center NLRI prefixes. Because the APJ data center has a local DX connection and will advertise the same NLRI prefix to AWS, AWS should see the path through the US DX connection as a worse path than the one to APJ, and vice versa. The APJ data center does the same with US NLRI prefixes.

When AWS public services must decide how to route traffic to a data center and have multiple options, they will generally pick the one with the lowest AS_PATH. The customer controls what path AWS will prefer by advertising the NLRI via BGP with attributes tuned for this.

This is not pictured, but the customer could also opt to not provide redundancy for the other data center, in which case the NLRI prefix for the other DC could be filtered from advertisement, or the NLRI could be tagged with a BGP community that keeps it from being exported beyond the local zone, as discussed earlier.

From the AWS advertisement perspective, the customer is free to accept some or all of the AWS prefixes and can filter out what is not needed by only accepting NLRI prefixes that have a desired BGP community.

What if the goal is to load balance connectivity over multiple Direct Connect VIFs? For this, let's look at a scenario with multiple Direct Connect VIFs attached to the same DX gateway and then a VPC with a private workload. How can traffic be directed such that the traffic from the VPC will use both VIFs? *Figure 4.23* offers a solution:

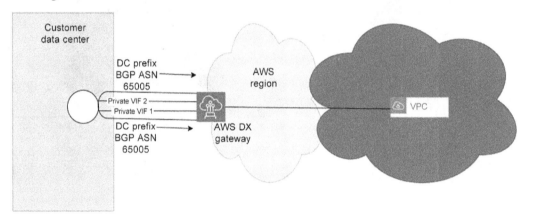

Figure 4.23: Using private VIFs to load balance

Because the BGP AS_PATH and NLRI prefix advertisement are the same, and there are no other attributes to consider, AWS will load balance traffic from the VPC to DX over both VIFs. Generally, load balancing is only possible if all attributes match, including the length of the NLRI prefix being advertised.

Using this same diagram, it's possible that if one VIF were to be preferred over another, the BGP AS_PATH could be manipulated to use a different length, or the BGP community could be set on the NLRI prefixes advertised over DX to AWS for low, medium, or high priority, which sets the AWS preference on which VIF to use when sending traffic back across DX.

Of course, on the customer side, local preference can be used without the need for BGP communities to influence traffic from the customer to AWS. This is simply a matter of setting local preferences on the routes coming in from AWS on the VIF you want to be the primary VIF. Care should be taken that the same VIF is preferred if not load balancing so that AWS and the customer premises pick the same VIF to prefer both ways.

Suppose that Trailcats wants to prefer certain paths over the regions to which it has a DX connection. It could modify the BGP NLRI that it advertises to AWS over different BGP peers with AWS preference BGP communities, ensuring that a certain path back was preferred. This would allow Trailcats to lower the latency of connectivity to AWS public services while still having resiliency against failure.

BGP and Transit VIFs

Using BGP with a transit VIF really means advertising and receiving routes from one or more AWS Transit Gateway instances. This is like how BGP peering works with private VIFs, with an important set of differences. Remember that TGW is a more intelligent routing construct and has things that a VGW does not have, such as multiple route tables, selective route propagation, and much higher route limits. This allows for far more scalable and configurable routing patterns. In the case of Trailcats, by connecting DX to TGW, Trailcats gains resiliency and a much higher route table limit – up to 10,000 routes instead of only 100. If Trailcats wants to selectively route traffic through TGW by segmenting route tables, this is possible with TGW but impossible with a VGW or DX gateway.

> **Note**
>
> One thing to be aware of, and a possible way AWS may seek to trick you in the exam, is that while up to three TGW instances can be attached to one DX gateway, TGW instances cannot be attached to the DX gateway if the CIDRs they advertise to the DX gateway overlap with each other.

TGW can be connected to up to 20 DX gateways, though each DX gateway may only be connected to at most three different TGW instances. In addition, a single DX gateway can connect multiple transit VIFs, though this does not extend the number of TGW instances with which a single DX gateway may connect.

Because, from the TGW perspective, a DX gateway is just another TGW attachment, it can granularly propagate routes from other attachments to a DX gateway or not, as the requirements demand. However, a single transit VIF cannot advertise more than 100 routes to a DX gateway from the customer side, despite TGW allowing up to 10,000 routes. To take advantage of that scale, it is necessary to use TGW Connect over the transit VIF to TGW. This will be covered in more detail in *Chapter 5, Hybrid Networking with AWS Transit Gateway*.

> **Note**
>
> Despite AWS TGW supporting up to 10,000 routes, when a DX gateway is used to terminate a transit VIF and connect TGW directly, the limit is still enforced at 100 routes. TGW Connect using GRE allows route scaleup to 1,000 routes. TGW limitations will be explored further in *Chapter 5, Hybrid Networking with AWS Transit Gateway.*

Direct Connect and Bidirectional Forwarding Detection

Direct Connect is automatically enabled to use **bidirectional forwarding detection** (**BFD**), but it needs to be configured on customer-side equipment. BFD is a layer 2 fault detection mechanism that verifies the reachability of the other endpoint much faster than traditional BGP failure mechanisms will detect a connectivity failure. Commonly, BFD is established between the customer equipment and the AWS router and the BFD health is tied to BGP, so BGP will fail more quickly if a connectivity loss occurs. This quickens how failover can happen if a problem occurs. Normal BGP timers would not detect a loss of connectivity until a keepalive timer fails for 90 seconds (3 times the keepalive timer of 30 seconds). This can be tuned to a minimum of 1-second keepalive and a 3-second hold time, but BFD allows for sub-second fault detection, quickening BGP neighbor teardown and route withdrawal in a failure.

Direct Connect and IPv6 Connectivity

IPv6 support does not functionally change most of the features and limitations previously discussed but does come with a few extra limitations and features. Here is a rollup of IPv6-specific features and limitations:

- A private/transit VIF can have one IPv4 and one IPv6 BGP session over the same VIF. This means that while IPv4 and IPv6 are supported on the same VIF, only one BGP neighbor is supported for each address family. Adding IPv6 does not functionally change this limitation.

- VPN connections built over Direct Connect must still use IPv4 addressing.

- You cannot run an IPv6 BGP session over an IPv4 VPN session with Amazon; however, certain NVAs could support such a configuration.

Summary

In this chapter, you have covered the details of what an AWS Direct Connect private circuit is, how to create one, and all the different types of DX VIFs that can be utilized. The constructs that can terminate DX in the cloud were also discussed, including VGWs, TGW, and DX gateways. BGP is a large part of hybrid cloud connectivity, and so you covered all of the ways that BGP can be used with DX and the best practices based on the VIF type. Finally, you discussed IPv6 and Direct Connect. AWS adds more and more IPv6 functionality every day to its offerings.

There are a lot of features, limitations, and caveats to consider and many of them will show up in the exam, so please spend some time rereading and memorizing the most important ones from this chapter, especially the notes that call out specific limitations that may be used to trick you in the exam. If you remember the different DX VIF types and when to use each, along with BGP and how it is used with each VIF, you should be well prepared. The next chapter will cover hybrid cloud networking and AWS Transit Gateway.

Exam Readiness Drill – Chapter Review Questions

Apart from mastering key concepts, strong test-taking skills under time pressure are essential for acing your certification exam. That's why developing these abilities early in your learning journey is critical.

Exam readiness drills, using the free online practice resources provided with this book, help you progressively improve your time management and test-taking skills while reinforcing the key concepts you've learned.

HOW TO GET STARTED

- Open the link or scan the QR code at the bottom of this page

- If you have unlocked the practice resources already, log in to your registered account. If you haven't, follow the instructions in *Chapter 16* and come back to this page.

- Once you log in, click the START button to start a quiz

- We recommend attempting a quiz multiple times till you're able to answer most of the questions correctly and well within the time limit.

- You can use the following practice template to help you plan your attempts:

Working On Accuracy		
Attempt	Target	Time Limit
Attempt 1	40% or more	Till the timer runs out
Attempt 2	60% or more	Till the timer runs out
Attempt 3	75% or more	Till the timer runs out
Working On Timing		
Attempt 4	75% or more	1 minute before time limit
Attempt 5	75% or more	2 minutes before time limit
Attempt 6	75% or more	3 minutes before time limit

The above drill is just an example. Design your drills based on your own goals and make the most out of the online quizzes accompanying this book.

First time accessing the online resources? 🔓
You'll need to unlock them through a one-time process. **Head to** *Chapter 16* **for instructions**.

Open Quiz	
https://packt.link/ansc01ch4 OR scan this QR code →	

5
Hybrid Networking with AWS Transit Gateway

As architectures grow to fit the needs of an enterprise, so do the networking needs of the architectures. With on-premises networks and multiple VPCs, sometimes spread over multiple regions, direct connections between nodes can become complex very quickly. **AWS Transit Gateway** (**TGW**) is a scalable service designed to simplify network connectivity in a hybrid cloud environment by acting as a hub to route traffic. It enables organizations to centralize and simplify their networking, eliminating the need for complex peer-to-peer VPC connections or point-to-point VPN tunnels.

This chapter focuses on AWS TGW, its components, and how it works. Then, it will cover the integration options and design patterns for hybrid cloud networking, including connecting to resources inside and outside the cloud.

Here's a brief rundown of the topics covered:

- Transit gateway components
- Transit gateway attachment types
- Transit gateway route table overview
- Transit gateway and dx gateway designs
- Transit gateway and inspection VPC designs

By the end of this chapter, you will have an understanding of what AWS TGW is and how it can be used to connect different networks together into a cohesive, unified solution. You will know how to manage the route tables to ensure connectivity or segmentation.

We will start with an overview of AWS TGW.

AWS TGW Basics

At a high level, an AWS TGW is a managed AWS router that can connect many different AWS networking constructs together in a hub-and-spoke topology. AWS realized that VPC peering had scale limits and manual route management issues and that the virtual private gateway construct lacked sufficient routing control and intelligence as well. The cloud-native connectivity constructs were proving unable to deal with the demands of businesses that wanted better connectivity options for critical applications. AWS introduced TGW to provide a managed, scalable, performant network construct able to deal with the challenges posed by increased user complexity.

TGW also extends routing intelligence by providing multiple route tables and simplifying the exchange of routes across resources connected to it. This improves operational scale and security options as route segmentation is now possible. TGW also simplifies connectivity between resources by providing a central intelligent hub to connect them together.

A basic TGW deployment might look something like this:

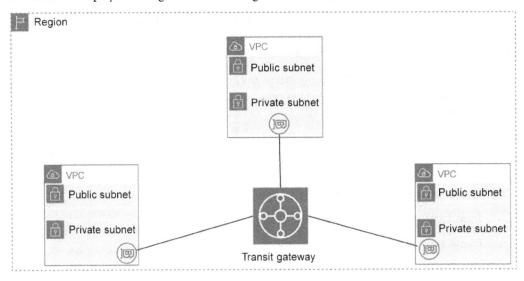

Figure 5.1: Basic AWS TGW deployment

In this example, TGW is just offering basic connectivity between three VPCs and their resources. An advantage of using TGW instead of VPC peering is that the route tables can be updated without manual intervention by route propagation on the TGW, something that will be covered in the *Transit Gateway Route Tables* section. Another advantage is the ability to provide selective connectivity via TGW route table manipulation, something else that will be covered in this chapter.

AWS TGW Components

A TGW instance is a managed AWS resource, which means you as the consumer are not responsible for things such as creating an EC2 virtual machine instance, configuring it, ensuring it is deployed in a fault-tolerant and highly available way, or managing its software lifecycle. This allows you to simply use it as a tool and leave all the details to AWS, but it also means a loss of granular control and being siloed into specific deployment models. There is always a trade-off.

The following sections will break TGW down into its components to explain how to consume it. TGW itself has already been explained, so the focus will be on what parts of TGW can be configured.

TGW Attachments Overview

For traditional network engineers, it may be helpful to think of a TGW attachment as a network cable being plugged into the TGW. The attachment serves as a target network object for the purposes of routing packets through the TGW. In conjunction with TGW route tables, complex routing decisions can be made by a TGW connected to many kinds of AWS network resources with a lower administrative overhead than manually connecting these constructs.

Under the hood, each attachment uses a different type of connectivity to connect these resources. For example, to attach a VPC to the TGW, VPC subnets must be selected so that an **elastic network interface** (**ENI**) can be deployed because that is how TGW handles connectivity with VPCs. With a VPN attachment, the connectivity is based on the same Site-to-Site VPN connectivity that is used to connect a customer gateway logical construct (customer equipment) to a **virtual private gateway** (**VGW**), except it connects to a TGW instead. All of this will be further explained in the next section.

TGW Attachment Types

What is an attachment in the context of TGW? From a TGW perspective, an attachment in a network construct associates some other AWS network resource with the TGW. AWS networking tends to be object-oriented in configuration, and this attachment is an object that has been built to be used as a network target for traffic coming from or destined for the attached resource via the TGW. For example, a TGW VPC attachment connects to AWS VPCs, as shown in *Figure 5.2*:

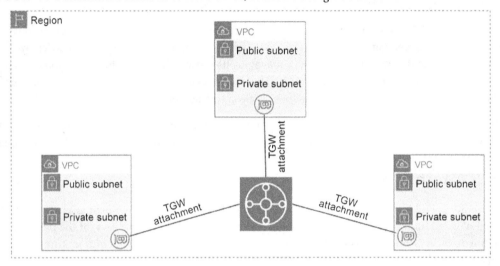

Figure 5.2: Basic TGW architecture with attachments

The preceding example shows TGW attached to three VPCs. The TGW attachment is just an object that serves to reference TGW and (in this case) VPC routing. Attachments currently come in five varieties:

- VPC attachment
- Direct Connect attachment
- VPN attachment
- Peering attachment
- TGW Connect attachment

The next sections will discuss each of these attachments.

VPC Attachments

The VPC attachment is the simplest of the TGW attachment types. When a TGW attachment to a VPC is created, an ENI will be created in one or more subnets (and **Availability Zones (AZs)**) of that VPC. Here is an example of this type of deployment:

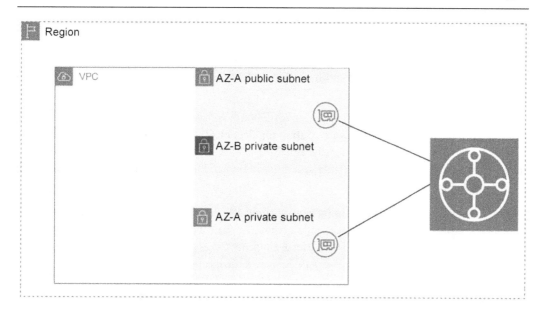

Figure 5.3: TGW VPC attachment example

In this example, the TGW has been attached to the VPC in two of three subnets. As previously explained, this attachment involves deploying an ENI into each subnet (and Availability Zone) desired. VPC attachments can be deployed into private or public subnets and can exist within the same availability zone. It may or may not be necessary to have more than one TGW attachment to a VPC, depending on the deployed resources and high availability requirements. Generally, the following rules govern how routing with a VPC works when using a TGW attachment as the next hop:

- TGW attachments can be the target of a subnet route table if and only if the subnet exists within the same availability zone as the TGW attachment ENI. For example, in *Figure 5.3*, the AZ-B private subnet cannot use the TGW attachment even though the VPC is attached.

- Public and private subnets can use a TGW attachment. There is no restriction on the type of subnet; it only has to be within the same AZ at the attachment ENI.

As an example, suppose that an application VPC exists with four subnets: two public and two private, with two of each in the same AZ but the other two in a different one. Within the private subnets, a TGW attachment gets deployed. Any resources within the public subnets could only utilize the TGW attachment of the private subnet that exists within their own AZ. Because of this, the application would only be resilient within its own AZ and would need to be scaled in a way that covers multiple AZs for true resilience.

> **Note**
>
> This is a hard rule that you should be aware of for the exam. A scenario may be presented that involves connecting a subnet in another AZ to a TGW attachment, and you should be ready to eliminate this option.

Attaching a VPC to a TGW comes with routing implications that will be covered in the upcoming *TGW Route Tables* section, but generally, attaching a VPC to a TGW will propagate routes from the VPC into the main TGW route table unless a custom one has been defined for the attachment. However, the reverse is not true.

One more detail that may be important when attaching a TGW to a VPC, especially if using security appliances within the VPC, is appliance mode support. Appliance mode modifies the normal behavior of a TGW. In normal operation, a TGW uses AZ affinity to keep traffic within the same AZ when routing into and out of a TGW. This can cause issues with inspection VPCs where security appliances are deployed because inter-VPC inspection can be impacted if the source and destination flows are not in the same AZ. Consider the following diagram:

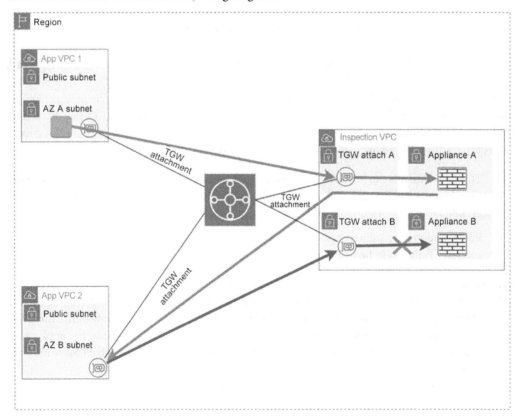

Figure 5.4: TGW flow failure without appliance mode

As seen in *Figure 5.4*, because TGW tries to keep traffic within the same AZ by default, the initial flow from **App VPC 1** has no issues – the flow hits the firewall in the inspection VPC in **AZ A** and is forwarded back to the TGW attachment for delivery to **App VPC 2**. The trouble happens when **App VPC 2**'s workload in subnet B responds to the packet and the TGW attachment routes it to inspection VPC subnet B where the other firewall has not seen the state of this flow and drops it.

The correct way to solve this problem is to activate appliance mode on the TGW attachments to the inspection VPC. This causes the TGW to use the same TGW attachment for ingress and egress flows regardless of the AZ. This state tracking is only good for a single TGW, however. If multiple TGWs are attached to the inspection VPC, the flow state details are not shared between these TGWs, so flow symmetry is no longer guaranteed.

> **Note**
>
> TGWs do not share flow hash data with each other, so any exam question answer option that involves flow symmetry through multiple TGWs attached to an inspection VPC is an incorrect option choice.

Direct Connect Attachments

Direct Connect and TGW were covered in *Chapter 4, AWS Direct Connect*, but in this section, the focus will be on the TGW and not the **Direct Connect (DX)** gateway. As a refresher, the DX gateway is used in conjunction with AWS DX to provide connectivity to one or more TGWs within or across regions. For simplicity's sake, a diagram from *Chapter 4, AWS Direct Connect*, is reprinted here:

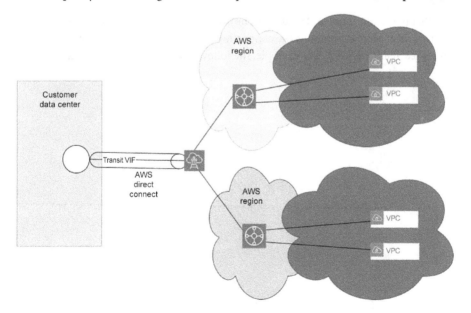

Figure 5.5: Basic DX gateway with Transit VIF connected to TGWs

Figure 5.5 shows how a DX gateway is often used with DX and TGW to subdivide connectivity. One TGW can be connected to up to 20 DX gateways, with an important limitation: the TGWs connected to a DX gateway cannot advertise the same CIDRs. Moreover, a DX gateway does not provide connectivity between any of the TGWs connected to it, only the Direct Connect endpoint using the DX gateway. As discussed in *Chapter 4, AWS Direct Connect*, this could cause bottlenecks between TGWs using the Direct Connect endpoint if proper route filtering does not take place or the TGWs are not peered with each other.

When a TGW is connected to a DX gateway, the routes from the DX gateway are propagated into the TGW main route table unless a custom route table is defined, in which case it is propagated there. This is not a **Border Gateway Protocol** (**BGP**) connection between the TGW and the DX gateway, nor the TGW and the equipment terminating the Direct Connect Transit VIF. A BGP connection is established between the DX gateway and the equipment using Direct Connect. More details can be found in *Chapter 4, AWS Direct Connect*.

TGW associations with DX gateways also use allowed prefix lists to govern what is advertised by the DX gateway to the on-premises location or provider extending that connection. This is configured on the DX gateway, but the behavior changes based on it being connected to a TGW rather than a VPC/VGW directly. For TGW associations, this allowed prefix list governs exactly what prefixes are advertised from the cloud side to the on-premises/provider side regardless of what prefixes exist in the cloud and are reachable via the TGW. This prefix limit is 200.

For example, a TGW attached to a DX gateway that is attached to Direct Connect with a Transit VIF could advertise 20.0.0.0/8, and that is the route that would be received, even if the TGW had no route or attachment connected to any resource within that CIDR range.

Practically, that makes no sense and would not be done, but it's important to understand that this is overriding behavior. More commonly, the connection would advertise a supernet or default route of the resources reachable by TGW. For example, the following figure shows a supernet use case:

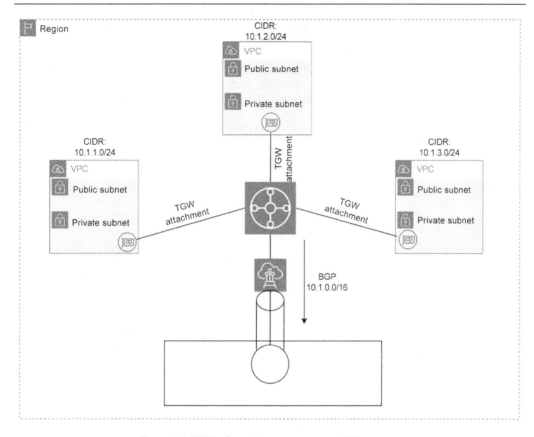

Figure 5.6: TGW advertising summary to a DX gateway

Figure 5.6 shows a TGW and DX gateway connecting an on-premises data center to three VPCs. In the diagram, the TGW advertises 10.1.0.0/16. You could also advertise complete nonsense to the DX gateway from the TGW attachment. This is not a normal use case; you would need a specific reason for that to make sense. The important detail here is that the DX gateway can be made to advertise anything, not only CIDRs that are reachable from the TGW.

> **Note**
>
> When you associate the DX gateway with a TGW, the allowed prefix list is not optional. The TGW does not automatically propagate any routes to the DX gateway; it is not dynamic. Also, the TGW allowed prefix list cannot overlap with any other TGW allowed prefix list attached to the same DX gateway, as mentioned before. Lastly, the total number of allowed prefixes per DX gateway/TGW association is 200.

This is not to be confused with the TGW ability to reference a prefix list for TGW attachments. The TGW attachment prefix list is a created object used to simplify repetitive route table entries. Some are created and managed by AWS directly concerning their services, and custom ones can be created. The prefix list reference is a different feature than the allowed prefixes when attaching a DX gateway to a TGW.

VPN Attachments

Of all the TGW attachments, the VPN attachment might be the simplest and most straightforward of them all. Instead of the normal Site-to-Site VPN option of connecting a **customer gateway** (**CGW**) to a VGW attached to a VPC, TGW gives the option to attach a TGW to a CGW instead. The setup steps are the same as a customer would take when attaching a VGW to a CGW, binding the TGW attachment to a CGW representing the equipment.

After attaching, the customer can download the VPN configuration to apply to their own equipment for connecting the VPN to TGW. This sets up a primary and redundant tunnel from the TGW to the customer equipment. The routing between the TGW and CGW can be BGP or static. If using BGP, the TGW supports no real traffic engineering, though the customer equipment can use BGP AS_PATH prepending to specify the path preference toward the TGW or use any available and supported BGP traffic engineering methods from the on-premises equipment on ingress traffic from the TGW.

Peering Attachments

TGWs are Region-locked, and that means an organization that uses multiple Regions will need multiple TGWs to connect resources. In fact, this can get quite messy when it comes to things such as cross-account sharing, peering, and routing. Consider *Figure 5.7*:

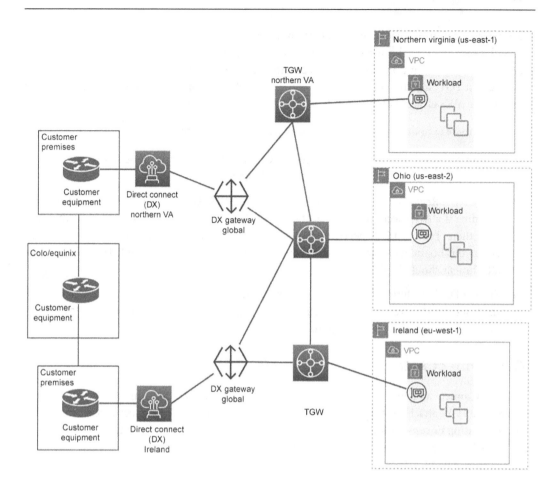

Figure 5.7: Complex TGW peering scenario

As shown in *Figure 5.7*, things can get very messy when peering TGWs together and attaching to DX gateways from a route propagation perspective. *Figure 5.7* shows DX in the Ireland and Northern Virginia Regions connecting customers through global DX gateways and TGWs to VPCs in Ireland, Northern Virginia, and Ohio. The DX gateway considerations have been discussed and so will not be repeated, but when peering a TGW to another TGW, the route tables of each TGW must be amended to point to the next hop of a TGW peering attachment in a manner very similar to VPC peering. This makes TGW peering a static affair.

TGW peering can use IPv4 or IPv6 peering, and when setting up a TGW peering attachment, a proposal is sent and must be accepted to create the attachment. This is required whether the TGWs exist in the same account or across AWS accounts. Cross-account TGW peering is different than cross-account TGW attachment sharing and will not be automatically accepted even if both accounts belong to the AWS Organization.

In addition, there are other limitations to a TGW peering attachment:

- Not all AWS Regions support TGW peering. Check out this link to find out more: `https://aws.amazon.com/transit-gateway/faqs/`.

- Some AWS Regions are automatically added to an AWS account and others are opt-in only, meaning the AWS account must select the AWS Region to allow resources to be created and used there. From a TGW peering perspective, the TGW peering attachment can be created across this boundary. Before removing the regional opt-in setting (if needed), the TGW peer attachment should be deleted.

- From a DNS resolution perspective, resources attached to a TGW in one Region cannot use the Route 53 DNS resolver in another Region to resolve either public or private IP addresses of VPC resources in another Region.

TGW Connect Attachments

The TGW Connect attachment is a bit of an outlier among the attachments in that this attachment uses another TGW attachment type as an underlay connection. The TGW Connect attachment utilizes **Generic Routing Encapsulation** (**GRE**) tunnel technology on top of the underlying TGW attachment with a third-party device using BGP as the common routing protocol. This allows TGW to handle routing more dynamically with third-party network appliances.

This is a bit complex, so a diagram will be useful here:

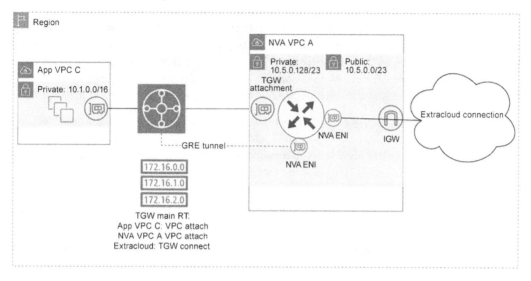

Figure 5.8: TGW Connect architecture

Figure 5.8 shows two VPCs (**App VPC C** and **NVA VPC A**) connected via a TGW. **NVA VPC A** has both private and public subnets, with connectivity to external networks via an internet gateway. A GRE tunnel provides secure communication between **App VPC C** and the NVA, while the route table indicates the networks reachable through the TGW. *Chapter 6, Connecting Third-Party Networks to AWS*, will cover a lot more deployment options using these TGW options. For now, the important things to understand about the TGW Connect attachment are as follows:

- The TGW VPC attachment to the VPC where the network virtual appliance is deployed is only used for underlay traffic to allow the TGW Connect GRE tunnel to be built over the top, like many other tunnel technologies.

- The route table of the TGW will contain the VPC CIDR of the attached VPC via the TGW VPC attachment, but it will learn any BGP routes via the TGW Connect attachment. These are different attachments, and it is possible to split the route tables and adjust route table propagation, but the TGW should never be set to prefer the route to the attached VPC (in the example case, 10.5.0.0/16) over the TGW Connect attachment or it will cause a recursive tunnel routing issue. For this same reason, the VPC CIDR (or at least the subnet where the GRE tunnel endpoint exists) should never be advertised over the GRE tunnel via BGP to the TGW.

- When creating a TGW Connect attachment, AWS assigns a /29 CIDR from the assigned TGW CIDR and does not use the IP of any TGW attachment ENI. This means that the IP of the GRE tunnel endpoint on the TGW is not drawn from the VPC attachment. Typically, when a TGW is created, or edited, the CIDR associated with a TGW can be applied.

- The CIDR associated with the TGW can be an IPv4 /24, /23, or /22 block of IP addresses. For IPv6, this could be /64, /63, or /62. It cannot include the space AWS typically uses for tunnel IPs from the 169.254.0.0/16 range and should be unique across the network, not overlapping other VPCs or on-premises networks. It is from this assigned range that the TGW Connect peer IPs will be drawn.

- AWS attempts to create two GRE tunnels per TGW Connect attachment with the remote device or devices. The IPs will be drawn from the /29 subnet the TGW chooses for the Connect attachment.

- BGP peering with a TGW using TGW Connect attachment requires using BGP Multi-Hop on the device peering with the TGW.

- That was a lot of information about TGW Connect. Remember that a TGW Connect attachment is an overlay tunnel technique that provides more bandwidth throughput than IPsec tunneling and BGP dynamic routing, and allows sharing more routes than are typically allowed by a VPC route table with a TGW.

- In the case of Trailcats, the fictional company utilizing AWS for cloud networking for this certification guide, a TGW Connect attachment would make the most sense when connecting its resources not connected to TGW using a third-party network device, such as a router or firewall. Trailcats could use this TGW Connect attachment to create an underlay attachment connection to the VPC with the third-party NVA, as in *Figure 5.8*, and over that TGW attachment connectivity, create an overlay GRE tunnel for the data plane and exchange of routes. This allows the BGP dynamic routing protocol to handle route propagation instead of having to statically edit native cloud route tables and improves the management of Trailcats resources.

- In this section, we covered all the different attachment types that TGW offers to serve as a hub of network connectivity. TGW is the workhorse of AWS cloud networking, offering the most granular routing experience with the most supported connectivity options. The next section will cover TGW route tables and discuss how they are managed and populated both statically and dynamically.

TGW Route Table Overview

TGW route tables are the secret weapon for managing a complex AWS network infrastructure. When a TGW is created, by default, a main route table is created that covers the TGW and any attachments that are created. This is optional, however. For complex routing architectures, the TGW can have multiple route tables, each associated with a different TGW attachment if desired. In this way, it is possible to provide segmentation or connectivity between resources attached to a TGW as desired, based on the route table of the attachment.

Commonly, the TGW route table is used just like a subnet route table, providing basic routing to and from the various TGW attachments. But if needed, each different attachment could have a different route table and allow different routing decisions. This will be explored further in the upcoming *TGW Route Propagation Overview* section.

TGW Route Tables

TGW route tables are remarkably like VPC subnet route tables from an administration perspective. Generally, routes are added to a TGW route table in the same fashion as route tables are attached to a VPC subnet, with a CIDR and the next hop declared. However, TGW route tables have a much larger scale than a subnet route table. In fact, a TGW can carry up to 10,000 routes, whether dynamically learned via BGP, statically configured, or a mix of the two. This is a soft limit that can be increased by contacting AWS, but the upper limit of that quota is not published.

Here are some other limits associated with TGW route tables:

Details	Default Limit	Can the Quota Be Increased?
Route tables per TGW	20	Yes
Total combined routes per TGW	10,000	Yes
Dynamic routes from TGW Connect peer to TGW	1,000	Yes
Routes advertised from TGW to TGW Connect peer	5,000	No
Static routes for a prefix to a single attachment	1	No

Table 5.1: TGW route quota limits

Despite TGW enforcing limits, the scale is far beyond what is achievable by other network constructs in AWS outside of third-party virtual network appliances. It is important to know the quota limits for the exam, however.

Note

Generally, it's unnecessary to memorize tables and numbers, but this is an exception. Know the TGW limitations for the exam, as a common trick question involves dynamic routes that exceed the quota limits.

Each TGW route table can be associated with one or more TGW attachments. Commonly, this is used when grouping resources together or connecting VPCs with a similar purpose. Consider the following diagram:

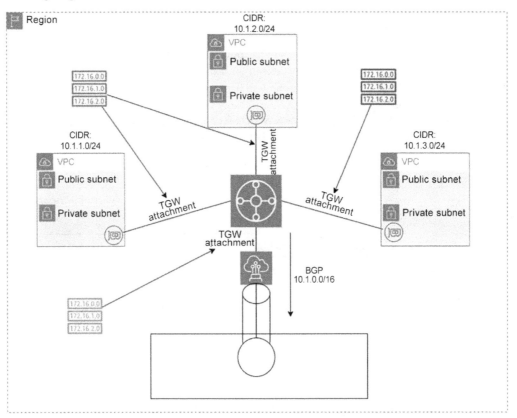

Figure 5.9: TGW with multiple route tables associated

In *Figure 5.9*, two VPCs are sharing the same route table. This is common if the VPCs need connectivity to the same resources. Remember that a route table associated with a TGW attachment will be consulted when a packet comes into the TGW from the attachment, not when a packet is leaving a TGW for an attachment.

Prefix lists have been mentioned before, and we can also use them as a reference for TGW route tables. They can be either AWS-configured prefix lists or custom lists, which helps reduce the manual effort associated with building and maintaining one or many route tables for your AWS deployment.

TGW Route Propagation Overview

Under normal circumstances, adjusting the routes within route tables is an entirely static affair. For example, when peering two VPCs together, resources in one VPC will not be able to reach the resources in another VPC without first creating routes directing traffic for the other VPC at the VPC peering object. This requires editing route tables in both VPCs and attaching them to subnets.

TGW makes this slightly easier, though it is still not a truly dynamic process. For traditional network engineers, it may at first look like TGW is using some form of routing protocol to propagate routes, but this is a dangerous expectation to set. What is really happening is that TGW can automatically program the routes for an attached resource into its local routing table, whether that is the main table or a specific one for the attachment used. However, the attached resource will not receive any sort of route from the TGW in turn.

TGW route propagation can be a very complex topic, and some diagrams will be needed. This will be further explored in the next section.

TGW Route Table Propagation

TGW route table propagation is a feature that allows route tables to be populated based on AWS automation. The TGW, by default, has only a single route table to worry about, and so route propagation is a simple method by which routes are automatically programmed into the route table when attaching a resource to the TGW, like this:

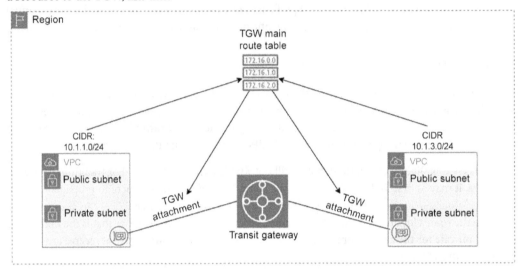

Figure 5.10: Basic TGW route table propagation

Figure 5.10 shows two VPCs with public and private subnets, connected via a TGW for routing and interconnectivity. As mentioned before, this is not a two-way street. The resource attached to the TGW does not have its own route table (if it has a route table) updated by the TGW with routes that the TGW is using automatically. In the case of VPC attachment, the subnet route table that receives the TGW attachment has no routes injected into the route table. Commonly, these are spoke VPCs that are part of a hub-and-spoke network architecture where the spoke VPCs connect to a central hub VPC. These spoke VPCs typically don't need to have routes injected. Remember that for almost all resources attached to a TGW, the TGW is the only way to connect to other resources, so programming extra routes would be a waste of time, not to mention the quota limitations are much lower on VPC route tables. *Figure 5.11* illustrates how VPC CIDRs are shared with a TGW and how manual route propagation works:

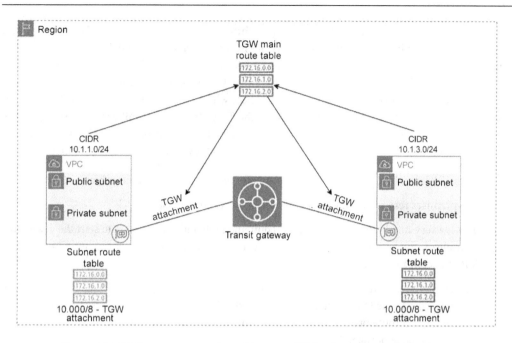

Figure 5.11: TGW route propagation with manual VPC subnet route table addition

In *Figure 5.11*, you can see the same image as in *Figure 5.10* but with added routing tables. Route propagation reaches its full potential in complex routing scenarios. Consider a scenario where the goal is to only allow certain resources to communicate. With TGW route propagation, handling complex routing scenarios is made much simpler. Start with a complex setup, such as this:

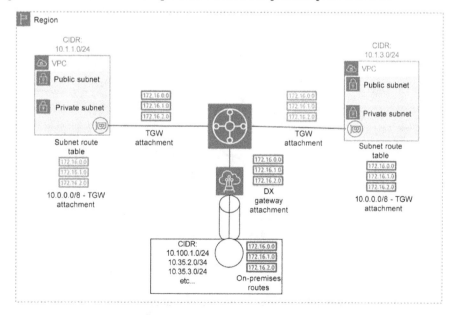

Figure 5.12: TGW complex route table deployment

Figure 5.12 expands on the previous network architecture by incorporating a DX gateway and on-premises connectivity, in addition to the two VPCs connected via a TGW. The main benefit of route table propagation is to simplify this process, especially in the case of dynamically learned routes over a VPN or a TGW Connect peer, where there may be far too many routes to handle manually. Remember that route table propagation also influences what routes the TGW will advertise to other peers over a VPN or TGW Connect as well. Since the TGW total route limit exceeds the number of routes that can be advertised to peers, it's important to plan the route propagation carefully. Consider how a TGW route propagation strategy could be used to achieve the following requirements:

- Each VPC should be able to reach on-premises resources but not each other

- On-premises resources should be able to reach resources in each VPC

- This is a very simple and common set of requirements, but it's a good place to start the discussion. If Trailcats, the fictional enterprise, had a security requirement that would not allow the two VPCs to communicate, this is one way to accomplish that goal.

- Remember that TGW does not advertise routes to a DX gateway and that the DX gateway will only advertise to the DX consumer a prefix list that is manually configured, as discussed earlier. Unlike normal route propagation behavior, this is a special case that overrides other types of TGW route table propagation.

- However, the reverse is not true. TGW will propagate routes learned from the DX gateway to whatever route tables are selected for propagation, with those routes coming from the customer/provider via BGP. Given the previously mentioned requirements, the route propagation strategy should be something like this:

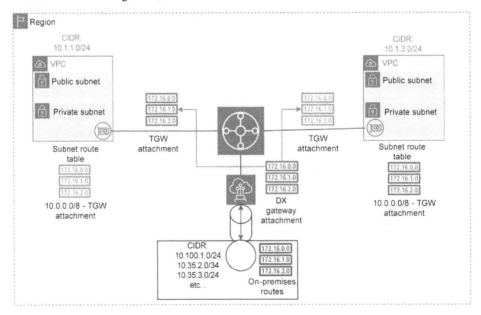

Figure 5.13: VPC and DX route propagation via TGW

Figure 5.13 shows the same as *Figure 5.12* but with additional arrows and labels showing the flow of traffic and how routing is configured for on-premises connectivity.

Several steps are needed to adjust the route tables so that there will be no unintended communication:

1. The route tables attached to the TGW attachment for each VPC should be different, and the other VPC attachment should not be propagated to ensure there is no communication between the VPCs.

2. Route table propagation should be turned on for the route tables attached to the TGW attachment for each VPC, and the DX gateway attachment should be propagated into them. This ensures that the TGW attachment will know where to send any packets coming from a VPC destined for the on-premises resources.

3. A suitable CIDR should be advertised from the DX gateway to the customer/provider device across DX. In this case, with only two VPCs, it could be summarized as `10.1.0.0/22`, for example. Note that, as stated before, this is a manual process and there is no requirement that CIDRs advertised should be reachable via the TGW; it just rarely makes sense to do otherwise.

4. To prevent the customer/provider equipment from becoming transitive and allowing the two VPCs to communicate using it as a hub, care must be taken when advertising routes from the equipment to the DX gateway. For example, advertising a flat `10.0.0.0/8` CIDR would encompass both VPC CIDRs and traffic would be attracted to the customer equipment in the absence of a more specific route in the VPC route table or the TGW attachment route table. There is another solution to this specific scenario, but this pitfall is very common to see in real life and potentially in the exam.

5. As an optional failsafe, a static route can be added to each route table attached to the VPC attachment on the TGW, using the CIDR of the other VPC, and with a next hop set to blackhole. This means that if the blackhole route is the most specific one, traffic destined for the other VPC will be dropped when entering the TGW.

Figure 5.14 illustrates this concept:

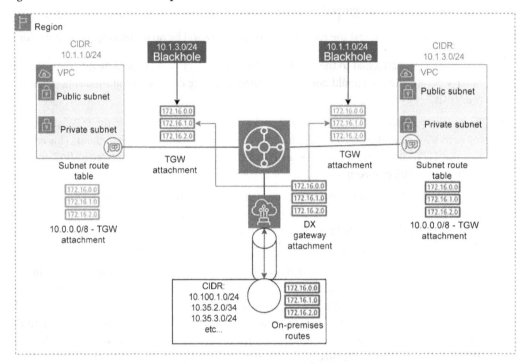

Figure 5.14: TGW route propagation with static blackhole routes

This may be preferable to route filtering on the DX equipment end, but it is a manual process and could also end up being forgotten down the line if requirements change. Managing route propagation between TGW and VPCs, as well as on-premises, is a design consideration that often requires careful attention to the route tables, the subnets in which there are TGW attachments, and DX route propagation with TGW. The next section explores more of the latter.

TGW and DX Gateway Design

Of all the hybrid cloud networking options, the combination of TGW and DX gateways tends to be one of the most confusing. There are a lot of caveats for both TGW and DX gateways, and when they are combined, there are even more. The biggest ones have been covered in their respective sections, so in this section, the focus will be on common design patterns and things to consider when planning to interoperate them.

When designing TGW and DX gateway integrations, the most important thing to remember is that TGW does not advertise routes to a DX gateway attachment; the DX gateway must be configured with the prefixes to advertise across DX. What's more, the advertised CIDRs cannot overlap other TGWs attached to the same DX gateway, so if redundancy in CIDR advertisement is needed, more DX gateways (and transit VIFs) are needed as well. This is easier to show than tell because it sounds far more complex than what will be deployed:

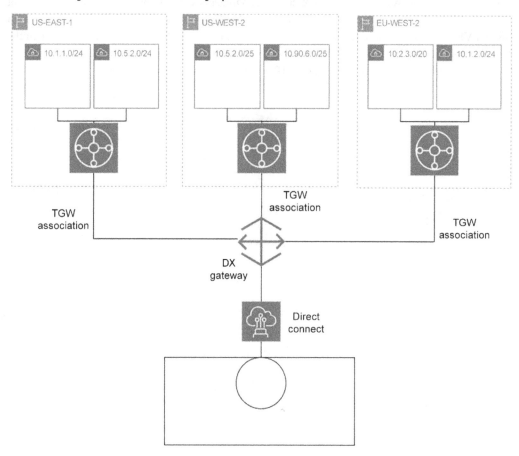

Figure 5.15: Multi-Region TGW with a DX gateway

In *Figure 5.15*, there are no problems. Each TGW attachment to the DX gateway is free to use a non-overlapping CIDR advertisement to the customer/provider equipment across DX. Proper network IP schema design has simplified the connectivity for hybrid connectivity.

What if the environment is a lot messier? Many brownfield environments will not be so cleanly summarized and will need more granular control of DX gateway prefix advertisements to be able to consolidate connectivity. Remember that a DX connection can only have a single transit VIF and that this VIF can connect to a single DX gateway. From a design perspective, it may be critical to connect TGW to that DX gateway, and this can only be done with non-overlapping prefix advertisements.

Consider instead a messy brownfield where network engineers were not involved with the deployment but are now trying to provide connectivity. Assume that Trailcats has two different applications in different VPCs, and each has a new requirement to communicate with database servers that are in the Trailcats on-premises data center. If they set up a DX gateway to connect the VPCs to the data center, this will not work because of the overlapping IP addresses shown in *Figure 5.16*:

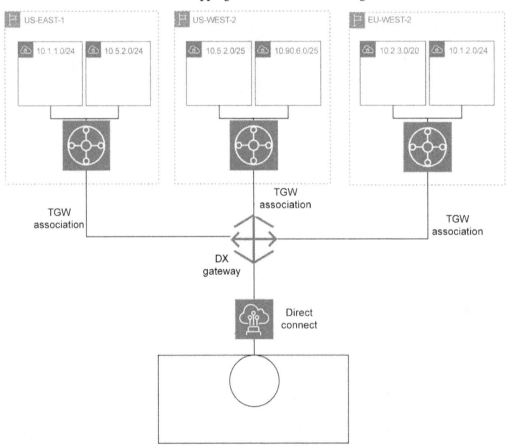

Figure 5.16: Multi-Region TGW with a DX gateway – IP overlap

As shown in *Figure 5.16*, two of the VPCs attached to the DX gateway would have the same VPC subnet address if this were possible to configure, but AWS will not allow VPCs to be attached to the same DX gateway with overlapping subnets.

There are multiple ways to resolve this problem without introducing a DX gateway, such as NAT, but none can be accomplished with only a DX gateway. Because the IP prefix advertisement from the DX gateway to the customer/provider connected to Direct Connect is static and cannot overlap other TGW attachments to the same DX gateway, one of the attachments would have to remove the overlapping prefix.

If Trailcats were having this issue, the simplest way to deal with this problem without requiring adjusting the IPs in one of the two VPCs would be to use a private Direct Connect VIF to connect to one of the overlapping VPCs instead, with or without another DX gateway. That wouldn't be enough to ultimately solve the problem if all cloud resources needed to communicate with all on-premises resources, though. This sort of messy brownfield design is what causes most of the headaches with cloud networking. If we assume this overlap must be resolved, it's almost always simpler to migrate the workloads than to try to implement a network address translation solution. In the real world, though, this could be a critical workload, or the app could be moved to the cloud without any modifications, leaving no flexibility for changes. In *Chapter 6, Connecting Third-Party Networks to AWS*, many different design patterns will be explored – some to resolve problems just like this. The following figure expands on some of the options available to cloud network designers:

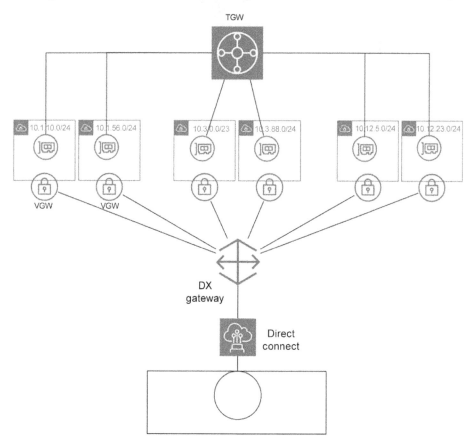

Figure 5.17: DX gateway private VIFs with TGW connecting VPCs

Figure 5.17 shows another design pattern, this one using a TGW only to connect VPCs and localize extra-cloud connectivity to Direct Connect. Often, the expected traffic patterns inform the design. In a design where TGW ends up being the hub, there is usually an equal amount of cross-traffic expected from an east-west and north-south perspective.

When using a pattern like the one shown in *Figure 5.18*, the bulk of traffic is expected to be north-south, from workload to extra-cloud, and a minimal amount is expected to be east-west between VPCs. Care must be taken when designing cloud networks to understand not only what resources must communicate, but to what extent and how things may change in a failure scenario.

Until now, the focus has been on a single DX gateway and a single DX location. How complex can things get when connecting across multiple locations and DX locations with redundancy? Consider the following figure, where a DX gateway and a TGW are both being used for connectivity:

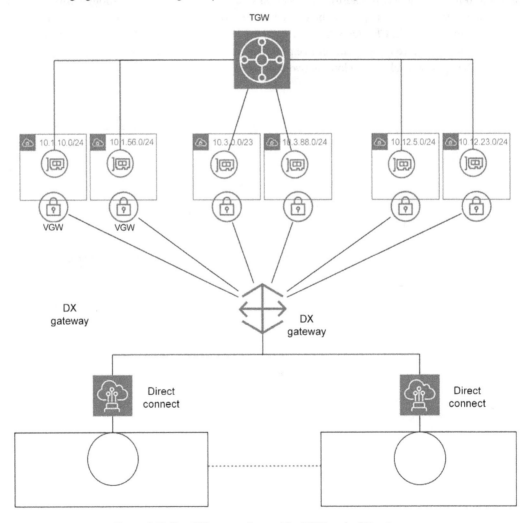

Figure 5.18: Two DX connections with a TGW and a DX gateway

There is a very large number of ways to mix and match designs, and they can't all be represented here. However, redundant designs pose challenges in providing the best path optimization while avoiding route looping. In *Figure 5.18*, for example, there are multiple points where traffic looping could happen if poor route design is used, but this design can also provide redundancy in case of single failures.

For example, if each DX location is connected to the other (which is very common in DX location designs), it would be possible for traffic to go from either DX location to any of the VPCs even if one of the DXs were to fail completely. This would require some creative routing on the customer/provider side but is very possible. Of course, inside the cloud, the VGW and TGW are already redundant within AWS. The TGW would continue to offer east/west connectivity to VPC workloads, and the working DX could be used to move traffic into and out of the cloud. There are a lot of design patterns involving other kinds of third-party connectivity, which will be covered in the next chapter.

TGW Inspection VPC Design

With the advent of **Gateway Load Balancer** (**GWLB**), the recommended TGW inspection VPC design has changed a bit to include that, but here is the inspection VPC design without GWLB first. Note that the TGW is being used to control the connectivity and the routing from east-west flows between VPCs to force inspection by a security appliance – in this case, AWS Firewall. The following figure depicts an inspection VPC design using AWS Firewall:

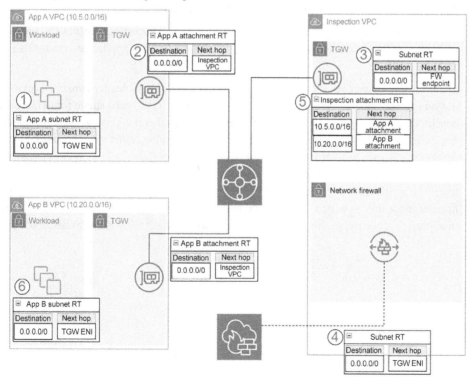

Figure 5.19: TGW with an inspection VPC design using AWS Firewall

Understanding the route table flows is key to understanding how security appliance insertion/redirection works in the cloud. The traffic flow is shown in *Figure 5.19*. Each step in the flow of traffic between VPCs is marked:

1. A workload in **App A VPC** wants to send traffic to the workload in **App B VPC**. The first step is consulting the subnet route table, which points all traffic at the TGW attachment.

2. Traffic ingresses the TGW VPC attachment, and the custom route table is consulted. Traffic coming from the VPC destined anywhere is redirected to the inspection VPC.

3. Once traffic comes into the TGW subnet in the inspection VPC, all traffic is forwarded to the AWS Firewall endpoint in the other subnet. That endpoint has a backend PrivateLink connection to the AWS Firewall service that inspects the traffic and, if allowed, forwards it to the endpoint/subnet.

4. The subnet route table points all traffic at the TGW attachment.

5. As traffic comes into the TGW attachment for the inspection VPC, the destination attachment is found, and the traffic is forwarded to **App B VPC**. The subnets in **App B VPC** use local routing for the VPC to deliver the traffic to the workload.

6. When the workload responds, it happens in the same fashion, starting with the subnet route table directing traffic to the TGW ENI. Not pictured is the TGW attachment route table for **App B VPC**, which would be identical to that of **APP A VPC**.

7. This is a simple east-west flow between two VPCs, but it perfectly identifies the method by which inspection is accomplished in the cloud, where security appliances cannot be in line with the traffic.

8. The most important concept to understand here is that route tables are consulted on ingress, not on egress, especially with TGW attachments. If the TGW had a single route table for all attachments, traffic between VPCs would never be inspected as the TGW would simply forward the traffic between the attachments. Custom route tables enable traffic steering based on ingress destination flows. This is true of east-west and north-south flows, as well as cloud ingress and cloud egress flows from other sources and destinations. Also remember that this diagram is highly simplified; a real design would include Availability Zone redundancy and multiple firewall endpoints, in which case it would make sense to run the inspection VPC attachment in appliance mode to pin the full traffic flow to a single Availability Zone, as discussed earlier.

9. Here is a similar design but utilizing GWLB and appliances with the GENEVE protocol:

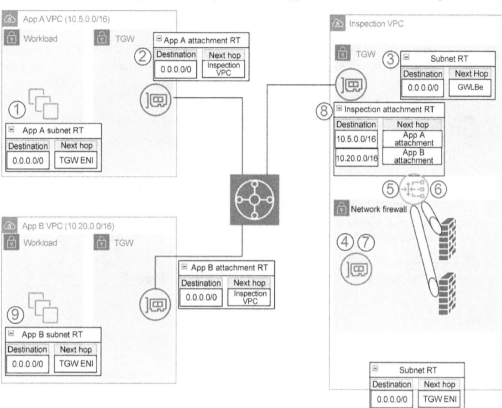

Figure 5.20: TGW inspection VPC using GWLB

In *Figure 5.20*, the addition of GWLB enables the preservation of the original traffic details as it is inspected by the security appliance with the addition of GENEVE. The steps are similar to the previous example, with very specific differences:

1. As before, the traffic starts with a workload sending it to an **App B VPC** workload. The subnet route table directs traffic to the TGW ENI.

2. The traffic is forwarded to the inspection VPC, as before, based on the TGW attachment route table.

3. This time, instead of the AWS Firewall endpoint, the traffic is sent to a GWLB endpoint, which is a special ENI tied to the GWLB.

4. The traffic enters the GWLB endpoint and uses AWS PrivateLink to take traffic to the GWLB.

5. Using a five-tuple hashing algorithm, the GWLB selects a security appliance that is healthy and uses the GENEVE tunnel protocol to forward the traffic for inspection, encapsulating the traffic in the tunnel with the GENE header.

6. The security appliance inspects the traffic and, if it is allowed, forwards it to the GWLB using the GENEVE tunnel. This allows the security appliance to inspect traffic without needing to know the routes for the traffic's destination and preserves the original source/destination information without the need for NAT scenarios, which can make inspection more difficult. When the GWLB receives the traffic back from the security appliance, the GENEVE encapsulation is removed.

7. The GWLB endpoint references the subnet route table and forwards the allowed traffic to the TGW attachment for delivery.

8. Using the TGW attachment route table, the traffic is sent to **App B VPC**.

9. Using local routing, traffic is delivered to the destination workload. If the workload responds, this process will happen in reverse as traffic goes back. Importantly, TGW appliance mode should be used to ensure symmetry across Availability Zones as GWLB uses AZ affinity as well.

10. The key takeaway from this traffic flow is to understand how each route decision feeds into the next, and the route table manipulation required for this traffic flow.

> **Note**
>
> For up-to-date best practice reference architectures with TGW using the AWS Well-Architected Framework, visit `https://docs.aws.amazon.com/vpc/latest/tgw/what-is-transit-gateway.html`.

Summary

TGW hybrid networking can be a complicated solution with many caveats. Throughout this chapter, the options and limitations have been conveyed using diagrams and packet flows, but this is just the tip of the iceberg. AWS is constantly updating its best practice designs and architectures as new features are released.

In this chapter, we covered TGW at length, including the components and how to use route tables to effectively steer traffic based on the attachment and desired traffic pattern. After reading this chapter, you should be able to integrate TGW with other cloud and extra-cloud networks to meet business requirements. The next chapter will expand on networking, covering how to connect third-party networks to AWS.

Exam Readiness Drill – Chapter Review Questions

Apart from mastering key concepts, strong test-taking skills under time pressure are essential for acing your certification exam. That's why developing these abilities early in your learning journey is critical.

Exam readiness drills, using the free online practice resources provided with this book, help you progressively improve your time management and test-taking skills while reinforcing the key concepts you've learned.

HOW TO GET STARTED

- Open the link or scan the QR code at the bottom of this page

- If you have unlocked the practice resources already, log in to your registered account. If you haven't, follow the instructions in *Chapter 16* and come back to this page.

- Once you log in, click the START button to start a quiz

- We recommend attempting a quiz multiple times till you're able to answer most of the questions correctly and well within the time limit.

- You can use the following practice template to help you plan your attempts:

Working On Accuracy		
Attempt	Target	Time Limit
Attempt 1	40% or more	Till the timer runs out
Attempt 2	60% or more	Till the timer runs out
Attempt 3	75% or more	Till the timer runs out
Working On Timing		
Attempt 4	75% or more	1 minute before time limit
Attempt 5	75% or more	2 minutes before time limit
Attempt 6	75% or more	3 minutes before time limit

The above drill is just an example. Design your drills based on your own goals and make the most out of the online quizzes accompanying this book.

First time accessing the online resources? 🔒
You'll need to unlock them through a one-time process. **Head to** *Chapter 16* **for instructions**.

Open Quiz https://packt.link/ansc01ch5 OR scan this QR code →	

6

Connecting Third-Party Networks to AWS

This chapter will focus on connecting third-party networks to AWS and managing the complexity of using different types of intracloud and extracloud connectivity. The goal of this chapter is to show how different native and non-native connectivity options interact with AWS networks. In the simplest terms, a **third-party network** is any network that is not inside the AWS cloud. Except for AWS Outposts, an on-premises deployment of AWS infrastructure, this covers on-premises networks, networks within collocated data centers, branch offices, and even other clouds. In all cases, AWS has a rigid set of integration options with any of these networks that differs only based on how the third-party network chooses to connect to AWS.

Here's a brief rundown of the topics covered:

- What is a third-party network?
- Third-party connectivity options
- Redundant connectivity to third-party networks
- Route optimization for redundant connections

By the end of this chapter, you should be able to understand what options exist to connect a third-party network to AWS and be able to compare and contrast them, design for resilient connectivity, and optimize routing between AWS and any third-party network.

Defining Third-Party Networks

In terms of this certification guide, and AWS in general, a **third-party network** could be considered any network that exists outside of the control of yourself or your organization. A third-party network is not classified by the way in which it connects to your network but by the operational control and responsibility of the network and its traffic. It is rarely worth sparing a thought for what sort of deployments, routing protocols, and designs a third-party network has; only consider what traffic it is sending to and receiving from your own network, and how to integrate the connectivity between the two.

To this end, before deciding what sort of connectivity is appropriate, you need to ask a few questions, such as this: Who is connecting and what do they need access to? The next section will answer these questions.

Identity of Third Party

When considering the best connectivity options for bringing in a third-party network, the very first consideration should be based on what organization needs connectivity. Is it a vendor? A partner? Home users? Executives? Consultants? All the above?

Consider *Trailcats* and their business needs. They may need to allow another company to connect to their cloud resources for some business purpose. This other business may need to access a Trailcats database or offer some service to Trailcats privately that requires a secure connection.

The level of trust extended, and the level of access required, are the very first considerations when planning a third-party network connection. Not all connection methods offer the requisite amount of control and accountability for networks using them. For example, a simple **site-to-site (S2S) virtual private network (VPN)** between a **customer gateway (CGW)** and a **virtual private gateway (VGW)** attached to a single VPC offers simple, basic connectivity to a particular set of cloud resources (those within the VPC). When setting up a VPN, the integrity is maintained by the encryption of the data, and the authenticity of the connecting network is based on mutually agreed-upon information, whether it be pre-shared keys or digital certificates. If these mutual authentication methods work, the third-party network is allowed to connect without further scrutiny.

This works when connecting to a third party that needs limited access to specific resources, with whom you have a good enough relationship to share basic connectivity using simple authentication methods. Commonly, this is used in **business-to-business (B2B)** connectivity, where one business needs to share resources with another privately.

Contrast that with something like an AWS PrivateLink connection. Commonly, the level of trust between the two organizations using this type of connectivity is very low (basically nonexistent) unless a custom PrivateLink service is created just for that connection. In that case, AWS itself brokers the connectivity by allowing a consumer to connect to the exposed service tied to PrivateLink. Other than that connectivity, no access is implied, and authentication is handled entirely by AWS.

In both previous cases, trust was established through the sharing of information or brokering through AWS. What if other requirements exist, such as a business partnership that requires high throughput or a more complex relationship of multiple VPCs? For this, we often turn to account sharing using network resources such as VPC subnets or a shared **transit gateway** (**TGW**). The drawback of sharing these resources is it requires a higher level of trust between parties because the level of implied access is higher.

This is the sort of thought process that goes into designing third-party connectivity in AWS (or anywhere else, really). Always aim to adhere to the principle of least privilege and least access wherever possible as requirements dictate.

Third-Party Access Requirements

The level and depth of access needed often prescribe the type of connectivity for third-party entities as well. For example, AWS Client VPN allows a remote worker to access resources, but there is an explicit allow methodology associated with the resources a remote VPN user can access. With AWS Client VPN, an administrator creates a target subnet for VPN users to access in a VPC as the landing point for connectivity. That subnet is all the access implicitly provided. All other access to resources must be explicitly provided and allowed by way of the Client VPN subnet route table and authorization rules allowing the traffic to reach the other networks.

Because a remote User VPN is inherently less safe than an on-premises connection, things such as identity, authentication, and authorization are very important. It is especially important to define what resources such a user can access. AWS provides a high level of granularity to client access VPN because Client VPN is a less trusted method of connectivity. In the case of Trailcats, the fictional example enterprise used in this book, a user accessing Trailcats resources from a Trailcats office using Trailcats equipment and coming from the Trailcats trusted network can be assumed to be more trustworthy because of the inherent difficulty in impersonating a Trailcats employee, gaining access to a Trailcats workstation in the office, and connecting to cloud resources. Contrast this with the far less secure method of VPN connectivity, which can come from any untrusted network, such as a coffee shop or library. With a VPN, often the only safeguard in place is user credentials, which are far easier to steal than physically accessing a building with a trusted network.

You may have heard of the term **zero trust access** (**ZTA**) before. This principle of least privilege based on the connection, IP address, or any other factor is an easy concept to understand but exceedingly difficult to implement without business impact. Often, the principle of ZTA is tied directly to identity and a policy based on that identity's level of privileged access. There is no trust extended based on what type of connectivity is used, whether it be from the corporate headquarters or over the internet. AWS Client VPN is a good example (though not complete) of ZTA principles as it allows user identity to be authenticated based on an identity store such as **Active Directory** (**AD**), and authorization to resources granted or restricted based on the identity and its membership in groups of interest that are granted access.

With most AWS third-party connectivity options, this level of granularity is not provided because the connectivity itself is at a different layer and needs to allow broader connectivity. A poor example of ZTA principles would be a third-party S2S VPN, where whole networks gain access to each other with no regard for the identity of the users or workloads crossing the VPN. Once the VPN tunnel is established, trust at the network layer is extended as far as routing will carry the traffic. As mentioned in the previous section, this is sometimes easier than trying to enforce ZTA principles at the network layer. Allow the traffic in with a certain level of trust and rely on upper-level processes to be the enforcement point. In these cases, ZTA principles need to be enforced at the application layer and not on the network layer. For end-to-end ZTA to be complete, this is arguably the correct layer of enforcement.

Third-Party Connectivity Options

This section covers third-party connectivity options, how they connect to AWS, and the level of trust required to facilitate connectivity. Some of this information may overlap with previous chapters, but the focus in this chapter is on the third-party connectivity aspects of the solutions and less focused on how the solution itself works.

AWS PrivateLink

PrivateLink is an AWS technology powering multiple services, some that are AWS services themselves and others that are **security-as-a-service** (**SaaS**) partner services. The concept is straightforward: create an interface or gateway endpoint in your environment that provides private AWS-managed connectivity to something on the other side, whether that is an AWS or third-party service. This diagram illustrates the basic idea:

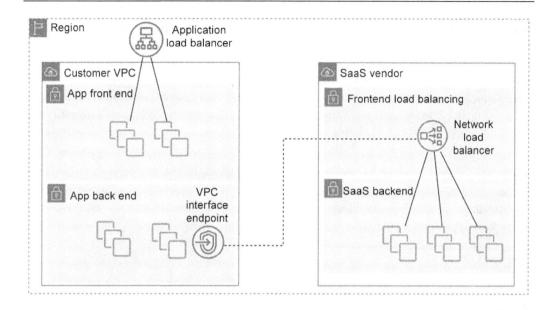

Figure 6.1: AWS PrivateLink SaaS vendor connectivity example

In *Figure 6.1*, our Trailcats company is connecting to a third-party vendor that is offering some sort of data transformation service. Suppose that Trailcats users can upload photos via the Trailcats website, and these photos will be altered in some way for the user to download. Maybe an AI image generator will modify the image to have a funny meme or add a funny face. Trailcats itself does not possess this AI technology; it consumes the services of another vendor that exposes its service via AWS PrivateLink.

The PrivateLink story starts with a vendor, whether AWS or a third party, exposing some sort of service to PrivateLink. Commonly, this is done via the AWS PrivateLink marketplace but can be created using the endpoint service, after which the service provider can use the marketplace to advertise and register the service or share privately with customers through other means. The service must be accessible via the consumer to use, which is where the conversation about trust and depth of connectivity begins.

The "Who's connecting?" question can be more complex than other AWS connectivity options and so it bears some discussion. If the service is listed on the PrivateLink marketplace, the answer is usually "Any AWS customer." If the service is unlisted and shared privately with customers by another means, the answer is, "Anyone you allow as a service provider." Note that just exposing the PrivateLink service does not provide connectivity to a consumer; it is merely the first step in providing access unless you grant access to everyone explicitly. This is the same for PrivateLink connection requests. When creating the service endpoint, the service provider must decide whether to allow connection requests automatically or require explicit acceptance. An AWS PrivateLink marketplace service operates on a scale that almost certainly requires auto-acceptance and a grant of access to everyone, while a private service is more likely to be controlled by the provider.

In short, the trust level is highly configurable. However, the access is the same regardless of trust extension. The VPC interface endpoint terminates a load balancer (network or gateway) in the service provider's environment, which connects to the provider's backend workloads. This connectivity is well understood and bears no further explanation, except that the return data will go back through PrivateLink to the consumer VPC via the endpoint. The Gateway Load Balancer often comes with a SaaS offering much in the same way you might construct an Inspection VPC yourself. As with other **elastic network interface** (ENI)-based objects, the route tables must be updated to direct relevant traffic into the ENI for this service to be consumed.

This entire conversation is different if you are the service consumer, which will almost always be the case, as most businesses consume more services than they offer. In that case, the consumer is accessing, not providing access, and the trust level is less relevant as PrivateLink is not a two-way service that gives access to the service provider.

S2S VPN

As mentioned earlier, the S2S VPN is perhaps the oldest and simplest example of third-party connectivity. It's a simple idea: through a controlled choke point, allow a third-party entity to access some resource within the private network without exposing that resource to other parties. AWS provides basic S2S VPN connectivity through a single VPC using **virtual private gateway** (**VGW**), or (potentially) multiple VPCs using TGW.

Setting Up a S2S VPN

The following steps are a guide to creating a simple S2S VPN to connect a third-party network to a VPC in AWS. The third-party device is not important; it just needs to be capable of using **IP security** (**IPsec**) and **Border Gateway Protocol** (**BGP**):

1. **Create the CGW**: Remember that the CGW is just an AWS representation of the actual on-premises device making the VPN connection to AWS, including things such as the public IP of the device and a BGP **Autonomous System Number** (**ASN**) if BGP will be used. In the VPC section of the AWS console, select Customer Gateway and supply the relevant information for the device outside of the cloud that will make the VPN connection, specifically any BGP ASN and the public IP of the device that will be making the VPN connection.

2. **Create the target gateway (VGW or TGW):** Because a S2S VPN can connect to a VGW or TGW, you must choose. A VGW gives access to a single VPC and a TGW gives access to any resources reachable via the TGW attachment. In the AWS console, still in the `VPC` section, select `Virtual Private Gateways` and supply a name for the VGW as well as a custom BGP ASN, or default to let AWS assign its own BGP ASN.

3. You must attach the VGW to a VPC to access the resources within that VPC. These steps do not cover creating a VPC, but you can attach a VGW to any VPC after it is created. After the VGW is created, under the `Actions` menu of the `Virtual Private Gateway` section, select `Attach to VPC` and select a VPC to attach to the VGW. This will make the VPC resources available to a VPN connection that terminates on the VGW.

4. **Configure routing:** Routing options are heavily dependent on things such as the use case for connectivity. The amount of access needed is determined by the answer to the "What do they need access to?" question. It can be static or dynamic with BGP and has limitations and caveats:

   - The VGW has a limit of 100 routes that can be learned by the VPN. The VGW cannot advertise anything other than the full **Classless Inter-Domain Routing (CIDR)** of the VPC CIDR across the VPN to the **customer gateway (CGW)**.

   - The TGW has a limit of 10,000 routes, which is far more useful. These routes can be propagated or not to other attachment route tables as desired, and the routes that are advertised from the TGW attachment to the CGW are based on the attachment route table. In this exercise, we will not be using the TGW, but it is important to understand the difference in limitations between the two options.

   - Select `S2S VPN Connections` and create a VPN connection. You will enter a name for the VPN, then select `Virtual Private Gateway` for the target type, and the VGW you created in *Step 2* from the list. `Customer Gateway` should be set to `Existing`, and then select the gateway you created in *Step 1*. For routing options, select `Dynamic`. Leave the other options blank and select `Create VPN Connection`.

5. **Update the security groups (SGs) of workloads**: To allow access inbound to cloud resources, any SG associated with workloads must have rules added to allow resources to access them as needed. Creating a workload is not part of this exercise, but to edit the SG of one, you go to the EC2 section of the AWS console and select `Security Groups`. From there, you can create an SG or edit one. *Figure 6.2* shows a sample SG where the VPN resources are using an IP address range of `10.100.10.0/24` and a rule is needed to allow traffic from that network to the workload:

Figure 6.2: Sample SG creation for access over S2S VPNCreate

6. **Create the VPN connection**: This is where the details of the VPN connection are set up. The configurable settings are as follows:

   - `Name`.
   - `Virtual private gateway` / `Transit gateway`.
   - `Customer gateway` (`New` or `Existing`).
   - `Routing options` (`Static` or `Dynamic` with BGP).
   - Tunnel inside IP version (IPv4 on VGW or TGW, IPv6 on TGW only).
   - (Optional) Specify allowed IPv4 / IPV6 CIDRs allowed to communicate over tunnels.
   - (Optional) Specify IP tunnel IP from `169.254.0.0/16` (IPv4) or `FD00::/8` (IPv6).
   - (Optional) Specify **internet key exchange** (**IKE**) pre-shared key (IKEv1 and IKEv2 supported).
   - (Optional) Advanced options such as dead peer detection, timeouts, logging, phase 1/phase 2 algorithms and lifetimes, and so on.

- (Optional) Advanced option to determine which side can initiate the tunnel negotiation. The default is Add, requiring the CGW to initiate the connection. Can only be set to Start as long as CGW has a public IP to have AWS perform tunnel initiation. Here is a screenshot showing the configurable options:

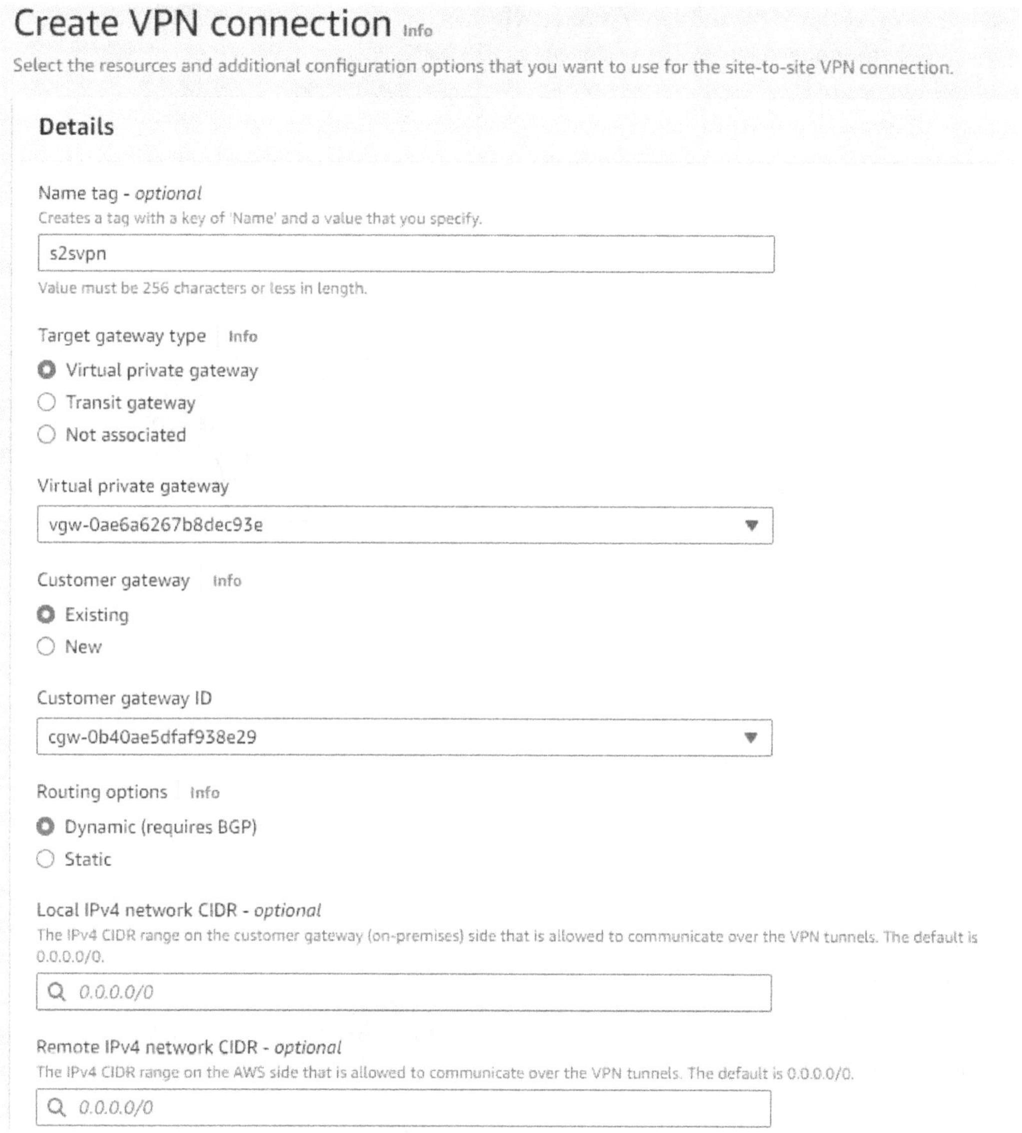

Create VPN connection Info

Select the resources and additional configuration options that you want to use for the site-to-site VPN connection.

Details

Name tag - *optional*

Creates a tag with a key of 'Name' and a value that you specify.

```
s2svpn
```

Value must be 256 characters or less in length.

Target gateway type   Info

- ● Virtual private gateway
- ○ Transit gateway
- ○ Not associated

Virtual private gateway

```
vgw-0ae6a6267b8dec93e                                    ▼
```

Customer gateway   Info

- ● Existing
- ○ New

Customer gateway ID

```
cgw-0b40ae5dfaf938e29                                    ▼
```

Routing options   Info

- ● Dynamic (requires BGP)
- ○ Static

Local IPv4 network CIDR - *optional*

The IPv4 CIDR range on the customer gateway (on-premises) side that is allowed to communicate over the VPN tunnels. The default is 0.0.0.0/0.

```
Q  0.0.0.0/0
```

Remote IPv4 network CIDR - *optional*

The IPv4 CIDR range on the AWS side that is allowed to communicate over the VPN tunnels. The default is 0.0.0.0/0.

```
Q  0.0.0.0/0
```

Figure 6.3: Creating a VPN connection

7. **Download the configuration file**: The configuration file to use for configuring the CGW can be generic or have specific vendor configurations for supported vendors. The file will include pre-shared keys, algorithms, addresses, and other VPN-specific settings discussed in the previous step. These settings must match the AWS side or the tunnel will not be negotiated.

8. **Configure the CGW and verify connectivity**: On the AWS side, under the VPC console, and VPN, the tunnel status will be visible under S2S VPN.

Here is a diagram showing a VPN connecting to a VGW:

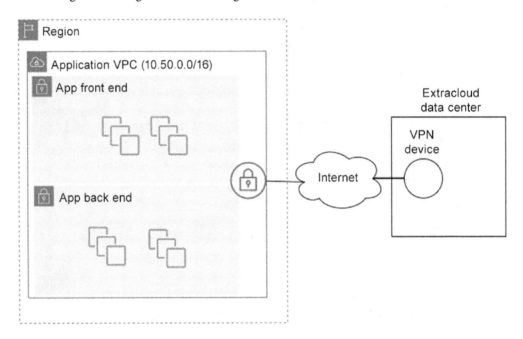

Figure 6.4: Basic S2S VPN connectivity to a VGW attached to the VPC

As a result, resources connected to the CGW would have access to the resources attached to the VPC associated with the VGW, and vice versa, as long as BGP is established, and the proper routes are exchanged.

This is a very basic configuration, giving connectivity to a single VPC. Because VPC peering is intransitive, even static routing cannot give the VPN access to other VPCs in that way (or vice versa). If the VPC were attached to a TGW, with some creative static route addition on either side of the VPN, traffic could use the VPC in a transitive fashion, but the use cases for this would be slim instead of just landing the VPN directly on the TGW. Perhaps a "clean room" VPC with inspection **network virtual appliances** (**NVAs**) would be a good example.

> **Note**
>
> AWS S2S VPN with VGW does not support IPv6. Only S2S VPN with TGW supports IPv6, and then only for the inside tunnel address, not for the outside public address. If you see an exam question involving connecting IPv6 VPCs or workloads and options for S2S connectivity, be sure to use S2S VPN with TGW, not VGW.

The TGW VPN attachment is briefly discussed next but is set up in largely the same fashion, substituting TGW for VGW. Here is the same use case diagram, but with a TGW VPN attachment instead:

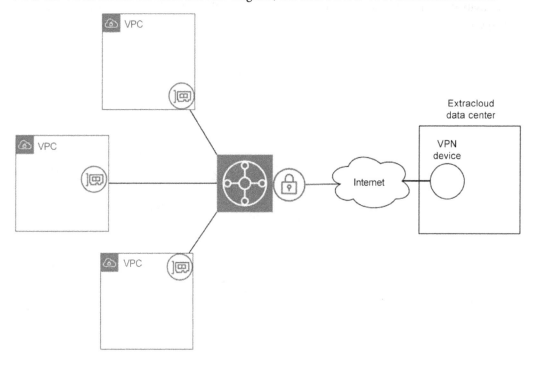

Figure 6.5: Basic S2S VPN connectivity to TGW attachment

In *Figure 6.5*, the connectivity is very simple. Remember that each TGW attachment can use a custom route table and propagate routes from each attachment to other attachment route tables. This makes the sharing of dynamic routes much simpler, as well as use cases such as segmentation.

Which to use depends on the answer to "What do they need access to?". If the third party only requires access to a few resources in a single VPC, a VGW may be sufficient and the right choice to easily restrict access. If there are multiple resources in different VPCs, a TGW with selective route propagation could be the correct choice. Care must be taken to ensure that route propagation is selective and that supernets, such as 10.0.0/8 or 0.0.0.0/0, do not allow traffic to traverse unintended paths and gain access to resources by choosing unintended, unsecured paths.

S2S VPN is redundant and highly available if the CGW equipment is also redundant. The S2S VPN configuration will include two tunnels that connect to different AWS infrastructures to protect against hardware failure in AWS impacting the VPN. On the third-party side, redundancy is achieved if these tunnels also terminate to different equipment using different connectivity.

Speed across a VPN is often a limiting factor. A common problem with S2S VPN performance, outside the limiting factor of the IPsec encryption overhead, is the latency between a CGW and the AWS infrastructure terminating the VPN connection. AWS offers a VPN Accelerator feature for this using the Global Accelerator AWS service. This service is a distributed infrastructure of AWS on-ramp connectivity that can lessen latency between a CGW and AWS so the VPN can take advantage of the high-speed AWS backbone faster.

Here is an example diagram:

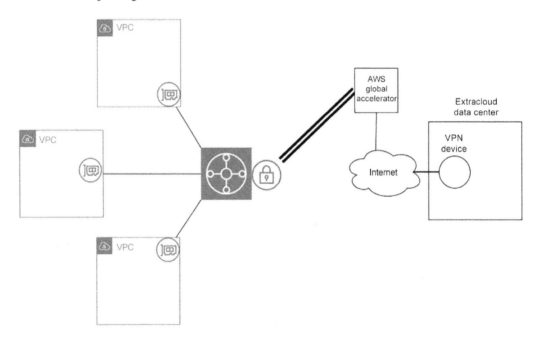

Figure 6.6: Accelerated VPN using AWS Global Accelerator

Figure 6.6 shows a simple AWS Global Accelerator deployment where a data center is connecting to a geographically closer AWS point of presence and using the AWS backbone to optimize traffic flow to cloud resources. By connecting to a Global Accelerator location that is geographically closer to the CGW, the VPN can use the high-speed AWS backbone much sooner, lowering latency. This does not increase throughput, however, and comes with some restrictions:

- The accelerated VPN is only supported on TGW, not VGW

- A **Direct Connect** (**DX**) public **virtual interface** (**VIF**) cannot use the accelerated VPN

- Acceleration cannot be turned on and off; the S2S VPN must be recreated to activate or deactivate the feature

- NAT-T support is required for acceleration, so be sure your CGW setup supports it and that the NAT-T setting is on in S2S VPN

- IKE and rekeys must come from the CGW side

- S2S VPN using certificate-based authentication will only work if the CGW supports IKE fragmentation due to a lack of support for this in Global Accelerator

These limitations are better understood in the context of the wider goal: to reduce latency by minimizing time spent on a non-AWS network. The accelerated VPN feature is intended only for that purpose, and the complexity of handling this under the covers requires a less primitive form of connectivity, which is why VGW is not supported.

The other requirements are often harder to overcome because the CGW equipment connecting to Global Accelerator must support them. This often means the equipment must be more modern and have a more advanced feature set available, which could be a limiting factor, depending on the equipment in use.

> **Note**
> This is another situation where it will be helpful to memorize the limitations and caveats associated with an AWS service. In the exam, there may be questions regarding accelerated VPNs posed along with some requirements that are incompatible to trick you.

In general, remember that the accelerated VPN can only improve latency, not throughput. In summary, the accelerated VPN is a limited option for improving latency in some situations where an on-premises location may be geographically far from the resources in the cloud.

TGW

A lot of the specifics of connecting networks to TGW have been covered elsewhere in this guide. What is worth covering in this section is how easy it is to share connectivity with a third party. In *Chapter 3, Networking Across Multiple AWS Accounts*, different cross-account sharing techniques were explained, but to refresh your memory, it will be reprinted here:

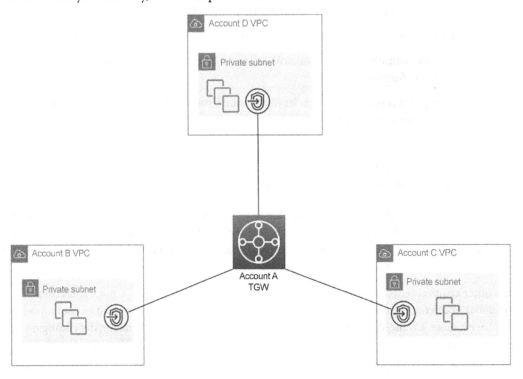

Figure 6.7: Shared TGW architecture

As shown in the diagram, TGW attachments can be shared to give connectivity to resources hosted in different AWS accounts. The question of "Who's connecting?" is based on explicit attachment sharing with other AWS accounts and their acceptance of that share. This also addresses the question of, "What do they need access to?", which can be controlled by tightly managing TGW route propagation. Even with this, usually, an Inspection VPC would be a good intermediate stop between a shared TGW attachment and its destination. Because of the way TGW is designed, it is perhaps the most frictionless way to provide connectivity across accounts within the cloud, but that comes with the possibility of unintended access based on improper route segmentation. Also, as stated, simple route segmentation is an insufficient security posture that may necessitate the use of security appliances or services for inter-attachment inspection.

For cases where multiple third-party accounts will be using the same TGW for connectivity, there may be other considerations, such as third-party security requirements. It is important to adhere not only to your own security posture but also to ensure the third parties using your connectivity communicate their requirements as well so that a decision can be made about what type of connectivity satisfies both parties.

VPC Peering

Peering VPCs together is by far the simplest form of third-party connectivity. It is so limited, however, that the use case for doing it is almost nonexistent. Cross-account VPC peering works on a proposal/acceptance model like most cross-account AWS services; so, just like them, the trust model is based entirely on whether you trust the third party enough to allow VPC peering to be created. Since VPC peering is intransitive, the question, "What do they need access to?" is limited to resources within the single peered VPC.

There is very little to say beyond that, though it should be noted that outside of the initial proposal/acceptance trust relationship, full trust is extended from a networking perspective. By default, if subnets are using the main VPC route table, and if a route to the other VPC is added, full connectivity is allowed between resources at the network layer. There are multiple ways to limit the access to resources if this is undesirable:

- SGs
- **Network access control lists (NACLs)**
- Custom route tables

If some resources but not others should be able to access resources across the VPC peering, those options can restrict access. These options exist in all third-party connectivity, but usually, there are better restriction options available in the connectivity solution. In the case of VPC peering, because of its simplicity, there no higher-level security features are available than those that are tied to the VPC and instance.

SGs are the last line of defense for an instance or endpoint, but depending on the requirements, it's possible that a single SG can't be used on multiple instances, or at least between application roles of instances, necessitating manual upkeep of many SGs to take care of restrictions.

NACLs are another way to tackle restrictions in a more scalable way, but often a cloud environment is not mature enough to take advantage. For example, consider a deployment involving multiple VPCs and subnets where different applications are deployed:

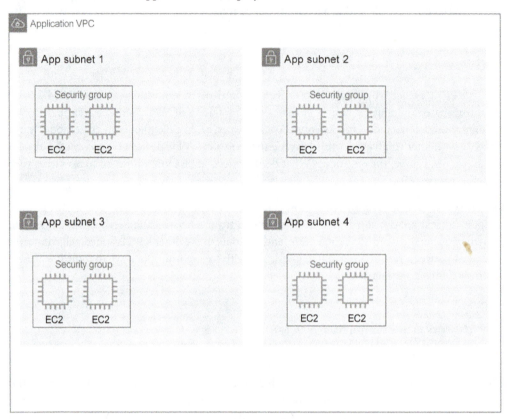

Figure 6.8: Basic VPC with subnets

In this case, imagine that each app subnet has an SG attached to similar EC2 instances using similar access rules. Of course, in this case, a NACL would be just as effective, but that will be covered next. When peering VPCs, the VPC peering mechanism offers no security, as discussed. The best way to restrict resources between peered parties is through route table manipulation, NACL, and finally SGs. Each has a limitation on how many entries can be used so a defense-in-depth strategy is often the right one depending on the level of access required. In *Figure 6.8*, the easiest way to restrict access if needed would be to edit SGs because they cover multiple resources. However, the application deployment in this VPC is also better than average. More often, what is inherited is legacy deployments where developers have no understanding or need to understand things such as segmentation and network-level security.

If Trailcats grows organically over time instead of with proper VPC design and application segmentation in place, it may end up with poorly planned permits that make it much harder to enforce security policy, as shown in *Figure 6.9*:

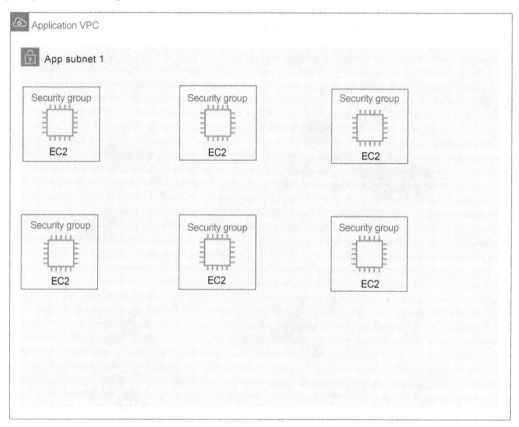

Figure 6.9: Common (poor) VPC design

Looking at this messy deployment, with multiple application resources deployed into the same subnet, there's no good way to control access other than editing every single SG. *Figure 6.9* shows a common result from when the cloud was new and seen more as a playground for app development than a production-grade design. Over time, apps tended to go into production without cleaning up the environment. That leaves network engineers with few options to secure them. It's not scalable to edit individual SGs for all access levels, but other options will not exist in an environment like the one in *Figure 6.9*. The most scalable and granular way to control access in a VPC peering situation looks like this:

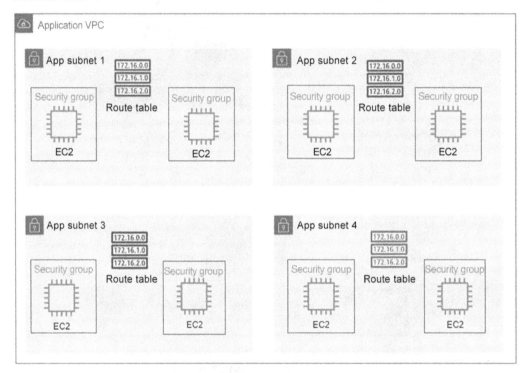

Figure 6.10: Scalable access control design for VPC peering

The design shown in *Figure 6.10* takes an increasing granularity defense in depth security posture. Based on requirements, all access, some access, or no access can be allowed to individual workloads or each subnet. Custom route tables offer route-level segmentation for an entire subnet if full access needs to be blocked. For example, if the third-party VPC is 10.10.0.0/24, a subnet route table can simply omit that route to the peered VPC. Care must be taken to ensure that supernets such as 10.0.0.0/8 do not attract traffic to an NVA that might reside in another subnet that does have access to the peering, or that this supernet does not point at the VPC peer as well. Routing is based on the longest match, and in the absence of an exact match, a less specific match can work to attract traffic in a way you do not intend.

NACLs can restrict traffic into and out of a subnet based on requirements. Remember that a NACL is *not* a stateful entity and that rules must be created to allow traffic into and out of a subnet. Access control at this level allows scalable control for multiple workloads that reside in a subnet without having to manage multiple SGs to achieve the same results.

The last line of defense is the SG itself. SGs are not particularly scalable but they are extremely granular. The pinhole approach works best with SGs, in situations where you may have very specific connectivity requirements for only specific workloads. Consider a scenario where a requirement exists that one workload in a subnet must be accessed via a third party. If this were the only requirement, it would make no sense to try to provide access control at other layers (RT and NACL) for a single access control. Remember that the goal is to balance scalable access control and granular access control. This is not true security as there is no inspection of traffic; that is something that needs to be handled differently.

The main concept to take from this discussion of VPC peering access control is the way to balance scale and granularity, providing access control at the right level based on the access requirements.

DX

There is not a lot of third-party connectivity through DX. The third-party aspect for DX is about whether the DX consumer needs to use an AWS DX Partner to deliver connectivity to them over a dedicated circuit or if the consumer will be using their own equipment to terminate that connection. This scenario turns the question of "Who's connecting?" and "What do they need access to?" on its head. The trust required is also turned around, as you are instead requiring the third party to deliver connectivity to you.

However, the fundamental questions still apply. Depending on the third-party provider and service offering, they might be delivering the full VIF directly to your equipment or just connectivity to the VIF over a Layer 3 **wide area network** (**WAN**) connection. For the purposes of answering the "Who's connecting?" question, your organization may have dedicated VIFs or share them between organizational units, and this influences the question of what they'll need access to.

For the exam, understand that the DX is, for all intents and purposes, over the control of your organization, but from a physical perspective, a third-party provider might be engaged to provide that physical layer and/or Layer 2 VIF connectivity if your own equipment is not in an AWS DX location to receive the cross-connect.

Client VPN

A **remote access VPN** is an access method almost as old as remote connectivity itself. The premise is very simple: allow some users authorized to connect to private resources access to them via an untrusted medium while keeping the data secure. With AWS Client VPN, users can connect to a native AWS VPN concentrator service and gain access to an authorized list of resources based on the level of authorization configured for that user. The following figure shows a basic end user connecting to Client VPN over the internet:

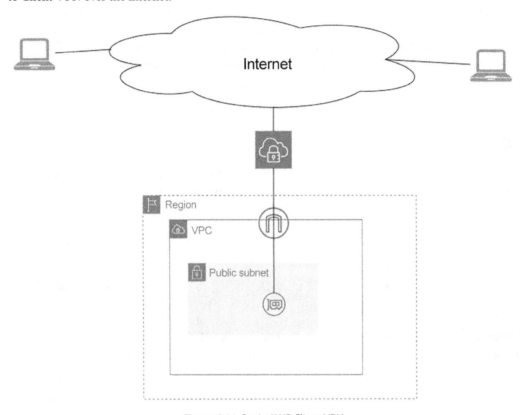

Figure 6.11: Basic AWS Client VPN

Figure 6.11 depicts a very basic setup where Trailcats wants a remote user to connect over the Internet to resources inside a Trailcats VPC.

The question of "Who's connecting?" is of utmost importance in a User VPN setup because trust is extended solely based on the user. Commonly, a User VPN solution includes validating a user against an authentication identity store and authorization database, and AWS is no exception.

To authenticate users, AWS offers the following options:

- AD authentication
- Mutual cert-based authentication
- Single sign-on using **Security Assertion Markup Language (SAML)** based authentication

You can also mix methods, such as using mutual cert-based authentication with SAML or AD-based authentication.

AD authentication assumes you have an AD built that is accessible by AWS to identify and authenticate a remote user. The AD can exist on-premises (with AWS AD Connector) or it could be cloud-based AWS Managed Microsoft AD. In either case, when using this method, you can also enable **multi-factor authentication (MFA)** if it is configured for use with your AD.

Mutual authentication is based on certificates and trust chains. Without getting fully into **public key infrastructure (PKI)**, the general idea is that a certificate is generated by a trusted party and uploaded to AWS to use as a server-side certificate by AWS. The trusted party must also be trusted by the clients connecting to the VPN, and the reverse is also true. Both AWS and the client must be using identity certificates issued by a party that each trusts to accept them. This is the "mutual" in mutual authentication. Each client receives its own digital certificate for identity, and this can be revoked if a user leaves the organization. When connecting, AWS checks the digital certificate for trust to verify the identity is allowed to connect. The client checks the AWS server certificate to ensure it is connecting to the real VPN and not a threat actor's middleware or other false entity.

Single sign-on using SAML federation is becoming the most common option as more and more applications move to a SaaS model and logging into each becomes a real chore. The idea is simple: create an **identity provider (IdP)** that can be used for authentication across any number of applications and then tell those applications to identify users by sending the users to them, keeping the IdP in one place, and not having to juggle usernames and passwords across hundreds of apps.

The AWS documents on **federated user authentication** for User VPN can be found here and should be used as a reference for the following figure: `https://docs.aws.amazon.com/vpn/latest/clientvpn-admin/federated-authentication.html`.

Here, you can see the steps for User VPN connectivity using SAML authentication:

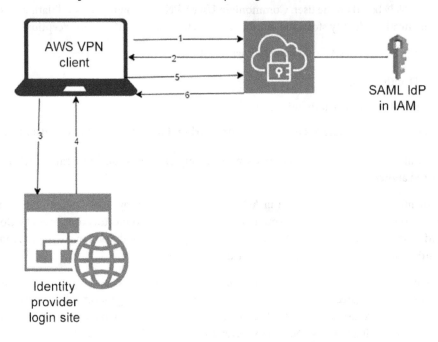

Figure 6.12: AWS Client VPN with SAML authentication

Figure 6.12 shows the sequence of events when a user attempts to connect to the VPN with SAML authentication:

1. The user opens the AWS Client VPN and initiates the connection to the AWS VPN endpoint.

2. The VPN endpoint responds with the IdP login site info based on what is configured in the IAM SAML provider for the VPN.

3. The VPN client opens a popup browser to the IdP login website to authenticate the user against the identity store used by the IdP. The identity store itself could be AD or some other identity database.

4. If the login is successful against that identity store, the IdP sends a SAML assertion back to the requesting client.

5. The VPN client provides SAML assertion information to the AWS VPN endpoint.

6. The AWS VPN endpoint validates the assertion and provides or denies access to the user.

This method of authentication comes with a lot of caveats due to the third-party authentication methods in use. The exact list of caveats is large and can be examined in full at the URL provided previously, but here are a few of them:

- Only Apple Safari, Google Chrome, Microsoft Edge, and Mozilla Firefox are supported for IdP authentication.

- Connection to the VPN endpoint must use the AWS-provided VPN client.

- A Client VPN endpoint only supports a single IdP.

- MFA is supported if the IdP itself supports it. What this means is that the AWS VPN client is limited in its capabilities and operating system support. If your organization is not able to meet these requirements, it may be better to go with a third-party VPN service. In that case, however, you often must deploy extra virtual instances to serve as the VPN termination points instead of relying on the AWS infrastructure alone. This is a more complex solution that will require upkeep and patching, but due to the limitations of AWS User VPN, it is often unavoidable.

> **Note**
> If an exam question requires choosing the correct authentication method for Client VPN, and MFA is a requirement, you can immediately rule out mutual authentication.

After dealing with the question of "Who's connecting?" and how they are connecting, the next thing to consider is "What do they need access to?" and how to authorize it appropriately.

This part is much simpler. AWS Client VPN endpoints can have SGs attached to them restricting traffic at the endpoint level, just like all other endpoints. In addition, Client VPN supports authorization rules as well. These rules act like a firewall granting access but do so via client policy and profile instead of network-level rules.

As an example, a client authorization rule can be created that allows access to a certain VPC CIDR based on the identity of the user or the groups to which they belong. The group membership and policy can be based on AD groups or granted directly to the user. This is not a replacement for inspection and true firewall security, however; it is only access control. If an attacker were to compromise an account and authenticate with certain access levels, AWS Client VPN would do nothing to stop any attacks they launch.

One last thing to cover with AWS Client VPN is **split-tunnel mode**. This is a setting that ensures that only AWS-destined traffic is sent through the AWS Client VPN. When the user connects to the VPN endpoint, the routes in the Client VPN endpoint route table are pushed to the client, and only traffic destined for those prefixes will be sent to the AWS tunnel. This is not a default setting; by default, all client traffic will be sent to AWS, including internet-bound traffic, which is often undesirable. Use split-tunnel mode when using something like endpoint or host security on the client or if the client's traffic is not of concern outside of AWS-destined traffic. Use the default setting if there is a security mandate to inspect *all* traffic from a client.

NVA/Overlay Networks

This section will focus on how NVAs and overlays enable connectivity to third-party networks in a different way to native services. Of course, specifics vary from vendor to vendor, but in general, third-party connectivity using NVAs will be using some level of native services in the middle as the NVAs themselves live in VPCs, inside subnets, and are subject to the same rules as any other workload.

The best NVA connectivity methods will be a mixture of native and NVA solutions, using each where they make the most sense. NVAs offer a large amount of granularity, scale, and familiarity, while native services do not require the care and upkeep associated with virtual appliances. Consider the following NVA connectivity pattern between an organization and a third party using NVAs:

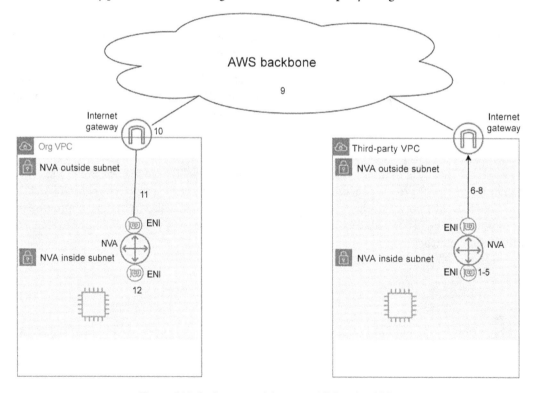

Figure 6.13: Basic connectivity across VPCs using NVA

In this diagram, connectivity between two VPCs is established without using a native service such as TGW or VPC peering. Instead, the NVA is given an Elastic IP address for the public subnet to allow it to communicate beyond the VPC. Using the AWS backbone, these two NVAs can communicate using those addresses. Commonly, the two NVAs establish a tunnel protocol to secure communications across VPCs in this way. However, the native networking services had to help make this happen.

Here is a walkthrough of this connectivity, as shown in *Figure 6.12*. See if you can separate the NVA networking from the native cloud networking to understand how they interact better:

1. A workload in a third-party VPC sends a packet to the org VPC. The destination IP in the packet is the CIDR of the org VPC. This packet uses the default gateway of the workload (the VPC router).

2. The private subnet route table is consulted, and a route is added for the org VPC with the next hop of the ENI attached to the NVA in the private subnet.

3. The packet comes into the ENI of the NVA and a check is performed before the packet is passed to the NVA. This check often results in network engineers graying prematurely or tearing their hair out in frustration because it is hidden in a little sub-menu of the ENI itself and not evident from the NVA or instance level: the source/destination check.

4. By default, an ENI is configured to only accept packets directly addressed to it as a security measure and drop packets intended for other destinations. This means that an NVA, whose job it is to route packets in transit 99% of the time, would be useless. To fix this, you must remove this default setting on the ENI itself. It is in a sub-menu, and the silent drop nature of the check can cause extreme frustration if you forget.

> **Note**
>
> This setting is far more likely to cause problems in real deployments, not certification exams, but there may be a question that involves taking a series of steps to make an NVA operational that includes this.

5. The NVA itself must deal with the packet from here. Depending on what sort of NVA it is, the packet might be inspected or otherwise manipulated before forwarding. Importantly, the NVA has its own route table and has no knowledge of the subnet route tables to which its ENIs are attached. Commonly, an NVA will have **Dynamic Host Configuration Protocol** (DHCP) for its interfaces to simplify IP and default gateway assignment, a large departure from on-premises network appliances. In these cases, the NVA default gateway is the VPC router IP of the subnet in which the ENI is deployed.

6. Assuming the packet is forwarded, the NVA forwards the packet to its next hop, which is usually the VPC router. In this scenario, however, the next hop is a tunnel interface. Because of this, the NVA encapsulates the original packet source and destination inside a new tunnel header where the source is the NVA's public interface, and the destination is the public interface of the NVA in the other VPC.

7. This new outside source/destination will be used for routing decisions until the other NVA de-encapsulates the packet. The packet leaves the ENI of the NVA in the public subnet, and the subnet route table is consulted for how to reach the next hop IP.

8. There are a few options to direct traffic here, but for simplicity and ease of use, usually, a default route pointing at the **internet gateway (IGW)** is used to route the packet. The IGW translates the private address of the NVA ENI to the public EIP associated and sends the packet to the AWS backbone.

9. The packet makes its way across the AWS backbone. Remember that though public IP addresses are in use here, the packet is kept on the AWS private backbone, not sent across the internet because both source and destination are AWS resources.

10. When the packet arrives at the IGW of the org VPC, the EIP associated with the org NVA is translated from public to private address, and the packet is sent to the local route table.

11. The local subnet route table uses local VPC routing to send the packet to the public subnet ENI of the NVA.

12. When the NVA receives this packet, it de-encapsulates the tunnel address and finds the original destination is reachable from its private interface ENI via its NVA route table.

13. The packet is sent from the NVA to the workload.

Return traffic happens in the same fashion. Note that to make this happen, a synergy of native cloud networking and NVA networking was required to make this connectivity happen. What does this mean to our questions? Who's connecting, and what do they need access to?

The answer to both questions is a resounding, "It depends." Much like a standard S2S VPN, the trust is based on shared tunnel setup details. This means each organization must configure its respective NVA and share connection information with the other party. The process of setting up this connection validates the "Who's connecting?" identity question. As to the, "What do they need access to?" question, the answer is very dependent on what the NVA itself can access. Often, the NVA handles some network routing or inspection for the workloads in its own VPC or other VPCs. What is reachable depends mostly on the routing rules of the NVA and any security policy in play. Often, this is a full security solution, not access control. Just like on-premises, a security NVA could be inserted between a third party and workloads to inspect the traffic.

NVAs also do a lot of heavy lifting when there is a requirement for granular traffic engineering. AWS offers some native options, of course, some of which have previously been discussed. However, when it comes to granularity and feature completeness, an NVA cannot often be beaten. The next section discusses ways to use that granularity for redundant traffic engineering, along with what options exist at the native level.

Redundant Connectivity to Third-Party Networks

Resilience and redundancy are important parts of maintaining network connectivity, and the same is true when dealing with third-party connections. Failure is always a possibility, and the ability to recover from it so that business can continue is critical to companies. There are a lot of ways to offer redundancy to third-party connectivity, and unsurprisingly, all of them are based on the connectivity method. This section will cover design patterns for redundant connectivity with a focus on the redundancy itself, not traffic engineering. Some of these redundancy options are built into the AWS native offers, and others require extra effort.

AWS PrivateLink

PrivateLink works on the concept of endpoints (Interface or gateway) being deployed in the customer VPC to provide connectivity. From this perspective, redundancy with an AWS PrivateLink connection is as simple as deploying more endpoints into other **Availability Zones (AZs)**, or other VPCs as design requirements dictate. The other end of that connection will still go to a provider load balancer with backend resources; so, redundancy outside of the consumer is covered by the provider. From a route redundancy perspective, each PrivateLink endpoint is deployed into a subnet, and that subnet route table can be configured to send traffic to its respective endpoint. If the workloads are balanced across AZ failure of a single AZ would impact only one PrivateLink endpoint and the subnets using it.

Redundancy is built into the fabric of the **AWS S2S VPN** solution. The most likely point of redundancy failure is on the third-party connectivity side. While VPN is often abstracted in diagrams to a single connection, here is a diagram of the actual deployment:

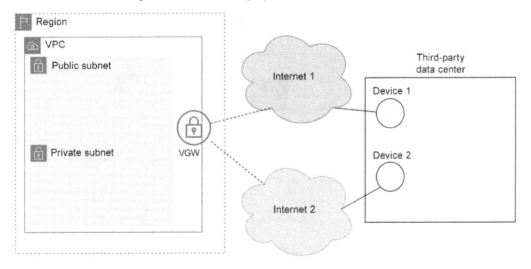

Figure 6.14: Redundant S2S VPN deployment

As shown in *Figure 6.14*, the AWS S2S VPN service is attached to a VGW, but that construct is intrinsically redundant. When the S2S VPN is created, AWS allocates two public IPs in different AZs to attach to the VGW for VPN tunnel endpoints. The rest of the redundancy requirements are on the third-party connection. In an acceptably redundant setup, the third party would be diversifying their connectivity to AWS and hardware as well.

TGW

Redundant connectivity with TGW involves mixing connectivity types in most cases, though there are also design patterns that involve cross-region or multi-TGW connectivity for redundancy. TGW is regionally redundant under the covers per AWS, so most of the concerns about redundancy have to do with the connecting party rather than the TGW.

Here is a common redundant TGW connectivity pattern from a single site utilizing S2S VPN and DX:

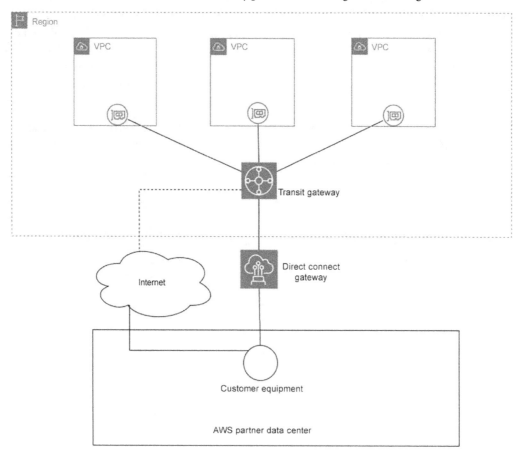

Figure 6.15: Redundant connectivity to TGW using VPN and DX

In this scenario, connectivity is maintained over two separate paths, using the AWS DX and the internet. This provides path redundancy, but not physical layer redundancy. Here is something that's more diverse and redundant:

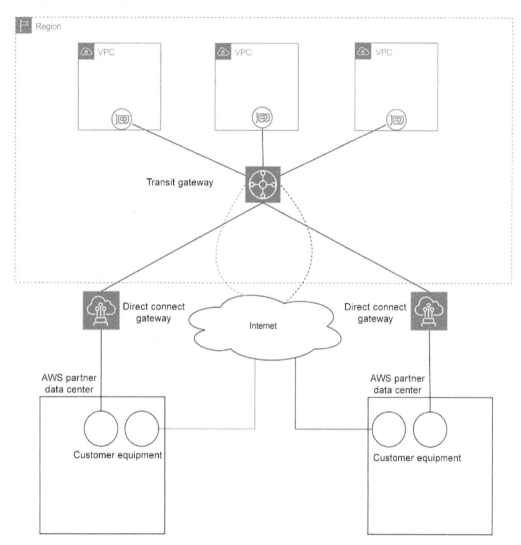

Figure 6.16: Improved resiliency with TGW using diverse equipment and paths

Addressing physical resiliency is often costly, but this scenario does offer a far better uptime than *Figure 6.15*. In this case, any physical connection or piece of equipment could fail without impacting connectivity. In fact, an entire AWS Partner data center could go down without impacting connectivity to the cloud if the applications in the colo data center are also redundant, or if there is backend connectivity (not pictured) to an on-premises data center.

The complexity of resiliency using TGW has more to do with the routing exchange and traffic engineering options than connectivity itself. These will be discussed in the upcoming section on route redundancy.

DX

Resiliency with DX is a question of money, geography, and space in a data center. As discussed in *Chapter 4, AWS Direct Connect*, **DX** is a geographically located physical layer solution to provide access to AWS via a private, dedicated circuit. There are limitations to this type of solution, and resiliency is one of them. Considerations for a DX connection between a customer in an AWS partner facility and the AWS cloud include the following:

- Physical data center location relative to other DX locations and cloud presence
- How many DX ports are available and whether they are on different AWS equipment in a data center
- Whether the DX partner cross-connect shares infrastructure with other DX cross-connects you have
- Whether there is diverse customer or provider equipment connecting to the DX ports
- Whether that equipment is in separate racks, with separate power and cooling

If Trailcats decides to use a provider to extend the DX to their own premises, the list changes. Other than the DX location itself, all of these become the provider's responsibility. Trailcats' responsibility as a customer of the provider is to ensure they have sufficient power, cooling, equipment, and cable diversity in their own facility instead. It is still important to verify with the provider that they have physical layer redundancy per leased DX port and that diverse paths exist to the extent possible within the AWS DX location, provider equipment, and the handoffs to customer WAN connection.

The customer WAN connection is often the provider's responsibility to the extent that the service continues to work. This means the provider needs to ensure that connectivity is available from the customer demarcation point through their lines and equipment. This can often be a sticky situation and not clearly understood; in some buildings, especially multi-tenant buildings, the provider is only responsible for providing connectivity to the building via some equipment on the outside or in the basement. The customer has to engage the building facility management to take care of the connectivity to their location within the building, and it would be the building facility management that fixes any issues from provider equipment to the customer equipment.

In the case of Trailcats, suppose its office is in a multilevel office building, and it leases an office from the owners of the building. If there are connectivity problems to the cloud, Trailcats would have to notify the building operators, who would have to investigate their own equipment inside the building up to the demarcation point where the provider's equipment resides. Commonly, this is in some sort of telecom room in the basement, but it could be anywhere that the provider installs their equipment for the building.

If this equipment is fine, the building operator must engage the provider, and if the provider's equipment is fine, they must engage their cloud service provider. Trailcats has a single point of contact for this scenario, and it is why many small companies opt for that kind of DX deployment, handing the entire responsibility over to a partner provider.

The redundancy of a remote facility is unlikely to be an exam question, but these are all important considerations when designing resilient connectivity to the cloud. Here's a common provider DX scenario that involves multiple physical locations and redundant connectivity:

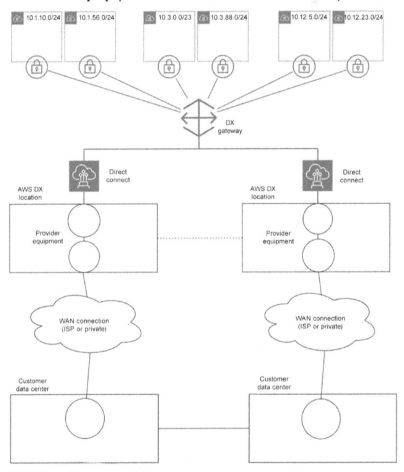

Figure 6.17: Provider to customer DX extension with resiliency

Note that, in *Figure 6.16* and *Figure 6.17*, the circles and connections are representative of multiple racks, power, cooling systems, cables, and network equipment to ensure physical layer redundancy. *Figure 6.16* shows data-center- and DX-level redundancy using multiple geographical areas. For simplicity's sake, the links are illustrative and not singular. Physical redundancy is best delivered via geographical separation to ensure that no weather, power, or disaster event can impact a business.

Moving on from physical layer redundancy, DX VIF redundancy is a simpler matter. Of course, there is no VIF redundancy without DX redundancy, but this has more to do with routing resiliency. Consider the following:

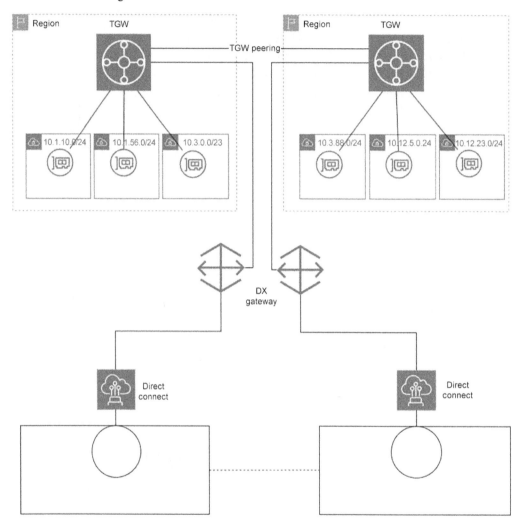

Figure 6.18: DX and TGW resilient architecture

Figure 6.18 focuses on cloud-level redundancy with multiple TGWs connected to DX gateways. Physical layer redundancy is important, but you also need redundant cloud constructs to connect with. From a route advertisement perspective, this architecture offers great resiliency. Notwithstanding the need for traffic engineering, which will be discussed in the upcoming section on route optimization, the on-premises networks can be safely advertised from each DX to the TGW. If desired, a supernet (such as 10.0.0.0/8), subset, or full route table (if what is under the route limit) for both customer data centers can be advertised over both DX to TGW. This allows several resiliency scenarios, including using the cloud as a backbone in the event of connectivity loss between the two data centers. This may or may not be a desirable outcome, so care must be taken when designing route resiliency.

In real deployments, often, the available options drive the design. For example, many organizations import routes from Microsoft 365 traffic steering, but this bloats the route tables of enterprise routers by a large amount. This is not an issue for the on-premises world, but these prefixes should not be advertised to cloud resources because the route quota limits will be quickly overrun, causing an impact when BGP sessions begin flapping.

In addition, from the cloud side, full connectivity to all the cloud resources can be achieved even in the event of a single DX failure or a whole AWS DX location data center failure. Private VIF resiliency is partly about physical diversity and partly about routing advertisement resiliency.

Client VPN

There's actually very little to say about Client VPN resiliency because it's almost entirely managed by AWS. When setting up Client VPN, you can associate multiple subnets (and therefore AZs) with the Client VPN endpoint. This provides local resilience for the connectivity of clients to AWS resources. Of course, you can also set up multiple Client VPN endpoints in AWS in different locations, not only for resiliency but to improve the latency of client connections. The resiliency of authentication and authorization infrastructure (and connectivity to that infrastructure) is not directly part of the Client VPN solution, but the service cannot operate without access to those identity stores. Make sure when designing an AWS Client VPN implementation that the resiliency and connectivity to those identity stores are considered.

NVA/Overlay Networks

As covered earlier in the chapter, there is a lot to unpack in the design of an NVA networking solution. Because these NVAs are your responsibility, you must design connectivity resiliency and instance resiliency as well. For the purposes of connecting to third-party networks, the main concerns when designing for NVA resiliency are instance resiliency and trust resiliency.

Trust resiliency and instance resiliency go hand in hand. Network resiliency is often the result of the previous two. Because of the inherent fragility of a single instance compared to a managed AWS service, care must be taken to ensure the fault domain of a single instance cannot impact the entire solution. However, the trust boundaries of instances with a third party are often very prescriptive and specific. This means that if an instance fails and must be replaced, the same trust extended to the previous instance may not extend to the new one. Moreover, in designing for resilience and involving multiple instances in that design, the complexity of redundant connections to third parties can become logarithmically complex. The next sections will cover the resiliency considerations in detail.

Instance Resiliency

The hardest part of dealing with NVA resilience is instance failure recovery. Often, the NVA is created with no configuration and must be configured after booting. This makes automatic recovery challenging, but there are a few common ways that vendors deal with NVA resiliency and build it into the product:

- **Bootstrap configuration options**: When creating the NVA, or on boot, the instance can connect to a saved configuration held in an AWS S3 bucket to restore its configuration or receive enough configuration to connect.

- **A software-defined networking (SDN) controller**: This out-of-band controller handles the administration of the NVAs, whether that be deployment, destruction, upgrades, or configuration management. A trust relationship is established between the controller and the NVA during boot or when onboarding the device to the controller, which will then handle life cycle management, failure detection, and recovery.

- **Manual restoration**: This is certainly the least likely to be used in a production network, but it is, of course, an option to manage instance resiliency in a lab or development environment. This involves launching a replacement NVA instance, logging into it, and configuring it manually following a failure.

- **Auto Scaling group (ASG)**: ASGs are an AWS-managed way to handle NVA resiliency by using a service instead of an SDN controller. This involves the detection of failure or some other trigger condition that causes AWS to create a new NVA instance. That instance will have bootup options and could be using a bootstrap configuration or be a dedicated instance AMI built with the configuration preloaded.

Note that an ASG is commonly used for scaling instances in and out based on demand and some other considerations need to be taken. For example, if an ASG triggers the creation of a brand new NVA instance with new IP addresses because the old one still exists, the trust relationship with the third party might have a problem depending on how trust was established.

Automation might also be needed on the subnet route tables to direct workload traffic to a new NVA instance as well if the ENI is not migrated. When an NVA fails, its network interfaces become unassociated, but the route tables in the VPC subnets may still be directing traffic toward them. In this case, some action must be taken to adjust the route tables and direct traffic to the new ENI. More commonly, a **Network Load Balancer** (**NLB**) can be used as the target for the subnet route tables, and the backend of that NLB is the ENIs of the NVA instances. This ensures that if an NVA fails, routing is not affected.

However, if there is a design by which routing neighbors are set up between those NVAs or an AWS service (such as TGW), that may not be possible. BGP neighbor packets do not take kindly to being load balanced between NVAs. This delicate balance of routing and automation is an ongoing area of complexity that NVA vendors are seeking to improve.

Trust Resiliency

Speaking of trust, depending on how an NVA instance is connecting to or accepting a connection from a third party, the replacement or addition of NVA instances can cause some problems. For example, in the case of an NVA failure, there may be a problem with trust transference as a result of replacing an NVA, as shown in the following figure:

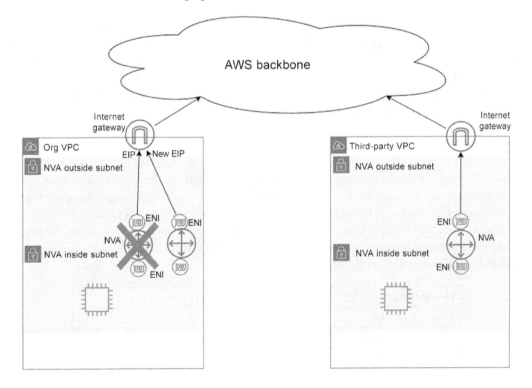

Figure 6.19: NVA replacement trust problem

After the events shown in *Figure 6.18*, when the NVA is replaced, without sufficient automation in play, a scale-out or NVA replacement event could break the trust relationship with a third party by creating a new instance with a new EIP that is not trusted by that third party for building tunnels. A more effective approach to NVA instance failure would be a workflow that detects failure and migrates the EIP to the new instance rather than assigning a new one to the new instance. This is often handled automatically by SDN controllers and requires more work without them. An AWS Lambda script could act based on a CloudWatch alert to perform the EIP migration. In this way, the third party does not need to be alerted to make changes on their side to restore connectivity.

Route Optimization for Redundant Connections

The last section of this chapter focuses on how to use traffic engineering to improve the resiliency and coverage of redundant connections. Different third-party connectivity services come with different options, and you will cover each of them in this section. This is not an exhaustive list of all possible options, as by blending multiple tools, there are many ways to accomplish the goal; however, the most common options should be covered in the following subsections.

AWS PrivateLink

PrivateLink doesn't exchange routes with the remote end, so the only real route optimization is making sure that AZ affinity is kept with subnet route tables. For example, an AWS PrivateLink endpoint deployed in a subnet in AZ A should not be used by workloads in other AZs unless that is the only endpoint. This prevents cross-AZ charges. Commonly, a PrivateLink deployment includes redundant endpoints in multiple AZs so that AZ affinity can be kept by workloads consuming the services connected to PrivateLink.

S2S VPN

S2S VPN is much more interesting in terms of traffic engineering. There are a lot of rules when determining things such as path preference given multiple route options. For example, static routes to an identical destination as those propagated by VGW using learned routes from an S2S VPN are preferred. What this means is that if two identical routes appear in an AWS route table, the origin of the route becomes the tiebreaker when determining which is used. Consider the following table, which shows a sample AWS route table:

Destination	Route Target
10.1.0.0/16	Local
10.200.1.0/24	vgw-12345 (propagated)
10.200.1.0/24	igw-6789 (static)

Table 6.1: Sample AWS route table with identical routes

In this case, the static route to the IGW would be preferred to the learned route from the S2S VPN. A VGW learning the same route from multiple destinations must prioritize a connection to send traffic. Remember that a VGW is attached to a VPC and can receive multiple types of connections and BGP advertisements from each. Though the CIDR may be propagated to the VPC route tables, ultimately, the VGW must decide which connection to send traffic for a prefix. The preference is as follows:

- BGP propagated routes from a DX

- Manually added static routes for a S2S VPN (since BGP is not used on S2S VPN)

- BGP propagated routes from S2S VPN

- If multiple S2S VPN BGP neighbors are advertising the same routes, AS_PATH is compared and the shortest is preferred

- When AS_PATH matches and the first AS is the same across multiple paths, MED is compared, and the lowest MED is preferred

A VGW cannot do ECMP for multiple VPN tunnels to the same destination. If ECMP is preferred, use S2S VPN on TGW instead. Since AWS performs maintenance on tunnel endpoints from time to time, they use the MED to set tunnel preference on the remote end. For a third-party connection, the preference is to let AWS use MED as the tiebreaker to select the best tunnel. AWS also recommends you use devices that can allow asymmetric traffic, if possible, such as a router instead of a firewall. However, third parties often have their own standards and preferences that cannot be adjusted to meet this recommendation. The health of a tunnel endpoint is the highest preference for AWS to choose which path to use in an S2S VPN, but no probes or health checks are performed outside of the endpoint availability itself.

There's a lot to unpack here. Generally, the recommendation is to let AWS pick which S2S tunnel to use in a particular connection, if possible, but this assumes the third party is using a device that allows asymmetry in routing, such as a router and not a firewall. Firewalls require full traffic symmetry to inspect the session flow in both directions and will drop half-seen sessions, so path preference is absolutely required for them.

Note that this is only for a single S2S VPN. When connecting multiple S2S VPNs, the question of which tunnel is less important than which VPN connection to use at all. Consider the following:

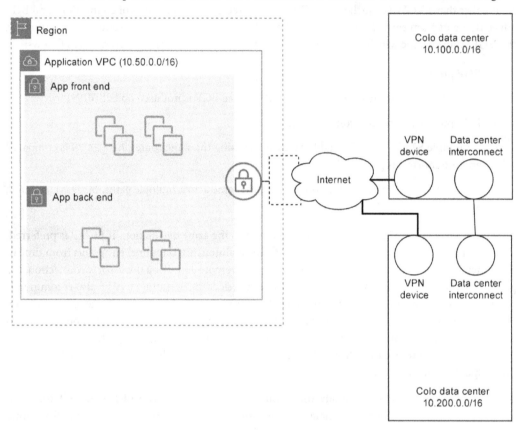

Figure 6.20: Redundant S2S VPN

In this case, there may be many considerations in play here. Are these VPN connections from each data center using BGP or are they statically configured? Is the intent to provide backup connectivity for each data center to the cloud in the case of a single VPN device failure? What prefixes should be advertised and with what attributes to achieve optimal traffic flow?

If we assume that the VPN is using BGP for dynamic route advertisement, it's possible to use either the longest match (meaning more specific route prefixes), BGP attributes, such as AS_PATH, or both to accomplish preferential routing. The route preference was covered earlier, but the longest match generally beats everything. Here's a possible solution to allow basic traffic engineering to take place over the VPN.

Suppose that, in *Figure 6.19*, no route preference is applied to prefixes received from the VGW for the attached VPC as BGP will use the extra AS_PATH added by each data center device to prefer the closest path to the cloud. For example, if the top **Data Center** is using BGP 65001, and the bottom 65002, when the top DC receives the route prefixes from the VGW over the VPN, it will append its own local AS (65001) before advertising it over the **data center interconnect** (**DCI**) to the other data center. Assuming the other data center is also receiving this same prefix from the VGW locally, the BGP AS_PATH will be shorter and therefore preferred to the one received over DCI. However, should the VPN fail in one of the data centers, a path to the cloud is still advertised from the other. When advertising route prefixes from the data centers to the VGW, there are two good options for traffic engineering:

- Use the longest prefix length as a traffic engineering tool. From the top DC, advertise 10.100.0.0/16 and 10.0.0.0/8. This ensures that in the case of a VPN failure in the bottom DC, traffic will be attracted to the top DCI and can flow across the DCI to the bottom DC. Do the same for the bottom DC, except advertise 10.200.0.0/16 and 10.0.0.0/8.

- Use BGP AS_PATH prepending. This involves artificially adding BGP AS_PATH hops to a prefix using a BGP route map. In the top DC, allow both 10.100.0.0/16 and 10.200.0.0/16 to be advertised to the VGW but add 3 or 4 more AS_PATH to just the 10.200.0.0/16 prefix. Do the same in the other DC for 10.100.0.0/16.

Complex traffic engineering is not supported by S2S VPN using VGW. For more complex options, a TGW is required.

TGW

The route optimization for redundant connections to TGW often relies entirely on how the external resource is connected and what options exist for those connections. TGW often extends the support of route preference and traffic engineering. In the previous section, some of the differences between S2S VPN on VGW and TGW were discussed, such as the route limitations and lack of segregated route tables. Where things tend to become interesting with TGW is combining multiple connectivity types along with their caveats and limitations to accomplish the route preference objectives.

Here is a very common redundant connectivity pattern for TGW:

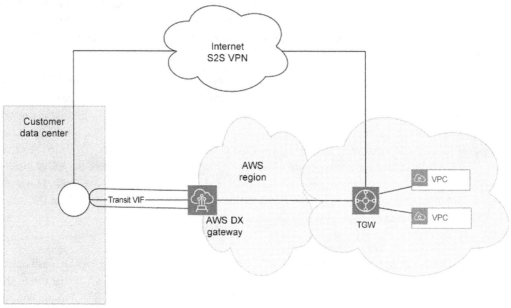

Figure 6.21: Basic redundant connectivity pattern utilizing DX, TGW, and S2S VPN

This diagram makes it look less than redundant, but there would be multiple customer data center routers or firewalls terminating the DX and S2S VPN. The DX is not redundant, but that is why the S2S VPN is used as a backup connection. The main consideration for this type of design pattern is whether the S2S VPN has enough bandwidth to support the traffic should the DX fail, making the S2S VPN the backup link and not a primary one.

From a routing perspective, the traffic engineering options are like those with VGW, except for IPv6 support and the ability to do equal-cost multipathing, utilizing multiple VPN tunnels. This is a configurable feature for the TGW, which enables the ability to load balance across multiple tunnels so long as all attributes are the same.

Discussed earlier are route preferences based on the origin of the routes. This must be considered when deciding how to engineer traffic across multiple TGW connections. In this case, if the goal is to use the DX as the primary connection and use ECMP for the backup VPN (increasing the bandwidth), there are two things that must happen:

- The path over the DX to the DX gateway must be preferred via a longer match or shorter BGP `AS_PATH` advertised from the data center to DX, and the return traffic advertised from the DX gateway should be the same as the customer equipment; a longer match, a shorter BGP `AS_PATH`.

- The route prefixes should not be preferred when advertised over the S2S VPN, by prepending BGP ASN, or by advertising less specific prefixes in both directions. To achieve ECMP on the customer side, it's important to make sure each tunnel receives identical prefixes and BGP attributes, and that BGP ECMP is configured. By default, BGP uses a tiebreaker to pick the best path.

The other thing to remember is that from a VPC perspective, native route tables cannot have overlapping CIDRs. This means that route propagation chooses what is used by the VPC and where the traffic will flow. This would only be a concern if the VPC has a S2S VPN to a VGW attached directly to the VPC, as well as a TGW attachment, and must pick which route to propagate to the route table. If the VPC is attached to a TGW directly, and not also attached to a VGW using a S2S VPN (as in *Figure 6.19*), this is not a concern as the VPC will only use TGW and TGW is making the destination routing decision.

As discussed earlier, the goal is to advertise more specific routes to the cloud via DX and to summarize over the S2S VPN. This allows the TGW route table for the VPC attachment to know multiple destinations to on-premises without having to pick one best option. Should there be a reason to advertise the same routes to TGW over DX and VPN, the goal is to lengthen the `AS_PATH` of the S2S VPN and allow TGW to propagate only the DX route into the VPC attachment route table, replacing it with the S2S VPN route only in case of failure.

In all cases, the question is "What is the desired primary and backup path?", followed by "How do you ensure this is achieved while still providing resiliency for each path?" The answers to these questions rely heavily on architecture and applications.

DX

Route optimization for redundant DX connections relies very heavily on what sort of VIF is deployed and how many DX locations are involved. As with TGW, traffic engineering with DX is very dependent on what backup connectivity is deployed, but with the added complexity of a VIF type on top of it. A large amount of time was spent discussing DX routing and traffic engineering in *Chapter 4, AWS Direct Connect*; so, for a refresher, please consult it if you don't remember how basic routing with DX works, as we will now expand on that when designing resilient connections. *Figure 6.21* was discussed in the previous section, and the only complexity that needs to be added here is another DX location and the resulting challenges.

In the example shown in *Figure 6.22*, a single DX gateway *could* have been used as they are global constructs but remember that DX gateways have a limitation: they cannot be attached to two different TGWs with the same prefix advertisements. That is a problem since the TGW peering across regions is meant to provide backup connectivity if a DX fails. This necessitates the extra DX gateway so that prefixes can be advertised in a fully redundant manner, as follows:

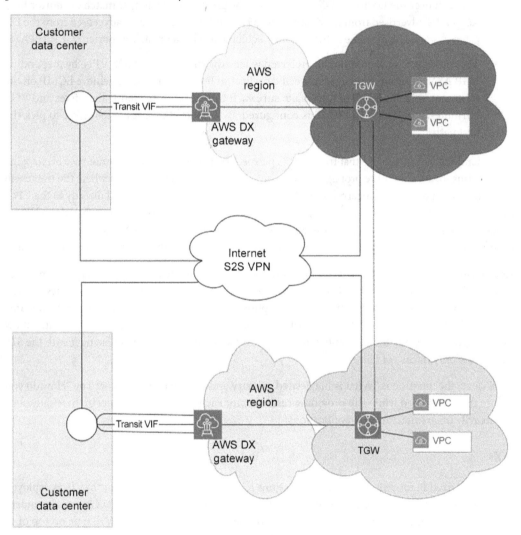

Figure 6.22: Resilient DX locations with DX gateways, TGWs, and S2S VPN

Figure 6.22 shows a resilient design using multiple data centers, DX circuits, DX gateways, and AWS regions. This is a fully redundant design, but it introduces a lot of complexity from a routing perspective. In a scenario like this, it is important to remember some of the caveats in play when considering resiliency:

- The TGW peering attachment is not a dynamic attachment; routes must be explicitly advertised across a TGW peering attachment. This means that new routes added to a DX won't end up being advertised across to the other region's TGWs by default. For this reason, the common TGW peering strategy is to only use supernets such as `10.0.0/8` or `192.168.0.0/16`, for example.

- The DX gateway is similarly not advertising routes learned from the TGW; it is instead configured to advertise routes statically with the destination being the TGW attachment.

- As stated earlier, route preference between DX and S2S VPN landing on a TGW can be controlled by a shorter and longer prefix length, by BGP `AS_PATH` manipulation, or both.

In general, the routes originating from outside the cloud will almost always be more numerous, have longer prefix matches, and be more intolerant of summarization, but this is not always the case. Especially in environments where developers turned proof on concepts into production workloads, the likelihood is high that little time was spent optimizing IP address management. The native cloud constructs are far less capable of route scale than the extra cloud and third-party connections are likely to be. So, with this in mind, here's a summary of the best practices with routing design:

- Summarize routes being advertised to the native cloud constructs, if possible. This is important for preventing any problems with hitting route quotas inside the cloud. Exceeding route quotas results in BGP sessions being reset and routes being dropped.

- Use longer prefix advertisements for traffic engineering purposes while keeping in mind route scale limits. Because each traffic prefix is different, any time you summarize or allow longer match prefixes through, it results in more routes to advertise.

- Use BGP `AS_PATH` if the route scale is too high to use longer prefix matching, or if summarization is impossible but will fall under any scale limitations. BGP will prefer a lower `AS_PATH` length over longer ones, and with BGP route maps, you can engineer the advertisements to the cloud in a way that prefers the path you want.

- AWS uses the BGP attribute MED on VPN tunnels to prefer one tunnel over another and is expecting both tunnels to be usable. If possible, use devices that are tolerant of asymmetric routing, and if not, use BGP `AS_PATH` prepending to ensure that AWS uses the tunnel you expect to the device you expect. This ensures that traffic symmetry is kept, which is most important if the terminating on-premises device is a firewall that needs the full traffic session to come through to prevent the accidental denial of traffic.

If the DX is connected to a single VPC using a private VIF to VGW connection, and the backup connection is an S2S VPN to the same VGW, there is no way to use both VPN tunnels for traffic. In this case, you should do the following:

- Advertise longer prefixes or shorter BGP `AS_PATH` over the DX to the VPC

- On the customer side, consider setting inbound BGP local preference on routes received from the DX peer, or adding BGP `AS_PATH` entries to routes received over the S2S VPN as the advertisement will be identical (the VPC CIDR)

Public VIF resiliency and traffic engineering were covered extensively in *Chapter 4, AWS Direct Connect*. You can refer to that chapter if needed. Essentially the chapter comes down to a combination of BGP communities and prefix length manipulation.

NVA/Overlay Networks

In designing a network using NVA or overlay technology, the main goal is to use the tools given to you by the NVA vendor to create the most effective hybrid network possible, leaning into NVAs' strengths and minimizing their weaknesses.

The main strengths of NVAs are as follows:

- NVAs generally have far more route optimization options, multiple routing protocols, and a granular level of control.

- NVAs can handle larger route scales, usually an order of magnitude more than native. This may be required if route summarization is otherwise impossible.

- NVAs, especially overlay NVAs, offer unparalleled levels of visibility into traffic flow. For solutions that can make routing decisions based on this data, it is a great strength.

The main weaknesses are as follows:

- Because NVAs are EC2 instances, they suffer from all the problems an EC2 instance may suffer from, including host failure, artificial data throughput limitations based on instance size, and lack of high availability unless the NVA supports it with other NVAs.

- NVA life cycle management becomes necessary, and upgrades and configuration are required to a high degree. Even with controllers managing the NVAs, this is a much heavier investment of time and effort than is required in a native offering.

- For all the strengths of NVAs, they cannot exist without the native cloud underneath, and they will have to interact with native cloud constructs. This means higher complexity and, eventually, NVAs still must be aware and adhere to native cloud limitations around routing, route scale, and BGP feature support.

The following is a scenario that is like something you may see in the exam.

Given the following business requirements, determine whether an NVA solution makes sense. Suppose Trailcats needs to establish a new DX connection with a new AWS region in which they are deploying new cloud applications. This new region is part of an acquired company that has a lot of sprawls in its VPC IP address allocation and a lot of overlapping IP addresses between VPCs. The VPCs in the region have no need to communicate with each other, only the on-premises networks. The DX will operate at 10 GB per second and can connect either to an AWS transit gateway or a VGW in a dedicated VPC in which a network NVA could be deployed to establish overlay routing.

Trailcats would like to have the least administrative overhead possible in managing this new connectivity, but the on-premises networks must be able to communicate with any of the VPCs in the newly acquired AWS region. What solution should they choose and why?

There is a lot of information given to sway your decision one way or the other, so it's important to break down the requirements and see whether they disqualify any solution.

First, the new region has a lot of VPC IP address sprawl. This is a bit of a distractor as an exact number is not supplied, and TGW can scale up to 10,000 routes. Unless the requirement explicitly exceeds that number, there is no reason to use this information to disqualify one of the options.

Second, the requirements mention overlapping VPC IP addresses but caveats that the VPCs do not need to communicate with each other. The on-premises resources need to be able to communicate with any VPC, however. This information is useful, and buried within this is an implicit, not-specified requirement that will lead you to the correct answer.

Third, the requirements mention DX speed. Most NVAs can handle 10 GB connectivity, so this is not a clear disqualifier and is in fact a distractor. Any information supplied that can apply to any of the options is probably distracting you from the real information you need to focus on, which is information that will allow you to pick the best option.

Lastly, Trailcats requests that the solution be the least administratively burdensome. This is a clear attempt to get you to pick the AWS TGW as managing NVAs is much more operationally difficult. However, this is just a part of the requirement, as whatever solution is chosen must be capable of solving the problem. In short, this qualifier comes last, after ensuring that the solution works in the first place; if there remain multiple options, this requirement can break any ties.

In this case, the disqualifier is buried in understanding the most critical requirement: the on-premises resources need to be able to communicate with any VPCs. This means that network address translation will be needed between on-premises and the cloud, and this is best handled by an NVA.

If that doesn't make sense, remember that while a NAT gateway can be used to handle translating between private resources, each VPC would need a NAT gateway for this to work, and the TGW would need a lot of manual route table manipulation. The NVA can do NAT in a much more scalable way using advanced networking features, which makes it a less burdensome choice.

Summary

Working with third-party networking in AWS can be highly complex. In addition to the usual design challenges, there's an added layer of difficulty in determining how much trust to grant to third parties while designing with the principle of least privilege. AWS provides numerous options to secure and extend connectivity across different solutions, and it is your responsibility to choose the option that best meets your needs while safeguarding the security and integrity of your organization. In this chapter, we covered third-party connectivity options, including redundant connectivity to third-party networks and how to optimize routing for redundant connections. After reading this chapter, you should be able to select the best connectivity options for third-party networking from the AWS connectivity options based on requirements.

In the next chapter, you will learn the basics of AWS Route 53.

Exam Readiness Drill – Chapter Review Questions

Apart from mastering key concepts, strong test-taking skills under time pressure are essential for acing your certification exam. That's why developing these abilities early in your learning journey is critical.

Exam readiness drills, using the free online practice resources provided with this book, help you progressively improve your time management and test-taking skills while reinforcing the key concepts you've learned.

HOW TO GET STARTED

- Open the link or scan the QR code at the bottom of this page

- If you have unlocked the practice resources already, log in to your registered account. If you haven't, follow the instructions in *Chapter 16* and come back to this page.

- Once you log in, click the START button to start a quiz

- We recommend attempting a quiz multiple times till you're able to answer most of the questions correctly and well within the time limit.

- You can use the following practice template to help you plan your attempts:

Working On Accuracy		
Attempt	Target	Time Limit
Attempt 1	40% or more	Till the timer runs out
Attempt 2	60% or more	Till the timer runs out
Attempt 3	75% or more	Till the timer runs out
Working On Timing		
Attempt 4	75% or more	1 minute before time limit
Attempt 5	75% or more	2 minutes before time limit
Attempt 6	75% or more	3 minutes before time limit

The above drill is just an example. Design your drills based on your own goals and make the most out of the online quizzes accompanying this book.

First time accessing the online resources? 🔒

You'll need to unlock them through a one-time process. **Head to *Chapter 16* for instructions**.

Open Quiz	
https://packt.link/ansc01ch6 OR scan this QR code →	

7

AWS Route 53: Basics

It's not DNS

There's no way it's DNS

It was DNS

The above Haiku is a popular internet meme reflecting the truism that **Domain Name System (DNS)** failures and misconfigurations are the root cause of many application and service outages. DNS is a fundamental building block upon which the internet rests. Why is this so? Can you imagine the need to know the IP address of a website so you can visit it in a web browser? If it's an IPv4 address, it might be somewhat feasible, but what about an IPv6 address that is 16 bytes long? For example, to reach the Trailcats website using the bare IP address, you would have to input `192.0.2.1` for IPv4 or `2001:db8:f000:g000:3333:4444:5555:ff11` for IPv6. The DNS is the translation layer of numbers to names, making the Internet accessible to humans.

The DNS is an essential underlying technology, and its misconfiguration represents a frequent cause of outages. This is because DNS configurations involve numerous settings and parameters, and a mistake in any part of the configuration can lead to failure. Human error is a major factor as DNS management often requires manual inputs, and even small mistakes such as typos, incorrect entries, and forgetting to update records can snowball into widespread problems. As a network engineer, it's table stakes to understand DNS principles and operations. The first part of this chapter will provide you with that baseline level of understanding. After establishing this vital background information, the chapter will focus on AWS Route 53.

In this chapter, you'll focus on the following topics:

- History of DNS
- DNS fundamentals
- Amazon Route 53
- Domain registration
- Routing traffic to your domain
- Route 53 resolver
- Conditional forwarding
- Route 53 health checks

By the end of the chapter, you will have a solid understanding of how DNS works and what makes Amazon Route 53 a compelling option for providing DNS services. You'll learn how to register domains with Route 53 and route traffic to your AWS resources using alias records, public hosted zones, private hosted zones, and Route 53 routing policies. You will learn how Route 53 Resolver provides DNS resolution for your **virtual private cloud** (**VPC**) resources, and you'll get an introduction to conditional forwarding and health checks. This chapter, along with *Chapter 8*, *AWS Route 53: Advanced*, will equip you with the skills to build and troubleshoot sophisticated hybrid cloud DNS architectures, preparing you for the exam and real-life applications.

History of DNS

The predecessor of the modern internet was the U.S. Department of **Defense Advanced Research Projects Agency** (**DARPA**) funded experimental wide area network called ARPAnet. ARPAnet came online on October 29, 1969, with 4 test sites and was decommissioned on Feb 28, 1990. Its purpose was to allow government contractors to share expensive computing resources.

Very early on, it was recognized that having a directory of hostnames to network address mappings made the network more usable. Stanford Research Institute maintained a text file called `HOSTS.TXT` containing all the ARPAnet addresses to name mappings.

By the early 1980s, ARPAnet had grown to the point that a simple text file was no longer feasible, which resulted in RFC 882 and RFC 883, the initial RFCs for DNS.

DNS Fundamentals

Computers use Internet Protocol addresses to connect and exchange information. The challenge is that humans don't remember sequences of numbers well, but they're very good at remembering sequences of words. The DNS allows humans to tell computers where to go by using words. The DNS is a distributed database containing all the information needed to convert a human-readable name into the numerical address of a computer connected to the internet. Without the DNS, the internet would be nearly impossible to use. With this context set, it's important to understand some key terms. The following short paragraphs provide definitions of key concepts, which will be discussed in detail starting with the next section.

Databases need an organizational structure, a way to search for and navigate to the record you're looking for. The records contained within the database should have a consistent format, and there must be a way to request and receive those records.

The domain namespace is the formal term for the organizational structure of the DNS. The domain namespace is put together in such a way that the individual parts of the database can be distributed across the internet.

Resource records (**RRs**) are contained within the database partitions, and they serve a variety of purposes. The ultimate function of most RRs is to resolve a domain name to an IP address, but some records serve other helpful functions.

Name servers hold database records. They respond to requests for RRs and exchange database information with other name servers.

Forwarders are a special type of DNS server that doesn't store database records; instead, they forward requests to other servers that pass the responses back. They're useful for routing requests to different DNS servers in more complex designs.

Resolvers, sometimes called clients, request RRs from DNS servers. When you type a domain name into a web browser, your computer's operating system has a DNS resolver that requests the needed DNS records from a DNS server.

With that brief overview out of the way, it's time to dive into the details, beginning with the domain namespace.

Domain Namespace

The domain namespace is logically structured as an upside-down tree with a dot at the root. Each subsequent node in the tree is delegated authority by the node above it. After the root are the **top-level domains** (**TLDs**) such as .org, .com, .net, and so on. After TLDs are where organization-controlled domains typically begin. As the tree goes from the top down, the domains are concatenated with a dot and read from left to right, as shown in the following figure:

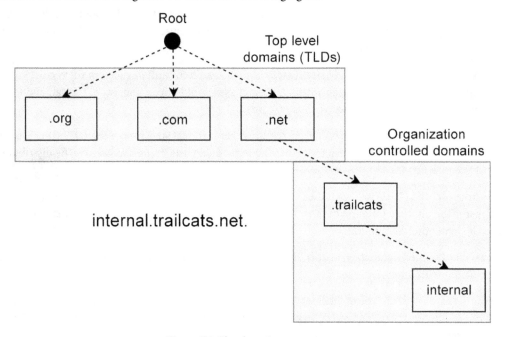

Figure 7.1: The domain namespace

The **Internet Corporation for Assigned Names and Numbers** (**ICANN**) maintains the complete list of TLDs. Delegation of the TLDs is spread amongst several organizations called registries. The registries maintain the records for all the domains registered under the TLDs they control. For example, the US-based infrastructure provider and domain registrar Verisign operates the .com registry.

Each domain has a master file of records called a zone file, which contains RRs for the domain. Name servers store these files and make their information available for use.

The label given to a domain is called a domain name. A **fully qualified domain name (FQDN)** specifies the full set of domain labels starting with the root domain—for example, `www.trailcats.net`. Relative domain names don't contain all the labels. A special type of name called a hostname is a name given to a computer, and this is the most commonly used form of relative domain name. For example, say you have a computer called Siamese. It is considered a relative domain if you refer to it by just that name.

In practical terms, if you type `siamese` in a web browser, whether you can reach that computer depends on whether your computer is configured with a search domain. A search domain would be a domain that your computer can query to see if it has a record for the name Siamese. For example, imagine your computer is configured with a search domain of `trailcats.net`. When you type `siamese` in your web browser, your computer appends "siamese" to `trailcats.net`, making the query `siamese.trailcats.net`. This is a handy feature when a large number of computers are in the same domain your computer is in. You can use only the hostname, and the computer fills in the rest under the hood.

If it's still confusing- think about it like this: Pretend for a moment your full name is Billy Bob Smith. People in your family just call you Billy and it's fine. Now imagine you've moved to another town and your friend wants to search for you in a directory so they can call you on the telephone. There are bound to be a lot of Billys in the telephone directory. So, they'll search using your full name, which is likely to be unique. This is exactly how relative and fully qualified names work.

> **Note**
>
> To prevent the client from using relative paths, use a FQDN, which includes the period at the root of the domain tree. **Example:** `www.trailcats.net.`

Name Servers

Name servers hold subsets of the domain space database, one file per domain, and respond to requests from resolvers. They use pointers to other servers for parts of the domain tree for which they do not have information.

> **Note**
>
> **Domain Information Groper (DIG)** is a command-line tool for querying DNS servers. It's useful for troubleshooting and learning. It's included with Mac and Linux operating systems. For Windows, you'll need to install **Windows Subsystem for Linux (WSL)** and install a Linux distribution. You are encouraged to try out the commands used to generate the example output. It will facilitate learning and retention.

There are 13 logical root name servers managed by several organizations, including ICANN, NASA, and Verisign. The TLD registries manage the TLD DNS servers. Using DIG, you can look up the root name servers as shown in the following figure:

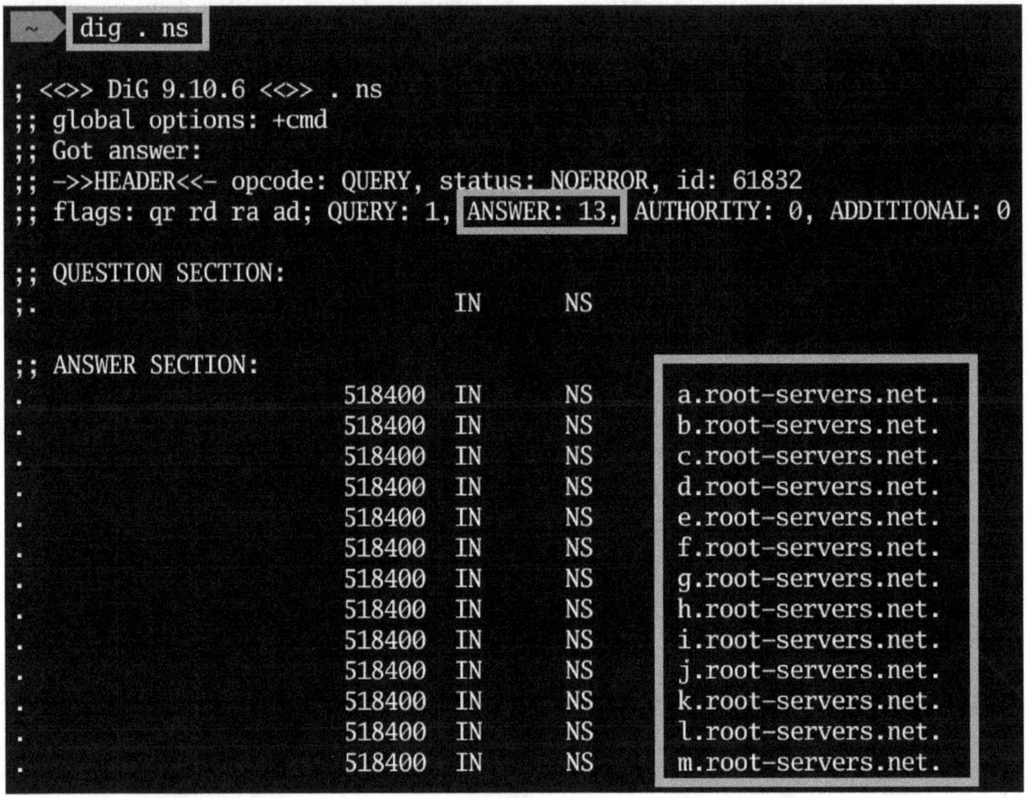

```
~   dig . ns

;  <<>> DiG 9.10.6 <<>> . ns
;; global options: +cmd
;; Got answer:
;; ->>HEADER<<- opcode: QUERY, status: NOERROR, id: 61832
;; flags: qr rd ra ad; QUERY: 1, ANSWER: 13, AUTHORITY: 0, ADDITIONAL: 0

;; QUESTION SECTION:
;.                              IN      NS

;; ANSWER SECTION:
.                      518400   IN      NS      a.root-servers.net.
.                      518400   IN      NS      b.root-servers.net.
.                      518400   IN      NS      c.root-servers.net.
.                      518400   IN      NS      d.root-servers.net.
.                      518400   IN      NS      e.root-servers.net.
.                      518400   IN      NS      f.root-servers.net.
.                      518400   IN      NS      g.root-servers.net.
.                      518400   IN      NS      h.root-servers.net.
.                      518400   IN      NS      i.root-servers.net.
.                      518400   IN      NS      j.root-servers.net.
.                      518400   IN      NS      k.root-servers.net.
.                      518400   IN      NS      l.root-servers.net.
.                      518400   IN      NS      m.root-servers.net.
```

Figure 7.2: Using DIG to retrieve the list of root name servers

In *Figure 7.2*, the dig command is used to query the root domain. The highlighted portions show that 13 records were returned, followed by the list of the 13 root name servers. These records are then used to query TLDs.

In the next example, DIG is used to look up the name servers or the .org TLD, which are learned from the root name servers from *Figure 7.2*:

```
~  dig org. ns

; <<>> DiG 9.10.6 <<>> org. ns
;; global options: +cmd
;; Got answer:
;; ->>HEADER<<- opcode: QUERY, status: NOERROR, id: 31410
;; flags: qr rd ra ad; QUERY: 1, ANSWER: 6, AUTHORITY: 0, ADDITIONAL: 0

;; QUESTION SECTION:
;org.                          IN      NS

;; ANSWER SECTION:
org.                  3531    IN      NS      a0.org.afilias-nst.info.
org.                  3531    IN      NS      a2.org.afilias-nst.info.
org.                  3531    IN      NS      b0.org.afilias-nst.org.
org.                  3531    IN      NS      b2.org.afilias-nst.org.
org.                  3531    IN      NS      c0.org.afilias-nst.info.
org.                  3531    IN      NS      d0.org.afilias-nst.org.
```

Figure 7.3: Using DIG to retrieve the list of .org name servers

In *Figure 7.3*, DIG is used to enumerate the name servers for the .org TLD. Six results are returned. One of these servers may then be queried to learn the name servers for all .org domains (for example, ietf.org).

A name server may carry out multiple roles:

- **Primary name servers:** Primary name servers maintain the master authoritative copy of records for that domain in a zone file. This server is referenced in a special domain RR called **Start of Authority (SOA)**.

- **Secondary name servers:** Secondary name servers maintain a read-only copy of the zone file, which is periodically transferred from the primary server.

- **Resolvers:** Resolvers are a particular type of DNS server that performs recursive queries on behalf of clients.

- **Forwarders:** Forwarders forward queries to other servers. This is particularly useful in organizations that have a discontinuous DNS structure. For example, there may be corporate DNS servers for internal use and Route 53 DNS servers for applications running in AWS. Forwarders provide a way to glue these discontinuous domain architectures together.

From a network perspective, DNS communication happens on UDP and TCP port 53 (DNS queries normally use UDP, with TCP primarily used for zone transfers between servers).

DNS Zone Delegation

DNS zone delegation is the process of dividing a DNS namespace into smaller, more manageable parts, known as zones. This is typically done to distribute the administrative and operational responsibility of the DNS namespace to different entities. For example, suppose `trailcats.net` is managed by the IT department and the organization wants to delegate `tabby.trailcats.net` to the marketing department. `trailcats.net` will include name server records for `tabby.trailcats.net`, pointing toward the name servers that are managed by the marketing department. These are called **glue records**, and you can see how they work by studying the recursive query process, which is described in the next section.

The Recursive Query Process

The recursive query process is a method used by DNS resolvers to obtain the final answer to a DNS query. DNS resolvers perform the recursive query process on behalf of clients. The resolver walks the DNS tree, starting at the root zone, and working its way to the DNS server that holds a zone file for the domain (remember, the zone file is the database of all the RRs in the domain).

The following mental exercise steps through the recursive query process. For this exercise, the assumption is that the client and resolver have empty caches. As a practical matter, unless it has just booted up, the resolver will have most TLD name servers already cached, so the step of querying the root name servers is rarely necessary:

1. When you type www.`trailcats.net` into a web browser and press *Enter*, behind the scenes, your computer issues a query to the resolver.
2. The resolver checks its cache.
3. The cache is empty.
4. The resolver queries a root name server.
5. The root name server will return the name server records for the TLD – in this case, `.net`.
6. The resolver queries a `.net` name server.
7. The `.net` name server returns the name server records for `trailcats.net`.
8. The resolver queries a `trailcats.net` name server for www.`trailcats.net`.
9. The `trailcats.net` name server returns the IP address of www.`trailcats.net` to the resolver.
10. The resolver passes the query result on to the client computer.

Figure 7.4 illustrates the recursive query process. The resolver walks the DNS tree from the root domain to the Trailcats DNS server, which is able to answer the resolver's query.

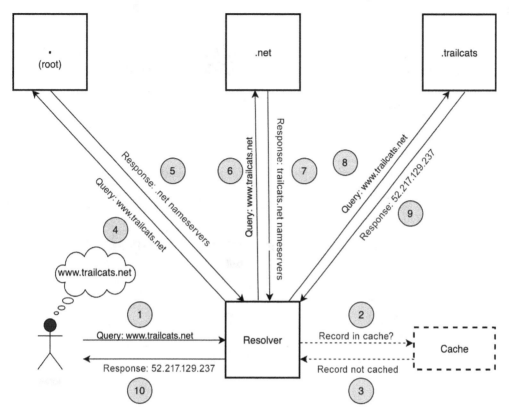

Figure 7.4: The recursive query process

Data Structure and Resource Records

Information for a domain is held in a plain text file called a zone file. The term zone refers to a zone of authority, the person or organization controlling the domain. For day-to-day practical purposes, a zone and a domain can be considered synonymous. It may seem strange to build a database out of plain text files. However, consider that the DNS was designed to replace hosts.txt when the practice was for human administrators to create and maintain database records by hand. Most modern DNS services store DNS zones internally in a binary format and output a plain text version for human readability and data transfer with other systems.

The data in the zone file is organized into RRs, which contain information about a domain name, such as its IP address or mail server. A valid zone file must have a **SOA** record, which defines who hosts the authoritative copy of the zone file, and a default **time-to-live** (**TTL**) entry for RRs that don't have a TTL specified. **Note**: TTL is expressed in seconds.

The important RRs to know about are listed in *Table 7.1*. Knowledge of these RRs is important information that you will need to know for the exam. They will be covered in detail later in the section.

Record Type	Description	Notes
SOA	Start of Authority	Defines who hosts the authoritative zone
A	Hostname to IPv4 mapping	The most basic RR type
AAAA	Hostname to IPv6 mapping	The most basic RR type
CNAME	Canonical Name	Multiple names for the same IP
MX	Mail Exchanger	Prioritized list of mail servers
TXT	Text record	Used for multiple purposes
NS	Name Server	Name server to IP mappings
PTR	Pointer	Used for reverse DNS
DNSKEY	Key signing keys	Used to validate RRSIG records
RRSIG	RRSET Digital Signature	Digital signature of an RR set
DS	Delegation Signer	Trust chain between parent/child zones

Table 7.1: DNS RR types

All RRs have a common top-level format, as shown in *Table 7.2*. Some types will have additional fields.

Name	TTL	*Class	Type	Record Data

Table 7.2: DNS RR format

> **Note**
> The class will always be IN for internet.

Using the DIG utility to look up the name servers for `trailcats.net` returns the name server records for the domain, as shown in the following figure:

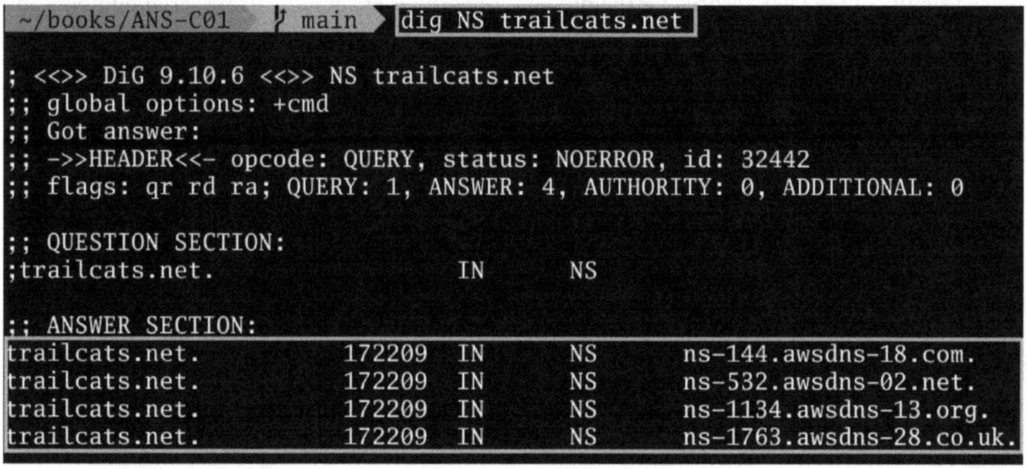

```
~/books/ANS-C01    ⎇ main    dig NS trailcats.net

; <<>> DiG 9.10.6 <<>> NS trailcats.net
;; global options: +cmd
;; Got answer:
;; ->>HEADER<<- opcode: QUERY, status: NOERROR, id: 32442
;; flags: qr rd ra; QUERY: 1, ANSWER: 4, AUTHORITY: 0, ADDITIONAL: 0

;; QUESTION SECTION:
;trailcats.net.                    IN        NS

;; ANSWER SECTION:
trailcats.net.          172209  IN        NS        ns-144.awsdns-18.com.
trailcats.net.          172209  IN        NS        ns-532.awsdns-02.net.
trailcats.net.          172209  IN        NS        ns-1134.awsdns-13.org.
trailcats.net.          172209  IN        NS        ns-1763.awsdns-28.co.uk.
```

Figure 7.5: DIG output showing RR format

In *Figure 7.5*, notice how the records in the answer section of the output exactly match the RR format shown in *Table 7.2*. This standards-based display output is a helpful aspect of DIG.

Now that you have a grasp of the RR format, you will learn about each of the record types depicted in *Table 7.1*.

SOA

The SOA record defines the authoritative name server. SOA records have a special format that may not be easily readable in DIG output. Therefore, the following is a representation of an SOA record in a different format that's more readable (this is `BIND` format – `BIND` is an older DNS server that used to be in popular use):

```
$TTL 86400
@   IN  SOA ns-144.awsdns-18.com. awsdns-hostmaster.amazon.com. (
        1               ;Serial
        7200            ;Refresh
        3600            ;Retry
        1209600         ;Expire
        86400           ;Negative response caching TTL
)
```

In reviewing the output of the SOA record, you will note it contains a large number of fields. The reason there is so much information is that the record has features that ensure the integrity, consistency, and manageability of the DNS records for the domain. Integrity is ensured by using serial numbers and listing the authoritative servers for the domain. Consistency is accomplished through values that control when other servers should request updates from the primary name servers. Manageability is accomplished by setting the default TTL for records, allowing the domain authority to control how long resolvers cache records. In short, the SOA record is the primary source of information about the domain.

The following list describes each of the fields in the SOA record shown in the BIND record format example:

- @: The operator that has the special meaning of "this domain."
- SOA: The record type, which is SOA.
- ns-144.awsdns-18.com: The authoritative name server.
- awddns-hostmaster@amazon.com: The administrator email address for the domain.
- **Serial number**: Secondary DNS servers use the serial number to determine whether the file has been updated.
- **Refresh timer**: The refresh timer is the polling interval used by secondary servers to check the serial number of the zone file.
- Retry: Retry is how often the secondary server should reattempt contact if the primary server is unreachable.
- Expire: Expire is how long a secondary server should continue to service requests for the zone in the event the primary server is unreachable. If the timer expires, the server will stop responding to requests from the domain.
- **Negative response TTL**: Negative response TTL is how long a resolver is supposed to cache negative responses to queries for the domain. For example, if tabby.trailcats.net was queried and there is no record of that name, the client should wait to ask again for 86,400 seconds, or 24 hours.

> **Note**
> If you ping a hostname that doesn't exist, your client will cache the negative response for the TTL listed in the SOA. This means that even if you fix the problem and create the record, the client will not check again, potentially for up to 24 hours. Therefore, it is a good idea to clear the client's DNS cache before testing again.

A Record

The A records represent hostnames. The value data will be an IPv4 address.

An A record for `www.trailcats.net` looks something like this:

Name	TTL	Class	Type	Data
www	3600	IN	A	192.0.2.200

Table 7.3: DNS A record

AAA Record

The AAA record is identical to an A record but for IPv6 addresses. The only difference is the data field will contain an IPv6 address.

CNAME

The **Canonical Name (CNAME)** or record is an alias that points to an A record. This is a useful technique when you want multiple hostnames to resolve to the same IP address or point to a different domain altogether.

For example, pretend you have `trailcats.net`, `www.trailcats.net`, `blog.trailcats.net`, and `trail.trailcats.net` all pointing to the same web server, `192.0.2.200`. If you were to use A records, you would see something like this in the `trailcats.net` zone file:

Name	TTL	Class	Type	Data
@	3600	IN	A	192.0.2.200
www	3600	IN	A	192.0.2.200
Blog	3600	IN	A	192.0.2.200
Trail	3600	IN	A	192.0.2.200

Table 7.4: Multiple records for the same server

In *Table 7.4*, if the server gets a new IP address, all four records need to be updated. In real life, there can be hundreds or even thousands of records that point to one server.

All four records must be updated if you move the website to a server with a different IP address. It is more work, and there's an increased possibility of errors. A better option would be to use a CNAME for three of the records, as shown here.

Name	TTL	Class	Type	Data
@	3600	IN	A	192.0.2.200
www	3600	IN	CNAME	@
Blog	3600	IN	CNAME	@
Trail	3600	IN	CNAME	@

Table 7.5: CNAME records

In *Table 7.5*, if the server gets a new IP address, only one record needs to be updated, the A record.

MX

The **mail exchanger** (**MX**) record points to the mail servers for a domain. The data field has two parts: a priority and an IP address or A record. CNAME records are not allowed. For priority, a lower value is better. Priority is used for resiliency if an MX record is down or for load sharing if the priority is the same across multiple servers.

Name	TTL	Class	Type	Priority	Data
@	3600	IN	MX	10	mx1.trailcats.net
@	3600	IN	MX	20	mx2.trailcats.net

Table 7.6: MX record

In *Table 7.6*, there are two MX records. Note the priority field. This is to provide redundancy. If mx1.trailcats.net were to become unavailable, a server attempting to send email to trailcats.net would then try mx2.trailcats.net.

TXT

The **text** (**TXT**) record is used to store metadata for various purposes. The most common use is spam reduction through the use of **Sender Policy Framework** (**SPF**) and **DomainKeys Identified Mail** (**DKIM**).

Name	TTL	Class	Type	Data
@	3600	IN	TXT	"attribute1=value1, attribute2=value2"

Table 7.7: TXT record

In *Table 7.7*, the data field could contain arbitrary information. This is most often used to prove domain ownership as only the owner of a domain has the ability to create a text record.

NS

The **name server (NS)** records are the name servers for the domain.

Name	TTL	Class	Type	Data
@	3600	IN	NS	ns-144.awsdns-18.com

Table 7.8: NS record

The name server records tell the internet where to go to find a domain's IP address. They are also the glue that ties the DNS together.

PTR

The **pointer (PTR)** record is used to perform a reverse lookup of an IP address to find its domain name. Reverse lookups are used by the traceroute program and for anti-spam checks, among other things. For it to work, there must be an `in-addr.arpa` domain for the IP subnet the IP address belongs to. For a PTR record, the IP address is concatenated with `in-addr.arpa` in reverse order.

Name	TTL	Class	Type	Data
61.135.231.54. in-addr.arpa	3600	IN	PTR	s3-website-us-east-1. amazonaws.com

Table 7.9: PTR record

PTR records are often used by anti-spam filters and logging. For example, if a mail server attempting to send mail to `mail.trailcats.net` announces itself as `mail.cnn.com`, `mail.trailcats.net` might perform a PTR lookup on the IP address of the server to verify that it resolves to mail. cnn.com. If it does not, the connection will be rejected.

DNSKEY

DNSKEY records hold the signing keys for the domain. There are two types of DNS keys: the **zone signing key (ZSK)** and the **key signing key (KSK)**. The ZSK signs groups of related RRs and that signature is stored in a **resource record signature (RRSIG)** record. A hash of the KSK is stored in the parent domain as a DS record. This second record is used to verify the ZSK is authentic.

The data portion of the DNSKEY has four fields:

- Flags
- A fixed value of 3
- Algorithm identifier
- Public key

The flags are 256 for a ZSK and 257 for a KSK. The algorithm identifier is an 8-bit number that determines the algorithm used to generate the key. The key field is the public key itself. To identify the algorithm being used, visit https://www.iana.org/assignments/dns-sec-alg-numbers/dns-sec-alg-numbers.xml.

Table 7.10 shows the format of the DNSKEY record.

Name	TTL	Class	Type	Flags	Fixed	Algorithm	Key
trailcats. net	3600	IN	DNSKEY	256	3	13	{public key}

Table 7.10: DNSKEY record

RRSIG

The RRSIG record contains the digital signature for an **resource record set (RRset)**. An RRset is a grouping of records with the same name and type. The RRSIG record has a large number of fields. They are the type covered, algorithm used, number of labels in the owner name, original TTL, expiration and inception dates, the key tag, the signer's name, and finally, a Base64 representation of the key.

As it's not feasible to show a representation of an RRSIG record in table format, *Figure 7.6* shows it as console output:

```
trailcats.net.          5     IN    RRSIG A 13 2 5 20230807020300
20230807000255 53358 trailcats.net.
/3R7jV2730SRAhPkjJ/0p0CTwHxOlxHmALkKczEM9b1eZqI+fC+9bFKs
2PNKYMXr85ZL65ZfIf3/BiNFexgnMQ==
```

Figure 7.6: RRSIG record

DS

The DS or delegation signer record is a hash of the key signing key that is stored by the parent domain. The purpose of the DS record is to create a chain of trust so the DNSKEYS can be verified as authentic. The fields are tag, encryption algorithm, digest algorithm, and signature.

Table 7.11 shows the format and fields of the DS record:

Name	TTL	Class	Type	Tag	Encryption	Digest	Digest
trailcats. net	3600	IN	DS	256	13	5	{signature}

Table 7.11: DS record

Amazon Route 53

AWS Route 53 is a globally resilient, scalable, highly available DNS service. It includes numerous helpful features beyond a traditional DNS service at an attractive price point.

For example, in the early days, Trailcats used a traditional DNS service with a single pair of servers in North America. As the business expanded to other continents, this architecture resulted in high latency, poor application performance, and frequent service disruptions for European and Asian customers. Trailcats switched to Amazon Route 53, which solved all of these problems. AWS's global network reduces latency by routing DNS queries to the nearest server, improving website load times. Latency-based routing directs customers to the closest web server, ensuring optimal performance. Health checks and automatic failover minimize downtime by redirecting traffic around failed servers. In this section, you'll learn about these features and more.

Route 53 provides three main services:

- **Domain registration**: Route 53 makes domain registration straightforward with its web-based management console, allowing users to easily search for and register domain names. Additionally, seamless integration with other AWS services and automatic DNS configuration simplify the process, enabling quick setup and management of domains.

- **Route traffic to your domain**: Route 53 offers a variety of traffic routing features, including public and private hosted zones, alias records, routing policies, rule-based conditional forwarding, and numerous other features.

- **Provide health checks on resources**: Route 53 health checks monitor the health of endpoints by sending automated requests to verify their status and availability. If an endpoint fails, Route 53 can automatically route traffic to healthy alternatives, ensuring continuous service availability. Additionally, integration with CloudWatch, a monitoring and management service for AWS resources, provides detailed monitoring and alerting, helping to quickly identify and address issues.

The following sections will cover domain registration and management, traffic routing, conditional forwarding, and health checks. *Chapter 8, AWS Route 53: Advanced*, will take a deeper dive into Route 53 features and capabilities.

Domain Registration

To create a website or application that runs on the internet, you must register a domain name for it. As it happens, domain registration is one of the three main services that Amazon Route 53 provides, and it does it in an intuitive way that handles most of the details for you. Still, it's a good idea to understand how the registration process works, both for the exam and for the real world. Domain registration involves the following elements:

- The registrant (you)
- The registrar (Route 53, GoDaddy, etc.)
- The registry operator (Verisign, IANA, Identity Digital, etc.)
- The DNS provider (which is often but not always the registrar)
- The TLD zone (operated by the TLD registry)

Domain registration involves the following steps:

1. The registrant verifies that the domain name is available.
2. The registrant submits a registration request to the registrar along with payment.
3. DNS is provisioned:

    I.   If the registrar is also the DNS host, they will create a zone file with all the needed records, load the zone file on their DNS servers, and add the DNS server information to the registration request.

    II.   If a third-party DNS host is used.

4. The zone file must be created and added to their DNS servers:

    I.   The registrant must communicate the DNS server names to the registrar.

5. The registrar sends the information to the registry operator.
6. The registry operator adds the information to the Whois database.
7. The registry updates the TLD zone file with the domain name and the name servers for the new domain.

The diagram in *Figure 7.7* depicts the process in a more visual way.

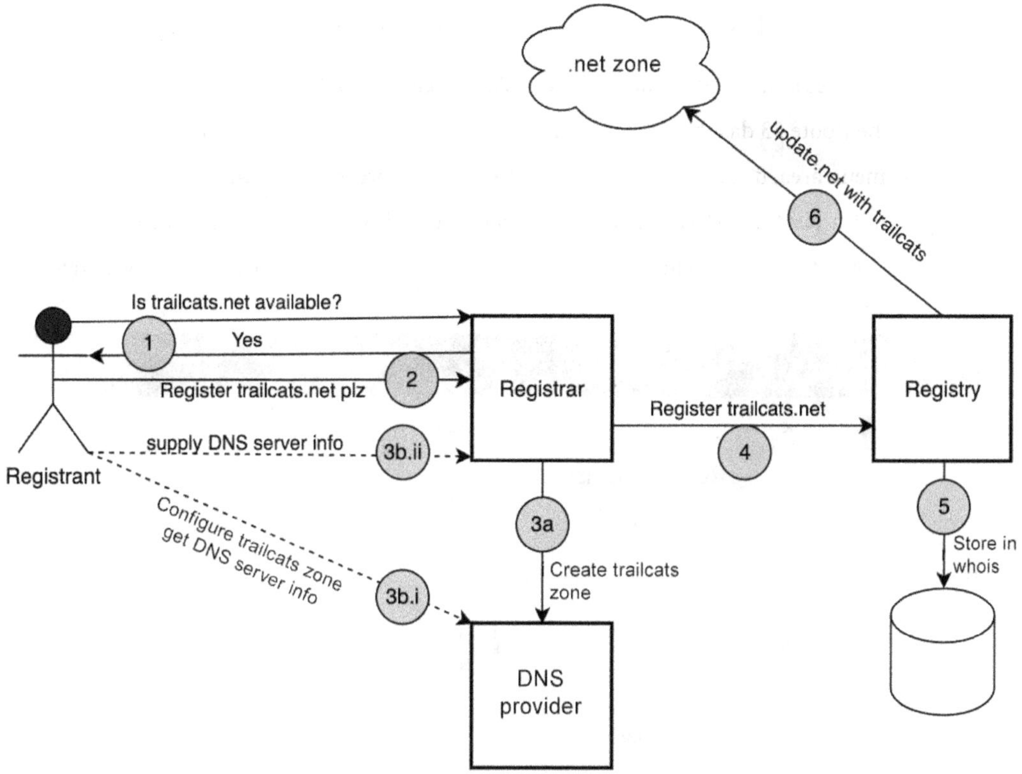

Figure 7.7: Domain registration process

As can be seen from *Figure 7.8*, using the registrar's DNS simplifies the registration process as the registrar handles the coordination and provisioning on behalf of the registrant.

> **Note**
>
> Even if you plan to use a third-party DNS provider, use the registrar's servers initially, then change DNS servers after registration. It's a lot less complicated.

Example 7.1: Creating a Domain Registration

The following steps show how you would create a new domain using the AWS Management Console:

1. In the services search bar of the AWS console, search for Route 53.

2. Open the Route 53 dashboard.

3. In the menu area, under `Domains`, select `Registered domains`.

4. Click on `Register domains`. The domain availability search bar will appear.

5. Search for a domain and click `Select` next to one or more that are available. See *Figure 7.8* for an example.

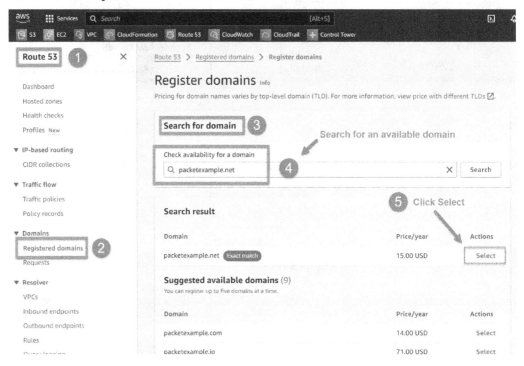

Figure 7.8: Finding and selecting an available domain in the AWS Route 53 console

6. Click `Proceed to checkout`.

7. Select payment options and fill out contact information for Whois.

8. Agree to the terms and conditions and click `Submit`.

If you had just created a new domain for your company, you would be able to monitor the status of your registration by clicking on `Requests` under the `Domains` menu option. After a few minutes, processing should be complete, and you can begin managing your new domain.

Once a domain has been registered with Route 53, its properties can be viewed under `Registered domains`. Route 53 will assign four DNS servers, which can be viewed or changed along with numerous other properties, such as contact information, auto-renewal, and **Domain Name System Security Extensions (DNSSEC).** More on this in *Chapter 8, AWS Route 53: Advanced*).

Figure 7.9 depicts the domain details for a registered domain.

Figure 7.9: Registered domain details page

The page depicted in *Figure 7.10* provides some useful information at a glance, including items such as 1. `Expiration date`, 2. if registration is configured for auto-renew, 3. if DNSSEC is configured, and 4. name servers for the domain.

Routing Traffic to Your Domain

This is where Route 53 begins to add value over a basic DNS service. Route 53 provides a standard DNS service, then it layers additional features and functionality. Features introduced, including alias records, public hosted zones, private hosted zones, routing policies, Route 53 Resolver, rule-based conditional forwarding, and health checks, will be revisited with a deeper dive in *Chapter 8, AWS Route 53: Advanced*.

Alias Records

Alias records are a proprietary extension to DNS that enables Route 53 to route traffic to AWS resources including but not limited to S3 buckets, load balancers, VPC interface endpoints, CloudFront distributions, and Elastic Beanstalk. The benefit is that the alias records automatically respond to changes in the underlying resources. For example, if the public IP address of a load balancer changes, Route 53 will return the correct IP address without operator intervention.

> **Note**
> Make sure you have a crystal-clear understanding of the difference between CNAME and alias records.

Alias versus CNAME records

CNAME and alias records are similar in that they are both pointers to another record. However, there are differences, and it's essential to be aware of them.

Alias records point to AWS resources. Whenever there is a need to create a record that points to an AWS resource, alias should be the first choice. Alias records cannot point to non-AWS resources. For example, there is no way to create an alias record that points `www.trailcats.net` to `google.com`. Also, because alias records are a Route 53 proprietary type, they can only be created in Route 53 hosted zones.

CNAME records can point to any other domain name with one exception: the naked domain name they are defined in. For example, there is no way to create a valid CNAME record for `trailcats.com` in the `trailcats.net` zone. However, a CNAME that points `www.trailcats.net` to `trailcats.org` is perfectly valid. Alias records do not have this limitation.

> **Note**
> If you are asked to create an RR in a hosted zone that points to the naked domain, CNAME will never be a valid answer.

Public Hosted Zones

A public hosted zone is exactly what it sounds like. It's a zone that is accessible from the public internet. When you register a domain through the Route 53 registrar, a public hosted zone is automatically created for you. If the domain is registered with another registrar, you must manually create the zone and then update the DNS records with your registrar to point to the Route 53 DNS servers.

Exercise 7.1: Create a Public Hosted Zone

In the following exercise, you will create a public hosted zone using the AWS Management Console:

1. In the services search bar of the AWS console, search for Route 53.
2. Open the Route 53 dashboard.
3. Select `Hosted zones` from the menu.

4.   Click Create Hosted Zone. The Create Hosted Zone dialog will appear.

5.   Enter a domain name (you can use trailcats.net if nothing comes to mind).

6.   Make sure the public hosted zone is selected.

7.   Click Create hosted zone. Following this, the Hosted zone details page will appear and there will be two records already created:

    I.   Name server record with four name servers

    II.   SOA record

A newly created public hosted zone should look something like the example in *Figure 7.10*.

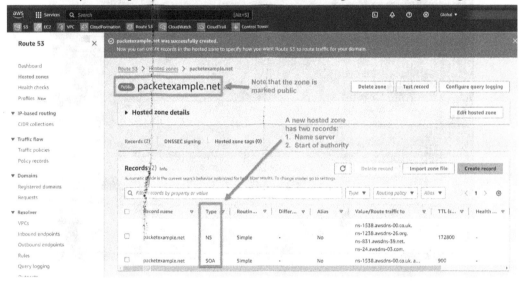

Figure 7.10: Newly created public hosted zone

Here's how to perform verification using DIG:

1. Look up the TLD name servers using `dig ns <tld> +short`.

2. Dig the NS records for your domain referencing a TLD NS using `dig @<tld.ns> ns <domain> + short`.

    In *Figure 7.11*, *steps 6* and *7* are followed for `.net` and `trailcats.net`.

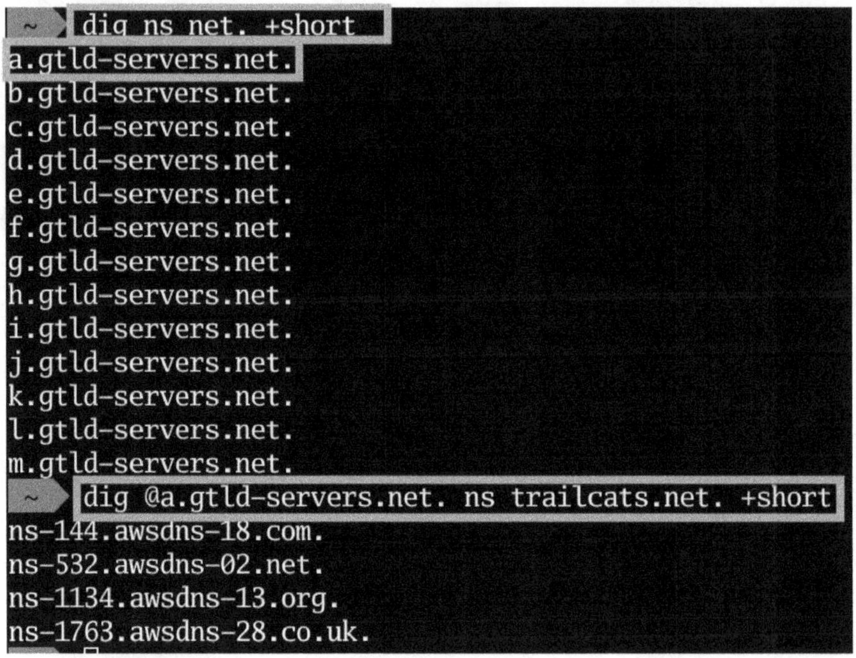

```
~  dig ns net. +short
a.gtld-servers.net.
b.gtld-servers.net.
c.gtld-servers.net.
d.gtld-servers.net.
e.gtld-servers.net.
f.gtld-servers.net.
g.gtld-servers.net.
h.gtld-servers.net.
i.gtld-servers.net.
j.gtld-servers.net.
k.gtld-servers.net.
l.gtld-servers.net.
m.gtld-servers.net.
~  dig @a.gtld-servers.net. ns trailcats.net. +short
ns-144.awsdns-18.com.
ns-532.awsdns-02.net.
ns-1134.awsdns-13.org.
ns-1763.awsdns-28.co.uk.
```

Figure 7.11: Using DIG to verify name server changes

In this exercise, you learned how to create a public hosted zone in Route 53, recognize important details about the zone on the Route 53 dashboard, and verify that the DNS information has propagated to the internet TLD servers by using DIG.

Private Hosted Zones

AWS private hosted zones are used for domain resolution within VPCs. This provides DNS services for private infrastructure where the names of things should not be available to the public, or for VPCs that do not have access to the public internet. For example, it would make no sense to publicly expose a database for a web app over the public internet, nor provide information about it, such as its hostname, through a public DNS.

There are a few important things to know about working with private hosted zones with Route 53:

1. Private hosted zones need to be explicitly associated with the VPCs where they are to be employed, to make the zone accessible to resources within the VPC. This is done by editing the private hosted zone settings, as shown in *Figure 7.12*.

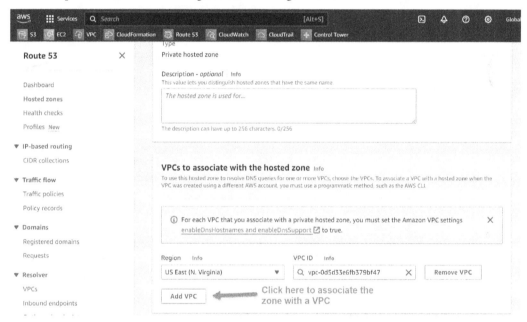

Figure 7:12: Associating a private hosted zone with a VPC

2. In addition to creating the VPC association, the `enableDnsHost` and `enableDnsSupport` VPC settings must be enabled for the private hosted zone to work with the VPC. Without them, your private hosted zone won't be able to provide DNS resolution for the resources within your VPC, and instances in the VPC won't get host names.

This is done by editing the VPC settings as shown in *Figure 7.13*:

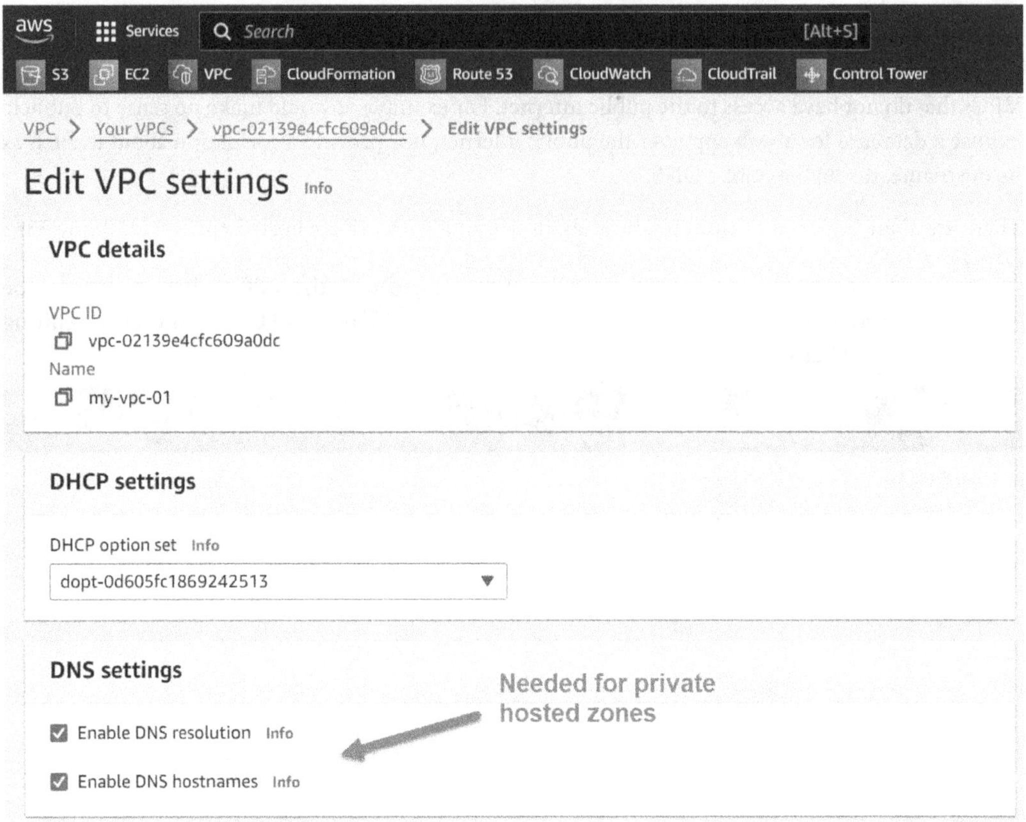

Figure 7.13: Configuring VPC settings for a private hosted zone

3. Private hosted zones do not support delegation. For example, if you have a private hosted zone called `trailcats.net`, you cannot create a new zone called `accounting.trailcats.net` and link them together with name server records.

> **Note**
>
> Route 53 supports split DNS. You can configure public and private zones for the same domain name (i.e., `trailcats.net`) and the private zone will be used in any VPC where it is attached.
>
> It's important to know that Route 53 Resolver will not attempt to recursively resolve a privately hosted zone. Here's an example: You have `trailcats.net` private and public hosted zones. There's a record for `www.trailcats.net` in the public hosted zone, but not the private one, and there's an EC2 instance in a VPC with the private hosted zone attached. If the instance tried to resolve `www.trailcats.net`, Route 53 Resolver would return a non-existent domain result.

Exercise 7.2: Create a Private Hosted Zone

In the following exercise, you will create a private hosted zone using the AWS console:

1. In the services search bar of the AWS console, search for Route 53.
2. Open the Route 53 dashboard.
3. Select Hosted zones from the menu.
4. Click Create hosted zone. The Create hosted zone dialog will appear.
5. Enter the domain name and make sure Private hosted zone is selected.
6. Under VPCs to associate with hosted zone, choose us-east-1.
7. Select the default VPC.
8. Click Create hosted zone.
9. The end result should look similar to *Figure 7.14*:

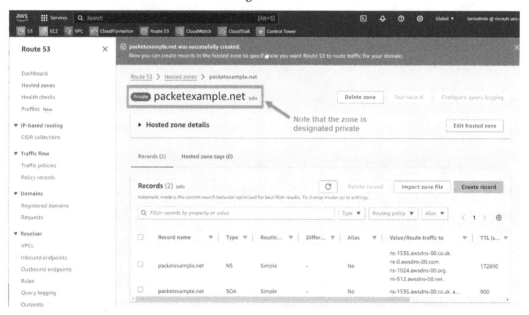

Figure 7.14: Newly created private hosted zone

Exercise 7.3: Create an A Record in the Private Hosted Zone

In the following exercise, you will create a DNS A record using the AWS console:

1. In the services search bar of the AWS console, search for Route 53 and open the Route 53 dashboard.
2. Select `Hosted zones` from the menu.
3. Click on the private hosted zone you created earlier.
4. Click `Create record`.
5. Select `Simple routing`, then click `Next`.
6. Click `Define simple record`.
7. Enter a name – for example, wwwin.
8. Under `Record type`, select `A record`.
9. Under `Value/Route traffic to`, select `IP address` or another value.
10. In the `Value` field, type the following:

    ```
    1.1.1.1
    1.1.1.2
    1.1.1.3
    1.1.1.4
    ```

11. Click `Define simple record`.

    You should have a table that looks something like this:

Name	Type	Value	TTL
www.trailcats. net	A	1.1.1.1 1.1.1.2 1.1.1.3 1.1.1.4	300

Table 7.12: A record example for a hosted zone

12. Click `Create record`.

The result should look similar to *Figure 7.15*:

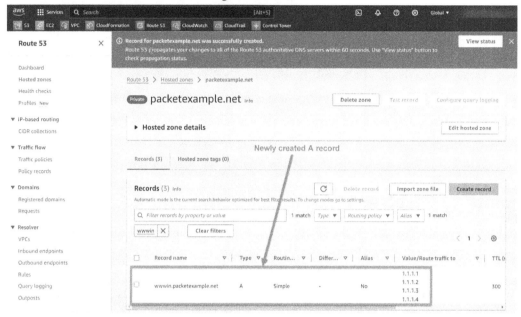

Figure 7.15: An A record in a private hosted zone

In this section, you learned about public and private hosted zones, including how to configure them. In the next section, you will learn about Route 53 routing policies.

Route 53 Routing Policies

Route 53 provides a robust set of methods for routing traffic to resources in AWS. This chapter covers simple routing only. There's a full review of the routing options in *Chapter 8, AWS Route 53: Advanced.*

Simple routing is the most basic DNS record. It consists of a name, record type, one or more values, and a TTL. *Table 7.13* shows an example of such a record:

Name	Type	Value	TTL
tabby.trailcats.net	A	1.1.1.1	300
		1.1.1.2	
		1.1.1.3	
		1.1.1.4	

Table 7.13: Simple RR example

Route 53 Resolver

Route 53 Resolver provides a DNS resolution service to VPCs and on-premises resources. Through a combination of endpoints and routing rules for conditional forwarding, Route 53 Resolver integrates on-prem resources in hybrid cloud designs.

Within a VPC, Route 53 Resolver is always the +2 address of the CIDR range. The Resolver is also available at 169.254.169.253. For example, if you have a VPC with a CIDR of 172.31.0.0/16, Route 53 Resolver will be 172.31.0.2.

Figure 7.16 illustrates where AWS will place the Route 53 Resolver endpoint in a VPC using the +2 rule.

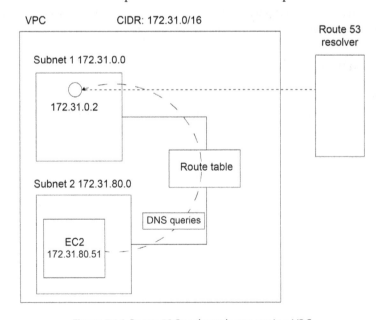

Figure 7.16: Route 53 Resolver placement in a VPC

Figure 7.16 shows that resources in other subnets within the VPC will have their DNS client traffic routed to the subnet where the Route 53 Resolver endpoint exists.

As you may recall from the section on private hosted zones, if a private hosted zone is attached to the VPC, Route 53 Resolver will return records for that zone. If there is a public hosted zone with the same name, Route 53 Resolver will not return records from the public hosted zone.

Conditional Forwarding

Route 53 supports conditional forwarding through Resolver rules. Conditional forwarding instructs resolution requests for a specific domain (for example, `finance.trailcats.net`) to a set of target IP addresses of remote DNS servers with the information for that domain. This is useful for connecting with other organizations as well as hybrid cloud integration where VPC resources need to resolve the addresses of on-premises resources, such as a database.

The following facts are important to know about conditional forwarding with Route 53:

- Its configuration scope is at the Region level. This stands in contrast to other Route 53 services.

- It requires inbound or outbound endpoints to be created. If you want to use the forwarders from a VPC that doesn't have the endpoints deployed, you'll need to ensure routing is in place.

- You must explicitly attach the forwarding rule to the VPC where it is to be used.

Conditional forwarding will be covered in greater depth in *Chapter 8, AWS Route 53: Advanced*.

Health Checks

Route 53 can perform health checks on resources and send alerts on failure, as well as being used to trigger the automatic failover of DNS records. The simplest way to use them is to associate them with RRs in a failover routing configuration.

Route 53 has three types of configurable health checks it can perform: the status of an HTTP(S) or TCP endpoint, calculated (combination of health checks), and the status of an Amazon CloudWatch alarm.

The endpoint status check tests to see if an endpoint, such as a web server, mail server, and so on, is online. The check can use HTTP, HTTPS, or TCP. The default interval is 30 seconds, with a 10-second interval available for an extra charge. TCP is a simple connection test; the HTTP(S) test can poll for a specific resource, such as a file. The HTTP(S) check status can also perform string matching on the response body.

By default, the endpoint status check uses health checkers in eight different AWS Regions. This is configurable, allowing you to select a subset. Imagine, for example, that Trailcats has a dedicated website in US-East-1 for North America and a separate website for other continents. It wouldn't make sense to test connectivity to the North America website from those other continents.

Creating an endpoint health check is straightforward. The following two examples will show a basic TCP health check and then a more elaborate HTTP check with customized regional health checkers.

From the Route 53 console, select Health checks, and then select Create health check, as shown in *Figure 7.17*:

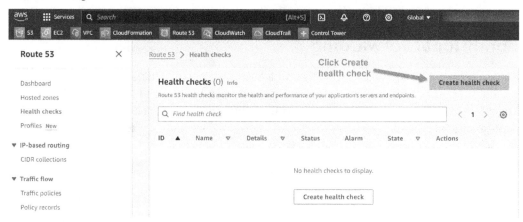

Figure 7.17: Creating a health check

In the case of a simple TCP health check, you could configure a simple TCP connection test as shown in *Figure 7.18*:

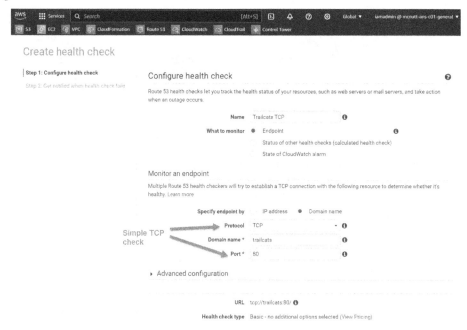

Figure 7.18: TCP health check

The TCP health works with a wide variety of applications, but it's a simple connection check. It can't tell you whether the application listening on the TCP is functioning as expected.

For websites or web applications, the HTTP check is better because you can test whether the server is working by requesting a specific resource or response in the message body. In *Figure 7.19*, an HTTP health check is configured with the health checker list customized to only test from North America. The screenshot has been cropped to fit the page:

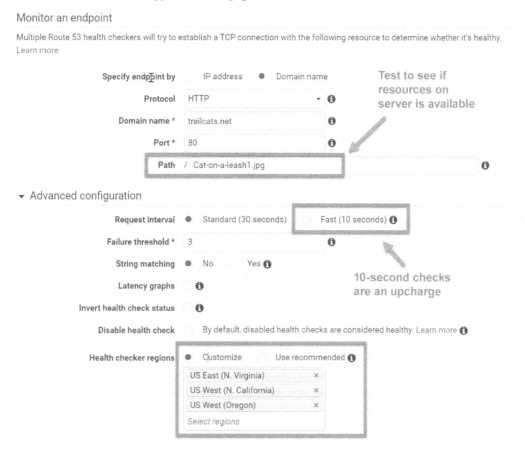

Figure 7.19: HTTP health check

In this example, the HTTP check is configured to retrieve `Cat-on-a-leash1.jpg`. This is a better check than the TCP check because it verifies the web server is functioning and returning content.

A calculated health check takes two or more health checks and calculates health based on selectable logic. The logic types are as follows:

- X of Y, where Y is the total number of checks
- All of the checks
- At least one check

A good use for calculated health checks is complex health monitoring. Calculated health checks can provide a more comprehensive health status.

As an example, imagine that Trailcats has a three-tier application consisting of a database server, an application server, and a web server. Health check 1 monitors the availability of the application server, health check 2 monitors the availability of the database server, and health check 3 monitors the status of the web server. A calculated health check can ensure that all three components are operational before considering the application healthy.

CloudWatch alarm checks monitor the status of CloudWatch Metrics, a monitoring service that collects and tracks metrics. The alarm check allows Route 53 to mark resources as unhealthy on the basis of a CloudWatch alarm being triggered.

As an example, imagine that Trailcats has an application where performance is critical, and if the application is running slowly, traffic needs to be rerouted to a redundant instance in another Region. In this example, you can set an alarm if the application response time exceeds 500 milliseconds over a 5-minute period. If the alarm triggers, the application is marked unhealthy and traffic will be rerouted.

Summary

For both the exam and your career as an AWS networking specialist, you should be able to design DNS solutions in public, private, and hybrid systems. In this chapter, you have covered what the DNS is, a little bit about its history, its architecture, how the protocol operates, and the major types of RRs such as A, CNAME, MX, and SOA. You have also seen how the DNS is a distributed database, how the **TLDs** are managed by registries, and how the subdomains of the TLDs are owned by individual organizations. The section on DNS components is foundational information and if you studied it carefully, you will have noticed that the subsequent information in the chapter was much easier to understand and absorb. This was by design, and you will find that that foundation will continue to serve you for years to come.

In the Route 53 portion of the chapter, you covered Route 53's major features, which are domain registration, traffic routing, and health checks. You learned how to register and manage domains with Route 53 and about some proprietary features of Route 53, such as alias records and private hosted zones. You have learned how Route 53 Resolver and private hosted zones integrate with AWS VPCs, and how conditional forwarding ties discontinuous domain trees together using a combination of rules with inbound and outbound endpoints in VPCs. You learned about Route 53 health checks and how they can be used to monitor your resources and only route traffic to them if those resources are healthy.

Chapter 8, AWS Route 53: Advanced, will build on what you have learned, creating advanced hybrid cloud architectures.

Exam Readiness Drill – Chapter Review Questions

Apart from mastering key concepts, strong test-taking skills under time pressure are essential for acing your certification exam. That's why developing these abilities early in your learning journey is critical.

Exam readiness drills, using the free online practice resources provided with this book, help you progressively improve your time management and test-taking skills while reinforcing the key concepts you've learned.

HOW TO GET STARTED

- Open the link or scan the QR code at the bottom of this page

- If you have unlocked the practice resources already, log in to your registered account. If you haven't, follow the instructions in *Chapter 16* and come back to this page.

- Once you log in, click the START button to start a quiz

- We recommend attempting a quiz multiple times till you're able to answer most of the questions correctly and well within the time limit.

- You can use the following practice template to help you plan your attempts:

Working On Accuracy		
Attempt	Target	Time Limit
Attempt 1	40% or more	Till the timer runs out
Attempt 2	60% or more	Till the timer runs out
Attempt 3	75% or more	Till the timer runs out
Working On Timing		
Attempt 4	75% or more	1 minute before time limit
Attempt 5	75% or more	2 minutes before time limit
Attempt 6	75% or more	3 minutes before time limit

The above drill is just an example. Design your drills based on your own goals and make the most out of the online quizzes accompanying this book.

First time accessing the online resources? 🔒

You'll need to unlock them through a one-time process. **Head to** *Chapter 16* **for instructions**.

Open Quiz	
`https://packt.link/ansc01ch7`	
OR scan this QR code →	

AWS Route 53: Advanced

As you learned in *Chapter 7, AWS Route 53: Basics*, Amazon Route 53 is a robust and versatile **Domain Name System (DNS)** web service within **Amazon Web Services (AWS)**. However, Route 53 offers far more functionality than basic DNS. Route 53 provides advanced features designed to optimize performance, enhance security, and streamline management across complex environments.

This chapter explores these advanced capabilities, focusing on five key topics for the exam and deploying DNS within your AWS environment:

- Conditional forwarding
- Route 53 routing policies
- Sharing services between accounts
- Route 53 logging and monitoring
- **DNS Security Extensions (DNSSEC)**

By the end of this chapter, you will be able to stitch discontinuous DNS architectures together into a functional whole to deploy DNS security to protect the integrity of your DNS records. Mastering these topics will prepare you both for the exam and for real-world design and deployment of sophisticated DNS architectures using Route 53.

Conditional Forwarding

Conditional forwarding is a way to stitch discontinuous DNS architectures together by forwarding queries to specific resolvers and overriding the normal process of recursive lookups. In Route 53, this is accomplished through a combination of endpoints to route query traffic and forwarding rules, which tell Route 53 where to send query traffic for external destinations. Each of these concepts will be covered in the following sections.

Route 53 Endpoints

To facilitate routing DNS traffic between Route 53 **private hosted zones** (**PHZs**) and on-premises networks, you create endpoints. Endpoints are groupings of elastic network interfaces in your VPCs that attach to Route 53 Resolver. There are two types of endpoints – inbound endpoints and outbound endpoints:

- **Inbound endpoints** allow on-premises DNS resolvers to resolve DNS names for PHZs. The on-premises resolvers forward queries to Route 53 Resolver via these inbound endpoints.

- **Outbound endpoints** allow Route 53 Resolver to resolve DNS names for on-premises DNS zones. Route 53 Resolver uses resolver rules to determine where to forward queries via the outbound endpoints.

> **Note**
>
> For outbound endpoints to forward traffic, you must configure forwarding rules that reference them.

For availability reasons, Route 53 requires you to specify a minimum of two IP addresses for each endpoint. Ideally, you want to place these in separate Availability Zones. Each IP address is backed by an **elastic network interface** (**ENI**) attached to the Route 53 resolver. It might be tempting to place endpoint interfaces in the environment liberally. However, the costs of doing so could be significant. A small number are typically deployed per Region in a shared services VPC.

Forwarding Rules

Forwarding rules are how you tell Route 53 Resolver where to send queries that should be handled outside of Route 53. When you specify a domain name in a forwarding rule, all subdomains will also be forwarded.

There is an exception/override to the forwarding rule, which is the system rule. The system rule is used to override specific subdomains from being forwarded. For example, you have a forwarding rule for `corp.trailcats.net`, and the business has decided to migrate `accounting.corp.trailcats.net` to AWS. You can create a system rule for the accounting domain name that overrides the forwarding rule for `corp.trailcats.net`. This would cause `accounting.corp.trailcats.net` to be resolved by Route 53 instead of being forwarded.

Forwarding rules are the final component for conditional forwarding. In order to configure a forwarding rule, you need to have a few things in place first:

- The VPCs from which queries will be forwarded

- An outbound endpoint to forward queries

- The IP address(es) of the target resolver(s)

Optionally, you may want to perform query logging and share the forwarding rule with other accounts, so you will also want to have that information available.

Figure 8.1 depicts a typical conditional forwarding setup in a hybrid cloud architecture.

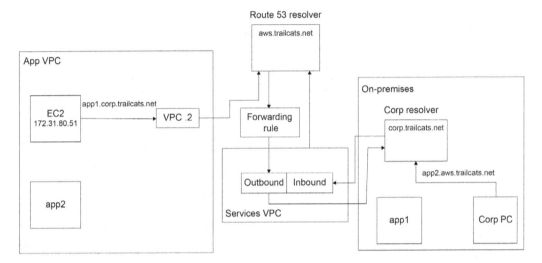

Figure 8.1: Conditional forwarding in hybrid cloud architecture

In this design, a services VPC has outbound and inbound endpoints configured to handle DNS query routing. In the outbound direction, Route 53 is configured to forward all queries for `corp.trailcats.net` to the corp resolver via the outbound endpoint. In the inbound direction, the on-premises resolver forwards queries for `aws.trailcats.net` to the Route 53 Resolver via the inbound endpoint.

Route 53 Routing Policies

Route 53 DNS records are different from normal DNS records because they offer more control over how traffic is directed. Normal DNS records just link a domain name to an IP address. In Route 53, routing policies let you send traffic based on things such as whether a server is healthy, where the user is located, or which server can respond the fastest. For example, if a server goes down, failover routing automatically sends users to a backup server. Latency and geolocation policies make sure users are directed to the closest or fastest server, giving better performance than normal DNS.

When configuring a DNS record in Route 53, selecting a routing policy is the first step when configuring a resource record in a hosted domain. The policy selected is based on what your goals are for that resource. Except for simple routing, all routing policies support the use of health checks.

The following is a summary of the routing policies available in AWS:

- **Simple routing** is a single-record response without complex rules. It's the most basic routing policy for a DNS record. It consists of a name, record type, one or more values, and a **time to live** (TTL).

  Table 8.1 is an example of a simple routing record. In the example shown, three IP addresses are returned, and the record has a TTL of 300 seconds or 5 minutes:

Name	Type	Value	TTL
`tabby.trailcats.net`	A	`1.1.1.1` `8.8.8.8` `8.8.4.4`	`300`

Table 8.1: Simple resource record example

  The format of the simple record looks like a normal DNS record that you would expect to encounter on any DNS server. All of the other records have special fields that are proprietary to Route 53 routing policies.

- **Failover routing** refers to the return of a record in response to a query. Route 53 can use health checks as a basis for whether to employ this policy for a given query. Route 53 supports active/ active and active/passive failover.

  In active/passive failover, primary and secondary records are tied to the main record. It's a typical pattern to point to a static page if the main web application is unhealthy. A classic example is the Twitter (now X) "fail whale," which was discontinued in 2013.

  Table 8.2 is an example of a failover record with primary and secondary entries:

Name	Type		TTL
`tabby.trailcats.net`	A		`300`
Record ID	Failover Type	Value	Health Check
1	Primary	`{Elastic Load Balancer)`	`{Health Check ID}`
2	Secondary	`{S3 bucket}`	None

Table 8.2: Active/passive failover example

In *Table 8.2*, the configuration shown will route traffic to an AWS elastic load balancer as long as the health check passes. If the health check fails, Route 53 will return a static web page hosted in a **Simple Storage Service (S3)** bucket.

- **Latency routing** is used to optimize the user experience by directing them to a healthy resource with the lowest latency. For this routing method to make sense, you should have the same resource available in multiple Regions. The latency calculation is not in real time; it is based on the Region where the resource is located combined with a latency database that AWS maintains on the backend.

Table 8.3 is an example of a website with copies in three different AWS Regions. The user will be directed to the region with the lowest latency:

Name	Type		TTL
tabby. trailcats.net	A		300
Record ID	Region	Value	Health Check
1	us-east-1	{us-east-1 ELB}	{Health Check ID}
2	ap-east-1	{ap-east-1 ELB}	{Health Check ID}
3	eu-north-1	{eu-north-1 ELB}	{Health Check ID}

Table 8.3: Latency routing example

In *Table 8.3*, someone in the United States would be directed to the us-east-1 load balancer. A user in the Philippines would be directed to the ap-east-1 load balancer. A person in France would be directed to the eu-north-1 load balancer. If ap-east-1 and eu-north-1 became unhealthy, all users would be directed to us-east-1. This provides resiliency as well as improved performance.

- **Weighted routing** allows you to choose the percentage of traffic you want to send to each healthy resource. This is useful for simple load balancing and for testing new software versions. The way it works is all the weight values in the records are summed up and then each resource's individual weight is divided by the sum to determine what percentage of traffic it will get. If the weight is set to 0, the resource will get no traffic.

In *Table 8.4*, the resource weights add up to 100. So, resource A would get 45%, resource B would get 45%, and resource C would get 10%.

Name	Type		TTL
tabby.trailcats.net	A		300
Record ID	Weight	Value	Health Check
1	45	{Load Balancer-1}	{Health Check ID}
2	45	{Load Balancer-2}	{Health Check ID}
3	10	{Load Balancer-3}	{Health Check ID}

Table 8.4: Weighted routing example

In *Table 8.4*, Load Balancer-3 might be a development version of a web application. With this design, Trailcats can test the new version on a small subset of its traffic. If there is a problem with the new code, only 10% of users would be affected.

- **Multivalue routing** splits the difference between simple and weighted routing. It allows you to return up to 8 records that pass a health check.

In *Table 8.5*, the addresses of all three load balancers would be returned if they passed their health check:

Name	Type		TTL
tabby.trailcats.net	A		300
Record ID		Value	Health Check
1		{Load Balancer-1}	{Health Check ID}
2		{Load Balancer-2}	{Health Check ID}
3		{Load Balancer-3}	{Health Check ID}

Table 8.5: Multivalue routing example

In *Table 8.5*, all load balancers would be returned to the user (assuming a passed health check), and it would be up to the user's computer to decide which one to use.

- **Geolocation routing** is used mainly for localizing language (for example, German for Germany) or for restricting content by location (for example, restricting a US-based user from watching a video on a Canadian website). It uses a hierarchy of tags from most to least specific. If there is no match, no record is returned. The order of hierarchy from most to least specific is state/subdivision, country, continent, and default.

Table 8.6 depicts a record that uses geolocation routing with three possible options – Alaska, United States, and a default:

Name	Type		TTL
tabby.trailcats.net	A		300
Record ID	Location	Value	Health Check
1	Alaska	1.1.1.1	{Health Check ID}
2	United States	8.8.8.8	{Health Check ID}
3	Default	{static S3 website}	{Health Check ID}

Table 8.6: Geolocation routing example

In *Table 8.6*, users in Alaska would be routed to 1.1.1.1. Users in the United States would be routed to the server at 8.8.8.8. Users outside of these locations would receive a static web page informing them they are in a service-restricted area.

> **Note**
> Remember that geolocation is **not** about routing the user to the closest resource. It's about restricting what resources users can access or performing localization, such as setting the language.

- **Geoproximity routing** is distance-based, with an additional control called bias, which can be used to expand or reduce the coverage scope of a rule. The distance metric can be either an AWS region or latitude/longitude coordinates. Geoproximity routing can only be configured by creating a traffic flow policy.

In *Figure 8.2*, a traffic policy has been configured for booking.trailcats.net that returns a different result based on four Regions that have been defined on the geoproximity map:

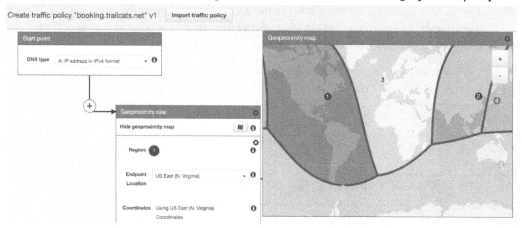

Figure 8.2: Geoproximity routing example

In *Figure 8.2*, the Regions are used to define where to route the traffic. For example, Region 1 directs traffic to the us-east-1 AWS region. This allows for the direction of traffic with some precision, using a graphical tool.

Route 53 Logging and Monitoring

Logging and monitoring are essential components of operating any information technology solution. Without logging and monitoring in Route 53, you wouldn't know whether something went wrong, such as a misrouted request or a failed DNS resolution. For example, if your website suddenly becomes unreachable, you'd struggle to identify the issue without monitoring in place. Troubleshooting would be nearly impossible, as you wouldn't be able to track DNS query failures or pinpoint where the problem lies. In the event of a security breach, incident responders would have no way to trace the attacker's activity or timeline without DNS logs to see what domains were queried and when, making it difficult to respond effectively.

The mechanisms for monitoring and logging in Route 53 include CloudWatch for DNS health monitoring and logging, CloudTrail for logging API calls, and the Route 53 console for viewing domain status and audit logs.

CloudWatch has two integration points with Route 53: CloudWatch metrics combined with health checks and CloudWatch logging. CloudWatch metrics can be used to observe the status of health checks, act as a health check, and fire alerts if a resource becomes unhealthy. Route 53 metrics are enabled by default so configuring them is unnecessary. CloudWatch logging is available for public hosted zones and Route 53 resolvers. Public hosted zones support query logging to a CloudWatch log group, while Route 53 Resolver supports logging to log groups and S3 buckets.

You may recall from *Chapter 7, AWS Route 53: Basics*, that DNS resolvers cache all lookups. This means only a tiny fraction of DNS queries for your public hosted domains will be made against the Route 53 DNS servers. This is important to bear in mind when reviewing logs and statistics. DNS query metrics are logged to CloudWatch by default at no charge, so there's no need to configure it. However, for detailed information about DNS queries, you must configure query logging.

> **Note**
> All query metrics are stored in us-east-1.

Route 53 public DNS query logging is configured from the Hosted zone page in the Route 53 console. There are two requirements to enable query logging: a CloudWatch log group in the us-east-1 region, and Route 53 must be granted permission to write to the log group. If the log group and permissions aren't already configured, those tasks can be performed from the Configure query logging page.

Resolver query logging handles logging for PHZs and will log anything that hits the Route 53 resolvers from your environment. This includes queries that originate from a VPC, queries from inbound endpoints (forwarded queries from on-premises), queries from outbound endpoints (queries from VPC to on-premises), and Route 53 DNS Firewall rules. This helps with troubleshooting, security analysis, and general activity monitoring of DNS activity within your Route 53 environment. Logs can be sent to CloudWatch, an S3 bucket, or Kinesis (not covered in this book).

CloudTrail is integrated with Route 53 and logs all Route 53 API calls by default; no configuration is required. CloudTrail stores events for 90 days free of charge and Route 53 events are searchable in the CloudTrail console under Event history. You can also create a trail that writes log data to an S3 bucket to provide longer retention periods and get information into other tools. By default, trails have a global scope (i.e., they will collect log information from all Regions). The Amazon free tier includes one free trail for management events (S3 charges will apply, however).

By leveraging trails, it's possible to create a centralized log collection point for an entire organization and take automated actions when needed. A common use case for trails is to facilitate the ingestion of log data by third-party monitoring and security tools by retrieving the log data from the S3 buckets.

Monitoring Domain Registrations

In the Route 53 console under Registered domains, you can view the status of your domains, their expiration dates, and other details. The Requests option under Domains provides a log of various domain operations such as registrations, transfers, DNSSEC configuration, and so on. The Requests page, located below Registered domains in the console menu, contains an audit log of domain operations.

The Requests page shown in *Figure 8.3* displays a log of domain operation actions:

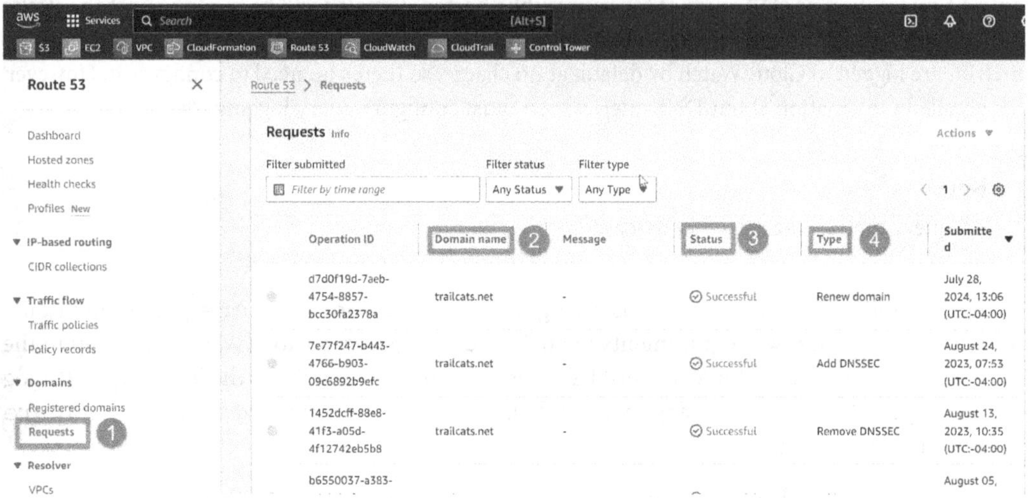

Figure 8.3: Route 53 domain operations log

The operations log is accessed from the Requests option (1). The Domain name column (2) shows the domain the operation was requested on. The Status column (3) shows whether the operation succeeded or failed, and the Type column (4) lists the specific operation. This log shows that trailcats.net was registered, DDNSSEC was configured, and the domain was renewed.

Public hosted zone logs record queries coming from the internet, and Route 53 Resolver logs record queries coming from your VPCs. In the following exercises, you will configure logging for public hosted zones and Route 53 Resolver.

Exercise 8.1: Configure Query Logging for a Public Hosted Zone

In this exercise, you will configure and verify query logging on a public hosted zone, which is used to monitor DNS queries from the public internet:

1. In the Route 53 console, select Hosted zones, then select a public hosted zone. Select View details. See *Figure 8.4* for the navigation steps:

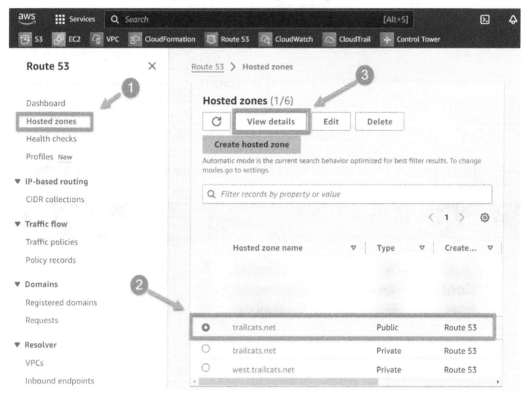

Figure 8.4: Selecting to view the details of a public hosted zone

2. Select `Configure query logging`, as shown in *Figure 8.5*:

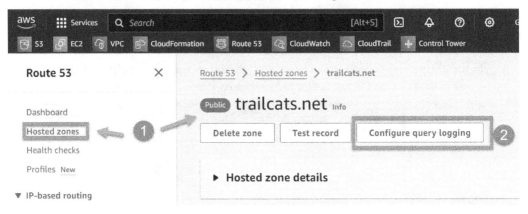

Figure 8.5: Configure query logging

3. In the `Log group` section, create a log group and name it `Route53PublicZones`. Click `Create`.

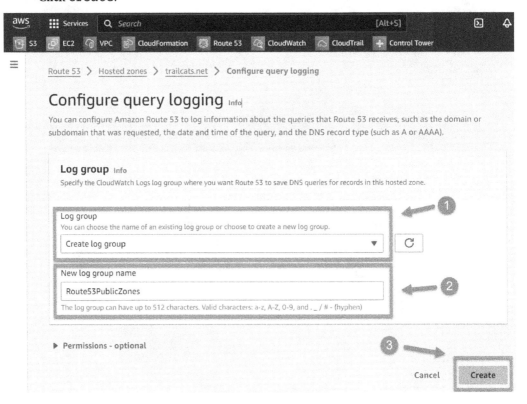

Figure 8.6: Creating the query logging

Query logging has now been configured for the hosted zone. Now, there will be a record of the DNS queries made for this domain.

Viewing the Query Logs

Once logging has been configured, it's a good idea to verify that logging is working as intended. To verify that logging is successfully configured, follow these steps:

1. Open the `CloudWatch` console. Under `Logs`, select `Log groups`, as shown in *Figure 8.7*, and select the log group you created.

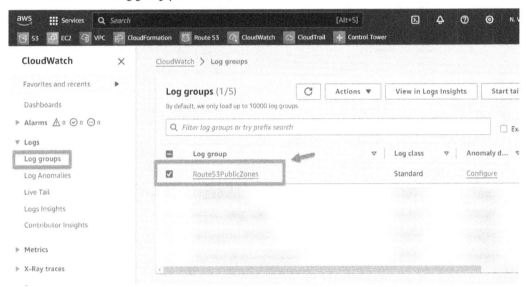

Figure 8.7: Finding and selecting a log group in CloudWatch

2. Under `Log streams`, there should be a stream called `route53-test-log-stream`, as shown in *Figure 8.8*. Select that log stream.

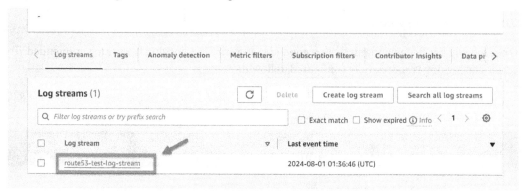

Figure 8.8: Identifying the test log stream

3. You should see a test event from Route 53, as shown in *Figure 8.9*:

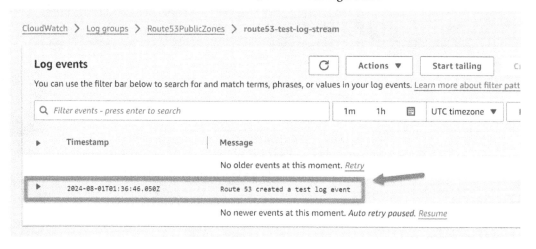

Figure 8.9: Test log event

In this exercise, you learned how to enable query logging for public hosted zones, and how to locate and view the logs in CloudWatch. In the real world, you perform these steps for all public hosted zones. This provides visibility into public DNS query activity for monitoring, troubleshooting, and security investigations.

In the next exercise, you'll configure Resolver Query Logging. This will log all DNS queries made by resources in your VPCs

Exercise 8.2: Configure Resolver Query Logging

In this exercise, you will configure Resolver Query Logging, which is used to monitor DNS queries within your AWS environment:

1. In the Route 53 console under Resolver, select Query logging. Note that the scope changes from global to regional. This is because Route 53 Resolver has a regional configuration scope.

2. In the Query logging configurations panel, select Configure query logging, as shown in *Figure 8.10*:

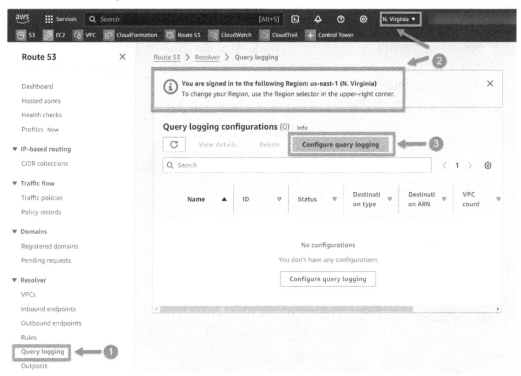

Figure 8.10: Configure query logging

3. In the Query logging configuration name field, enter ResolverQueryConfiguration, as shown in *Figure 8.11*.

4. For Query logs destination, select CloudWatch Logs log group, then select Create log group.

5. Call the log group `Route53Resolver`, as shown in *Figure 8.11*.

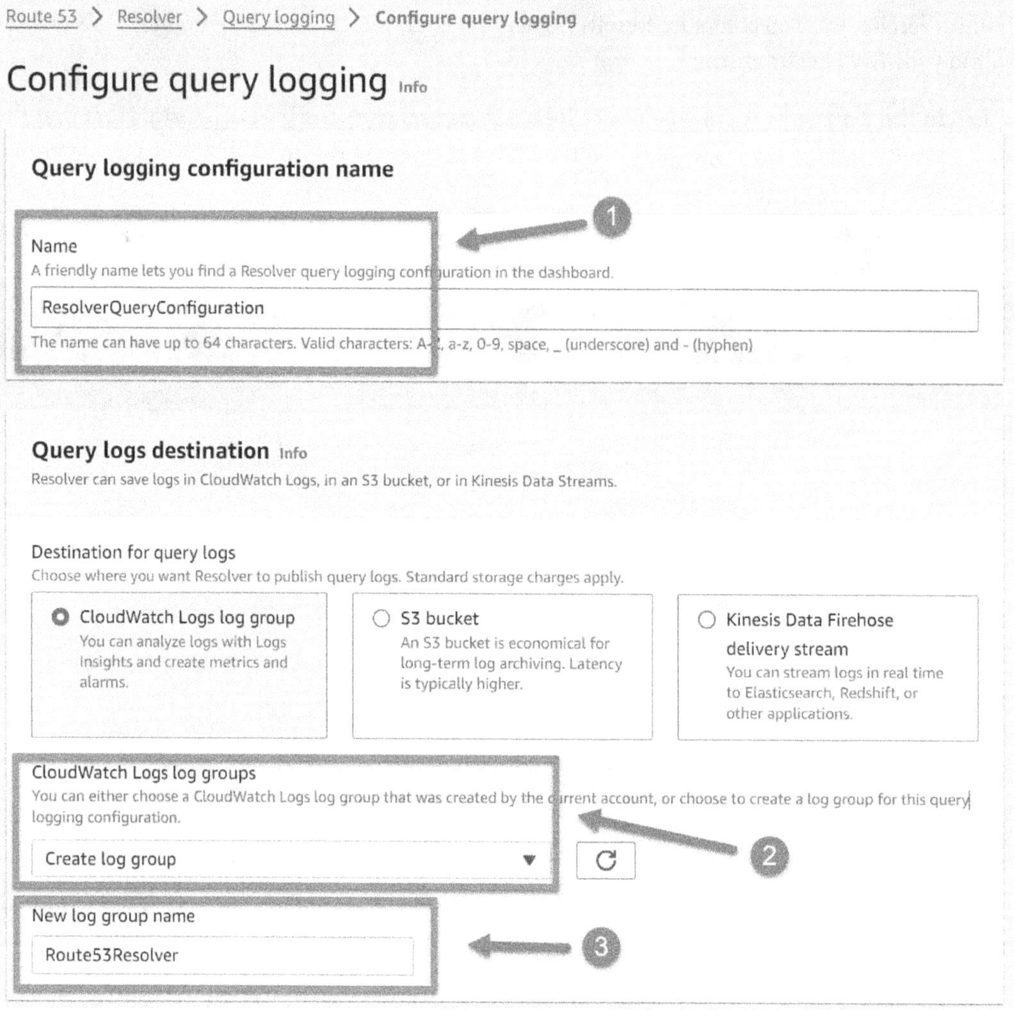

Route 53 > Resolver > Query logging > Configure query logging

Configure query logging Info

Query logging configuration name

Name

A friendly name lets you find a Resolver query logging configuration in the dashboard.

ResolverQueryConfiguration

The name can have up to 64 characters. Valid characters: A-Z, a-z, 0-9, space, _ (underscore) and - (hyphen)

Query logs destination Info

Resolver can save logs in CloudWatch Logs, in an S3 bucket, or in Kinesis Data Streams.

Destination for query logs

Choose where you want Resolver to publish query logs. Standard storage charges apply.

🔘 CloudWatch Logs log group
You can analyze logs with Logs Insights and create metrics and alarms.

⚪ S3 bucket
An S3 bucket is economical for long-term log archiving. Latency is typically higher.

⚪ Kinesis Data Firehose delivery stream
You can stream logs in real time to Elasticsearch, Redshift, or other applications.

CloudWatch Logs log groups

You can either choose a CloudWatch Logs log group that was created by the current account, or choose to create a log group for this query logging configuration.

Create log group ▼   C

New log group name

Route53Resolver

Figure 8.11: Configure query logging

6. Under `VPCs to log queries for`, select `Add VPC` or `More VPCs` and add them, as shown in *Figure 8.12*:

CloudWatch Logs log groups

You can either choose a CloudWatch Logs log group that was created by the current account, or choose to create a log group for this query logging configuration.

| Create log group ▼ | | C |

New log group name

Route53Resolver

VPCs to log queries for - *optional* (0)   Info   Remove   Add VPC

Resolver logs DNS queries that originate in the VPCs that you choose here. If you don't choose any VPCs, Resolver doesn't log any queries.

| Q Find resource |

‹ 1 › ⚙

| VPC ID | VPC name | Status | IPV4 CIDR | IPV6 CIDR | Owner |

No resources

You don't have any resources.

Add VPC ⬅ ①

Figure 8.12: Add VPC

7. Select at least one available VPC and click Add.

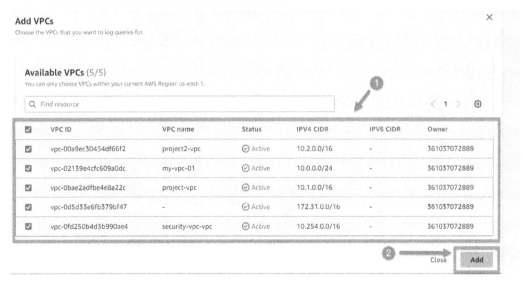

Figure 8.13: Select and add an available VPC

8. Click `Configure query logging` to commit the configuration, as shown in *Figure 8.14*:

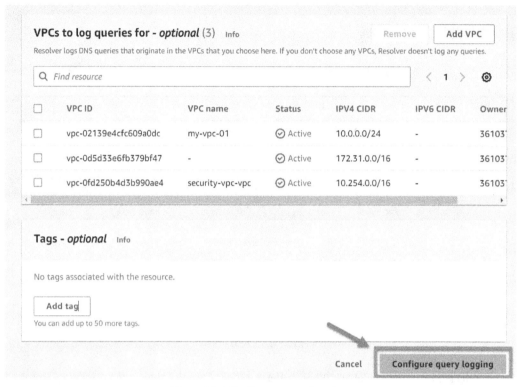

Figure 8.14: Configure query logging

You should see several success messages relating to the query logging configuration, and the logging configuration will appear with a status of Created, as shown in *Figure 8.15*:

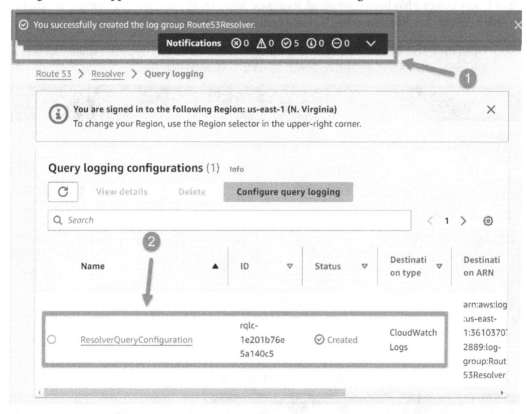

Figure 8.15: Created query logging configuration

You have now configured Query Logging for Route 53 Resolver. This will enable you to monitor and troubleshoot DNS queries made to Route 53 Resolver from within your infrastructure.

Viewing the Resolver Query Logs

Once Query Logging has been configured, it's a good idea to verify that query logging is working as intended. To view the Resolver query logs and verify the configuration, perform the following steps:

1. Open the `CloudWatch` console. Under `Logs`, select `Log groups`. You should see a new log group titled `Route53Resolver`. Select it to view the log group:

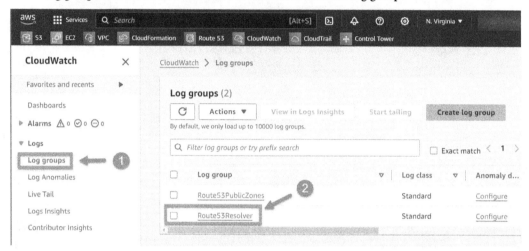

Figure 8.16: Viewing the Resolver query log

2. Locate the `Log streams` portion of the page (see *Figure 8.17*):

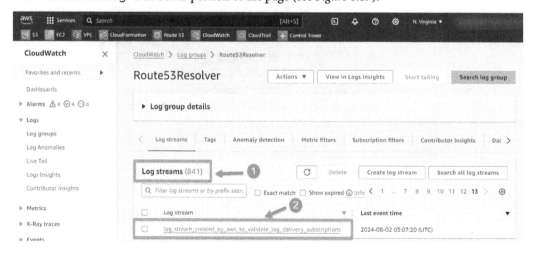

Figure 8.17: Locating the log streams

3. There should be a log stream that starts with `log_stream_created_by`. Select the log stream to view its contents, as shown in *Figure 8.18*:

Figure 8.18: Viewing the contents of a log stream

In this exercise, you configured query logging for Route 53 Resolver, attached the configuration to a VPC, and located the logs in CloudWatch. Query logging is important for visibility, troubleshooting, and security.

Sharing DNS Services between Accounts

Sharing AWS DNS services between accounts makes it easier to manage domain names and DNS settings in one place. This helps avoid repeating the same setup, saves money, and improves security by keeping control of DNS settings in a single location.

Imagine for a moment that as Trailcats grew into a global company, it organized its operations by **geographical regions (geos)**, with each geo responsible for budgeting and managing its own expenses. Under this model, each geo may want to have its own AWS accounts for billing purposes. Yet at the same time, the infrastructure needs to operate as a cohesive whole across those accounts. For reasons such as this, a best practice design pattern for architecting in AWS is to create logical separation at the account level, and then connect the accounts together in a hub-and-spoke topology with shared services placed in the hub account. In a hybrid design, the on-premises network would also be wired into the hub account.

Figure 8.19 illustrates a hub-and-spoke hybrid cloud architecture using accounts.

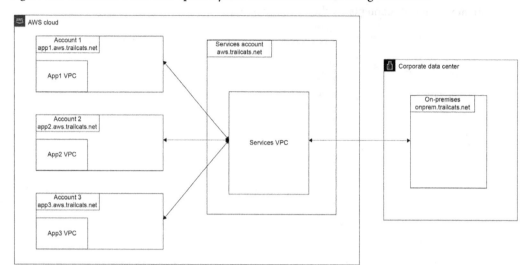

Figure 8.19: Hub-and-spoke architecture using accounts

In *Figure 8.19*, segmentation at the top level is accomplished through the use of separate accounts. Each application runs in a hub account, which is wired into a central services account or hub account. The on-premises environment is also wired into the hub account.

In order to operate as a cohesive entity, Route 53 provides tooling to create a unified DNS architecture across these types of designs. In this section, you will learn about those elements, how to use them, and where to place them.

These are the requirements for sharing DNS services between accounts:

- Query logging configuration through **AWS Resource Access Manager (AWS RAM)**
- Forwarding rules through AWS RAM
- Cross-account VPC to private hosted zone association
- Placement of Resolver endpoints in a shared services account
- Well-designed DNS naming structure

It's important to appreciate that query logging and forwarding rules have a regional scope. This differs from other aspects of Route 53 that have a global scope. This has implications that apply to multi-account designs and resource sharing between those accounts. Remember that if the design being discussed involves multiple Regions, you will need a separate configuration for each region.

Query Logging Configuration Sharing

Query logging configuration sharing facilitates logging to a central location. After you share a query logging configuration through AWS RAM, a service to share AWS resources securely across AWS accounts, you can attach it to VPCs in other accounts.

Figure 8.20 shows how shared query logging configuration is applied across accounts:

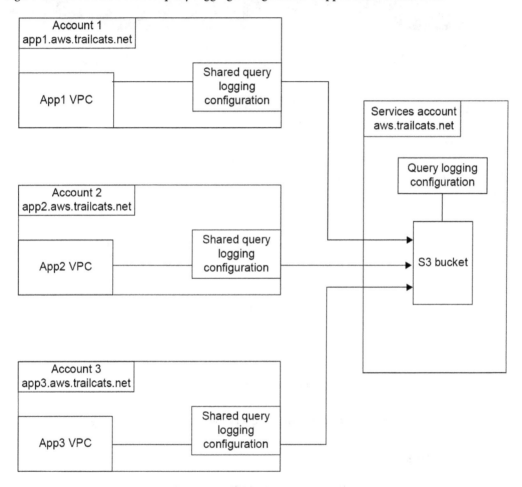

Figure 8.20: Query logging configuration sharing across accounts

In *Figure 8.20*, the query logging configuration in the central services account is shared with the hub accounts. This ensures logging consistency across the architecture.

To perform cross-account configuration sharing, there is a `Share configuration` option in the Resolver query configuration page, which cross-launches into AWS RAM, or you can go directly to the AWS RAM console.

Figure 8.21 depicts the `ResolverQueryConfiguration` page for `Route53Resolver`, which was created in *Exercise 8.21*:

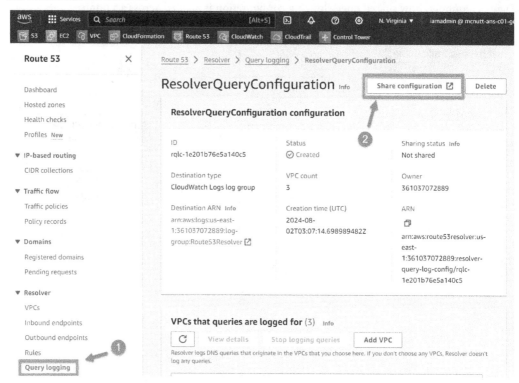

Figure 8.21: Share query logging configuration

In *Figure 8.21*, *step 2* is how you access the shared query logging configuration page.

Sharing Forwarding Rules

The purpose of sharing forwarding rules is to have a consistent policy for resolver forwarding across accounts. This is particularly useful in hub-and-spoke designs where inbound and outbound resolver endpoints are located in a central services VPC. The idea is to create the forwarding rules in the hub account and then share them with the spoke accounts.

Figure 8.22 shows how shared forwarding rules are applied across accounts:

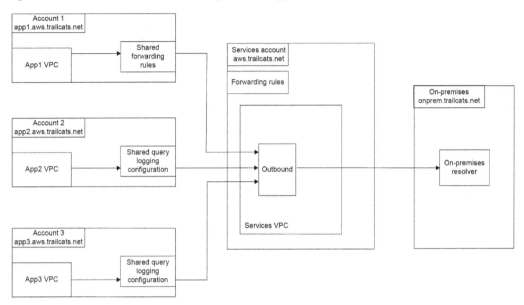

Figure 8.22: Sharing forwarding rules across accounts

In *Figure 8.22*, the forwarding rules configuration in the central services account is shared with the hub accounts. This ensures forwarding consistency across the architecture.

VPC to Private Hosted Zone Association in Another Account

The way Route 53 is architected, Route 53 **PHZs** are attached to VPCs. For example, if resources in app1.trailcats.net need to access resources in app2.trailcats.net, then you would associate app2.trailcats.net with the VPCs for app1.trailcats.net. There is a small wrinkle, however. Sharing PHZs between accounts is not supported through AWS RAM. However, PHZ association can be performed across accounts through the CLI or API.

> **Note**
>
> If you encounter a scenario that involves cross-account PHZ access, know that sharing cannot be done through AWS RAM, but it can be done through the CLI.

Creating a cross-account VPC to PHZ association is done through the CLI as follows:

1. In the account with the PHZ, create a VPC association authorization for the target VPC using the `create-vpc-association-authorization` argument.

2. In the account with the VPC, associate the VPC with the hosted zone using the `associate-vpn-with-hosted-zone` argument.

The full AWS CLI commands for the preceding need some additional parameters such as the hosted zone, the AWS region, and the VPC. The following shows what the commands look like with all the required parameters added:

- **Step 1**:

```
aws route53 create-vpc-association-authorization --hosted-zone-id
"Z02108366LMIYU328UF0" --vpc VPCRegion=us-east-1,VPCId=vpc-
06ac71e6c78fba7f0
```

- **Step 2**:

```
aws route53 associate-vpc-with-hosted-zone --hosted-zone-id
"Z02108366LMIYU328UF0" --vpc VPCRegion=us-east-1,VPCId=vpc-
06ac71e6c78fba7f0 --region us-east-1
```

Placement of Resolver Endpoints in a Shared Services Account

To scale unified DNS resolution across a complex architecture, centralized hub-and-spoke topologies are helpful for managing complexity and cost. Resolver endpoint placement in a central services account will simplify the design and eliminate the costs involved with having dedicated resolver endpoints in every account. A combination of shared forwarding rules and cross-account VPC associations facilitates the delivery of the query traffic to the central services account where the endpoints are deployed.

Figure 8.23 depicts how resolver endpoints may be placed in a central location for simplicity and cost containment:

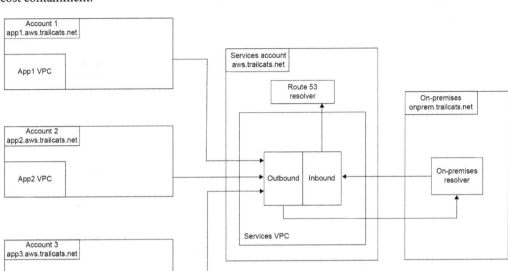

Figure 8.23: Resolver endpoint placement in central services VPC

In *Figure 8.23*, the Route 53 Resolver endpoints are placed in a central services account. This saves on cost and reduces complexity.

DNS Name Structure

To architect a unified DNS structure with PHZs and on-premises networks, it is essential to have a naming structure that aligns with DNS forwarding topologies. This simplifies configuration, facilitates troubleshooting, and reduces the possibility of errors.

Take `trailcats.net` as an example. This might be the public-facing domain, `corp.trailcats.net` might be the on-premises domain, and `aws.trailcats.net` might be the base AWS domain name, hosted in the shared services VPC. Each private application hosted could have its own domain name, such as `billing.aws.trailcats.net` and `inventory.aws.trailcats.net`. The on-premises DNS servers would be configured to send all `aws.trailcats.net` queries to the central services VPC inbound endpoint via forwarders. The spoke VPCs in AWS would forward `corp.trailcats.net` to the shared services VPC via shared forwarding rules. Finally, application-to-application (spoke-to-spoke) resolution can be done by sharing the PHZs between accounts and directly attaching them to the VPCs.

Figure 8.24 depicts a DNS name structure that aligns with the topology of the infrastructure:

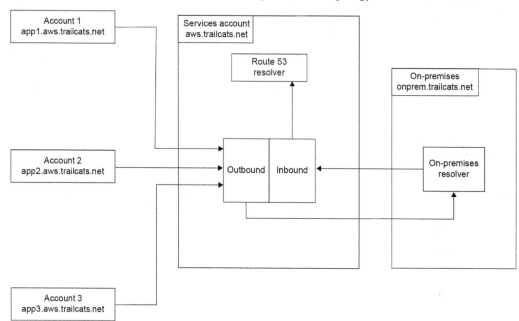

Figure 8.24: DNS naming structure in hybrid cloud design

In *Figure 8.24,* you can see how the DNS name structure of `aws.trailcats.net` is used for AWS resources, and `onprem.trailcats.net` is used for on-premises resources. This aligns with the DNS forwarding topology.

DNSSEC

DNS was invented in the 1980s for **Advanced Research Projects Agency Network (ARPANET)** and predates the internet. As ARPANET was a research network, security and hardening weren't concerns, which was also reflected in the design of other early protocols. Because of this, DNS is vulnerable to various attacks where an attacker can send the victim the IP address of a malicious website without the victim having any way to verify the information.

DNSSEC addresses this problem by providing authentication of the DNS server and maintaining the integrity of the records. This is done by using a public key infrastructure on top of DNS, where each level of the DNS hierarchy validates the level below it, starting at the root.

In order to make sense of how DNSSEC functions, you will learn about the underlying cryptographic concepts:

- **Public key cryptography**: Asymmetric encryption using public-private key pairs

- **Cryptographic hashes**: Fixed-length output from variable input

- **Digital signatures**: Verifiable proof of message authenticity

- **Chain of trust**: Hierarchical verification of trusted entities

With this foundation in place, you will be equipped to learn about how DNSSEC works, which is broken down into the following sub-topics:

- **DNSSEC records**: DNS records containing cryptographic keys that secure DNS responses

- **DNSSEC initialization**: Setting up the cryptographic keys and records for DNSSEC

- **DNSSEC record validation**: Process of verifying DNS record authenticity and integrity

- **Root signing ceremony**: A formal procedure to secure the root DNS keys

After working your way through this information, you will have a good understanding of DNSSEC. You will be well-prepared for DNSSEC-related questions in the exam, and to implement and support DNSSEC on AWS. Next up, you will dive into public key cryptography.

> **Note**
>
> For DNSSEC to make any sense, it's essential to know a few basic concepts about cryptography. Feel free to skip this section if you're comfortable with concepts such as public key cryptography, cryptographic hashes, and digital signatures. If it turns out you don't remember as much as you thought you did, you can always come back and read this later.

Public Key Cryptography

The primary purposes of public key cryptography are to provide authentication, confidentiality, integrity, and nonrepudiation. With DNSSEC, the primary properties of interest are authentication and integrity. The goal is to prove that resource records are unmodified and come from an authenticated source.

Public key cryptography is synonymous with asymmetric encryption. In asymmetric encryption, there are two related keys called a key pair. One is the private key, which is kept secret by the owner of the key pair; the other is the public key, which is shared with the world. Using mathematical algorithms, the key pairs have a special property where one can undo the operations of the other. For example, if some text is encrypted with the private key, it can be decrypted by the public key. The same principle works in reverse. If something is encrypted with your public key, only you, the private key holder, can decrypt it and read the message.

In *Figure 8.25*, the same cipher text is encrypted and decrypted twice. The difference is which key is for encrypting and which key is for decrypting. A public key is openly shared, and the private key is closely held by its owner.

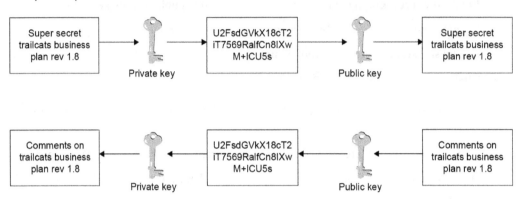

Figure 8.25: Key pair operations

In *Figure 8.25*, there are two examples of encryption and decryption showing that encryption/decryption works in both directions. In the first example, the text is encrypted with the private key and then decrypted with the public key. In the second example, the text is encrypted with the public key and decrypted with the private key.

Cryptographic Hashes

The purpose of a cryptographic hash is to prove message integrity, that is, to provide a guarantee that the message has not been altered. Here's how that works. A cryptographic hashing algorithm takes a block of text of arbitrary length as an input and then outputs a fixed-length value called a digest or a hash. Changing so much as a single bit in the input would result in a completely different hash value. This is commonly used to verify that data hasn't been corrupted at rest or in transit. This is very easy to test for yourself. Here are some example commands that work on Linux and Windows:

Linux:

```
echo "hello world" | md5sum
6f5902ac237024bdd0c176cb93063dc4
```

Windows:

```
echo "hello world" | md5
6f5902ac237024bdd0c176cb93063dc4
```

Figure 8.26 shows three blocks of text processed through the **Message Digest 5 (MD5)** hash algorithm:

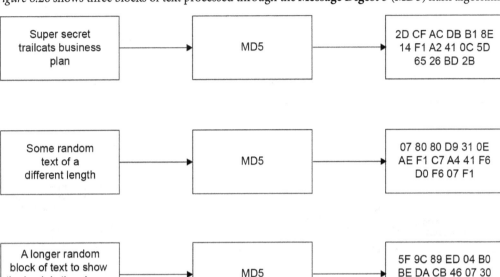

Figure 8.26: MD5 hash examples

Figure 8.26 depicts three different snippets of text, each of a different length, being fed into the MD5 algorithm. The length of the hash is always the same, regardless of the length of the input.

Digital Signatures

A digital signature is a hash of a message encrypted with a key from an asymmetric key pair (usually the private key). The idea here is that you can authenticate whom the message came from and that the message has not been altered. Here's how that works:

1.   The sender hashes the message.

2.   The sender encrypts the hash (normally with the private key). This is the signature.

3.   The sender provides the message, the hash, and the hashing algorithm.

4.   The receiver uses the public key to decrypt the hash.

5.   The receiver computes the hash of the message and compares it to the decrypted hash.

6.   If the hashes match, the message is valid.

Figure 8.27 illustrates the creation and validation of a digital signature:

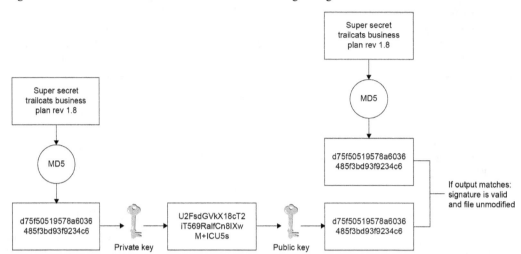

Figure 8.27: Digital signatures

In *Figure 8.27*, a block of text is hashed with MD5, then the hash is encrypted with a private key. The resulting ciphertext is decrypted by the recipient with the corresponding public key, revealing the hash. The recipient then hashes the original block of text and compares the computed hash with the decrypted hash. If the hashes match, the text was received unaltered.

Chain of Trust

A chain of trust is a hierarchical framework that authenticates using digital signatures, starting from a trusted root authority and progressing through one or more authorities to the end user. Each link in the chain is signed by the preceding authority, ensuring the validity and integrity of the links below it. This structure ensures that users can trust the identities and data encrypted or signed by these authorities.

Imagine that you just got a college degree and you're applying for a job. Why does that college degree have meaning to someone else? It is because the college is accredited, the teachers who work for the college are accredited, and accredited people develop the course material. The chain of trust starts at the top and flows down. Someone looking at your degree trusts what that certificate represents based on each link in the chain.

Figure 8.28 shows a university system as an analogy for the chain of trust concept. The accrediting body is the root of trust upon which everything else rests:

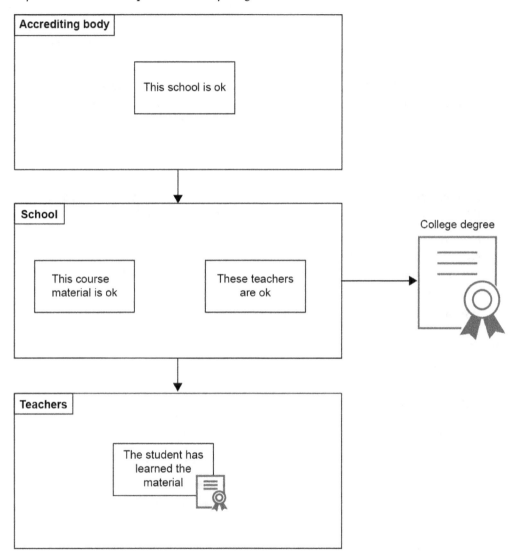

Figure 8.28: Chain of trust in education

Figure 8.28 depicts a chain of trust for a college degree. There is a root authority that grants authority to a sub-authority, which can grant to an authority below that, and so on. Validating a chain of trust involves starting at the leaf node and recursively working back toward the root, ensuring that all the digital signatures and hashes in the chain are validated.

DNSSEC Resource Records

DNSSEC has four types of records. RRSIG records contain digital signatures of the resource records, a DNSKEY record with the **zone signing key** (**ZSK**), a DNSKEY record with the **key signing key** (**KSK**), and a **delegation signer** (**DS**) record, which contains a hash of the KSK in the parent domain. The following list provides a definition of these records:

- **RRSIG record**: An RRSIG record is a digital signature of a **resource record set** (**RRset**). An RRSET is simply all of the records with the same name and type, for example, all of the MX records for a domain, or all of the A records for `www.trailcats.net`. The reason for doing this is so that signatures don't have to be computed for every single record every time there is a key rotation.

- **ZSK**: The ZSK is used to sign and validate RRSIG records. A hash is created of the RRSET, then encrypted by the private key in the ZSK pair. The public key is published in the zone in a DNSKEY record and is used by resolvers to verify the integrity of the RRSET.

- **KSK**: The KSK is used to sign the ZSK. Like the ZSK, the public key is stored in a DNSKEY record. You are probably wondering why there are two keys. The reason is the DS record. The idea here is to rotate the ZSK as frequently as desired without needing to update the DS record in the parent zone.

- **DS**: The DS record is the glue record for the trust chain in DNSSEC. It contains a hash of the KSK for the child domain. When a resolver performs chaining validation, it computes a hash of the KSK, retrieves the DS record from the parent domain, and checks to make sure the hashes match.

DNSSEC Initialization

Getting DNSSEC up and running for a domain requires the following:

1. Create the KSK and the ZSK key pairs for the zone and store them in DNSKEY records.

2. To set up the link in the chain of trust, the DNS provider will initiate a request to the upstream domain registry to create a DS record containing a hash of the KSK.

3. When the registry creates the DS record, DNSSEC is initialized and ready for operation.

4. Generate RRSIG records for each RRSET in the domain.

Figure 8.29 depicts DNSSEC in its fully initialized state.

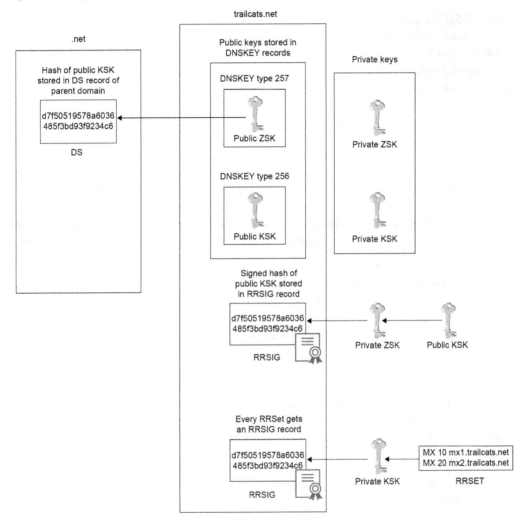

Figure 8.29: DNSSEC initialized state

In *Figure 8.29*, all of the elements needed for DNSSEC are in place. The upstream DNS provider has a DS record to validate the ZSK. There is a KSK to sign the RRSIG records. There is an RRSIG record for each DNS record set.

DNSSEC Validation

When a DNSSEC-capable resolver performs a lookup on a client's behalf, it will indicate that capability in the extended DNS header. The authoritative DNS server responds with the requested record and the RRSIG record. The resolver then requests the DNSKEY records for the ZSK and the KSK. If the hashes are validated, the resolver will request the DS record from the parent domain to verify the KSK hash. This process iterates recursively to the root domain.

Root Signing Ceremony

At this point, there's an obvious question you're probably asking yourself: How do I trust the root? What makes the root trustworthy is the security of its private KSK, which is held in a **hardware security module (HSM)** in two highly secured locations. The **domain signing key (DSK)** for the root domain is rotated every 3 to 4 months in an elaborate procedure called the **root signing ceremony**. In this ceremony, the DSK signing request is signed by the private KSK for the root domain. It's beyond the scope of this book to discuss it but it is quite fascinating. Details can be found at `https://www.iana.org/dnssec`.

Now that you understand how DNSSEC works, in the following exercise, you have the opportunity to get hands-on with setting up DNSSEC on Route 53.

Exercise 8.3: Set Up DNSSEC on Route 53

As DNSSEC is complicated, self-deploying has a high level of difficulty. A benefit of hosting with Route 53 is that the service handles all the complexity on your behalf. Because Route 53 handles all of the underlying details, this is one of those rare situations where increasing security is easy. To perform this exercise, you will need to have a domain registered with Route 53 and a corresponding public hosted zone. If you decide to step through the exercise, be sure to delete any resources you've created, as they will result in recurring charges:

1. In the `Route 53` console, navigate to `Hosted zones` and select your public hosted zone, as shown in *Figure 8.30*:

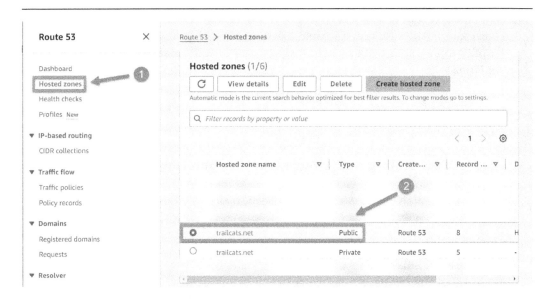

Figure 8.30: Selecting a public hosted zone

2. Select DNSSEC signing and enable it, as shown in *Figure 8.31*:

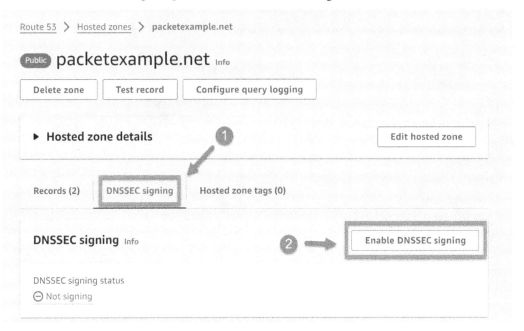

Figure 8.31: Enable DNSSEC signing

3. Use exampleKSK for the KSK name. Select Create customer managed CMK and name it exampleKSK. Then, select Create KSK and enable signing, as shown in *Figure 8.32*:

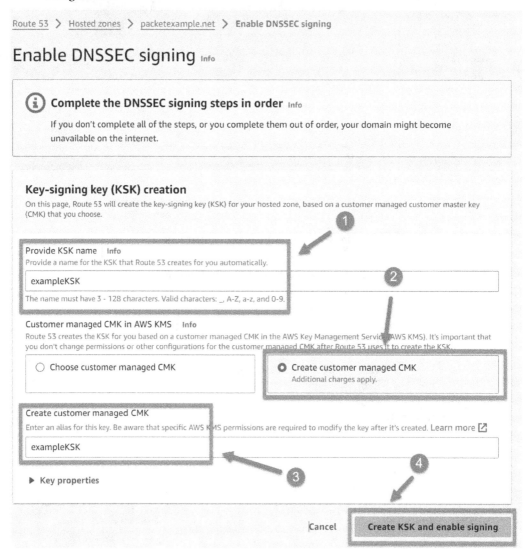

Figure 8.32: Create KSK

The remaining steps in the exercise will only work if the domain is registered with the Route 53 registrar. If you're using a domain that is registered somewhere else for the exercise, these steps should be skipped.

Now that the zone is signed, a DS record needs to be created:

1. On the DNSSEC signing tab of the public hosted zone, select View information to create DS record. Also, note the newly created KSK under Key-signing keys, as shown in *Figure 8.33*:

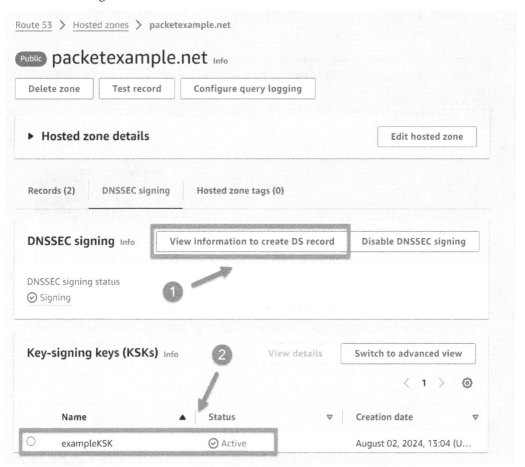

Figure 8.33: View information to create DS record

2. Select Establish a chain of trust, choose Route 53 registrar, and copy the public key, as shown in *Figure 8.34*:

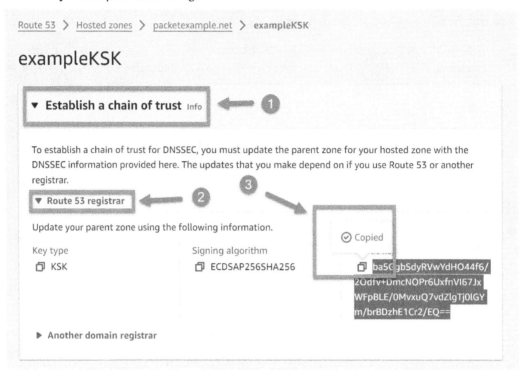

Figure 8.34: Copying the public KSK

3. In the Route 53 console, navigate to Domains and select Registered domains. Choose the domain for the PHZ that was configured in *step 1*.

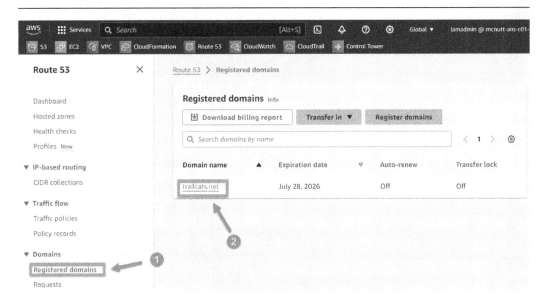

Figure 8.35: Selecting a registered domain

4. Select the DNSSEC keys tab and click Add, as shown in *Figure 8.36*:

Figure 8.36: Adding DNSSEC key to registered domain

5. Make sure the key type is 257 - KSK, paste the public key into the text box, and select Add, as shown in *Figure 8.37*:

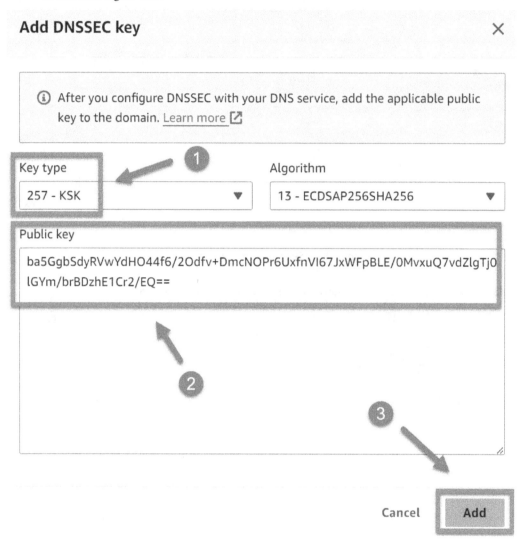

Figure 8.37: Adding a DNSSEC key

If Route 53 is your registrar, a DS record request will be generated and sent to the **top-level domain** (TLD) registry on your behalf.

6. Verify the creation of the DS record. It may take some time before the DS record is provisioned in the TLD name servers. Use the domain from *step 1*. In this example, ietf.org is used:

```
> dig ietf.org. ds +short +nocrypto
2371 13 2 [omitted]

> dig ietf.org. dnskey +short +nocrypto
256 3 13 [key id = 34505]
257 3 13 [key id = 2371]
```

In this exercise, you configured DNSSEC for a Route 53 hosted domain. You enabled DNSSEC, generated keys, and generated a DS record with a TLD registry. This hands-on experience will help you pass the exam.

Summary

For both the exam and your career as an AWS networking specialist, you should be able to design DNS solutions in public, private, and hybrid systems. In this chapter, you learned about the advanced DNS features and DNSSEC in Amazon Route 53. You learned about conditional forwarding, which uses inbound and outbound endpoints to integrate Route 53 with external DNS services. You learned about Route 53 routing policies, which provide a variety of methods to route traffic to your resources in AWS. You learned about Route 53 logging and monitoring, which is important for monitoring performance, troubleshooting, and security. You learned how to use cross-account sharing of Route 53 services and configurations to facilitate sophisticated multi-account architectures. And you learned about DNSSEC, and why it is an important element in protecting your DNS infrastructure. You have gained valuable skills for architecting DNS using Route 53, which will prepare you both for the real world and the exam.

In the next chapter, you will learn about elastic load balancing.

Exam Readiness Drill – Chapter Review Questions

Apart from mastering key concepts, strong test-taking skills under time pressure are essential for acing your certification exam. That's why developing these abilities early in your learning journey is critical.

Exam readiness drills, using the free online practice resources provided with this book, help you progressively improve your time management and test-taking skills while reinforcing the key concepts you've learned.

HOW TO GET STARTED

- Open the link or scan the QR code at the bottom of this page

- If you have unlocked the practice resources already, log in to your registered account. If you haven't, follow the instructions in *Chapter 16* and come back to this page.

- Once you log in, click the START button to start a quiz

- We recommend attempting a quiz multiple times till you're able to answer most of the questions correctly and well within the time limit.

- You can use the following practice template to help you plan your attempts:

Working On Accuracy		
Attempt	Target	Time Limit
Attempt 1	40% or more	Till the timer runs out
Attempt 2	60% or more	Till the timer runs out
Attempt 3	75% or more	Till the timer runs out
Working On Timing		
Attempt 4	75% or more	1 minute before time limit
Attempt 5	75% or more	2 minutes before time limit
Attempt 6	75% or more	3 minutes before time limit

The above drill is just an example. Design your drills based on your own goals and make the most out of the online quizzes accompanying this book.

First time accessing the online resources? 🔒

You'll need to unlock them through a one-time process. **Head to** *Chapter 16* **for instructions.**

Open Quiz	
https://packt.link/ansc01ch8	
OR scan this QR code →	

9

AWS Elastic Load Balancing

Load balancing is a technique for evenly distributing traffic across multiple servers that support an application. Modern applications need to work from any location, potentially serving traffic to millions of users concurrently. To manage this, most apps use several servers with the same data on them placed in strategic locations. For example, three servers in North America, three servers in Europe, and three servers in Asia.

AWS adds the concept of elasticity to this, meaning that the number of servers can dynamically grow or shrink to provide enough capacity to meet demand without overspending. In this chapter, you will learn about the fundamentals of load balancing and how AWS implements load balancing, focusing on three key topics:

- Load balancing fundamentals

- Load balancing in AWS

- AWS **load balancer** (**LB**) types

This chapter aligns with the exam domains of **Elastic Load Balancing** (**ELB**); specifically, that of load balancing fundamentals and AWS-specific LBs, such as the **Application Load Balancer** (**ALB**), **Network Load Balancer** (**NLB**), and **Gateway Load Balancer** (**GLB**). Learning about these topics will prepare you both for the exam and for the real-life design and deployment of scalable application architectures that leverage AWS ELB.

Load Balancing Fundamentals

Understanding load balancing is easier in the context of an example using our fictional enterprise, Trailcats. In the early days of `trailcats.net`, the site ran on a small EC2 instance, and performance was fine. Users were happy and recommended the site to their other hiking, cat-loving friends. Before long, trailcats.net had to be rehosted on a larger server; then again, and again. Eventually, Trailcats was on the largest EC2 instance Amazon offers, and demand was still growing.

The obvious solution was to have more than one server, but questions remained about how to spread traffic between the servers. DNS appeared to be a good solution, and that worked well initially. However, the site kept growing, and soon there were a dozen servers with even more needed.

In doing some research, the developers found out about LBs. They discovered LBs can spread the traffic across thousands of servers. Instead of large, expensive instances (scale up), LBs allowed them to use lots of small, low-cost instances (scale out) and reduce the number of servers needed during lower traffic periods (scale in). In addition to supporting scaling server resources out and in (elasticity), LBs can facilitate other useful things, such as taking unhealthy servers out of service gracefully and routing important users to dedicated servers.

The story of the `trailcats.net` scaling evolution, beginning with larger servers to the **Domain Name System** (**DNS**) to LBs, is typical in the lifecycle of an application. One of the benefits of using a cloud services provider such as AWS is that it is easy to start with LBs as part of the application design and save the trouble of rearchitecting the deployment as demand grows.

Put in more academic terms, LBs aim to improve the availability, performance, and scalability of applications. LBs accomplish this by routing incoming client requests to a list of healthy targets. When a target becomes unhealthy or unavailable, the LB will direct traffic to the remaining healthy targets. The details of how this happens depend on the type of LB being employed.

Types of LBs

LBs come in two broad categories: **NLBs**, which operate at the network and session layers, and **ALBs**, which operate at the application layer. Among NLBs, there is a special subtype provided by cloud service providers (such as AWS) that facilitates the inspection of traffic by security appliances called **GLBs**.

NLBs

NLBs operate at the network and transport layers (layer 3 and layer 4) of the **Open Systems Interconnection** (**OSI**) model. In most on-premises designs, the NLB presents a single frontend IP address representing the targets behind it. This is often called a **virtual IP** (**VIP**) address. The LB typically identifies traffic flows via a hash of the 5-tuple. This means taking the source IP address, source port, destination IP address, destination port, and protocol (i.e., **Transmission Control Protocol** (**TCP**) or **User Datagram Protocol** (**UDP**), and combining them to create a unique fixed-length value called a hash. When the load-balancing algorithm chooses a target, all traffic that matches the hash will be sent to the same target. This is efficient and fast.

Figure 9.1 depicts an architecture that uses an NLB:

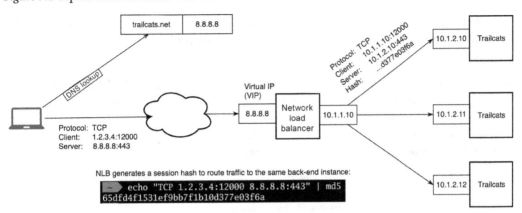

Figure 9.1: NLB operation

In this example, the 5-tuple of protocol, source address/port, and destination address/port is passed into the **Message Digest 5 (MD5)** hashing algorithm. The resulting hash is then assigned to a server in the pool. All traffic that matches the hash will be routed to the same server. The hash mathematically guarantees all traffic in the same flow goes to the same server.

During Trailcats' annual mega sale, an NLB would provide high performance by using an easily computed hash to distribute incoming customer traffic across multiple servers. Load balancing at the network layer not only preserves end-to-end encryption for secure transactions but also ensures broad compatibility with various applications, making it easy for the system to handle diverse non-web protocols. The LB's simplicity allows for the seamless insertion of third-party **network virtual appliances (NVAs)** such as firewalls when needed, ensuring that the entire shopping experience remains smooth, reliable, and secure, even under heavy demand.

A drawback of using NLBs is their limited ability to evaluate application layer information, which means they can only route traffic based on network and transport layer details such as IP addresses and ports. This limitation can lead to less optimal routing decisions, as the LB cannot consider the content or context of the data being processed. Additionally, the health checks performed by NLBs are typically basic, such as simple liveness checks using **Internet Control Message Protocol (ICMP)** or TCP connect tests, which might not fully assess the actual health or performance of the application, potentially leading to missed problems.

ALBs

ALBs operate at the session, presentation, and application (layers 5-7) layers of the OSI model. Being application-aware, they can route traffic using conditions such as URI paths, HTTP headers, and query parameters.

ALBs differ significantly from NLBs in their architecture in that they essentially act as reverse proxies. This means that ALBs terminate the connection on both sides rather than an end-to-end connection between the client and the server, potentially rewriting or adding information. This has implications from an information security standpoint. For this reason, an ALB would not be appropriate if there is a compliance requirement for end-to-end encryption.

Figure 9.2 depicts an architecture that uses an ALB:

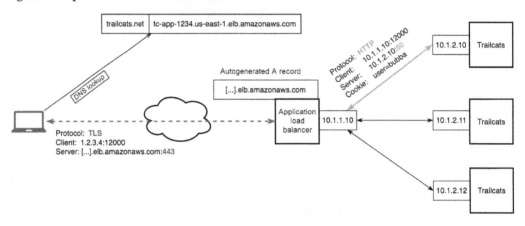

Figure 9.2: ALB operation

In this example, an ALB terminates the **Transport Layer Security** (TLS) session from the client and forms an HTTP connection with the server. The LB reads the user attribute from the HTTP header to route the connection to the correct server. Allowing the LB to handle the transport encryption with the client improves web server performance.

Having access to session information and payloads gives LBs tremendous power and flexibility. For this reason, modern web applications are often architected with this in mind, leveraging HTTP headers to signal control information that the LB can interpret. For example, they can route incoming requests to servers performing different functions, acting not only as a proxy but as an application-level router, getting the request to the appropriate component in the application.

Figure 9.3 depicts an architecture where an ALB is used as an application router:

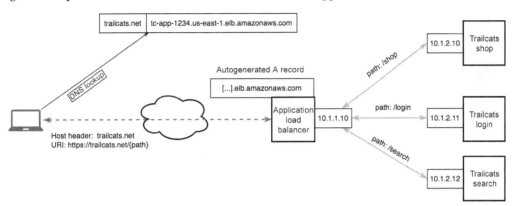

Figure 9.3: ALB path-based routing example

In this example, an application is divided into microservices. There is a shop, a login, and a search service. The LB evaluates the URL and routes traffic to the correct service based on the path.

For Trailcats, application load balancing offers significant advantages by providing flexible request handling, allowing the system to route customer traffic based on detailed application layer information such as URLs, cookies, and headers. This adaptability ensures that users are always directed to the optimal server based on their unique needs. Additionally, TLS termination enhances performance by offloading the decryption process from the servers, freeing up resources for faster processing. The LB can also manipulate headers, enabling custom routing and security policies tailored to Trailcats' unique requirements. Furthermore, application layer health checks go beyond basic liveness tests, providing a deeper assessment of server health, which ensures that only fully functional servers handle customer requests, resulting in a more reliable and responsive shopping experience.

While application load balancing offers advanced features, it also introduces several drawbacks. The complexity of managing and configuring ALBs can be challenging, especially as the system grows. ALBs are limited to a few protocols, which might restrict flexibility when integrating with other systems. The process of TLS termination, while beneficial for performance, breaks end-to-end encryption, potentially exposing sensitive data during transmission. Moreover, the loss of the client's IP header can complicate tasks such as logging, analytics, and security. Lastly, compared to NLBs, ALBs typically offer lower performance due to the additional processing required for inspecting and routing based on application layer data, which can lead to higher operating costs by requiring more compute resources.

GLBs

GLBs offer significant advantages in network architecture, particularly with traffic management and security. One of the key benefits is the simplified insertion of traffic inspection appliances, allowing for easy integration of security measures such as firewalls, intrusion detection systems, or other inspection tools without disrupting the network flow.

Additionally, GLBs preserve the IP header of packets, ensuring that the original source and destination addresses remain intact. This is crucial for maintaining transparency in traffic routing, enabling better tracking and monitoring without altering packet-level data. Historically, *service insertion* (a catchall term for placing traffic inspection appliances in the packet stream) has been a complex and complicated process. In offering GLBs, service providers such as AWS have abstracted away the complexity and made service insertion dramatically easier to implement.

Although considered to be NLBs since they route traffic at the network layer, GLBs have two unique properties that differentiate them. First, GLBs can only use NVAs as targets. Second, traffic between the GLB and the targets is encapsulated to preserve the header information and make the NVAs a transparent "bump in the wire."

> **Note**
> The NVA must support the overlay protocol for connection to the GLB (**Generic Network Virtualization Encapsulation (GENEVE)** in AWS).

Load Balancing in AWS

AWS ELB offers four flavors of LBs: an NLB, an ALB, a GLB, and a **Classic Load Balancer (CLB)**. Regardless of which LB you choose to implement, they all share the same basic set of constructs. The benefit is that once you understand these constructs, working with different LBs is consistent. Like most Amazon services, they are regional in their configuration scope, which is an important consideration in a multi-region deployment. In the next section, you will learn about these constructs and the features common to all LB types.

ELB Components

Elastic LBs have several key components that work together to manage and distribute incoming traffic. **Listeners** are the first point of contact for incoming requests. They are configured to check for specific ports and protocols (such as HTTP, HTTPS, TCP, etc.) and determine how the LB should handle the traffic.

Routing rules then come into play, defining how the incoming traffic should be directed based on conditions such as URL paths, host headers, or other factors. The routing rules ensure that the traffic is forwarded to the correct target groups. **Target groups** are sets of destinations, such as EC2 instances, containers, or IP addresses, where the requests are ultimately sent that manage the health checks and ensure that only healthy targets receive traffic. Together, listeners, routing rules, and target groups provide a flexible, efficient, and customizable system for managing and distributing traffic in a cloud environment.

ELB Common Features

Certain features are shared between all types of LBs in AWS and should be discussed together rather than be restated in each LB's section.

ALBs, NLBs, and GLBs have an attribute called **deletion protection**. Its purpose is to protect the LB from being inadvertently deleted. Deletion protection must be disabled to delete the LB, adding an extra measure of safety to infrastructure components.

Balancing allows traffic to be distributed evenly across all registered targets in all enabled availability zones, rather than being limited to targets in the same zone as the incoming request. This improves load distribution, enhances fault tolerance, and ensures that no single zone is overwhelmed with traffic.

When a target becomes unhealthy or is deregistered, connection draining allows existing connections to be gracefully completed within a specified timeout, ensuring that active requests are not abruptly terminated. The amount of time an LB will wait for connections to finish is called the **deregistration delay**.

Deregistration delay waits a configurable amount of time (default of 300 seconds) for existing connections to complete before deregistering the target.

These are the common components you will encounter regardless of the LB, which is important to know both in real life and for the exam. In the next section, you will learn about security policies, which listeners use to set TLS parameters.

Security Policies

AWS TLS security policies define the protocols and ciphers that ELB uses to encrypt traffic between clients and the LB, which is used by listeners for SSL/TLS connections. The policy defines the list of protocols (i.e., TLS version) and cipher suites the listener will accept in a negotiation with a client. At the start of a connection, the client sends the listener a list of supported suites, ordered from strongest to weakest. The listener compares the lists and picks the first match in the security policy.

The idea behind security policies is to give you the power to decide on the trade-offs you want to make between security and compatibility. From a security standpoint, this is important and deserving of attention in an actual deployment.

At a high level, TLS is designed to provide confidentiality, integrity, and authenticity. TLS lays out the rules, methods, order of operations, packet formats, and so on. Cryptographic security cipher suites do the heavy lifting of protecting your data via data encryption. A cipher suite refers to the common security algorithms used to encrypt and decrypt data; the cipher suites must be identical on both ends of a connection for encryption and decryption to work. TLS comes in four flavors – TLS 1.0, 1.1, 1.2, and 1.3. TLS 1.0 and 1.1 are deprecated. However, you may need to use them for compatibility reasons, so AWS allows you to include them in a security policy.

TLS Cipher Suites

The **cipher suites** consist of three main elements. The encryption algorithm, which provides confidentiality, will normally be the **Advanced Encryption Standard** (**AES**) or **Elliptic Curve Diffie-Hellman** (**ECDHE**). Next, the authentication algorithm is typically **Galois/Counter Mode** (**GCM**) or **Hashed Message Authentication Code** (**HMAC**); the last element is the hashing algorithm, which provides integrity and is usually some **Secure Hash Algorithm** (**SHA**) variant.

The naming convention is to string the elements of the cipher suite together like this: AES128-GCM-SHA256. This would be AES, 128 bits in length, GCM for authenticity, and SHA, 256 bits in length for integrity.

The default AWS security policy at the time of this writing is TLS-13-1-2-2021-06. This policy allows TLS 1.2 and TLS 1.3 an encryption size of at least 128 bits with an SHA-2 hash size of at least 256 bits. The strongest policy is TLS-13-1-3-2021-06 and allows TLS 1.3 only, with a select group of three cipher suites: AES-128-GCM-SHA256, AES-256-GCM-384, and ChaCha20-Poly1305-SHA256.

Now, you know what security policies are and how to deduce their contents through the AWS naming convention. In the next section, you will learn about techniques to preserve client information as traffic passes through networks and ALBs.

Client Information Preservation

Because LBs change the source IP address and possibly the port, targets need an alternative way to get this information. Network-based load balancing uses the proxy protocol, which inserts connection information into the payload. For HTTP/HTTPS application load balancing, this is accomplished through HTTP headers. There are three commonly used headers: X-Forwarded-For, X-Forwarded-Proto, and X-Forwarded-Port.

- **X-Forwarded-For**: X-Forwarded-For is an HTTP header an ALB or CLB can insert to provide the client IP address information to the web server. This is important because the LB acts as a middleman, so from the server's perspective, the connection is with the LB. X-Forwarded-For enables a web application to know the IP address of the actual client making the request. Because the LB terminates the connection, information about the client would be otherwise lost when the LB sends the connection to the target; this header preserves client information that may be needed by the application.

- **X-Forwarded-Proto**: X-Forwarded-Proto is an HTTP header an ALB or CLB can insert to let the web server know if the client is connected with HTTP or HTTPS. This is important for security reasons but also to indicate whether the payload is encrypted.

- **X-Forwarded-Port**: X-Forwarded-Proto is an HTTP header an ALB can insert to let the web server know what port the client used to connect. This may be important for the target application to know.

Proxy Protocol

The **proxy protocol** technically operates on layer 4 and is used by NLBs to track connection information. The LB injects extra data called a preamble into the packet payload that contains the IP address and port information of the client side of the connection with the LB.

There are two benefits to using the proxy protocol instead of the X-Forwarded-For header: It works with multiple protocols (not just HTTP), and if you need to implement unbroken TLS encryption to the backend application, it can be used to preserve that end-to-end encryption. There are two proxy versions: version 1 which is human-readable, and version 2, which is binary.

The human-readable version 1 message looks something like this:

```
PROXY TCP4 192.168.0.1 192.168.0.11 56324 443\r\n
```

The CLB supports proxy version 1, while NLB uses proxy version 2 (the non-human-readable version).

Integration with Other AWS Services

ELB integrates with several AWS services to enhance the functionality and efficiency of your applications.

With Route 53, ELB integrates route traffic based on policies such as geolocation or latency, ensuring that requests are directed to the most optimal LB in a global infrastructure. **AWS Certificate Manager** (**ACM**) integrates with ELB to automatically provision and manage SSL/TLS certificates for your LBs, simplifying the process of securing communications with HTTPS. EC2 Auto Scaling works closely with ELB to dynamically add or remove instances from the LB's target group based on demand, ensuring consistent application performance during varying loads.

For containerized environments, ELB integrates with **Elastic Container Service** (**ECS**) and **Elastic Kubernetes Service** (**EKS**) to distribute incoming traffic across containers, automatically adjusting to the scale of containerized applications as they grow or shrink.

A **web application firewall** (**WAF**) can be directly attached to ELB, allowing you to set up rules to filter and monitor HTTP/HTTPS traffic flowing through the LB, providing an extra layer of security against web threats.

Finally, Global Accelerator integrates with ELB by routing traffic to the best-performing LB across different AWS regions, improving application availability and performance on a global scale.

This section touches on some of the technical details of these integrations that you will need to know for the exam and the real world:

- **Route 53**: LBs receive autogenerated DNS A records in Route 53. Normally, you create a CNAME record in your Route 53 hosted zone that points to the autogenerated record. For example, a CNAME of `app.trailcats.net` points to `trailcats-app-123456789.us-east-1.elb.amazonaws.com`. This is important so that a user can use a friendly, easy-to-remember DNS name of the LB instead of needing to use an IP address or the AWS autogenerated DNS name.

- **ACM**: ACM integrates at the listener level of the LB configuration to provide certificates for the LB. If you are hosting your domain with Route 53, wiring ACM up to generate a certificate is simplified, as many details are handled for you.

- **EC2 Auto Scaling**: When you attach an LB to an autoscaling group, EC2 will automatically register and deregister targets with the LB.

- **ECS**: AWS LBs are supported for both EC2 and Fargate instance types. ALBs, in particular, are a great use for ECS as you have a lot of flexibility in routing traffic to instances based on criteria such as path, headers, cookies, and so on.

- **EKS**: When deploying a Kubernetes ingress controller, by default, AWS automatically instantiates an ALB to direct external traffic into the cluster. It is also possible to use an NLB with EKS by using the AWS LB controller. This is instantiated by using the Kubernetes service type of load balancer.

- **WAF**: To integrate a WAF web **access control list** (**ACL**) with an ALB, you associate the web ACL with the LB on the `Integrations` tab of the load balancer. The LB can be configured to either fail close or fail open if the WAF becomes inaccessible. If it's configured to fail close, the LB will fail to allow new connections; if it's configured to fail open, the LB will accept new connections even without security in place. This option is called WAF fail open and is located on the `LB attributes` page.

- **Global Accelerator**: A Global Accelerator can be created and integrated at the time the LB is created or later added under the `integrations` tab of the LB. Global Accelerator uses anycast addresses to place resources closer to clients.

LBs, by their nature, are meant to integrate with other services to provide value; the goal of load balancing is to offer resilient scalable connectivity, and this is best achieved by integrating with AWS services that have the same goal. The next section discusses LB types.

AWS Load Balancer Types

AWS offers four types of LBs, each with distinct features tailored to specific use cases.

The **ALB** operates at the application layer (layer 7) and provides advanced routing capabilities, including path and host-based routing, making it ideal for microservices and modern web applications.

The **NLB** functions at the network and transport layers (layers 3 and 4) and is designed for high-performance, low-latency applications, capable of handling millions of requests per second while maintaining ultra-low latencies.

The GLB integrates with third-party virtual appliances, simplifying the deployment, scaling, and management of these appliances by enabling seamless insertion into your network traffic flow.

Finally, the CLB, which is now deprecated, provided basic load balancing across multiple EC2 instances, operating at both the application and transport layers, but has since been superseded by the more feature-rich ALB and NLB.

In this section, you'll learn about each of these LBs and their important configurables.

AWS ALB

ALBs support only HTTP and HTTPS. Listeners can have multiple rules that are processed by priority. Each ALB can have up to 100 routing rules spread across its listeners. Rules can have up to five conditions each, meaning that there is ample opportunity within a rule to be inclusive or exclusive depending on the intent of the rule. At the time of writing, there are six condition types:

- The **host header** condition in AWS ALB allows you to route traffic based on the domain name included in the HTTP host header of incoming requests. This is particularly useful when you host multiple domains or subdomains on the same LB and need to direct traffic to different target groups based on the specific domain requested by the client.

- The **path** condition enables routing based on the URL path of an incoming request. For example, requests with paths such as `/images/*` or `/api/*` can be directed to different target groups. This condition is ideal for structuring traffic in applications with different services or content types, ensuring that the appropriate backend resource handles each request.

- The **HTTP request method** condition allows routing decisions to be made based on the type of HTTP method used in the request, such as `GET`, `POST`, `PUT`, or `DELETE`. This condition is useful when you need to handle different types of operations separately, such as read versus write requests, directing them to different servers or target groups optimized for each operation.

- The **source IP** condition lets you route traffic based on the IP address or CIDR block of the client making the request. This is particularly useful for scenarios requiring geo-blocking and security controls or when routing traffic from specific IP ranges to dedicated resources for compliance or performance reasons.

- The **HTTP header** condition allows for routing based on specific headers included in the request, such as User-Agent, Accept-Language, or custom headers. This condition enables granular routing based on the client's browser, language preferences, or other contextual information, tailoring the response to the client's needs or the characteristics of the request.

- The **query string** condition enables routing based on key-value pairs found in the query string of the URL. For example, requests that include `?version=2` or `?user=admin` can be routed to different target groups. This is particularly useful for A/B testing, serving different versions of an application, or handling special cases based on URL parameters.

Encryption and ALBs

In most cases, you will want to use TLS, which requires installing a certificate on the ALB. The ALB integrates with the ACM, which in turn integrates with Route 53. This makes generating and installing certificates on your ALBs easy. Be aware that this does break end-to-end encryption. If end-to-end encryption is required, the NLB is the correct choice.

AWS NLB

The AWS NLB operates on layers 3 and 4 of the OSI interconnection model. Like the ALB, the AWS NLB also supports the concept of listeners. However, NLB listeners do not have rules, as they only match the protocol and port and do not perform any packet inspection. Due to this simple nature, NLBs have higher performance and broader application support than ALBs.

NLBs can be assigned static IP addresses, which is useful for situations where whitelisting by IP is required; for example, customers need to create a firewall rule to allow traffic from your service.

NLBs can also be used with PrivateLink, which allows you to place an endpoint in a customer's **virtual private cloud** (**VPC**) without a VPC peering. This is a simple way to provide services without concerns over IP address conflicts.

CLBs

The AWS CLB has the functionality of both a network and ALBs. It works with EC2-Classic networking and VPC networking. Compared to the v2 LBs, it has several limitations that affect its scalability and utility. You would only use it to support EC2-Classic networking, which is deprecated.

The CLB has so many limitations, though. There are no target groups and no routing rules, EC2 instances are the only supported target type, and the LB supports only one SSL certificate per listener. The last limitation is the most impactful, as you would need to spin up a separate LB for each required SSL certificate, which adds cost and complexity.

GLB

The purpose of the AWS GLB is to simplify the deployment, scaling, and management of third-party virtual appliances, such as firewalls, intrusion detection systems, and deep packet inspection tools, in your network. It allows you to seamlessly insert these appliances into your network traffic flow, acting as a transparent gateway that directs traffic through them while maintaining the original IP headers. This enables consistent security and traffic inspection across your network while simplifying the operational complexity of managing these appliances at scale.

Imagine that Trailcats has grown so that a **chief information security officer (CISO)** is brought on board to ensure that Trailcats and its users' data are protected from attackers. One of the things CISOs do is think of all the ways someone may try to attack an organization and come up with countermeasures to those attacks. In the case of Trailcats, the CISO has determined that all traffic between the Trailcats servers and the internet should pass through a **next-generation firewall (NGFW)** for inspection and control in case something were to go wrong.

The challenge is that stateful firewalls were not architected with elasticity in mind. When traffic enters a firewall, the firewall adds the connection between the endpoints to a state table. In this way, it is able to allow return traffic without needing to create explicit rules to do so by consulting the state table to determine what connection the return traffic is associated with. A side effect of this, however, is that the state of the connection is tied to that specific firewall instance. If the firewall were to fail, all the connections would be lost, which is disruptive.

Figure 9.4 depicts how a stateful firewall allows for the implicit allowing of the return traffic from a flow that was initiated in the outbound direction:

Firewall rule		
Source	Destination	Action
TCP/10.1.1.1/any	TCP/1.1.1.1/443	Allow

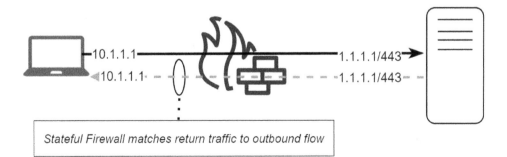

Stateful Firewall matches return traffic to outbound flow

Figure 9.4: Stateful firewall rules processing

As the session state is tracked as allowed traffic traverses the firewall, there is no need to create an explicit rule to allow the traffic that returns. This is automatically tracked in the firewall session table.

To provide high availability, firewalls have long had a concept called a **failover pair**, where a standby firewall receives all of the information about the connections on the active firewall, and should the primary firewall fail, the standby can step in without the connections being dropped. This is called **stateful failover**. A more elaborate but scalable technique called **clustering** can also be employed. In clustering, the firewalls share the traffic load, directing the traffic to a surviving member should a member fail or suddenly go offline.

The problem is that classic stateful failover and clustering both depend on layer 2 tricks to direct the traffic to the surviving members. The surviving firewalls spoof the Ethernet **media access control** (**MAC**) addresses of the failed firewall, tricking the upstream and downstream devices into directing traffic to them. As you likely know, the data link layer (layer 2) is not exposed to customers in public cloud environments, so these tricks do not work.

Figure 9.5 depicts a basic stateful failover design and illustrates why it does not work because MAC address-based forwarding doesn't exist in the cloud:

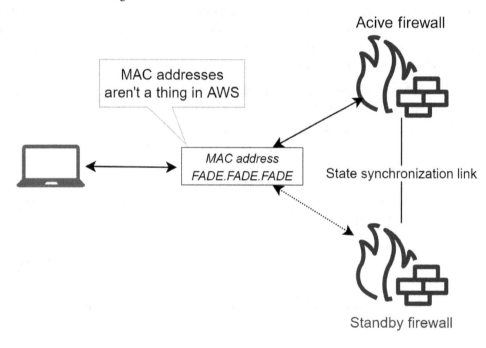

Figure 9.5: Why MAC-based stateful failover doesn't work in the cloud

Because MAC addresses are not tracked at the tenant/customer level, the firewalls a customer builds cannot use MAC-based forwarding. That data is simply not exposed to the firewall.

The most common method of getting around the problem of providing elasticity (scaling and resilience) for NGFWs in the cloud is a design sometimes referred to as an **LB sandwich**. In an LB sandwich, there is an outside LB and an inside LB with the firewalls placed in the middle. When the incoming traffic egresses the firewall, the firewall performs **source network address translation (SNAT)** on the traffic to ensure the return leg of the connection passes through the firewall where the connection was initiated. This design results in a lot of operational complexity and negates many of the benefits of using cloud service providers.

Figure 9.6 depicts a typical LB sandwich architecture:

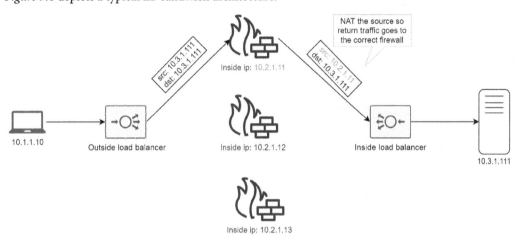

Figure 9.6: LB sandwich using SNAT

In the preceding figure, you may observe that when the packet from 10.1.1.10 leaves the firewall, the source address is changed to 10.2.1.11, the firewall's inside IP address. This ensures that the return traffic from the server comes back to the same firewall. When the firewall receives the return traffic, it will rewrite the packet's destination address to 10.1.1.10.

An obvious improvement would be to employ a cloud-native firewall architecture, and that is something of a work in progress. So far, cloud-service-provider-managed firewalls lack the features and efficacy of top-tier NGFWs. Recognizing this was causing customers a lot of pain, cloud providers came up with the concept of the GLB, which provides elasticity for their customers while still giving them the capability to run their NGFW of choice.

How AWS GLBs Work

The GLB lives in a dedicated security/service provider VPC. It connects to workload VPCs utilizing **Gateway Load Balancer endpoints (GLBe)** that are placed in a dedicated subnet of the workload VPC. IP routing gets traffic in and out of the GLB via the GLBe.

Figure 9.7 depicts a GLB traffic flow broken down step by step for flows initiated from a workload:

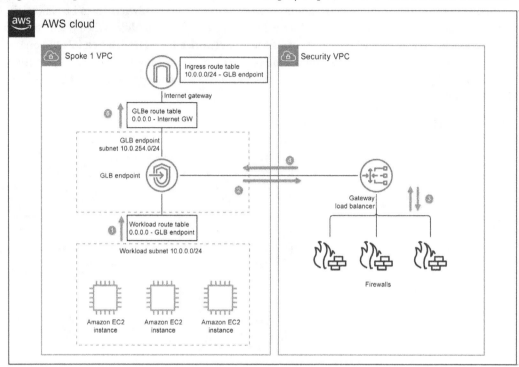

Figure 9.7: GLB *traffic flow in the egress direction*

The following steps illustrate what happens in *Figure 9.7*:

1. The routing table for the workload VPC points to the GLBe as the next hop for destination networks (normally, the default route, which would cause all egress traffic to be directed to the GLBe).

2. The GLB receives the traffic from the endpoint.

3. The GLB encapsulates the traffic in GENEVE packets, which preserves the original IP packet and header, and forwards the traffic to the NVAs. The GLB uses session stickiness to ensure that sessions are always routed to the same NVA instance (unless that instance were to fail, of course).

4. The NVAs process the traffic and return it to the GLB, which then forwards the packet to the originating GLBe.

5. The routing table for the subnet where the GLBe lives forwards the traffic to the internet gateway.

Like other AWS LBs, there is a listener bound to a target group. Health checks and cross-zone load balancing are supported. Traffic to the targets (the NVA instances) is encapsulated and sent to the targets on port UDP 6081.

Load Balancer Selection

With so many choices and so much configurability, selecting the best option for a given situation might seem less than intuitive. ALBs are the best choice most of the time.

Table 9.1 illustrates the requirements that would indicate the use of NLBs, GLBs, or CLBs:

Requirement	Application	Network	Gateway	Classic
Unbroken end-to-end encryption		x		
Performance		x		
Non-HTTP(S) protocols		x		
Static IP (used for whitelisting)		x		
PrivateLink		x		
Third-party security appliance			x	
Classic networking				x
If none of the above apply	X			

Table 9.1: LB selection

The preceding table matches the layer of networking features required against the list of available LBs. For the exam, it will be important to understand the capabilities and limitations of each type of LB to make a good choice when presented with a scenario that requires choosing the best one for a task.

Summary

In this chapter, you learned about load-balancing fundamentals and the principles of operation. Then, you learned about different types of LBs, including NLBs, ALBs, and GLBs, along with their benefits and drawbacks. You learned how AWS implements load balancing, which includes tight integration with core AWS services to facilitate ease of use. Lastly, you learned how to discern what LB to select for a given use case. After learning these topics, you are much better equipped to tackle designing and deploying architectures that utilize AWS ELB.

The next chapter changes the subject completely and covers security frameworks.Serei sil unit, Patus, C. Usque inatu maximur ehenteri publis, confit erbitus crevit ego mei sidiem denihil lessula vis condii parbi iame tem es cum nonsuni ium dum resendum ompro is? Gerfenestra rebes An no. Maricitrari iu se es fectus, caperte consuspio, esus enatienatua Senes! Nicae cre consin tem autermiustum perceris? Am sil consus, Cateatod Cato peri consilium med con se auctum oc, Catus fit. Marte auctandena, quodiondam, Ti. As ne clego vidius, nondact odius, te re patium aucepopopote ius cupio vid cre inatis horicavehem, co me iamquem ulocci publici intre consult odiconsim nos conlocae noniquam demusquam et factuam introx mus bons eo con tus ad acchum diusque eris, deliquosti, tum ne ad

Exam Readiness Drill – Chapter Review Questions

Apart from mastering key concepts, strong test-taking skills under time pressure are essential for acing your certification exam. That's why developing these abilities early in your learning journey is critical.

Exam readiness drills, using the free online practice resources provided with this book, help you progressively improve your time management and test-taking skills while reinforcing the key concepts you've learned.

HOW TO GET STARTED

- Open the link or scan the QR code at the bottom of this page

- If you have unlocked the practice resources already, log in to your registered account. If you haven't, follow the instructions in *Chapter 16* and come back to this page.

- Once you log in, click the START button to start a quiz

- We recommend attempting a quiz multiple times till you're able to answer most of the questions correctly and well within the time limit.

- You can use the following practice template to help you plan your attempts:

Working On Accuracy		
Attempt	Target	Time Limit
Attempt 1	40% or more	Till the timer runs out
Attempt 2	60% or more	Till the timer runs out
Attempt 3	75% or more	Till the timer runs out
Working On Timing		
Attempt 4	75% or more	1 minute before time limit
Attempt 5	75% or more	2 minutes before time limit
Attempt 6	75% or more	3 minutes before time limit

The above drill is just an example. Design your drills based on your own goals and make the most out of the online quizzes accompanying this book.

First time accessing the online resources? 🔒

You'll need to unlock them through a one-time process. **Head to** *Chapter 16* **for instructions**.

Open Quiz

https://packt.link/ansc01ch9

OR scan this QR code →

10
AWS CDN and Global Traffic Management

One of the most critical aspects of networks is not just connectivity, but the impact it can have on the end-user experience. If you have ever experienced a slow or intermittent internet connection, you can recall the frustration it causes. In the modern age, where network bandwidth is increasingly available, users have grown relatively impatient. For instance, if the e-commerce site for Trailcats fails to load the web store or a user's shopping cart within a few seconds, it could significantly reduce the likelihood of a successful transaction. This example highlights how the end-user experience can directly affect business outcomes. While a few lost sales might not seem critical, the consequences could be much more severe during high-traffic events such as a Black Friday sale.

This focus on enhancing the end-user experience has led to the widespread adoption of **content delivery networks** (**CDNs**). A CDN distributes and caches content as close to end users as possible, improving their interaction by reducing latency. By caching data in regional data centers worldwide, a CDN allows users to access content from the data center closest to them. AWS provides a robust CDN solution through two primary services: Amazon CloudFront and AWS Global Accelerator. This chapter will explore how these services can be utilized to improve application performance on a global scale.

In this chapter, you will cover two services that relate to the AWS CDN offerings:

- Amazon CloudFront
- AWS Global Accelerator

By the end of this chapter, you will be able to describe the core functionalities and benefits of Amazon CloudFront and AWS Global Accelerator. You will identify the appropriate use cases for Amazon CloudFront and AWS Global Accelerator in content distribution and global traffic management. You will learn how to apply Amazon CloudFront's caching and security configurations to optimize content delivery, protect distributed applications, and analyze the impact of using AWS Global Accelerator for improving network performance and availability for global applications. These objectives will guide your learning and ensure that you can effectively leverage AWS CDN services for optimal global application performance.

Amazon CloudFront

The primary service that enables the use of Amazon's CDN is **Amazon CloudFront**. In this section, we will introduce CloudFront and its key components, explore its caching behaviors, and discuss methods for securing CloudFront distributions and origins.

Amazon CloudFront is a **global** service that can enhance the performance and delivery of static and dynamic content of internet-based applications. This content can be things such as web pages (and their contents), images, or media files. When end users access this content, it can be served directly from a location that is closest to them.

For example, assume Trailcats has a static web page containing a large repository of cats modeling their merchandise, which they want to make available to end users. Amazon CloudFront can service this web page. Users downloading these files in the **United States** (**US**) would use a US-based AWS data center to access the content, while users in Australia would use an Australia-based one.

Understanding the orchestration of this content across all the regional AWS edge data centers is key for building content distribution on Amazon CloudFront. Next, you will look at some of the components of CloudFront.

Origins

The **origin** refers to the source of the content for a CloudFront distribution. It is the location from which CloudFront requests information when it initially receives inbound requests from remote users. Origins can be AWS resources or external servers that store or generate the content delivered by CloudFront.

Amazon CloudFront currently supports the following origin types:

- **Amazon S3 bucket**: A scalable object storage service used to store and retrieve any amount of data. When used as an origin, CloudFront can distribute content such as static web pages, images, and media files stored in an S3 bucket to users worldwide.

- **AWS Elemental MediaStore container or MediaPackage channel**: MediaStore is optimized for media streaming, providing low-latency performance for live video workflows. MediaPackage prepares and protects video for delivery. CloudFront uses these services as origins to efficiently stream live or on-demand video content to viewers.

- **Application Load Balancer (ALB)**: A load balancing service that distributes incoming application traffic across multiple targets such as EC2 instances, containers, and IP addresses. By setting an ALB as an origin, CloudFront can deliver dynamic web content and balance the load across multiple backend resources.

- **Lambda function (via Lambda@Edge, API Gateway, or direct URL)**: AWS Lambda allows you to run code without provisioning servers. CloudFront can invoke Lambda functions at the edge (Lambda@Edge) to customize content delivery. It can also use API Gateway endpoints or direct URLs to Lambda functions as origins for delivering dynamic content or APIs.

- **Custom origin (includes EC2 instances and/or web servers)**: Any HTTP server outside of AWS services, such as web servers running on EC2 instances or on-premises servers. CloudFront can use these custom origins to distribute content hosted on your own infrastructure to a global audience.

When CloudFront is configured to service a particular application, it forwards any requests for content that are not cached to the origin for retrieval. For example, if a viewer tries to load a web page that is part of a CloudFront distribution and the images or text on that page are not currently cached in CloudFront, an **origin fetch** occurs, and the request is forwarded to the origin for retrieval. You will read more about caching behaviors in the *Caching behaviors in Amazon CloudFront* section of this chapter.

Distributions

The configuration of a CloudFront **distribution** defines both the front-end configuration that is deployed via CloudFront and how the content is retrieved from its original location.

The front-end configuration of a CloudFront distribution determines several characteristics of the distribution. This includes details such as what **geographical locations** are used to service content, **access restrictions**, **security settings**, and **logging**.

When you configure a CloudFront distribution, AWS provides you with a default domain name to serve your application. This domain name follows a specific format: `<unique-id>.cloudfront.net`, where `<unique-id>` is a unique alphanumeric string assigned to your distribution. AWS automatically configures the **Domain Name System** (**DNS**) for this domain name to resolve to IP addresses that are advertised globally from all the geographical regions you have selected for your distribution. For example, a default domain name might look like `d1234567890abcdef.cloudfront.net`. You can also add an alternate domain name (CNAME) if you wish to use your own custom domain.

Since this domain name is used to serve as the front-end for the CloudFront distribution, any remote users or applications would need to reference the distribution domain name. For example, if Trailcats has a web application with an image located at `https://app.trailcats.net/images/picture.jpg`, once CloudFront serves the web app, the new location would be `https://d1234567890abcdef.cloudfront.net/images/picture.jpg`.

> **Note**
> Additional DNS configuration is required to associate a custom domain with a CloudFront distribution.

Behaviors

The traffic direction and caching related to a CloudFront distribution are controlled by something called a **behavior**. Each behavior for a distribution is associated with a specific origin. By default, each distribution will have a default behavior, labeled `default (*)`, and you can think of this as a catch-all policy for the CloudFront distribution. You can create more specific behaviors that will take precedence. This even allows you to separate distributions as **public** versus **private**, however, the public or private nature of a distribution is purely controlled by the behavior configured for the distribution.

For example, assume Trailcats has a CloudFront distribution for `store.trailcats.net`. They have a default behavior assigned to the distribution, which will handle all traffic for `store.trailcats.net/*`. This default behavior is configured to cache objects from an Amazon S3 bucket as the origin. On the website, there is a folder for a path of `/images/*` that Trailcats employees should be able to access. This can be serviced by a private behavior (separate origin) that requires further authentication through **signed URLs and cookies**.

Refer to *Figure 10.1* to see the separate behaviors that could be associated with public versus private distributions:

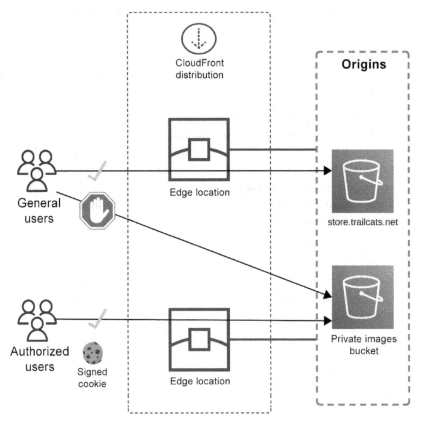

Figure 10.1: Public versus private distributions

In the preceding figure you can see that to keep content privately accessible, Amazon CloudFront allows you to use signed URLs or signed cookies. These mechanisms ensure that only authorized users can access restricted content. The key benefit of using **signed URLs or cookies** is the ability to tightly control access to content, such as preventing unauthorized downloads or sharing.

A **signer** is the entity responsible for creating and validating the signed URLs or cookies that grant access to the content. The signer can be your AWS account or a specific **trusted key group** associated with the CloudFront distribution. When a signer is configured for a CloudFront distribution, CloudFront will require viewers to present valid signed URLs or cookies to access the restricted content (or the subset defined by the distribution's behavior).

You can configure a signer for your CloudFront distribution in one of two ways:

- Add a **CloudFront key** as the account **root user** and assign the account as a trusted signer

- Create a **public-private key pair**, upload that public key to CloudFront, and assign the key to a trusted key group as a trusted signer (recommended by AWS)

Note

The use of trusted key groups is now the recommended method from AWS to associate a trusted signer with a CloudFront distribution/behavior. This simplifies the process by eliminating the need to use the AWS root account to configure the trusted signer.

CloudFront distributions can have either a single behavior or multiple. If a distribution has a single behavior that is public or private, then inherently, the entire distribution will be public or private. However, it is more common to see a mix of private and public behaviors within a distribution. If you again refer to *Figure 10.1*, you can see a single distribution configured with both private and public behaviors. This can be a flexible way to serve both public and private content.

Edge Locations

Amazon CloudFront can effectively deliver content using edge locations. An **edge location** is one of several regional data centers distributed globally, which are directly connected to the **AWS backbone network**. When a distribution is created, the designated **price class** determines which edge locations will service the CloudFront distribution. This means that the IP addresses tied to the domain name used for the distribution will then be advertised from all the chosen edge locations. There are around 400 AWS edge locations at the time of this writing.

Edge locations are in global **points of presence (POPs)** that have direct interconnections with public IPs. The public IPs tied to the configured CloudFront distribution will be advertised directly from these POPs. In addition, these edge locations are responsible for **caching** the content from the origin. As requests for content from the distribution are received, the content will be pulled from the origin and cached within these edge locations.

When a user makes a request for web content from a distribution, the DNS will automatically direct the user to the **closest** edge location in their geographic region from a latency perspective. When a request for web content is made to a distribution, the behavior depends on whether that content is already located in the cache of the receiving edge location or not:

- If the content is cached in the edge location, it will be immediately delivered to the user.

- If the content is not yet cached in the edge location, the request will be forwarded directly to the origin for retrieval using the AWS backbone. As the content is delivered to the user through the edge location, it will then be cached there as well.

Edge locations are responsible for more than just assisting with CloudFront; they also run several other AWS services that are deployed at a global scale. Here is a list of the services that run at every AWS edge location:

- Amazon CloudFront

- Amazon Route 53

- AWS Firewall Manager

- AWS Shield

- AWS **Web Application Firewall (WAF)**

- AWS **Lambda@Edge** (for use with CloudFront)

Regional Edge Cache

As stated, there are around 400 edge locations distributed across the world. Caching content around the globe at all these locations can be a task that requires optimization. AWS needs to make sure the most popular and sought-after content is cached in those edge locations. AWS has optimized the caching behavior by introducing a **regional edge cache**. There are 13 regional edge caches deployed around the world today. These regional caches contain a much larger caching capacity than the regional POPs.

When content is very popular, it will be stored at the edge location that it is requested from. However, as the content becomes less popular (meaning there are increasingly fewer requests for it from end users), the caching of this content will fall back to the regional edge cache. Once the content is requested again, the edge location will be able to pull it directly from the regional edge cache, without having to pull the content all the way from the origin.

Refer to *Figure 10.2* as an example of how regional edge caches work. This figure represents a static website hosted by Trailcats for the media files they provide to their global distributors on media. trailcats.com, which is part of a CloudFront distribution.

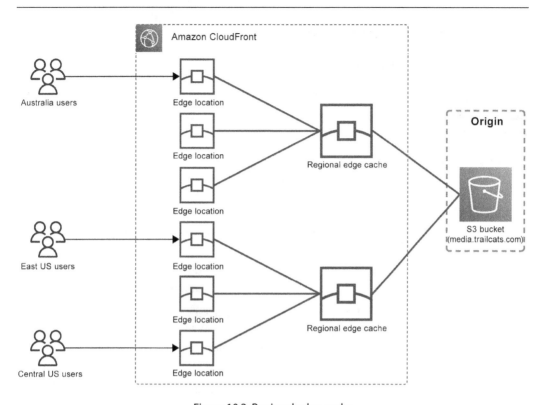

Figure 10.2: Regional edge cache

In *Figure 10.2*, you can see that the distributors (users) from Australia, the eastern US, and the central US all generate the requests directly to their local edge location. There may be a time of year when trade shows are very popular, and this content is constantly in demand from the distributors. During these periods, the content will likely stay within the edge location. When the trade show season dies down and the content is requested only a fraction of the time, the cached content will be removed from the edge location and will instead be stored in the regional edge cache.

Summary of Amazon CloudFront Components

Amazon CloudFront's components—including origins, distributions, behaviors, edge locations, and regional edge caches—form the backbone of its content delivery capabilities, enabling secure and efficient global content distribution. Understanding these elements allows you to build tailored solutions that can serve both public and private content with low latency and high scalability. By optimizing how CloudFront caches and delivers content, businesses can enhance user experience and improve overall performance.

With a clear understanding of CloudFront's core components, it's now essential to explore the key benefits that CloudFront provides. The next section will highlight the advantages of using CloudFront in real-life scenarios and discuss common use cases that demonstrate its value for modern CDNs.

Key Benefits of Amazon CloudFront

Amazon CloudFront's global network of regional edge locations is designed to optimize content delivery by improving performance, scalability, and security. This distribution model enhances the user experience by reducing latency and ensuring faster access to content, even in high-demand situations. Here are the key benefits that make CloudFront an ideal choice for delivering content efficiently and securely:

- **Low latency and higher transfer speeds**: As stated previously, CloudFront allows the global distribution of content to AWS edge locations, which can provide the lowest possible latency. Even at worst, if the content is not yet cached within an edge location, the user gains access to the AWS backbone network from the closest edge location to them geographically. This can greatly improve throughput and reliability, even when accessing the content all the way from the origin (assuming it is hosted within AWS).

- **Scalability**: Amazon CloudFront is built to automatically scale based on demand. When traffic volumes increase during peak times, CloudFront can consistently provide reliable access to content.

- **Security**: Amazon CloudFront provides best-practice security features such as **distributed denial-of-service** (**DDoS**) protection via AWS Shield, WAF integration, and **Secure Socket Layer/Transport Layer Security** (**SSL/TLS**) based encryption.

- **Cost efficiency**: Amazon CloudFront can be cost-effective in the sense that it is pay-as-you-go which does not require upfront payments or long-term commitments. You also pay for the services that you consume.

- **Integration with other AWS services**: Amazon CloudFront allows for direct integration with several other AWS services such as **Elastic Load Balancing** (**ELB**), S3, EC2, and Lambda, which can assist with interoperability within the AWS environment.

- **Customization and flexibility**: The use of CloudFront with Lambda@Edge can allow extensive customization of content by running code close to the end users. This can range from detailed analytics to personalizing content.

These key benefits can be some of the driving factors as to why you would choose to use Amazon CloudFront to be the CDN for your application. It is important to define your requirements and ensure that Amazon CloudFront meets the criteria to host the specific application in use. The next section will focus on the primary use cases for Amazon CloudFront.

Use Cases for CloudFront

Amazon CloudFront can address a variety of use cases for delivering content to remote users. It is widely used for various purposes across different industries. Next are some key use cases of Amazon CloudFront.

Static Content Delivery

There are many sites that are relatively static in nature. This means that they contain images and content files that are consistently accessed by remote users, but the content does not change or get updated on a frequent basis. This could be as simple as a personal blog with text and images or a complex company web page running entirely on JavaScript. This may be one of the most obvious use cases for Amazon CloudFront, as all the static content can be cached around respective edge locations and enhance the delivery of this content to remote viewers based on their location.

When this content is hosted on S3, there can be seamless integration options, such as enabling **origin access control** (**OAC**), which can easily secure access to your S3 content so that it can only be accessed via the CloudFront distribution.

For example, Trailcats has a web page where they publish maps and trail guides for global, pet-friendly trails. The site is hosted in the US, but many users access the content from Europe, Asia, and Oceania. By using Amazon CloudFront, Trailcats can significantly enhance the user experience by caching static files, such as maps and guides, at edge locations closer to these users. This reduces the time it takes for the content to load, as users don't need to retrieve it from the US-based origin server. Additionally, CloudFront leverages AWS's global network backbone to optimize data routing, ensuring faster and more reliable delivery of content, even for large files.

Dynamic Content and Streaming Media

In addition to static content, CloudFront can also assist with the delivery of dynamic content. This can often take the form of streaming media in some capacity to your users. Streaming media could be either pre-recorded (i.e., **video on demand** (**VOD**)) or live streaming via broadcast. VOD content can be cached at the edge similar to static content and live streaming content via CloudFront can assist with reducing load on the configured origin by storing media fragments at the edge in flight.

> **Note**
>
> More information about delivering VOD and live streaming via CloudFront can be found here: `https://docs.aws.amazon.com/AmazonCloudFront/latest/DeveloperGuide/on-demand-streaming-video.html`.

For example, assume Trailcats has a large library of videos with instructional videos on how to use their products when out on the trail. These videos can be somewhat large as they include hiking footage as well. Their VOD platform is fronted by a CloudFront distribution and allows these videos to be cached and consumed directly from AWS edge locations, avoiding incessant buffering and download times for their users.

Customize Content at the Edge

In a somewhat unique case, Amazon CloudFront can also allow the customization of distributed content within its edge locations using serverless code via **Lamba@Edge**. Lambda@Edge is a service that runs at each edge location and does not need to run within a VPC.

When content is being served from a CloudFront distribution, there is the possibility of four transactions in the communication path. Between the edge location and the remote user, there is a **viewer request** when content is requested from the edge location, and a **viewer response** when the edge responds with content or a custom response. Between the edge location and the origin, there is an **origin request** and an **origin response**, which operates in a similar fashion when a distribution needs to fetch the content from the origin. The origin request is made from the CloudFront distribution to the origin, and the origin response is from the origin back to the distribution.

Refer to *Figure 10.3* to see where **Lamba@Edge** functions can be used within a CloudFront distribution:

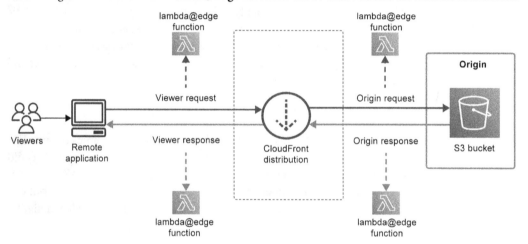

Figure 10.3: Lambda@Edge architecture

In theory, it is unlikely you would use all the points shown in the preceding figure on a single behavior. This is just to symbolize where **Lambda@Edge** functions can be invoked on incoming requests and responses.

Lambda@Edge allows for functions to be run during any or all these transactions for customization purposes. This could be used for things such as migrating between origins (directing traffic between multiple origins), serving specific content based on geolocation, or many more use cases.

Refer to *Figure 10.4* for an example of how **Lambda@Edge** can be configured on a behavior within a CloudFront distribution:

Function associations - *optional* Info

Choose an edge function to associate with this cache behavior, and the CloudFront event that invokes the function.

	Function type		Function ARN / Name	Include body
Viewer request	Lambda@Edge	▼	arn:aws:lambda:us-east-1:1234567	☐
Viewer response	No association	▼		
Origin request	Lambda@Edge	▼	arn:aws:lambda:us-east-1:0987654	☐
Origin response	No association	▼		

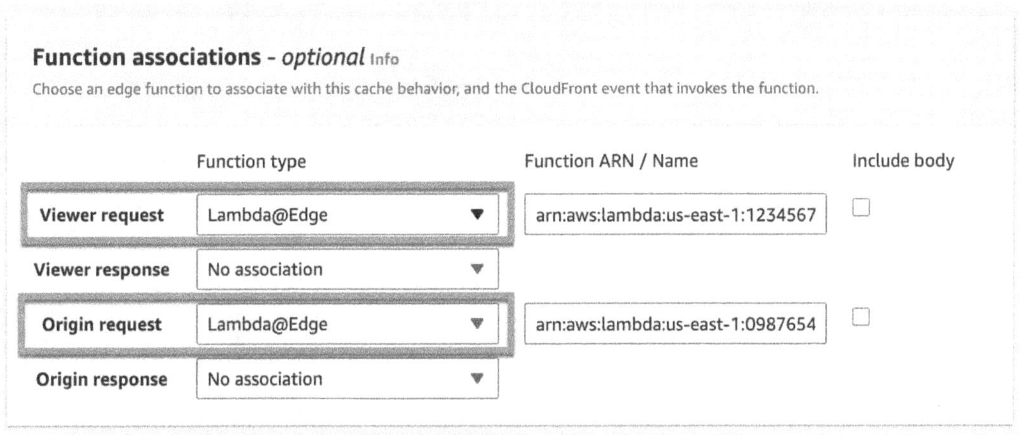

Figure 10.4: Lambda@Edge configuration

As you can see in the figure, you are also able to pass the body of the HTTP request through to the Lambda function if that data is required to run the function successfully. For example, Trailcats has a popular line of cat harnesses that say **Hello** in large text on them. They have updated the origin to include images of the product that include every major language written on the harness. A **Lambda@ Edge** function could be configured so that the origin request per image could be altered to pull the image with the language that matches what is most common in the country that the request originates from.

> **Note**
>
> Further reading about **Lambda@Edge** example use cases can be found here: `https://docs.aws.amazon.com/AmazonCloudFront/latest/DeveloperGuide/lambda-examples.html`.

Distributing Private Content

As mentioned previously, you can create specific behaviors within your CloudFront distributions that would be classified as public versus private. Using a private behavior within a distribution can be an effective way to distribute private content to users who require authentication before accessing specific content. Using a trusted signer, you can enable CloudFront application flows to ensure proper user authentication before accessing private content.

Behaviors within a distribution can be configured to **restrict viewer access**, which requires that viewers use either a signed URL or a signed cookie to view the content of the origin associated with that behavior. As mentioned before, AWS recommends associating a trusted key group with the behavior to enable a trusted signer.

Refer to *Figure 10.5* for an example of a Trailcats application that uses a combination of public and private distributions within a single application. This application needs to first authenticate users before providing them access to images of their upcoming products that reside within an S3 origin.

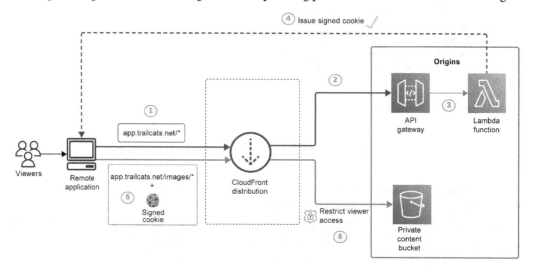

Figure 10.5: Private content distribution – signed cookie

In *Figure 10.5*, you can see that there is a default behavior (*1*) that services the application at app. trailcats.net/*. The default behavior for the distribution is associated with an API Gateway (*2*) that triggers a specific Lambda function (*3*), which is tasked with authenticating the user via some external means, which could be using **single sign on** (**SSO**) via Microsoft, Google, and so on. The Lambda function uses the keys within the trusted key group to issue a signed cookie (*4*) back to the application. The application then uses that signed cookie to access app.trailcats.net/ images/* (*5*). The behavior configured for this URL path has restricted viewer access enabled (*6*), with the same trusted key group assigned. Since the signed cooked was generated with a key from the trusted key group, the viewer will be able to view the content.

From this example, you can see how AWS customers can utilize a combination of distribution behaviors to distribute their private and public content.

Signed URLs and Cookies

As mentioned earlier in this chapter, a key method for securing private content within a CloudFront distribution is by using signed URLs or cookies. When restricting viewer access, you have the option to use either, and you may need to consider when to use one or the other. When evaluating which to choose, it is important to keep in mind the behavior of each of them:

- **Signed URL**: This provides access to a **single** object.
- **Signed cookie**: This provides access to several objects or **groups** of objects.

Generally, it is recommended to use signed cookies unless there is a specific reason that warrants the use of signed URLs. Historically, old browsers did not support signed cookies, which could be a deciding factor if you need to maintain that supportability for those browsers.

Caching Behaviors in Amazon CloudFront

This section about Amazon CloudFront is going to cover how caching works within a CloudFront distribution. It is important to understand how CloudFront will cache content in edge locations and how you may need to manipulate this behavior to avoid any issues with the distribution of up-to-date content. You will need to understand how caching is performed, **time to live** (**TTL**), and invalidations to ensure your content is effectively cached for your remote users.

How CloudFront Caches Content

As you've read up to this point, Amazon CloudFront caches content within edge locations to improve customer performance when accessing that content. When content is cached into these edge locations, the viewer of that content will get the content quickly and efficiently because they are getting it from a location much closer to them, hence improving the overall experience. This is also able to relieve potential strain on the origin of the content. Caching at the edge allows for far fewer requests to be sent all the way to the origin, therefore fewer resources will be consumed on the origin side.

The first time a specific piece of content (whether it be an image, video, or any kind of file) is accessed via a CloudFront distribution, it will need to be pulled from the origin of the distribution. As mentioned previously, this is referred to as an **origin fetch**. Remember, the requests for content always land at the closest serviceable edge location to the user. So, as the content is being pulled from the origin to provide to the user, the content is stored in the local cache of both the edge location and the regional edge location.

When content is requested from an edge location and that file is not currently stored within the cache, it is called a **cache miss**. Whenever content is requested and it is currently within the local cache, it is called a **cache hit**. Generally, you want to try and reduce the number of cache misses and increase the number of cache hits as much as possible. The more cache hits occur, the fewer requests need to be sent to the origin, which improves both efficiency and user experience.

Now, look at an example of how this works for some media content that Trailcats has published via a CloudFront distribution. A distribution has been set up for `media.trailcats.com`, which hosts some video content that is available for users to download. They have a file called `outdoor-kitty.mp4` that has just been posted on social media, so users are requesting to download it. Refer to *Figure 10.6* as you walk through this process:

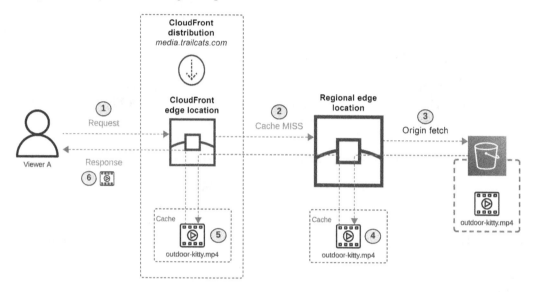

Figure 10.6: CloudFront caching – cache miss

The step-by-step process outlined in the figure is as follows:

1. Viewer A navigates to `media.trailcats.com` and requests to download the file. They are the first viewer to ever request this file via the CloudFront distribution. The request is sent to the edge location closest to Viewer A.

2. The edge location does not have this file stored within the local cache, so a **cache miss** occurs. The edge location then requests to see if the content is stored within the regional edge location, which has a much larger cache.

3. The regional edge location also does not have the file, so an **origin fetch** occurs, and the file is requested directly from the origin, which is an S3 bucket.

4. As the content is fetched from the origin to provide to Viewer A, it is also stored within the local cache of the regional edge location.

5. Additionally, the file is stored in the cache of the edge location.

6. The `outdoor-kitty.mp4` file is delivered to Viewer A.

As you can see, this process of a cache miss may require a series of steps and processing that all take time to complete. While each of these steps may only take a few milliseconds, that can still affect how fast the content is provided to the end user. However, if another user tries to pull down this same content shortly after, a cache hit will greatly improve this process. Refer to *Figure 10.7* to see how a cache hit will improve the user experience.

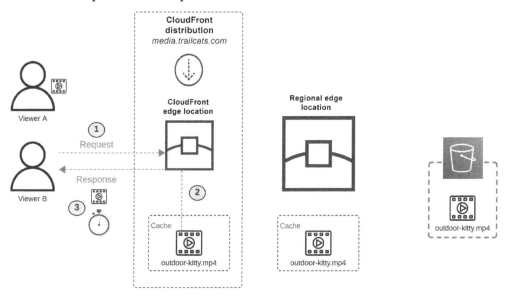

Figure 10.7: CloudFront caching – cache hit

Compared to a cache miss, you can see that a cache hit requires much fewer steps and in turn less time from the user's perspective. The steps for this cache hit are detailed here:

Viewer B requests the same file, `outdoor-kitty.mp4`. Viewer B is in a similar geographic area to Viewer A, so this request hits the same edge location.

The file is currently in the cache of the edge location, so it can be immediately delivered to Viewer B.

The file is delivered directly to Viewer B with reduced latency and increased performance.

> **Note**
>
> It is important to note that Amazon CloudFront only performs download or read caching of content. CloudFront does not offer any form of write caching.

Now that you've seen how Amazon CloudFront caches content within a distribution, you will next look at the methods you can use to influence and control the lifecycle of the content being cached.

TTL

By default, all objects that are stored within the cache of an edge location will have a TTL value associated with them. This is a value, defined in seconds, that determines how long an object is to be stored within the cache before it is considered expired.

After the TTL value has expired, the next request for that content will result in the edge location forwarding that request to the origin to verify whether it still has the latest version of that content or not. If CloudFront *does* still have the latest version of the file (i.e., it has not changed since it was last cached), then the origin will respond with a status code of 304 Not Modified, and the TTL value will be reset. If CloudFront *does not* have the latest version, then a status code of 200 OK will be sent by the origin, along with the latest version of the file.

The default TTL value for object caching within a distribution is 24 hours and is defined within a behavior. In addition to the default TTL, you can configure values for Minimum TTL and Maximum TTL, as shown in *Figure 10.8*.

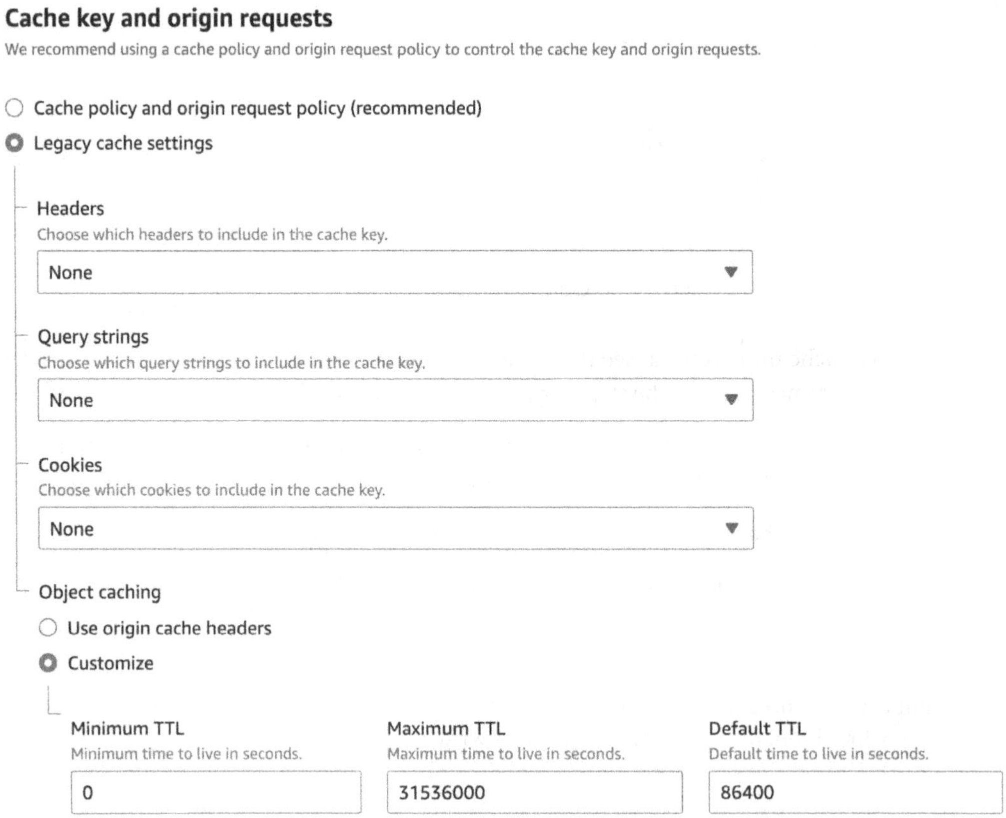

Cache key and origin requests

We recommend using a cache policy and origin request policy to control the cache key and origin requests.

○ Cache policy and origin request policy (recommended)

◉ Legacy cache settings

Headers
Choose which headers to include in the cache key.

None	▼

Query strings
Choose which query strings to include in the cache key.

None	▼

Cookies
Choose which cookies to include in the cache key.

None	▼

Object caching

○ Use origin cache headers

◉ Customize

Minimum TTL	Maximum TTL	Default TTL
Minimum time to live in seconds.	Maximum time to live in seconds.	Default time to live in seconds.
0	31536000	86400

Figure 10.8: Object caching settings

The configuration of a minimum TTL and maximum TTL only has an effect when the origin is configured to add specific HTTP headers on specific objects. These headers are `Cache-Control max-age`, `Cache-Control s-maxage`, or `Expires`. When either the `Cache-Control max-age` or `Cache-Control s-maxage` header is set on an object, this value will define the number of seconds before that object expires. The `Expires` header is used to specify a specific future date and time that the particular object will expire. However, the limit for all these headers is defined by the minimum TTL and maximum TTL value set within the behavior. For example, no value could be set by the `Cache-Control max-age` header that exceeds the configured maximum TTL.

Any objects retrieved from an origin that do not have these headers set will use the `Default TTL` value for the distribution.

Invalidations

While caching content can provide improved performance, it can also result in users not always getting the most up-to-date content.

For example, refer again to *Figure 10.7*. Assume that someone at Trailcats noticed there was a typo in some text within the `outdoor-kitty.mp4` video file, leading users to an incorrect website for more information. They immediately re-edited the video, fixed the typo, and uploaded it to the origin (S3). S3 automatically overrides the file with the new version. However, since the content is already cached in the edge location and is using a default TTL of 24 hours, that means users could continue downloading the incorrect file until it expires from the cache.

This can be resolved by creating an invalidation. An **invalidation** can be manually created within a distribution and essentially causes all the edge locations to purge the desired content from the local cache.

> **Note**
> Since an invalidation applies to all edge locations, it is not immediate. There will be a delay as the content is removed from all edge locations.

Invalidations are created on a per-distribution basis and can be created for single objects or groups of objects. For example, an invalidation could be created for only the `outdoor-kitty.mp4` object, as shown in *Figure 10.9*.

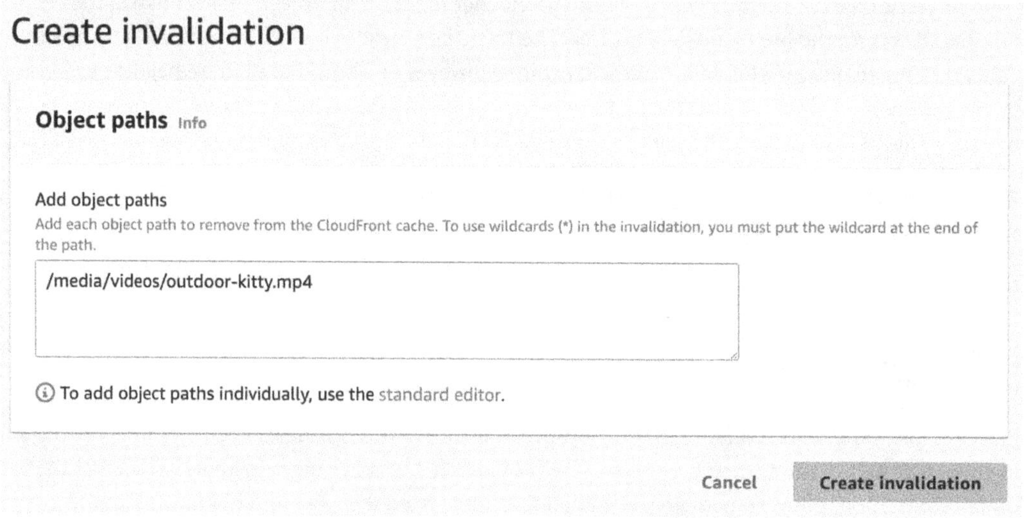

Create invalidation

Object paths Info

Add object paths

Add each object path to remove from the CloudFront cache. To use wildcards (*) in the invalidation, you must put the wildcard at the end of the path.

> /media/videos/outdoor-kitty.mp4

ⓘ To add object paths individually, use the standard editor.

Cancel   **Create invalidation**

Figure 10.9: Invalidation for a single object

However, if there are other objects that you wish to invalidate within the path, you could also use an invalidation for `/media/videos/*` to invalidate all the objects contained in that path. You could also create an invalidation for `/*` if you wanted to purge all objects for that origin.

> **Note**
>
> There is a cost associated with invalidations. The cost is the same whether you are invalidating a single object or a group of objects.

While managing content updates with invalidations is essential for ensuring users receive the most current information, it's equally important to safeguard the integrity and availability of that content from potential threats. This brings us to the next critical aspect: securing your CloudFront distributions and origins to protect against cyberattacks and unauthorized access.

Securing CloudFront Distributions and Origins

With the rise of sophisticated cyberattacks such as DDoS attacks, botnets, and data breaches, securing CDNs such as Amazon CloudFront has become more critical than ever to ensure the integrity, availability, and confidentiality of your data. Amazon CloudFront offers robust security features that help protect both your distributions and origins from these evolving threats, ensuring secure and reliable content delivery. This section will explore the key strategies and tools available to secure your CloudFront distributions, including configuring Amazon S3 as an origin with HTTPS, implementing secure origin access, leveraging AWS security services such as AWS WAF, Shield, and **AWS Certificate Manager** (**ACM**), and applying geographic restrictions to control content delivery. By understanding and utilizing these services, you can effectively fortify your CloudFront distributions against unauthorized access and attacks, ensuring a secure and reliable content delivery experience for your end users.

Amazon S3 as Origins with HTTPS

As previously mentioned, there are several AWS services that can be associated with an Amazon CloudFront distribution as an origin. One service that needs little explanation is ALB, which already is a front-end for web applications, but CloudFront can expand it to a globally efficient scale. One origin integration of particular relevance is Amazon S3.

As you may already be aware, S3 allows the hosting of a static website directly from a bucket that allows public access. This is a useful tool that allows simple websites to be hosted from S3 with only some HTML code running within an object in the bucket. However, one major drawback is that the hosting of a public website on S3 only supports the use of HTTP. It does not allow the use of the more secure HTTPS, which is a standard requirement today for almost any website.

Configuring a CloudFront distribution with an S3 origin allows seamless enablement of HTTPS on a static S3 website. With direct integration with ACM, you can assign a custom **SSL certificate** per distribution.

Within the CloudFront distribution settings, you can also adjust the viewer protocol policy, as shown in *Figure 10.10*:

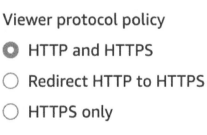

Figure 10.10: Viewer protocol policy

As shown, you have the option to support both HTTP and HTTPS, redirect all HTTP requests to use HTTPS, or strictly only support HTTPS.

> **Note**
>
> Configuring an S3 bucket as an origin for a CloudFront distribution does not automatically disable public access to the bucket. If public access to the bucket is allowed, users will still be able to reach the bucket directly using HTTP. To lock this down further, you will need to use OAC, which is covered later in this chapter.

Securing Origin Access

When using CloudFront to distribute content, you often want that to be the sole means of getting that specific content. This means you need to ensure that viewers cannot bypass the CloudFront distribution and access the origin directly. For example, if a distribution is configured with an S3 origin, you would need to secure the S3 bucket to ensure viewers do not directly navigate to the static web page for the bucket and download the content as this will not use the edge locations and caching provided by CloudFront.

This section will focus on securing access to your CloudFront origins, whether it be one that supports AWS native security controls or a custom origin.

OAC

As one would expect, securing origin access when the origin is an AWS native service can offer some added simplicity. AWS offers a feature called OAC, which can provide a simple way to restrict origin access to CloudFront only. OAC is a method that allows CloudFront to sign all requests (origin fetch) to an origin so that you can lock down the origin so that only those signed requests are permitted.

> **Note**
>
> Prior to OAC, AWS offered a feature through CloudFront called **origin access identity** (**OAI**), which provided similar functionality but lacked some of the enhancements that were brought in with OAC. OAC is now the recommended approach from AWS, but you may still see a reference to OAI on the ANS-C01 exam.

As of today, OAC supports the following origins:

- Amazon S3 buckets
- AWS Lambda functions (URL, not Lambda@Edge)
- AWS Elemental MediaPackage v2 or MediaStore

One of the most common deployments of OAC is for Amazon S3 origins. To be specific, this is when you are using the direct integration within a CloudFront distribution to use an S3 bucket and not referring to the URL of a static website hosted on S3.

OAI was the previously recommended way of securing S3 origins before OAC was introduced. You can consider these features to be essentially the same, but OAC offers additional capabilities that were not present in OAI. This includes security best practices, comprehensive HTTP methods support, SSE-KMS support, and coverage for S3 in all AWS regions.

Refer to *Figure 10.11* for a simplified view of how OAC works for S3 origins. In this example, Trailcats has a public-facing blog hosted on Amazon S3 that they publish through a CloudFront distribution.

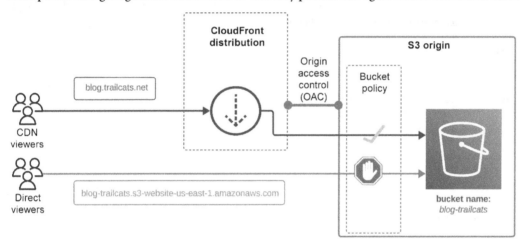

Figure 10.11: OAC for S3

As you can see in the diagram, some viewers are trying to access the bucket content both through the CloudFront distribution as well as the S3 bucket directly. The CloudFront distribution has enabled OAC within the origin settings, which means CloudFront will sign all the requests sent to the origin. Then, the bucket policy on the S3 bucket is configured to only allow GetObject requests from the distribution. Thus, any viewers trying to access the bucket directly will be denied because they are not coming from the correct source.

Exercise 10.1: Configuring OAC

In this exercise, you will learn how to configure OAC for an Amazon CloudFront distribution to secure your content hosted in an S3 bucket. By the end of this exercise, you will understand how to create an OAC, attach it to a distribution, and update the S3 bucket policy to allow access only through CloudFront.

Perform the following steps:

1. To update the origin settings, first navigate to your CloudFront distribution settings, then, on the `Origin and Origin groups` tab, select the origin that you want to update and click `Edit`. Under `Origin access`, select `Origin access control settings (recommended)`, as shown in *Figure 10.12*.

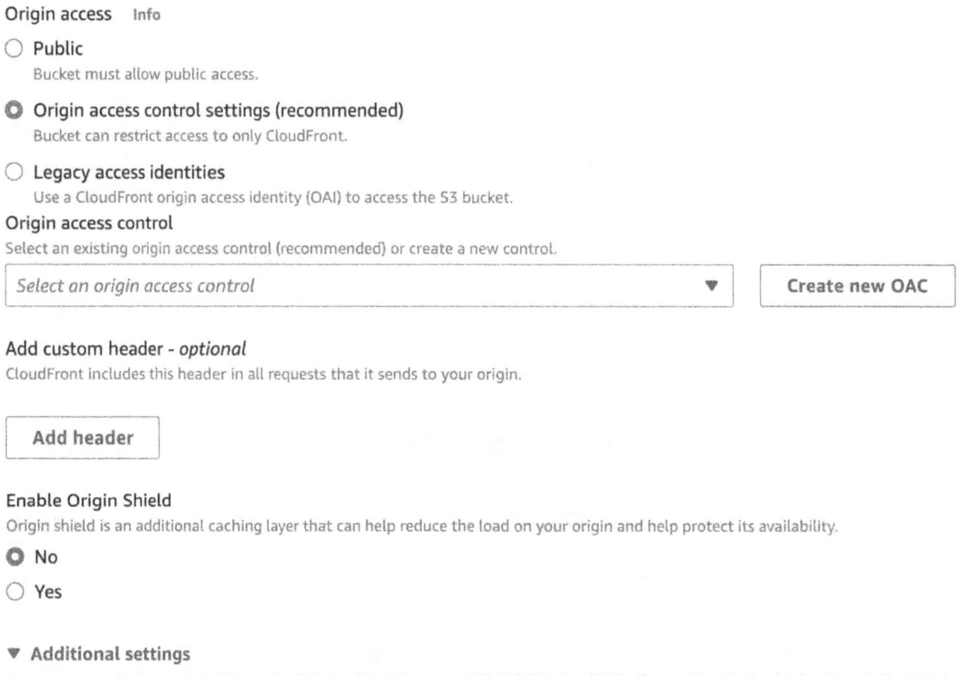

Origin access   Info

○ Public
   Bucket must allow public access.

● Origin access control settings (recommended)
   Bucket can restrict access to only CloudFront.

○ Legacy access identities
   Use a CloudFront origin access identity (OAI) to access the S3 bucket.

Origin access control
Select an existing origin access control (recommended) or create a new control.

| Select an origin access control ▼ | Create new OAC |

Add custom header - *optional*
CloudFront includes this header in all requests that it sends to your origin.

Add header

Enable Origin Shield
Origin shield is an additional caching layer that can help reduce the load on your origin and help protect its availability.

● No
○ Yes

▼ Additional settings

Figure 10.12: Enabling OAC on a CloudFront distribution

2. To create a new OAC, in the `Create new OAC` dialog, provide a name for the OAC. Under `Signing behavior`, choose whether you want CloudFront to sign requests or not. Typically, you would select `Sign requests`. For the authorization header behavior, choose whether to override the authorization header if needed. This is rarely required.

3.   Click Create, as shown in *Figure 10.13*.

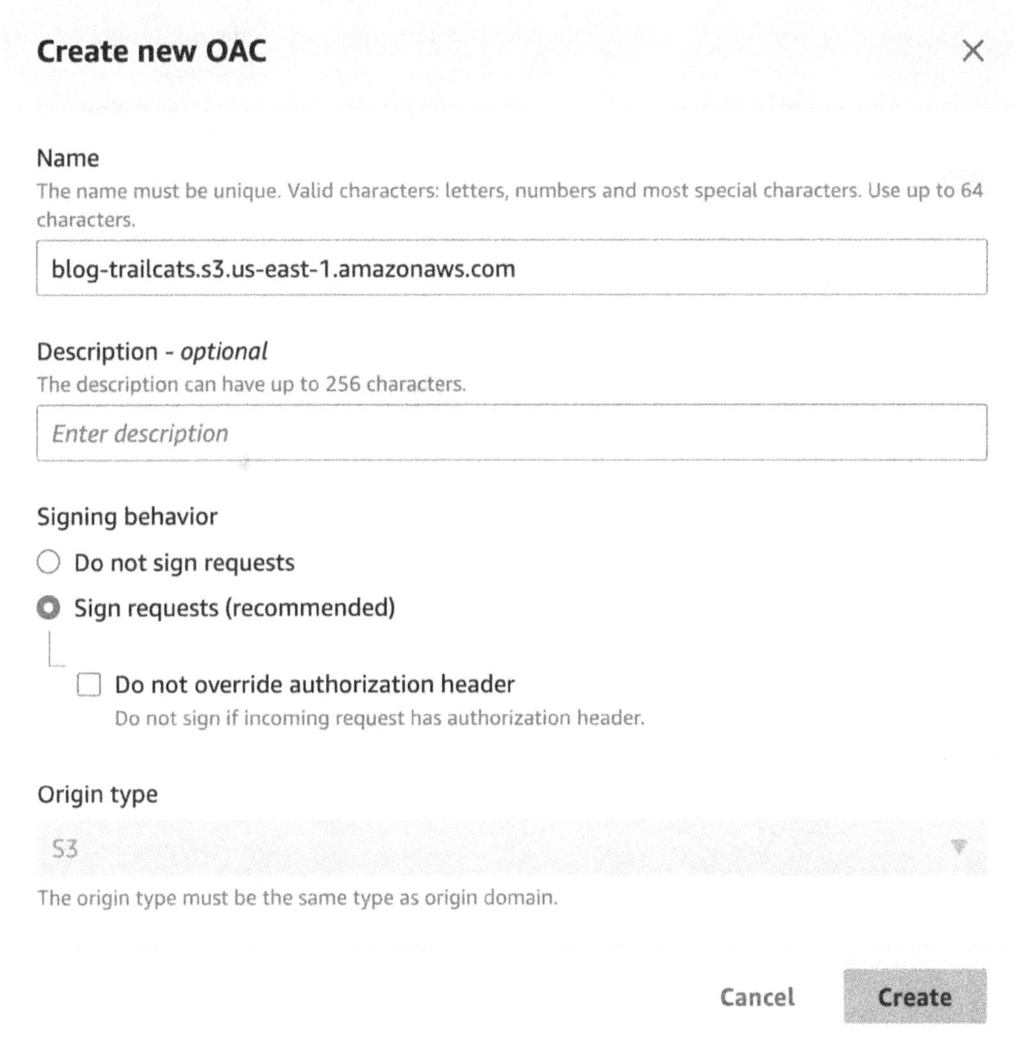

Create new OAC   ✕

Name

The name must be unique. Valid characters: letters, numbers and most special characters. Use up to 64 characters.

```
blog-trailcats.s3.us-east-1.amazonaws.com
```

Description - *optional*

The description can have up to 256 characters.

```
Enter description
```

Signing behavior

◯ Do not sign requests

🔘 Sign requests (recommended)

☐ Do not override authorization header

Do not sign if incoming request has authorization header.

Origin type

```
S3                                                                   ▾
```

The origin type must be the same type as origin domain.

Cancel   **Create**

Figure 10.13: Create new OAC

4.   To attach the OAC to the distribution, after creating the OAC, return to the origin settings in your CloudFront distribution. Select the newly created OAC and save your changes.

5. To update the S3 bucket policy, go to your S3 bucket that serves as the origin for your CloudFront distribution. Open the `Permissions` tab and select `Bucket policy`. Modify the bucket policy to allow access only from the CloudFront distribution's **Amazon Resource Name (ARN)**. The updated bucket policy should look similar to *Figure 10.14*, where only the specified CloudFront distribution is permitted to perform `GetObject` actions on the bucket.

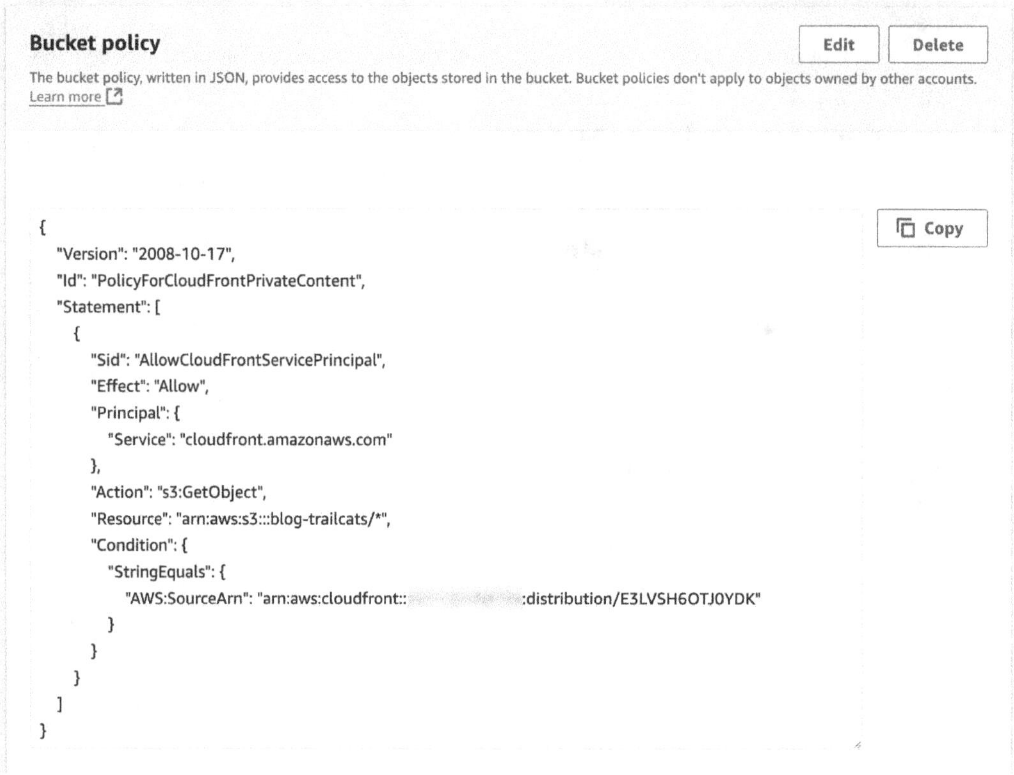

Bucket policy   Edit   Delete

The bucket policy, written in JSON, provides access to the objects stored in the bucket. Bucket policies don't apply to objects owned by other accounts.
Learn more

```
{                                                                        Copy
    "Version": "2008-10-17",
    "Id": "PolicyForCloudFrontPrivateContent",
    "Statement": [
        {
            "Sid": "AllowCloudFrontServicePrincipal",
            "Effect": "Allow",
            "Principal": {
                "Service": "cloudfront.amazonaws.com"
            },
            "Action": "s3:GetObject",
            "Resource": "arn:aws:s3:::blog-trailcats/*",
            "Condition": {
                "StringEquals": {
                    "AWS:SourceArn": "arn:aws:cloudfront::            :distribution/E3LVSH6OTJ0YDK"
                }
            }
        }
    ]
}
```

Figure 10.14: OAC S3 bucket policy

As shown in the figure, only the ARN of the CloudFront distribution will now be able to perform any `GetObject` actions on the Trailcats S3 bucket, thus locking down the access from any other source.

You have successfully configured an OAC for your CloudFront distribution and secured your S3 bucket by limiting access to only the authorized CloudFront distribution. This ensures that your content can only be accessed through the CloudFront distribution, enhancing the security of your origin.

Securing Custom Origins

In addition to securing access to origins that are native AWS services, you may also need to implement controls to secure access to custom origins. These origins could be web servers running completely outside of your AWS cloud environment, or even running on an EC2 instance within your own VPC.

There are two primary methods that can be used for securing custom origins and they can either be used separately or in a combination, depending on requirements.

Firstly, you can secure custom origins by using **custom HTTP headers**. Since all the communication between CloudFront and a custom origin is over HTTP/HTTPS, you have the option to specify a custom header to be added by CloudFront when fetching content from the origin. Then, you can manually set the web server to require this header to be present for any requests for them to be processed.

Another method you can use for securing custom origins is **IP-based filtering**. This method includes having the origin web server sitting behind a traditional network firewall that filters all inbound requests. AWS publishes all the public IPv4 addresses that are used by the edge locations, which would be the addresses used as the source to fetch content from custom origins. The network firewall can be configured to especially only allow inbound web requests to the origin server from these specified IP ranges.

> **Note**
> More details on the public IP address ranges used by Amazon CloudFront can be found at the following URL: `https://docs.aws.amazon.com/AmazonCloudFront/latest/DeveloperGuide/LocationsOfEdgeServers.html`.

Refer to *Figure 10.15* for a visual representation of both methods:

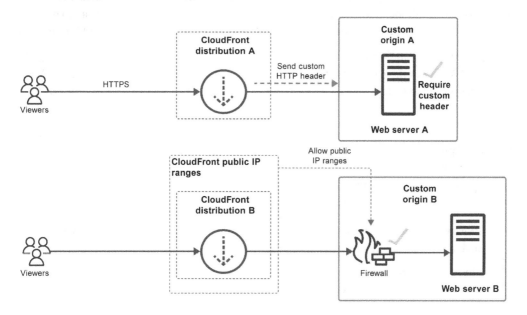

Figure 10.15: Securing custom origins

You will see in the figure that Web Server A only allows web requests using the custom header, while Web Server B is behind a firewall that is whitelisting the CloudFront public IP ranges. In the diagram, these methods are used separately across multiple distributions, but they could be used at the same time on the same origin if the security requirements call for such a design.

> **Note**
> When using a custom origin, you define what protocol should be used to access content from the origin. You can use HTTP-only, HTTPS-only, or match viewer. The match viewer option will ensure that all origin fetches will use the same protocol as the viewer request that was received.

While securing access to your origins is crucial for protecting the source of your content, it's equally important to defend your entire CloudFront distribution from external threats. To achieve this, AWS offers several integrated security services that enhance the protection of your CloudFront distributions against common attacks and unauthorized access. The next section will explore how services such as AWS WAF, AWS Shield, and ACM can be utilized to strengthen your security posture and ensure the reliable delivery of your content.

AWS Security Service Integrations

There are a few built-in integrations that can assist with providing security for your distributions and the data within them. Integrating security for a content distribution network is critical because of its presence on the public internet. The attack surface, or exposure to bad actors, is important to protect the availability of the content distribution as well as the integrity of the data in the content it is providing.

The three AWS integrations for security with CloudFront you should be aware of are AWS WAF, AWS Shield, and ACM. All these services have been covered in detail in *Chapter 12, AWS Security Services*, so this section will only detail the CloudFront integration options.

AWS WAF

AWS WAF protects web-based applications from many common attacks on the public internet. When AWS WAF is enabled for a CloudFront distribution, all traffic inbound to the distribution is first processed by AWS WAF and the assigned **web access control lists** (**web ACLs**). The use of these web ACLs can enable the following protections on your CloudFront distribution:

- **Prevent common web attacks**: This provides defenses for web applications against vulnerabilities such as SQL injection, **cross-site scripting** (**XSS**), and other common exploits
- **Source IP restrictions**: This limits inbound connections to only those from approved source IP addresses, enhancing access control

- **Geographic restrictions**: This allows for fine-tuned control over access based on the geographic location of the request

- **Bot control**: This mitigates traffic from known botnets and automated sources that could threaten the integrity of the application

- **Rate limiting**: This imposes restrictions on the number of requests from a single source IP over a specified period, reducing the risk of **denial-of-service (DoS)** attacks

When enabling AWS WAF on a CloudFront distribution, there are some initial settings you will define, as shown in *Figure 10.16*:

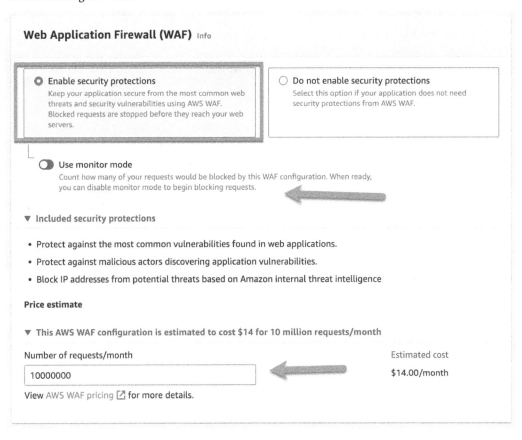

Figure 10.16: Enabling WAF on a CloudFront distribution

When creating an Amazon CloudFront distribution, you can select the `Enable security protections` option to integrate AWS WAF with your distribution. As shown in *Figure 10.16*, you have the `Use monitor mode` option if you wish to first monitor the logs of all the traffic that WAF blocks before enforcing the policy. You can switch this to enable blocking later from the `Security` page.

AWS Shield

AWS Shield is a service that focuses on protecting against DDoS attacks. AWS Shield is offered in two tiers: Standard and Advanced. All CloudFront distributions are automatically protected with AWS Shield Standard with no additional configuration required.

If you are an AWS Shield Advanced customer, you can enable AWS Shield Advanced on a CloudFront distribution by enabling CloudFront as a protected resource from the AWS Shield console and selecting the specific CloudFront distributions you would like to protect.

ACM and SSL

ACM is a **regional** AWS service that is used for the assignment or importing of SSL certificates within AWS. One of the most common services that utilizes ACM for certificate management is Amazon CloudFront. SSL/TLS certificates are often used not just for secure connectivity over HTTPS but also for the purpose of verifying ownership of a domain.

By default, CloudFront distributions support SSL without any special configuration. This is because the domain name assigned to the distribution, such as `d111122223333.cloudfront.net`, uses the default SSL certificate for `*.cloudfront.net`, which is managed by AWS. However, it is more common that a custom domain name will be used with your distribution, which does require an additional SSL certificate to be associated with the distribution for that custom domain name.

> **Note**
> For further details about SSL/TLS certificates, refer to the following link for more details: `https://aws.amazon.com/what-is/ssl-certificate/`.

When configuring CloudFront distributions, you have the option to add your own custom domain names and up to one custom SSL certificate. This certificate can be directly integrated from ACM. Refer to *Figure 10.17* for an example of the workflow to enable SSL using ACM on a CloudFront distribution:

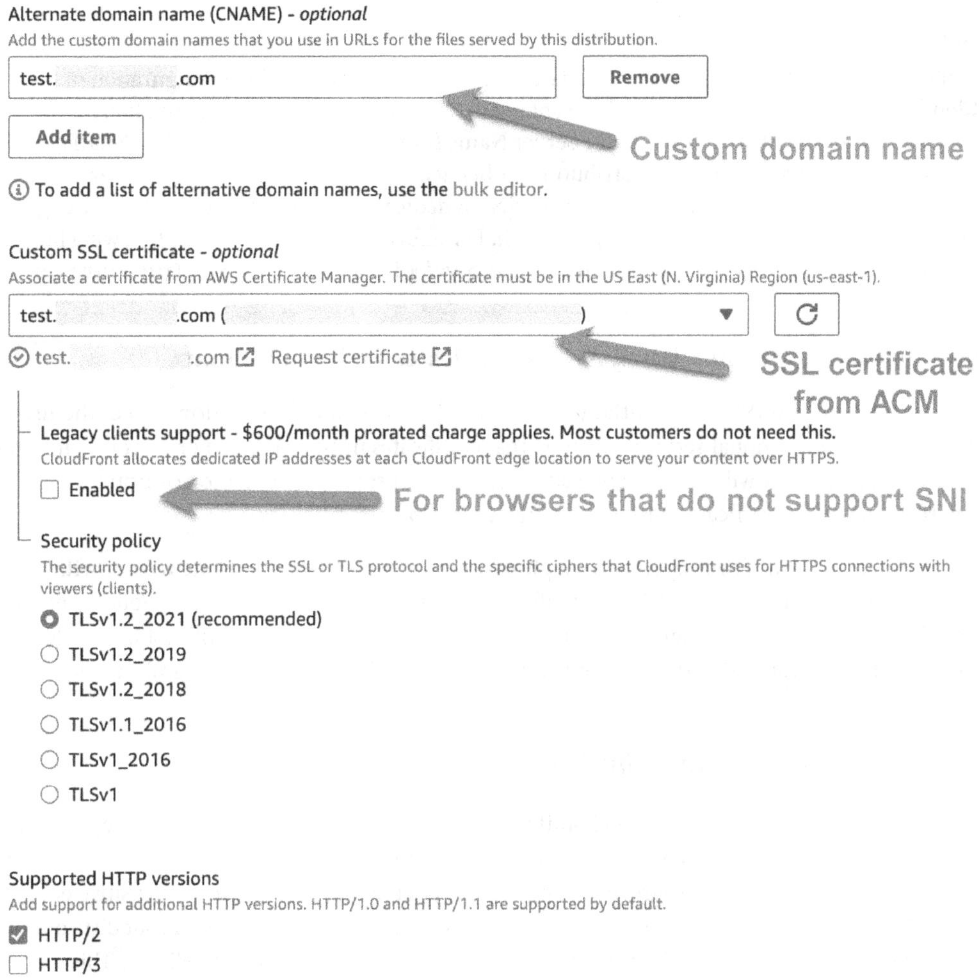

Alternate domain name (CNAME) - *optional*

Add the custom domain names that you use in URLs for the files served by this distribution.

| test.⬛⬛⬛⬛.com | Remove |

Add item

ⓘ To add a list of alternative domain names, use the bulk editor.

Custom domain name

Custom SSL certificate - *optional*

Associate a certificate from AWS Certificate Manager. The certificate must be in the US East (N. Virginia) Region (us-east-1).

| test.⬛⬛⬛.com (⬛⬛⬛⬛⬛⬛⬛) ▼ | C |

⊘ test.⬛⬛⬛.com ☑ Request certificate ☑

SSL certificate from ACM

Legacy clients support - **$600/month prorated charge applies. Most customers do not need this.**

CloudFront allocates dedicated IP addresses at each CloudFront edge location to serve your content over HTTPS.

☐ Enabled   For browsers that do not support SNI

Security policy

The security policy determines the SSL or TLS protocol and the specific ciphers that CloudFront uses for HTTPS connections with viewers (clients).

- ◉ TLSv1.2_2021 (recommended)
- ○ TLSv1.2_2019
- ○ TLSv1.2_2018
- ○ TLSv1.1_2016
- ○ TLSv1_2016
- ○ TLSv1

Supported HTTP versions

Add support for additional HTTP versions. HTTP/1.0 and HTTP/1.1 are supported by default.

- ☑ HTTP/2
- ☐ HTTP/3

Figure 10.17: SSL configuration for a CloudFront distribution

As you can see in the figure, you can configure your custom domain name and import the corresponding SSL certificate.

> **Note**
>
> You'll notice an option for `Legacy client support` within the SSL configuration of the CloudFront distribution. This option is meant to be used when viewers of the distribution may be using older browsers that do not support **Server Name Indication** (**SNI**). SNI enables CloudFront to determine which web server (distribution) is being accessed to provide the appropriate SSL certificate. For browsers that do not support SNI, a dedicated public IP address at every edge location to support this will cost $600 per month. For information on SNI, see the following link to read more: `https://en.wikipedia.org/wiki/Server_Name_Indication`.

Important Consideration – Using ACM Certificates with CloudFront

When working with ACM, it's essential to understand how regional limitations affect the use of certificates. ACM is a regional service, meaning that certificates must be created or imported into the specific region where they will be used. For example, if you are using an ALB in the `ap-southeast-2` region, the associated certificate must also be created in `ap-southeast-2`.

However, Amazon CloudFront operates as a **global** service, and its control plane resides within the `us-east-1` region. As a result, all ACM certificates that need to be integrated with CloudFront must be stored in the `us-east-1` region. Certificates stored in any other region will not be usable with CloudFront distributions. This distinction is crucial for ensuring successful integration and secure content delivery.

Amazon CloudFront Geography-Based Controls

When distributing content via Amazon CloudFront, security and compliance may require restricting access to specific geographic regions. For example, you may need to limit content distribution to certain countries due to licensing agreements or regulatory constraints or to protect sensitive data. By implementing **geographic restrictions**, you can control who can access your CloudFront distributions based on the viewer's location, adding an extra layer of security to your content delivery. This section will cover the built-in geographic restriction options in CloudFront and explore scenarios where you might use a **third-party geolocation** service for more granular control.

Geo-Restrictions

Within a CloudFront distribution, you are able to configure geographic restrictions that use an IP database to determine the source country of every viewer based on the source IP address. AWS reports that this database can determine this information with up to 99.8% accuracy. It is important to note that this service is only able to filter based on country. No other source criteria can be considered. Therefore, if more granular control is required, you may need to use an alternative solution.

CloudFront geographic restrictions can be configured either as an **allow list** or a **block list**. This means you have the option to either specify all the countries that are allowed to access the distribution and block the rest (allow list) or you can define what countries to block and permit the rest (block list). Any requests that are denied access to the distribution will receive an HTTP status code 403 (Forbidden) message.

Refer to *Figure 10.18* for an example of the configuration within CloudFront.

▼ **CloudFront geographic restrictions**

Restriction type
- ○ No restrictions
- ◉ Allow list
- ○ Block list

└ Countries

| Select countries | ▼ |

| Australia ✕ | United Kingdom ✕ | United States ✕ |

Figure 10.18: CloudFront geographic restrictions

In the preceding figure, you'll see that a distribution is configured with an allow list, and only the countries of Australia, the United Kingdom, and the US are permitted. Any requests from other countries will be denied.

Third-Party Geolocation

Should you need something more granular than just filtering based on country alone, you can use a third-party geolocation service. This setup is much more complex but offers a greater deal of flexibility.

To use a third-party location service, you will need to set up a **private CloudFront distribution** as covered previously in this chapter. The distribution would require either signed URLs or signed cookies to be used to access content within the distribution.

Once you have the private distribution in place, an application server will need to be deployed outside of the CloudFront distribution. This app server is responsible for the entire classification and approval process for all remote viewers. This means the app server could use something more accurate than the IP database offered through Amazon CloudFront and could even be more granular and allow or deny access to certain cities or regions within a country. The app server is responsible for issuing the signed URLs or signed cookies to the users, which they will then use to access the private distribution (like any normal private distribution).

One thing that is unique about using this option is that this method is not limited to just location. In theory, the app server could allow/deny access based on almost *any* defined criteria. This could be based on user attributes, authentication, or much more.

AWS Global Accelerator

Lastly, this brief section of this chapter will focus on AWS Global Accelerator. **AWS Global Accelerator** is a service from AWS that allows leveraging the **AWS global network backbone** to improve and accelerate the performance of public-facing customer applications running on AWS. This service has some similarities to Amazon CloudFront in that it uses much of the same infrastructure. However, one major characteristic that separates Global Accelerator from CloudFront is that it is not restricted to HTTP-based applications, nor does it provide any caching. Amazon CloudFront's primary purpose is content distribution, whereas AWS Global Accelerator is more like an on-ramp to AWS. The foundation behind Global Accelerator is to get global user traffic onto the AWS network as soon as possible, which has better performance, latency, and predictability compared to the public internet.

While the AWS infrastructure that Global Accelerator uses is very similar to Amazon CloudFront, the functionality of the service is quite different. The globally dispersed edge locations, in conjunction with Amazon's POPs, play a key role in the way Global Accelerator operates. For the scope of the *ANS-C01 exam*, you'll need to ensure you understand three key components of Global Accelerator: accelerators, listeners, static IP addressing, and endpoints.

Accelerators and Listeners

An **accelerator** is a component of AWS Global Accelerator that starts ingesting traffic at AWS edge locations destined for your public-facing applications. You control the behavior of steering traffic for accelerators by using listeners. A **listener** within Global Accelerator is somewhat like that of a listener that you would configure for a load balancer. The listener looks for incoming traffic on designated protocols and port ranges to forward them to your applications. Listeners can be configured to process traffic for TCP, **User Datagram Protocol** (**UDP**), or a combination of both protocols.

Accelerators come in two different types:

1. A **standard** accelerator can be used to direct traffic over the AWS network backbone to an array of AWS services such as ALBs, **Network Load Balancers** (**NLBs**), EC2 instances, or even Elastic IP addresses. While many of these services can be configured to be public-facing, putting them behind an accelerator can increase the global performance of your applications behind these services. You can also set weights on how much traffic goes to each endpoint.

2. A **custom routing** accelerator can be used in scenarios when you have specific remote users who need specific routing exceptions (i.e., you can map certain public IP addresses for remote users to specific ports/protocols on the accelerator, which will route that traffic to a unique endpoint, which may differ from the standard behavior of the listener).

Static IP Addressing

When an accelerator is configured, AWS assigns a set of **static IP addresses** that will be allocated and associated with the accelerator. AWS Global Accelerator can support dual-stack applications or IPv4 only. Accelerators configured for IPv4 only will be assigned two static IP addresses, while dual-stack accelerators will be assigned four static IP addresses (two IPv4 addresses and two IPv6 addresses).

> **Note**
> Custom routing accelerators do not support dual-stack IP addressing.

From a remote user perspective, these are the IP addresses that will be used to associate with the public-facing application they are accessing. This can be beneficial in scenarios where remote users need to allow public IPs to access them publicly. These static IP addresses are associated with an accelerator for its entire lifetime (i.e., they are only released once the accelerator is deleted).

These static IP addresses not only provide consistency and ease of management for your global applications but also enable efficient routing of user traffic. AWS Global Accelerator leverages a routing technique called **anycast** to optimize the performance and availability of these IP addresses on a global scale.

Anycast

AWS Global Accelerator can use these public IP addresses effectively at a global scale using a method called **anycast**. Anycast is a method of advertising the same IP prefixes from several locations to ensure anyone trying to access these IP addresses will always go to the closest point at which they are advertised.

The static IP addresses associated with an accelerator can be advertised from AWS out to the public internet at the several edge locations and **internet exchange** (**IX**) points that are used globally. This is how AWS Global Accelerator provides an effective on-ramp to the AWS network backbone for your applications closest to remote users.

For example, Trailcats has a mobile application that is serviced by an NLB residing in the us-east-1 region. There is also a standard accelerator associated with the application, with the anycast IPs directed to the NLB. Users in Ireland, Australia, and Japan can all see an improved user experience because their traffic enters the AWS network backbone at the closest advertisement point within their region.

AWS also publishes a speed test that you can use to see what the potential improvement would look like using AWS Global Accelerator. Refer to *Figure 10.19* for an example of the speed test, which can be found at the following URL: `https://speedtest.globalaccelerator.aws`.

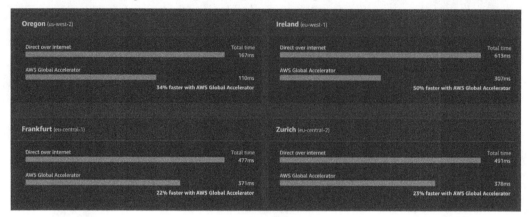

Figure 10.19: Global Accelerator speed test

The preceding speed test in *Figure 10.9* was run from Los Angeles, testing with a file size of 100 KB. You can see significant improvements in performance for several AWS regions, with data transfers happening as much as 50% faster to the `eu-west-1` region.

Endpoints

Simply put, **endpoints** are resources that an accelerator will send traffic to within AWS. As mentioned in the details about accelerators, these can be ALBs, NLBs, EC2 instances, or **Elastic IP address (EIPs)**. Endpoints can also have attributed weight values to adjust the amount of traffic that is sent to each endpoint by an accelerator. Endpoints can also be associated with endpoint groups, which allows for grouping resources on a regional basis.

Service Quotas

The ANS-C01 exam may expect you to have knowledge of some of the service quota limits for the topics covered in this chapter. This is because some service quotas have hard limits on the maximum values for each service quota. The following table includes some AWS service quotas you should be aware of for the exam.

Name	Default Limit	Adjustable?	Comments
Data transfer rate per CloudFront distribution	150 Gbps	Yes	
CloudFront distributions per AWS account	200	Yes	
Origins per CloudFront distribution	25	Yes	
Standard AWS accelerators per AWS account	20	Yes	
Custom routing accelerators per AWS account	10	Yes	
Port ranges per listener [AWS Global Accelerator]	10	Yes	

Table 10.1: CloudFront and AWS Global Accelerator service quotas

> **Note**
>
> These service quota details were pulled directly from the following AWS documentation pages:
>
> https://docs.aws.amazon.com/AmazonCloudFront/latest/DeveloperGuide/cloudfront-limits.html
>
> https://docs.aws.amazon.com/global-accelerator/latest/dg/limits-global-accelerator.html

Summary

In this chapter, you explored how to utilize the AWS global network for effective content distribution and application performance enhancement by leveraging two key services: Amazon CloudFront and AWS Global Accelerator. You gained insights into design patterns for using Amazon CloudFront to optimize content delivery across the globe, which is essential for ensuring low-latency access to web content and media. You also learned how to implement global traffic management using AWS Global Accelerator to enhance network performance and availability for applications hosted on AWS. Furthermore, the chapter covered integration patterns with other AWS services, demonstrating how CloudFront and Global Accelerator can work together to deliver a seamless user experience. By understanding these concepts, you have acquired crucial skills that are directly relevant to the AWS Certified Advanced Networking (ANS-C01) exam, as they reflect best practices for designing resilient, high-performing networks on AWS. These skills are also valuable in a professional context, enabling you to architect and manage complex global networks for enterprises using AWS, thus enhancing your career prospects in cloud networking and solutions architecture.

In the next chapter, you will learn about AWS security frameworks.

Exam Readiness Drill – Chapter Review Questions

Apart from mastering key concepts, strong test-taking skills under time pressure are essential for acing your certification exam. That's why developing these abilities early in your learning journey is critical.

Exam readiness drills, using the free online practice resources provided with this book, help you progressively improve your time management and test-taking skills while reinforcing the key concepts you've learned.

HOW TO GET STARTED

- Open the link or scan the QR code at the bottom of this page

- If you have unlocked the practice resources already, log in to your registered account. If you haven't, follow the instructions in *Chapter 16* and come back to this page.

- Once you log in, click the START button to start a quiz

- We recommend attempting a quiz multiple times till you're able to answer most of the questions correctly and well within the time limit.

- You can use the following practice template to help you plan your attempts:

Attempt	Target	Time Limit
Working On Accuracy		
Attempt 1	40% or more	Till the timer runs out
Attempt 2	60% or more	Till the timer runs out
Attempt 3	75% or more	Till the timer runs out
Working On Timing		
Attempt 4	75% or more	1 minute before time limit
Attempt 5	75% or more	2 minutes before time limit
Attempt 6	75% or more	3 minutes before time limit

The above drill is just an example. Design your drills based on your own goals and make the most out of the online quizzes accompanying this book.

First time accessing the online resources? 🔒

You'll need to unlock them through a one-time process. **Head to** *Chapter 16* **for instructions**.

Open Quiz

https://packt.link/ansc01ch10

OR scan this QR code →

11
Security Framework

In this chapter, you will learn about security frameworks as they apply to cloud networking. The topics you will learn about include the following:

- The cloud shared responsibility model
- Network segmentation
- Threat modeling
- Common security threats
- Mitigating security threats

You will gain an understanding of how cloud security frameworks are used in the real world, as well as how they come up in the exam. The objective of this chapter is to introduce security concepts and relate them to AWS cloud networking. In the exam, you may be presented with scenarios that require choosing security services based on the requirements of that scenario, so understanding threats and how to mitigate them is important.

For a company such as Trailcats, which relies heavily on cloud services for its operations, following cloud security frameworks is essential. These guidelines help Trailcats protect its data and applications by securing the network infrastructure. Ensuring that data moving to and from Trailcats' cloud servers is safe, using tools such as encryption, firewalls, and secure access controls, is essential to securing the environment.

By adhering to the **AWS Well-Architected Framework**, which is the AWS guidance and framework around secure cloud design and deployment, businesses can better protect their cloud networks from threats such as hacking and data breaches, ensuring their cloud-based resources remain secure and resilient against cyberattacks.

The Cloud Shared Responsibility Model

Fundamental to any discussion about cloud security is an understanding of the customer's and cloud provider's responsibilities, and where they differ, which is referred to as the shared responsibility model. The scope of responsibility between the customer and provider changes depending on what type of cloud service is being used.

Cloud services can be broadly categorized into three service models: **software as a service** (**SaaS**), **platform as a service** (**PaaS**), or **infrastructure as a service** (**IaaS**).

The shared responsibility model for SaaS services places most of the security responsibility on the provider. In the case of SaaS, Amazon (or a third-party SaaS provider) handles everything except for IAM and data protection. Examples of SaaS would be Amazon WorkDocs and hosted WordPress.

The shared responsibility model for PaaS services splits the responsibility for security between the customer and the service provider. In the case of PaaS, Amazon manages the operating system and platform, and the customer shares responsibility for the customer-exposed portions of the network infrastructure and communications paths. **Identity and access management** (**IAM**), securing the applications, and the security of the data are the customer's responsibility. Examples of PaaS would be DynamoDB and Amazon S3.

The shared responsibility model for IaaS services places much of the security responsibility on the consumer. The cloud provider is only responsible for the physical security of the hardware and the availability of the software that is used to create the IaaS workload. An example of IaaS would be EC2 instance types.

Figure 11.1 shows the division of responsibility between the cloud provider and the customer for IaaS:

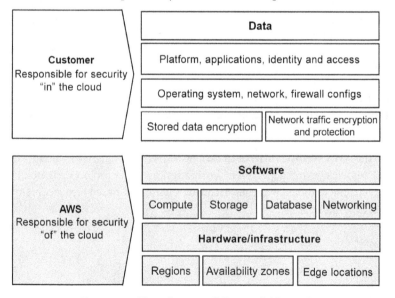

Figure 11.1: Shared responsibility model for IaaS

In *Figure 11.1*, you can observe that the division of responsibility for IaaS is somewhat complex. However, it's significantly less complex than managing everything, while still providing a large degree of control.

> **Note**
>
> In the cloud, IAM and data are *always* the customer's responsibility, regardless of the type of service.

As you can see, there's a clear trade-off here between responsibility and control. The more control the customer wants, the more they're responsible for from a security perspective. One of the appealing aspects of PaaS is the customer has a lot of flexibility while minimizing responsibility for the care and feeding of the underlying infrastructure, including security.

Network Segmentation

Network segmentation is a security practice that involves dividing a network into smaller, isolated segments or zones. This is important because it helps contain potential threats. If a cyberattack occurs in one segment, it is far less likely to be able to spread to other segments, limiting the potential damage inflicted. By controlling and restricting access between these segments, organizations can better protect sensitive data and critical systems. Network segmentation also makes it easier to monitor and manage security by helping to identify and respond to suspicious activity more effectively. Overall, it strengthens an organization's defense against cyberattacks.

In this section, you will learn how network segmentation is accomplished in AWS using **virtual private clouds** (**VPCs**), subnets, and firewalls. You'll also see a comparison of distributed and centralized models for firewall deployment.

Segmentation Basics

Zone-based security (sometimes referred to as macro-segmentation) is a way to provide a baseline level of network security by organizing resources by sensitivity/security level and placing them in dedicated VPCs. In the simplest form, the VPCs use **network access control lists** (**NACLs**) at the peering points, with a firewall at the network ingress/egress (often referred to as north/south/east/west) points. A more sophisticated approach is using firewalls between all VPCs and inspecting traffic moving in all directions, not just ingress and egress.

Good segmentation design is foundational to zone-based security (and network architecture in general), so although much of this section should be a review for most readers, it's worth taking a deeper look. In AWS, network segmentation generally happens at two levels – VPCs and subnets.

VPCs are a key component in network segmentation design. A VPC allows an organization to create isolated sections within a public cloud, each with its own network configuration. This isolation helps control traffic flow between different parts of the network, ensuring that sensitive areas are separated from less secure ones. Strictly speaking, the VPC as a construct is a hardline segment into which workloads can be deployed, and connectivity to other services and segments (VPCs) is a deliberate act accomplished through connectivity services such as VPC peering, **Transit Gateway** (**TGW**), or **network virtual appliance** (**NVA**).

When creating a network in AWS, it's important to pick the right type of subnet based on how you want your resources to connect with the internet and other networks. AWS offers different subnet types, each with its own level of access and security. Here's a simple breakdown of the main types and what they're used for:

- **Public subnet**: This subnet features a direct path to an internet gateway, allowing resources within it to access the public internet.

- **Private subnet**: This subnet lacks a direct route to an internet gateway. To access the public internet, resources here need a NAT device.

- **VPN-only subnet**: This subnet connects to Site-to-Site VPN via a virtual private gateway but does not have a route to an internet gateway.

- **Isolated subnet**: This subnet has no external routes outside its VPC. Resources here can interact only with other resources within the same VPC.

The key difference between the subnet types is routing. For example, you make a subnet private by not having a route to an internet gateway.

Subnets have NACLs that control inbound and outbound traffic for subnets. They act as stateless firewalls, allowing or blocking specific IP addresses or protocols based on set rules, helping protect resources by filtering traffic at the subnet level.

A common design is to place the user-facing portion of an application in a public subnet and put the backend components (such as business logic and stored data) in private subnets. For high-sensitivity resources, the isolated subnet would potentially be a good option. *Figure 11.2* depicts a basic segmentation hierarchy using VPCs and subnets:

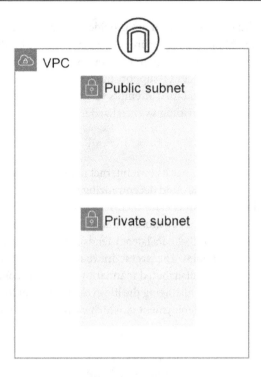

Figure 11.2: Network segmentation hierarchy

In *Figure 11.2*, the outer boundary is the VPC. At the next layer are public subnets, that is, subnets where traffic has a route to the internet. The innermost layer is the private subnet. Private subnets do not have a route to the internet.

While these constructs provide the basic building blocks of segmentation, most organizations will want to deploy firewalls and similar tools that provide a combination of visibility and control in order to detect and respond to threats (to keep things simple for the sake of discussion, the term firewall will be used here). What follows is a brief discussion of firewall deployment models in AWS.

Deployment Models

In AWS, firewall deployment models are crucial for managing and securing network traffic. Organizations can choose between distributed, centralized, or a combination of these models, depending on their security needs and architecture. Each approach offers different benefits in terms of control, scalability, and performance. Understanding each model, with its benefits and drawbacks, is important to select the best option for your cloud environment.

Distributed is more complex than centralized but provides better performance. Centralized is easier to manage and troubleshoot, but potentially represents a performance bottleneck if care is not taken. Another consideration is business organization. For example, if each business unit controls its own resources and there's no shared services IT support model in place, a distributed design may align better despite being less flexible. Combined attempts to split the difference, allowing local ingress/egress for performance while still providing a centralized inspection for east/west traffic.

Distributed Design

In a distributed deployment, each VPC has its own internet ingress/egress point protecting one or more public subnets. This use case centers around decentralizing the enforcement of security to move that enforcement and policy closer to the workloads themselves. The major benefit of this approach is that it saves on data transfer charges and can be more granularly applied to each VPC and each workload within that VPC. An added benefit is that data latency tends to be far lower because there is no need to pull data to a remote enforcement point. The largest drawback to such a design is that it will certainly cost more to deploy firewalls in this distributed manner by increasing compute and licensing costs, while also increasing the time cost of managing the life cycle of a fleet of firewalls. *Figure 11.3* depicts an example of a distributed firewall deployment in which each VPC has its own firewall:

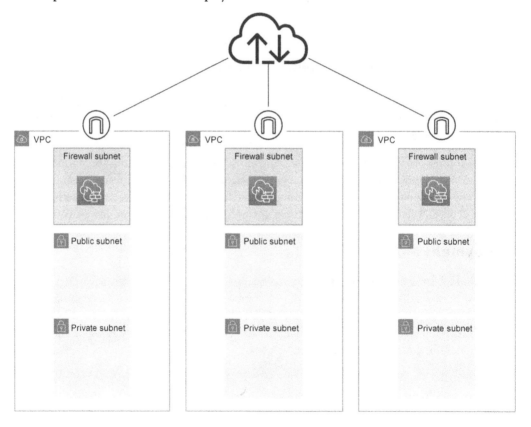

Figure 11.3: Distributed firewall design

In the preceding figure, each VPC has its own firewall that serves as the ingress and egress point for traffic entering or leaving the VPC. This allows for a granular policy to be deployed that can apply to each VPC individually.

Centralized Design

In a centralized design, a dedicated security VPC is created, and all inter-VPC traffic is routed through the security VPC, providing inspection and control for traffic flowing in all directions, that is, between VPCs, as well as ingress/egress. This allows for a much smaller firewall deployment at strategic network choke points, such as the security VPC being attached to a transit gateway. This approach saves on cost and time spent managing the life cycle of firewalls as well. The drawbacks to this approach include paying for data transfer that may ultimately be dropped or denied by the firewall and the extra latency introduced by routing traffic to remove enforcement point.

Figure 11.4 depicts a centralized design that uses a transit gateway in each VPC to ship traffic to a dedicated traffic inspection VPC:

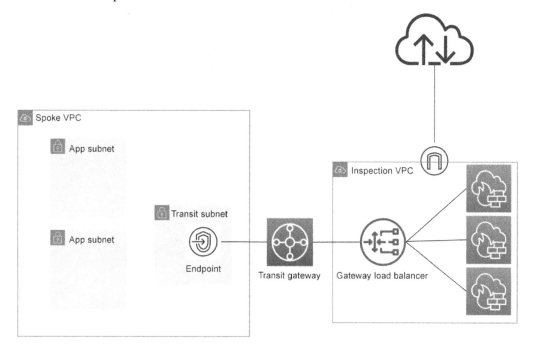

Figure 11.4: Centralized firewall design

In this figure, all traffic from a VPC will be routed to a remote security VPC for inspection by a firewall before being dropped or routed to its original destination.

Combination Design

In a combination design, east/west and on-premises ingress/egress traffic is routed through the security VPC. Each VPC then has its own internet ingress/egress point protecting its public-facing subnets. This approach aims to minimize the drawbacks of both prior designs and maximize the benefits. It is not entirely successful, however. From a benefits perspective, the traffic to and from the internet will not incur data transfer charges by being pulled to a security VPC. It will also have lower latency as the ingress and egress traffic is distributed. Most importantly, the firewalls deployed to the VPCs directly can be sized in a more appropriate way to focus only on internet traffic instead of all traffic, which may save on compute costs.

Centralizing the firewalls for east-west traffic only allows for lower data transfer charges and allows you to implement different security for east-west traffic itself instead of having a shared policy for both north-south and east-west traffic.

Ultimately, this approach increases traffic complexity by a large amount and should be carefully considered. Route tables will be much more granular to steer traffic to the appropriate firewall, and traffic patterns will be harder to troubleshoot.

Figure 11.5 depicts a combination design where ingress/egress traffic is routed to a VPC-attached firewall, and inter-VPC traffic is routed to a dedicated VPC for inspection:

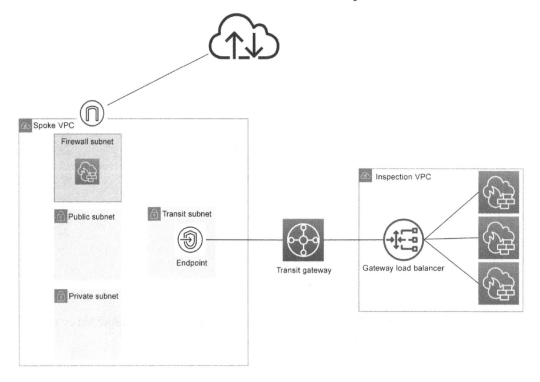

Figure 11.5: Combination design

The combination design is about traffic engineering to steer traffic to the most advantageous place for a security policy to be enforced based on the destination.

Threat Modeling

Threat modeling provides a clear understanding of the security threats to a system. It begins by mapping out the system's architecture to identify potential vulnerabilities and the specific threats that could exploit them. Once these threats are identified, the likelihood of each threat occurring and its potential impact is assessed. This assessment helps determine the level of risk associated with each threat. With a clear understanding of these risks, appropriate countermeasures can be implemented, such as patching vulnerabilities or using compensating controls to mitigate or eliminate the threats. Using threat modeling, organizations can proactively manage their security, prioritize risks, and ensure better protection against potential attacks.

Someone studying for a networking-oriented exam may question the relevance of threat modeling. It is listed as a blueprint topic, and in the real world, you may be expected to work with security teams as a subject matter expert, so knowing the basics of threat modeling will be useful.

Threat, vulnerability, likelihood, impact, and risk have specific meanings in cybersecurity and it's important to understand them to construct a working threat model:

- **Threat**: A threat is an undesirable event. It could be human or natural, malicious or accidental, for example, ransomware or flooding.

- **Vulnerability**: A vulnerability is a weakness that allows the threat to occur. It could be an easily guessed password or a building located in an area that frequently experiences severe weather.

- **Likelihood**: Likelihood is the probability that a threat will combine with a vulnerability to manifest into an event. For example, the likelihood of a flood is much higher in a coastal area than it would be on top of a mountain.

- **Impact**: Impact is the effect of the event. For example, a broom closet being broken into is a low-impact event. A flood in the data center where the main company database is located is a high-impact event.

- **Risk**: Risk can be understood in different ways, either by using detailed numbers or by making a general assessment. In simple terms, risk can be thought of as a combination of how likely something is to happen and how bad the outcome would be if it did. This basic idea of risk is what will be explored here.

Table 11.1 depicts a method of evaluating risk using a 5 x 5 matrix. Multiplying the *x* by the *y* axis (impact times likelihood) provides a score for quantitative analysis:

	Impact: 1 (Low)	Impact: 2 (Moderate)	Impact: 3 (High)	Impact: 4 (Very High)	Impact: 5 (Critical)
Likelihood: 1 (Rare)	Low	Low	Moderate	Moderate	High
Likelihood: 2 (Unlikely)	Low	Moderate	Moderate	High	High
Likelihood: 3 (Possible)	Low	Moderate	High	High	Very High
Likelihood: 4 (Likely)	Moderate	High	High	Very High	Critical
Likelihood: 5 (Almost Certain)	Moderate	High	Very High	Critical	Critical

Table 11.1: 5 x 5 risk matrix

In the preceding table, the risk levels in the matrix are as follows: *Low* risk requires monitoring but no immediate action. *Moderate* risk should be addressed but is not urgent. *High* risk needs attention soon and may require mitigation strategies. *Very High* risk demands immediate action, while *Critical* risk must be addressed as a top priority to avoid severe consequences. The matrix can be adapted based on specific organizational needs and the context of the risks being evaluated.

As an example, a common threat is the threat of a phishing attack, where an attacker may send an email to users to get them to click a link that installs malware onto their system. The likelihood of such an attack is *Likely* (4), on *Table 11.1*, and the impact is probably *Moderate* (2) for a standard user, but the threat could be *High* (3) or even *Very High* (4) depending on the user. Multiplying the matrix, we come up with a quantitative score of 8 (2*4), 12 (3*4), or 16 (4*4). This number gives guidance on the appropriate threat model and what level of countermeasure should be considered.

When thinking about risk, it's helpful to consider some real-world situations. For example, the risk posed by a poorly secured broom closet is low even though the likelihood of someone entering and taking something may be high. That's because the impact of stealing toilet paper is negligible. Therefore, it makes little sense to invest in countermeasures beyond a basic doorknob lock. On the other hand, losing the critical company database to a flood could be a critical risk because the impact is extremely high, even with a relatively low likelihood of occurrence. This risk could justify the expense of continuous replication to an offsite location, or even changing facilities altogether.

Threat Modeling Process

The process of threat modeling can be condensed into three main steps. First, it involves identifying and listing all potential threats to the system. This is called threat enumeration. Next, these threats are categorized and ranked based on factors such as their likelihood and potential impact. Two popular risk models are **STRIDE** and **DREAD**, which will be covered in the section on risk models. Finally, the organization decides on the most effective countermeasures and mitigation strategies to address the identified threats, ensuring that the system is better protected against potential risks. The next section will discuss the process.

Enumerating Threats

A good approach to threat enumeration is to break down the system and understand a few key aspects. Start by defining what a normal use case looks like, as this helps in identifying deviations that could signal a threat. Next, consider the attack surface, which includes all the points where an attacker could potentially gain access. It's also important to identify the assets that might be of interest to an attacker, as these are likely targets. Finally, evaluate the access rights required by different entities interacting with the system, since these rights can be exploited if not properly managed.

Using the concept of threat enumeration, let's consider an example with Trailcats, our fictitious company that operates an online platform for feline adventure gear.

First, Trailcats starts by defining what a normal use case looks like. For instance, a customer browsing the website, selecting gear, and making a purchase is a typical use case. Next, they identify the attack surface, such as the website's login page, payment processing system, and customer database. Then, they consider what assets an attacker might be interested in, such as customer credit card information or personal details stored in the database. Finally, Trailcats reviews the access rights required by entities interacting with the system, such as the permissions granted to customers, administrators, and third-party payment processors. By breaking down these aspects, Trailcats can better understand where potential threats might arise and take steps to mitigate them.

With cloud services, it's much easier for attackers to discover and enumerate some targets because they are associated with services that are designed by the cloud service provider to be public and highly available. Because of this, it's important to have an in-depth defense. You're not going to be able to stop an attacker from scanning you in the cloud, but you can get visibility, and you can control access to your resources.

Risk Models

To help organize and prioritize risks, it's helpful to use models. Two commonly used models are STRIDE and DREAD.

The STRIDE threat model is a framework developed by Microsoft to help identify and categorize potential security threats in a system. The acronym **STRIDE** stands for **spoofing, tampering, repudiation, information disclosure, denial of service, and elevation of privilege**. Each of these categories represents a different type of threat that could impact the security of a system. STRIDE is widely used in the field of cybersecurity for systematically analyzing threats and is especially useful during the design phase of software development to ensure security considerations are integrated from the start.

The DREAD threat model is a risk assessment framework also developed by Microsoft to evaluate and prioritize security threats. **DREAD** stands for **damage potential, reproducibility, exploitability, affected users, and discoverability**. Each of these categories helps assess how severe and likely a threat is, allowing organizations to prioritize which risks need the most attention. The model provides a structured way to score and compare threats, making it easier to determine where to focus security efforts.

STRIDE is ideal for identifying and categorizing different types of security threats during the design phase of a system. It's particularly useful for teams that want to ensure they are considering all possible threat vectors, such as identity spoofing, data tampering, or denial-of-service attacks. STRIDE helps in systematically thinking through potential vulnerabilities and ensuring that security is integrated from the beginning.

DREAD, on the other hand, is more focused on assessing and prioritizing the risks associated with identified threats. If a team has already identified potential threats (using STRIDE or another method), DREAD helps them evaluate which threats are the most dangerous or likely to occur. By scoring threats on aspects such as damage potential and exploitability, DREAD allows teams to prioritize their security efforts, focusing on the most critical risks first.

In summary, STRIDE is best for identifying and categorizing threats, while DREAD is more suited for evaluating and prioritizing the risks those threats pose. Together, they offer a comprehensive approach to threat modeling and risk management.

Common Attacks

This section outlines the common types of attacks that network engineers should be prepared to defend against. In each case, there will be an explanation of the attack, followed by a corresponding set of countermeasures for that type of attack. Since this exam focuses on networking, only network-relevant threats are discussed. For instance, while phishing is a prevalent attack vector, it will not be covered here.

As in the section on threat modeling, it's useful to explicitly define what some key terms mean, and in this case, those terms are attack and countermeasure.

An attack is any deliberate attempt to compromise the confidentiality, integrity, or availability of a system, network, or data. The goal of an attack is typically to steal data, disrupt services, gain unauthorized access, or cause harm to an organization's resources.

A countermeasure is a specific action, process, or technology implemented to prevent, mitigate, or respond to a security threat or attack. Countermeasures can include software tools, security policies, practices, and other controls designed to protect systems, networks, and data from unauthorized access, damage, or disruption. Examples of countermeasures are firewalls, encryption, intrusion detection systems, and logging.

The following list is non-exhaustive but shows many different types of attacks:

- Account compromise
- Exploiting inadequate permissions
- Port scanning
- Hijack attacks, which include the following:

  - **Man in the middle (MitM)**
  - DNS cache poisoning

- Application vulnerabilities, such as the following:

  - **Cross-site scripting (XSS)**
  - Cross-site request forgery
  - Command injection
  - SQL injection

Account Compromise

With cloud services, there are two different kinds of account credentials to be concerned with. The first is user credentials, and the second type is account keys, which are typically used to access API-based services. Attackers have many creative ways to gain access to credentials; it's safe to assume credential compromise will happen, so it's important to focus on early detection and minimizing impact.

One way for an attacker to establish persistence is to create account keys for an IAM user. As AWS allows up to two keys per IAM user, it provides an opportunity for an attacker to create an account key without being noticed.

The attacker will generally follow these steps:

1. Enumerate IAM users with `aws iam list-users`.

2. Enumerate account keys with `aws iam list-access-keys –user-name jsmith`.

3. Identify which IAM users have 0 or 1 keys.

4. Create a new key with `aws iam create-access-key –user-name jsmith`.

To enhance security, it's important to implement several key practices. First, make use of logging tools such as CloudWatch and CloudTrail, along with any specific logging capabilities provided by applications or services, to monitor and record activity. **Multi-factor authentication (MFA)**, though sometimes inconvenient, is a highly effective way to protect user accounts from being compromised. To secure APIs, consider using OAuth, an authorization protocol that uses short-lived tokens for accessing services, providing an additional layer of security. Regularly rotating account keys is another critical measure, as it reduces the risk and impact of compromised keys. Finally, applying the principle of least privilege through **role-based access control (RBAC)** ensures that users have only the permissions necessary for their job roles, and nothing more. It is also important to remove outdated permissions when someone changes roles to prevent unauthorized access over time.

Inadequate Permissions

The risks of unsecured S3 buckets are well known. To address this, AWS now makes S3 buckets private by default, but many still have weak permissions that are easily exploited. One reason for this is that it's easy to find and test S3 buckets. Since each S3 bucket must have a unique name, attackers can use public web addresses to check whether a bucket exists and then test its permissions. These addresses include `s3.amazonaws.com/BUCKETNAME` and `BUCKETNAME.s3.amazonaws.com`. Attackers often start by scanning an organization's website using tools such as **CewL** to create a list of possible bucket names. They then apply rules to generate variations of those names and use the list to check for existing buckets. Moreover, S3 buckets are sometimes exposed through website crawling and Google searches, especially when they're used to store static resources such as PDFs and images.

Another common issue related to permissions is that over time, organizations may unknowingly add sensitive items to those public S3 buckets, not recognizing the risk. This is also common with Windows file shares, FTP repositories, and other types of file storage systems.

> **Note**
> A popular tool for enumerating S3 buckets is Bucket Finder, available at `https://digi.ninja/projects/bucket_finder.php`. The CeWL wordlist generator is available at `https://digi.ninja/projects/cewl.php`.

By default, S3 access logging is turned off, which is typical for most AWS services. It's important to enable logging to track who is accessing your S3 buckets. Additionally, regularly scanning your assets for inappropriate access permissions or sensitive content helps prevent unauthorized access. Finally, ensure that all data stored, known as data at rest, is encrypted to protect it from potential breaches.

Port Scanning

Port scanning is still a popular and effective method of service discovery whereby an attacker will launch a program that attempts to connect on every possible service port, looking for open ports and running services to exploit. In the cloud, it comes with some added twists The most important thing to know is you cannot stop people from scanning you. This is how attackers can identify and exploit targets on cloud service providers such as AWS at scale:

1. Attackers spin up an EC2 instance in the same Region they're interested in scanning.
2. They retrieve the IP address ranges for the Region from `https://ip-ranges.amazonaws.com/ip-ranges.json`.
3. They use `masscan` to find instances listening on port `443`.
4. They use `tls-scan` to extract information from the x.509 certificates of their targets.
5. They use Nmap and other web vulnerability scanners to get a deeper look at interesting targets from the `tls-scan` step.

You might think that AWS would shut down entities performing large-scale scans, but the reality is that's very rare. It costs attackers pennies to scan for targets at an internet scale, so you can expect to be scanned regularly; if an obvious vulnerability is found, someone is going to take advantage.

If an attacker gains a foothold on an EC2 instance, they will surely use it as a pivot to scan private subnets for additional targets. Don't assume that because a resource lives in a private subnet, it's not subject to scanning.

> **Note**
>
> Some common programs that attackers use include the following:
>
> `masscan` is a high-performance port scanner. It can be found at `https://github.com/robertdavidgraham/masscan`.
>
> `tls-scan` scans web servers, extracting certificate information for later use. It can be found at `https://github.com/prbinu/tls-scan`.
>
> `jq` is a command-line tool for parsing JSON. It's quite helpful for working with information returned from cloud service providers. Find out more at `https://jqlang.github.io/jq/`.
>
> Nmap is the classic port and service scanner. It also supports service discovery and has built-in scripting. It can be found at `https://nmap.org/`.

Utilizing VPC flow logs is important for monitoring and analyzing traffic within your network. It's also important to implement **network security groups (NSGs)** and NACLs, which help control inbound and outbound traffic. Segmenting subnets with firewalls further enhances security because the firewall becomes a traffic inspection and policy enforcement point. Using tools such as TGW combined with Gateway Load Balancer can be particularly effective for scaling firewall usage. Regularly auditing your ACLs and NSG permissions ensures that access is appropriately restricted and aligns with security best practices.

Hijacking

Hijacking is a class of attack where the attacker responds to a service request purporting to be the legitimate resource. DNS cache poisoning is a common technique to redirect the victim to the attacker. In a variation of this attack, MitM, the attacker accepts the connection from the victim, then forms a connection to the intended resource, placing themselves in the data path. This allows the attacker to intercept and potentially alter data without the victim being aware. This type of attack is harder to pull off in the cloud than in an on-premises environment, but it can be done. The easiest way to accomplish an MiTM attack is to intercept traffic close to the end user and proxy communications to the service being hosted in AWS.

Implementing **Domain Name System Security Extensions (DNSSEC)** is important to protect against attacks targeting your domain name service. It's also essential to use transport encryption methods such as **Transport Layer Security (TLS)** and **Internet Protocol Security (IPSec)** to safeguard data as it travels across the network. Additionally, using a load balancer as a TLS termination point can offload encryption tasks from the server, meaning that the server itself doesn't need to handle encryption, which can simplify the setup and improve performance.

Application Vulnerabilities

Exploiting application vulnerabilities is a large field of study, but for the purposes of this book, the focus is on XSS, command injection, and SQL injection. As a network engineer, it's most likely not going to be your concern to fix the vulnerable code, but you may be asked to implement a compensating control to prevent the vulnerable code from being exploited while the problem is being addressed. So, you'll need to know a little about these attacks and how they work.

XSS is an attack against users by tricking a vulnerable web server into sending malicious scripts to a web browser that will execute them. The two most common forms are reflected XSS and stored XSS.

Reflected XSS is a directed attack, commonly delivered as embedded links in a phishing email. When the user hovers over the link, the domain will look legitimate and most likely is. The script/payload is delivered as a parameter in the URL, which is long and hard to read. Because the domain in the URL is the target domain, the user generally doesn't worry about it. This is a powerful form of attack because JavaScript can do things such as steal authentication cookies for other websites.

Stored XSS is when vulnerable websites accept malicious input from one user and display it to other users. Think product reviews, product descriptions, and so on. Anyone who browses to the page with the malicious content will end up executing the embedded script.

Command injection takes advantage of a web server that calls an external application to perform a function. For example, an application may be invoked by a server that resizes an uploaded image, but with command injection, that legitimate function can be replaced by different, malicious commands. The idea here is to manipulate the input, so the server executes the command of the attacker's choosing. The technique typically involves something called command stacking, which uses delimiters to send multiple commands in a single line.

SQL injection has similar characteristics to command injection. The idea is to manipulate the input to get the web server to pass queries to a backend database server. This typically involves using logic that is always true combined with operators and special characters.

Web application firewalls (**WAFs**) are a valuable tool for detecting and blocking attempts to manipulate input fields on your website where the website developer may have missed something. To protect your application, it's important to implement input sanitization, which ensures that only valid data is submitted by users. Using parameterized queries is another effective measure, as it prevents attackers from manipulating SQL queries to gain unauthorized access to your database. Lastly, an **object-relational mapper** (**ORM**) can be used to create an abstraction layer between your code and the SQL data source, reducing the risk of SQL injection attacks by managing database interactions in a secure way.

> **Note**
> WAFs are not a substitute for secure coding practices. They're a backstop. This isn't relevant to the exam, but in the real world, it's important to understand.

In this section, you learned about common attacks in cloud environments and how to defend against them. You'll find this information useful for the exam and the real world. The remainder of the chapter will cover encrypting data and securing DNS.

Encryption Methods for Data in Transit

In accordance with the shared responsibility model, AWS handles some data-in-transit encryption, while other flows are the customer's responsibility.

All data traffic between AWS data centers is automatically encrypted at the physical level. Similarly, data exchanged within a VPC and between interconnected VPCs across different Regions is also encrypted at the network level, provided that the Amazon EC2 instance types supporting this feature are used.

At the application level, the customer decides how to implement encryption. The most common option is TLS. AWS service endpoints accommodate TLS, enabling secure HTTPS connections for API requests. Load balancers are a great way to control the parameters of TLS encryption with minimal impact on the workload, assuming that end-to-end encryption is not a requirement.

In a hybrid cloud design, IPSEC may be used for transport encryption between the AWS environment and the on-premises environment. This ensures end-to-end encryption of the data between the on-premises egress point and the workload residing in AWS.

Security Methods for DNS Communications

DNS's design predates the internet, when security was not a top-of-mind concern. This makes DNS relatively vulnerable to attack as it wasn't designed with the idea of making it resilient in the face of threat actors.

DNS spoofing and cache poisoning are common techniques to facilitate MitM attacks, where the attacker masquerades as the legitimate service the user is trying to reach. Additionally, DNS can be a useful source of target discovery through the enumeration of DNS records. Attackers typically use tools and scripts to automate the process.

Segmentation, such as using split DNS or **private hosted zones** (**PHZs**), along with implementing DNSSEC are key strategies for enhancing network security. These approaches help control access to DNS information and protect against potential domain name service attacks. The following list explores each in more depth:

- **Segmentation**: Segmenting the DNS namespace between public and private zones makes target enumeration more difficult because attackers would need access to both and to be able to correlate them. Segmentation also makes it more difficult to discover the topology and architecture of your backend systems because an attacker may not be able to access different segments easily, which makes mapping difficult. This is often referred to as split DNS. In AWS, this is implemented using PHZs.

- **DNSSEC**: The purpose of DNSSEC is to authenticate DNS records. It does not encrypt records in transit or anything like that; it's strictly about authenticity. DNSSEC overlays a **public key infrastructure** (**PKI**) on top of the global DNS system with the root zone (.) acting as the root of trust through an elaborate root signing ceremony. Digital signatures are created for DNS records and are themselves stored as DNS records. These digital signatures allow DNS servers and clients to verify the authenticity of DNS records.

> Note
>
> **DNS over HTTPS** (**DoH**) encrypts DNS queries in transit and is gaining popularity. Since it's not covered in the exam, it wasn't covered in this section; however, as a networking professional, you should be aware of it. More information is available at https://www.rfc-editor.org/rfc/rfc8484.html.

Summary

In this chapter, you learned about the cloud shared responsibility model and how AWS implements it. Of particular importance is that the customer is always responsible for IAM and data, regardless of the service model. You learned about using network segmentation to divide the network into zones as a way of enhancing security. You also learned about threat modeling along with common attacks and their corresponding countermeasures, which will give security-focused engineers a playbook to start tackling threats in their own networks. You learned some of the most common categories of threats, which will help you organize the security features that AWS offers into a more understandable framework. On the exam, you will be better able to identify the appropriate security services to employ based on the threats mentioned. You learned about the mechanisms available to secure application flows and encrypt data in transit. Lastly, you reviewed DNSSEC and split DNS as methods for securing DNS communications. Knowledge of these topics will help prepare you for both the exam and the real world.

In the next chapter, you will learn about AWS security services.

Exam Readiness Drill – Chapter Review Questions

Apart from mastering key concepts, strong test-taking skills under time pressure are essential for acing your certification exam. That's why developing these abilities early in your learning journey is critical.

Exam readiness drills, using the free online practice resources provided with this book, help you progressively improve your time management and test-taking skills while reinforcing the key concepts you've learned.

HOW TO GET STARTED

- Open the link or scan the QR code at the bottom of this page

- If you have unlocked the practice resources already, log in to your registered account. If you haven't, follow the instructions in *Chapter 16* and come back to this page.

- Once you log in, click the START button to start a quiz

- We recommend attempting a quiz multiple times till you're able to answer most of the questions correctly and well within the time limit.

- You can use the following practice template to help you plan your attempts:

Working On Accuracy		
Attempt	Target	Time Limit
Attempt 1	40% or more	Till the timer runs out
Attempt 2	60% or more	Till the timer runs out
Attempt 3	75% or more	Till the timer runs out
Working On Timing		
Attempt 4	75% or more	1 minute before time limit
Attempt 5	75% or more	2 minutes before time limit
Attempt 6	75% or more	3 minutes before time limit

The above drill is just an example. Design your drills based on your own goals and make the most out of the online quizzes accompanying this book.

First time accessing the online resources? 🔒

You'll need to unlock them through a one-time process. **Head to** *Chapter 16* **for instructions**.

Open Quiz	
https://packt.link/ansc01ch11	
OR scan this QR code →	

12

AWS Security Services

Imagine you're planning a security system for your house. First, you assess potential risks—such as break-ins or fires—and create a detailed plan to address each one. You learned how to do this in the previous chapter. Based on this plan, you then choose specific tools, such as high-quality locks, cameras, and alarms, that best match the needs you identified.

Similarly, in this chapter, you will take the security plans you've developed and learn how to select and implement the right defenses to bring those plans to life effectively.

To that end, this chapter will cover the following topics:

- Controlling traffic
- Observability and monitoring
- Securing traffic
- Managing and auditing security configurations

By the end of this chapter, you will understand how to deploy traffic controls in layers, how to monitor these controls, how to secure your traffic with encryption, and how to monitor your configurations to ensure they function as intended. This will help prepare you for the implementation of security services both in the real world and for the exam.

Controlling Traffic

When managing traffic in AWS, a layered approach (placing controls starting at the network edge and ending at the host running the application) ensures robust security and smooth traffic flow across the online infrastructure. This strategy covers multiple layers, from the global level to the instance level, to protect and manage network traffic effectively.

At the global and regional levels, you can utilize AWS Shield and AWS **Web Application Firewall** (**WAF**) to defend against **distributed denial of service** (**DDoS**) attacks and other web-based threats that could impact web applications. Route 53, AWS's DNS service, is used to direct customer traffic efficiently to the appropriate resources, ensuring a fast and secure user experience. Meanwhile, CloudFront helps deliver content quickly to customers around the world, whether they're browsing product pages or streaming videos.

As we focus on **virtual private clouds** (**VPCs**) and subnets, AWS Network Firewall provides advanced traffic filtering and deep packet inspection, adding an extra layer of security to your network. To manage incoming traffic and maintain high availability, you can leverage Elastic Load Balancer, which distributes requests across multiple servers, ensuring that web applications remain responsive even during peak traffic periods. Finally, at the instance level, Trailcats uses operating-system-level controls to manage traffic, allowing only authorized access to applications.

By integrating these AWS services, you can effectively control traffic and safeguard your online presence from potential threats, ensuring a secure and seamless experience for users.

Layered Approach to Traffic Control

When traffic from the internet heads to resources hosted on AWS, it goes through several layers of AWS infrastructure before reaching the specific instance that runs your service or application. At each of these layers, AWS offers tools to decide whether the traffic should move to the next level. You can use these tools to build strong, layered protection for your applications and data. This section will help you understand which AWS tools or services to use in different situations, which is important for both the exam and the real world. *Figure 12.1* depicts these layers and the services, tools, and features at each layer:

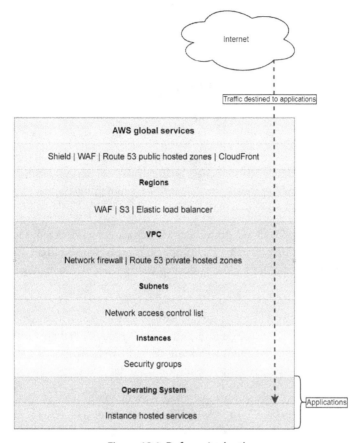

Figure 12.1: Defense in depth

Figure 12.1 provides a reference for how controls map to the layers of AWS infrastructure. Understanding where a control is relevant and the scope of its effects is fundamental.

In the following sections, you will work your way down the levels, starting at the global level and working down toward the instance level.

AWS Shield

AWS Shield is a managed DDoS protection service that operates at the network, transport, and application layers (layers 3, 4, and 7). It comes in two flavors: Standard and Advanced. Standard is provided automatically, free of charge. AWS Shield Advanced is a subscription service that provides the following:

- WAF at no charge (with some limitations).

- Extra protection for some types of resources.

- Health-based detection via Route 53.

- The AWS **Shield Response Team** (**SRT**), which requires a Business or Enterprise Support plan.

- **Economic denial of sustainability** (**EdoS**) coverage in the form of Shield Advanced Service Credits. The credits protect you from a large spike in your AWS bill due to a DDoS attack.

The headline features of Shield Advanced that you may want to memorize for the exam are the Shield Response Team and the Shield Advanced Service credits.

The AWS SRT is a group of security experts specializing in mitigating DDoS attacks. Available 24/7 to AWS Shield Advanced customers with Business or Enterprise Support plans, the SRT assists during DDoS events by analyzing traffic patterns, implementing custom mitigations, and providing guidance to maintain application availability and performance.

AWS Shield Advanced offers DDoS cost protection, which includes service credits to offset unexpected expenses resulting from scaling during a DDoS attack. If protected resources scale up in response to an attack, AWS may issue credits for the associated charges, helping to manage unforeseen costs.

AWS WAF

AWS WAF is a managed service that monitors and controls requests that are forwarded to your web applications. It can allow, deny, or count web requests using **web access control lists** (**web ACLs**). AWS WAF is deployable on API gateways, Application Load Balancers, and CloudFront distributions.

At the heart of a web ACLs are **rule conditions**, which determine what characteristics of the incoming traffic are being evaluated, such as IP addresses, HTTP headers, or the presence of specific strings in the request. Once a condition is matched, a **rule action** is triggered, which can either allow, block, or count the request. **Priority** plays a critical role in AWS WAF, as it determines the order in which rules are evaluated. Higher priority rules are assessed first, ensuring that more critical protections are applied before others.

You can also set up **custom responses** for blocked requests, allowing you to return specific messages or codes to users when their traffic is blocked by WAF.

For scenarios where controlling the rate of requests is important, **rate-based rules** are invaluable. These rules allow you to throttle traffic based on the number of requests from a single IP address, protecting your application from denial-of-service attacks or abusive traffic.

AWS WAF also offers **managed rule groups**, which are pre-configured sets of rules created by AWS or third-party vendors. These rule groups provide protection against a wide range of common threats, saving time and effort in creating custom rules from scratch.

Together, these elements allow you to build comprehensive, flexible, and scalable defenses for your web applications using AWS WAF web ACLs. The following section discusses these features in more detail.

An AWS WAF web ACL rule includes several key elements. It defines conditions, such as specific IP addresses, geographic locations, or particular HTTP requests, that determine which traffic the rule applies to. The rule then specifies actions, such as allowing, blocking, or counting the requests, based on those conditions. It can also include rate-based rules to manage excessive traffic from specific sources. These elements work together to control and filter web traffic according to the security policies set in the web ACL. *Figure 12.2* is an example of a web ACL with three rules:

Figure 12.2: Web ACL example

It's worth noting that at the time of writing, the WAF console is not as well organized as other consoles, and getting to the rule list in a web ACL requires some navigation. *Figure 12.2* depicts the navigation steps needed using numbers. Reviewing the preceding figure, the navigation steps are as follows:

1.   Search for WAF & Shield in the AWS console search bar.

2.   Under AWS WAF, select WebACLs.

3.   Select the web ACL of interest.

4.   Select Rules.

5.   The Rules table provides a summary of the rules in the ACL. The summary includes the rule's name, action, and priority, and a custom response (if one has been configured). The rule list doesn't show the conditions that will cause a rule to fire; you must edit the rule to review that information.

In the upcoming section, you'll take a closer look at web ACL rule elements to gain a deeper understanding of how they work, starting with rule conditions.

Rule conditions are used to match traffic for processing by the rule. The available conditions are extensive. Some example conditions are as follows:

- **Cross-site scripting**: Blocks malicious code injections
- **Country of origin**: Filters requests by country location
- **IP address**: Allows or blocks specific IPs
- **HTTP headers**: Checks information in request headers
- **JSON attribute/value pairs**: Examines JSON data for patterns
- **Strings within the message body**: Searches content for specific phrases

A single condition can be matched, or multiple conditions can be chained together and matched using **OR** logic (match any condition in the list) or **AND** logic (match all conditions in the list). For requests that don't match any rules, there is a configurable default action. *Figure 12.3* illustrates a country-of-origin condition, which filters web requests based on the requester's country, allowing or blocking access by location:

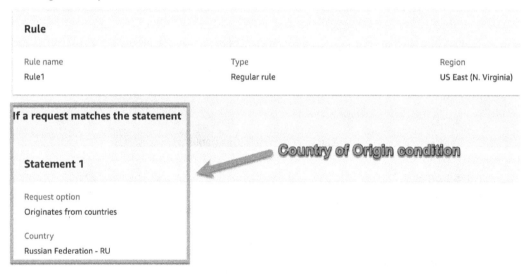

Figure 12.3: Rule condition example

Figure 12.3 depicts a web rule called rule 1. The matching condition for rule 1, called `Statement 1`, matches traffic originating from the Russian Federation. If this condition matches, it will cause the rule to fire.

Based on this configuration, if a web request from the Russian Federation hits the WAF, the rule and its associated action will be triggered.

Rate-based rules use a special type of rule condition that applies an action after a set number of requests are received within a specified time window. It's configurable to only count requests that meet match criteria, or to count all requests. This is different than the standard rule illustrated in rule conditions and rule actions. *Figure 12.4* illustrates a rate-based matching condition:

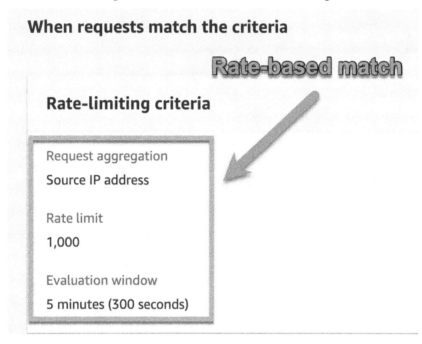

Figure 12.4: Rate-based match

In this example, if the same IP address makes 1,000 requests in a five-minute window, the rule and its associated action will be triggered.

Rule actions occur in a rule when required conditions are matched. These are the possible actions for a rule match:

- **Allow**: The request is allowed. A custom header can be added prior to forwarding.
- **Block**: The request is blocked. Optionally, a custom response can be defined, which consists of an HTTP response code and one or more custom response headers.
- **Count**: WAF counts the request but does not make a processing decision. WAF will continue to process rules in priority order.
- **Captcha and challenge**: WAF gives the user puzzles or challenges to verify the request is not coming from a bot or other form of automation.

Figure 12.5 illustrates a block action with a custom response:

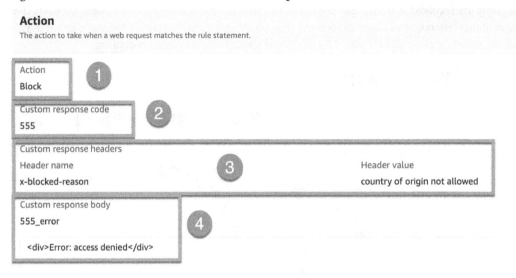

Figure 12.5: Web ACL block action

In *Figure 12.5,* the action (1) blocks the request and then returns (2) a custom response code. 3 is a custom response header and 4 is an informational message in the response body.

Managed rule groups are preconfigured groups of rules provided by AWS and other vendors. They can also be used alongside customer-created rules. Some managed rule groups are free and some must be purchased. *Figure 12.6* illustrates managed rule groups added to a web ACL:

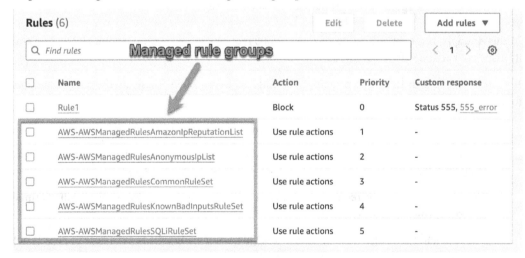

Figure 12.6: Managed rule groups

Figure 12.6 depicts a selection of free AWS managed rule groups added to a web ACL.

Priority is used to set the rule processing order and is configurable. WAF processes rules in the order they're listed in the web ACL. It evaluates each rule sequentially, starting from the top. If a request matches a rule's conditions, the specified action (allow, block, etc.) is taken, and processing typically stops. If no rule matches, the default action for the web ACL is applied. The count action is non-terminating, which means after counting the request, AWS WAF will continue processing the remaining rules.

Custom response is a configurable HTTP response code along with one or more user-definable headers. WAF custom response actions let you define specific responses when a rule blocks a request. You can customize the HTTP status code, add headers with details, and even include a response body with a message, providing more context or instructions for blocked users. This helps improve user experience and offers clear feedback when access is denied.

When configuring an AWS WAF web ACL, it's important to be aware that there are limits on several key elements. These limits help optimize performance and maintain stability. If you need higher limits, AWS offers ways to request increases through support for specific use cases. Each web ACL can have a maximum number of rules, managed rules, and rate-based rules, typically 1,000 per ACL. It also limits the number of IP sets, string-matching conditions, and custom response bodies you can configure.

The following list contains the configuration limits for WAF. Note that they're applied on a per-account, per-Region basis:

- 50 web ACLs
- 100 rules
- 5 rate-based rules
- 100 of each condition type
- 10 regex conditions
- 10 conditions per rule
- 10 rules per ACL

You've explored WAF, where it's used, what its major components are, and how web ACL rules work. In the next section, you'll briefly revisit Route 53 as it applies to the concept of providing traffic control.

Route 53

With Route 53, you can use public and private hosted zones combined with routing policies to control traffic flow both inside and outside your network.

Route 53 uses public and private hosted zones for traffic control by directing requests based on network visibility. Public hosted zones manage traffic for resources accessible on the internet, using routing policies to control global user access. Private hosted zones, on the other hand, manage traffic within a VPC, enabling DNS routing for internal resources.

AWS Route 53 routing policies enhance traffic control and security by directing users based on predefined rules while protecting resources. Geolocation policies restrict access to specific Regions, ensuring only authorized locations can connect. Weighted policies distribute traffic across multiple endpoints, reducing load on any single resource, which can mitigate risks. Latency-based and failover routing improve response times and redirect traffic to healthy, secure resources if issues arise. Combined with health checks and private hosted zones, Route 53's routing policies securely manage access to resources, minimizing exposure to unauthorized or high-risk traffic.

By combining zones with routing policies, Route 53 can efficiently control both public and private traffic flow, ensuring optimal resource access and security.

Detailed coverage of Route 53 can be found in *Chapter 7, AWS Route 53: Basics*, and *Chapter 8, AWS Route 53: Advanced*.

CloudFront

Like Route 53 public hosted zones, CloudFront distributions exist in the public-facing portion of AWS and thus are inherently accessible from the internet. The control points prevent direct access to the origin data, and geo-restriction controls where the origin can be accessed from.

To protect your origin, make sure that DNS always points to the CloudFront distribution's DNS name. For S3 bucket origins, configure the bucket's access policy to allow only CloudFront Origin Access Identity to access it. For custom origins, accept only authenticated requests, using methods such as certificates, cookies, or custom headers provided by the CloudFront distribution.

Geo restriction uses either an allow list or a block list of countries that can access the origin. Additionally, you can configure how blocked requests are handled and specify custom error messages for users whose access is denied.

Simple Storage Service (S3)

When it comes to security incidents (particularly sensitive data leaks), S3 needs no introduction. Like Route 53 and CloudFront, S3 lives in the public area of AWS infrastructure. S3 is associate-level knowledge, so this will largely be an overview and won't cover low-level details.

In the past, many high-profile security breaches (for example, Accenture, Verizon, the US Department of Defense) were due to the fact that public access was enabled by default. This is no longer the case. Unless there is a specific reason, do not enable public access to an S3 bucket. There are cases where public access is appropriate (for example, hosting a static website). However, there's a risk that, over time, more sensitive content might accidentally get added to that bucket. To prevent this, it's important to control who can add or change files by managing write access carefully. This is a good use case for examining a couple of options that can help with this from a network security perspective.

Imagine there is a static website hosted on an S3 bucket where the operations team manages the website, the marketing team needs to add content, and the security team needs audit access. Over time, having a single bucket policy to control all of this could get complicated, introducing the possibility of someone gaining the incorrect level of access. A couple of tools that can be used to reduce this risk are **S3 access points** and **bucket access conditions**.

S3 access points let you create multiple entry points to the same bucket, each with a unique URL and its own access policy. This approach allows you to use several straightforward policies instead of managing one complex policy for the entire bucket. You can also configure access points to only accept traffic originating from a specific VPC, and this cannot change once the access point is created. This eliminates the possibility of the access point ever being publicly accessible.

Here is an example of how you would create an access point for a bucket named `trailcats-bucket` and restrict access to a VPC named `vpc-1a2b3c`:

```
aws s3control create-access-point --name example-vpc-ap --account-id
123456789012 --bucket trailcats-bucket --vpc-configuration VpcId=vpc-
1a2b3c
```

Bucket access conditions may be employed within a bucket policy to control where the bucket can be accessed from. Here is an example of what an access condition that restricts access to a specific IP subnet might look like:

```
{
    "Version": "2012-10-17",
    "Id": "S3PolicyId1",
    "Statement": [
        {
            "Sid": "IPAllow",
            "Effect": "Deny",
            "Principal": "*",
            "Action": "s3:*",
            "Resource": [
                "arn:aws:s3:::DOC-EXAMPLE-BUCKET",
                "arn:aws:s3:::DOC-EXAMPLE-BUCKET/*"
            ],
            "Condition": {
```

```
        "NotIpAddress": {
            "aws:SourceIp": "192.0.2.0/24"
        }
    }
}
]
}
```

The bucket access policy shown is an example of a VPC endpoint policy. VPC endpoint policies are an access control feature of AWS PrivateLink. For purposes of review, recall that PrivateLink is a way to connect services to a VPC as though they were in that VPC by attaching an **elastic network interface (ENI)**, also known as an endpoint.

> **Note**
>
> A VPC endpoint policy is a JSON document. The principal argument is required. The other elements, such as action, resource, and condition, are dependent on the specific service. The list of services and the documentation for them can be found at `https://docs.aws.amazon. com/vpc/latest/privatelink/aws-services-privatelink-support.html`.

AWS Network Firewall

AWS Network Firewall is a stateful network firewall that includes an **intrusion detection system (IDS)** and **intrusion prevention system (IPS)** to enhance network security.

"Stateful" means that the firewall monitors and keeps track of the state of active connections passing through it. Unlike stateless firewalls, which analyze each packet independently, a stateful firewall considers the context of a connection, such as whether the packet is part of an established, new, or related session. This allows the firewall to make more intelligent decisions by recognizing whether a packet is a legitimate part of an ongoing communication or potentially malicious. Stateful firewalls are thus better at distinguishing between allowed traffic and unauthorized access attempts, providing stronger security by understanding the full conversation rather than isolated packets.

An IDS monitors network traffic for suspicious activity or known threats and generates alerts when potential security issues are detected. It's designed to detect and report, not directly block, malicious activity, which allows security teams to analyze and respond to threats manually. An IPS goes a step further by actively inspecting and filtering traffic. When it detects malicious traffic, the IPS can automatically block it, preventing threats from entering the network. IPS functionality is crucial for automated threat mitigation, reducing response time and risk of impact.

AWS Network Firewall integrates these systems to provide advanced protection, inspecting incoming and outgoing traffic to detect and prevent security threats in real time. This combination of IDS and IPS in a stateful firewall helps secure your AWS environment by actively monitoring, detecting, and blocking unauthorized or harmful traffic. AWS Network Firewall is placed at the VPC perimeter. It can inspect traffic to and from internet gateways, NAT gateways, VPN connections, and direct connections.

The stateful portion of the AWS Network Firewall is based on Suricata (an open source network threat detection engine), providing compatibility and portability with Suricata rules. It can perform domain name filtering and does smart protocol detection (for example, HTTPS on port 8000 instead of 443).

To ensure symmetric routing, firewall endpoints must go in their own subnets. This allows the firewall VPC endpoint to sit between the VPC egress and the resources to be protected. *Figure 12.7* depicts where the firewall endpoint is placed for a VPC with an internet gateway attached to it, and the routes needed to direct traffic:

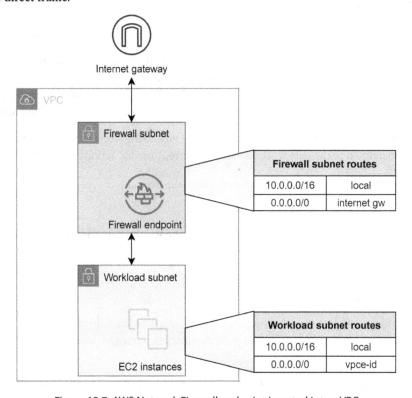

Figure 12.7: AWS Network Firewall endpoint inserted into a VPC

In *Figure 12.7*, a firewall endpoint is placed in a dedicated subnet to inspect traffic. Traffic in the workload subnet is routed through the firewall endpoint using specific routing rules, ensuring all outgoing traffic is inspected before reaching the internet. The firewall subnet routes external traffic to the internet gateway, while internal traffic stays within the VPC. This configuration isolates the firewall and enforces security for all outbound traffic.

AWS Network Firewall is made up of three key components: firewall appliances, firewall policies, and rule groups:

- **Rule groups** are collections of rules that determine how traffic is matched, inspected, and managed by the firewall. These rule groups can be either stateless or stateful. Stateless rules act as a prefilter for stateful rules. We'll discuss the differences between stateless and stateful rules in more detail later.

- **Firewall policies** define a set of rule groups and additional settings. The main purpose of a firewall policy is to apply the same traffic filtering rules to multiple firewalls. While a single policy can be applied to many firewalls, each firewall can have only one policy assigned to it.

- **Firewall appliances** are the actual workloads that connect to your VPCs and process traffic based on the firewall policy. These appliances are provided by AWS as a service and are designed to be highly available.

In the following passages, you'll explore how stateless and stateful rule groups work.

Stateless rule groups define standard network connection attributes with no additional context. In other words, each packet is processed with no knowledge of direction, existing connections, and so on. Packet selection employs a classic 5-tuple, which consists of five key elements: the source IP address, the destination IP address, the source port number, the destination port number, and the protocol type.

Stateless rules are evaluated by the firewall in order, rule processing stops after the first rule is matched and the action defined in the rule is taken. There are three possible actions for a stateless rule:

- **Pass**: A pass action permits the packet through the firewall

- **Drop**: A drop action discards the packet

- **Forward**: A forward action sends the packet to the stateful rule engine for further processing

The advantage of stateless rules is they have low processing overhead. It's a common design in firewalls to use stateless rules to make a quick decision on traffic before passing the remaining traffic to the more computationally expensive stateful rule engine.

Stateful rule groups inspect the packets with the context of a flow (a unique combination of the source IP address, destination IP address, source port, destination port, and protocol type). Stateful rules groups can have three types of rules:

- **Suricata rules** are the native rule type for the AWS Network Firewall stateful rule engine. The fields in a Suricata rule are action, header, and optional settings.

- **Standard rules** are easier to work with than Suricata rules. They're translated into Suricata rules by the rule engine. This is the type of stateful rule that an operator would typically create.

- **Domain list rules** allow or deny traffic based on domain names. An interesting feature of domain list rules is that the engine uses the **Server Name Indication** (**SNI**) field of the TLS negotiation to match hostnames on encrypted connections.

Stateful rules are processed by the firewall in order of action and stop at the first match, then the matching action is executed. The action order for stateful rules is pass, drop, and alert.

If there is no match, the default (catch-all) rule is to pass the traffic. This is testable, so make sure to commit this fact to memory.

> **Note**
>
> *ANS-C01* is not a Suricata exam, so you do not need to understand Suricata rules in depth. The main thing is to be able to identify that Suricata rules are supported and can be imported.

Now that you've explored the capabilities of AWS Firewall, you'll examine how AWS **Elastic Load Balancing** (**ELB**) can be used for traffic control.

Elastic Load Balancer

Basic traffic control through a load balancer (network or application) is simple. A load is protected by security groups and will only accept traffic that matches a configured listener. AWS **Application Load Balancers** (**ALBs**) can handle user authentication on behalf of an application by integrating with identity providers, such as Amazon Cognito or any provider compatible with **OpenID Connect** (**OIDC**). When a user tries to access an application, the ALB redirects them to the identity provider for authentication. Once authenticated, the provider issues a token, which the ALB verifies before allowing access to the application. This offloads the authentication process from the application itself, improving security and simplifying authentication management. ALB supports a wide variety of authentication methods to meet diverse security and access control needs, including OIDC, **Security Assertion Markup Language (SAML)**, **Lightweight Directory Access Protocol (LDAP),** Microsoft Active Directory, and Amazon Cognito.

Network Access Control Lists

AWS **network access control lists** (**NACLs**) are stateless, subnet-level security layers that control inbound and outbound traffic by allowing or denying specific IP traffic based on rules. By now, you're likely familiar with them, as they're considered fundamental knowledge. The goal here is to summarize key points regarding their use as part of a defense-in-depth strategy for traffic control.

NACLs are configured at the VPC scope and can be attached to any IP subnets within that VPC. There can be only one NACL per subnet, and rules are processed in sequence. Processing stops on the first match, so getting the rules in the correct order is important to prevent traffic from being inadvertently dropped. NACLs are stateless, meaning they evaluate on an individual packet basis. If no traffic is matched by a rule, the packet will be dropped by the default rule (explicit deny). Packet selection uses a classic 5-tuple (source IP address, destination IP address, source port, destination port, and protocol type).

As with NACLs, security groups are a fundamental element that you will have encountered the first time you configured an EC2 instance in AWS. The goal here is to summarize and solidify the key points as they relate to a traffic control strategy.

Security Groups

AWS security groups are stateful, instance-level virtual firewalls that control inbound and outbound traffic to resources based on specified rules.

Security groups have a VPC scope and can be attached to any ENIs. Each ENI can have up to five security groups attached to it. Rules are processed using a logical OR approach, meaning that if any rule matches and permits the traffic, it will be allowed to pass. As a result, the order of rules does not matter. Security groups do not have an explicit deny rule. They are stateful, which is useful as it simplifies rule creation by allowing for the return traffic of an allowed flow to automatically be allowed.

Instance-Hosted Services

Instance-hosted security services on AWS are tools and software that you can install directly on your EC2 instances to enhance security. These services can include endpoint detection and response software, log forwarders, and other security applications that monitor and protect the instance from threats. By running these services directly on the instance, you can add an extra layer of defense, customize security settings to fit your needs, and protect sensitive data within your cloud environment. This approach allows you to control and manage security from within the instance itself, offering flexibility and additional protection for your resources. It's important to remember, however, that AWS does not manage these services; they're your responsibility.

With the discussion of instance-hosted services, you've explored traffic control from the outer perimeter of AWS to the operating system of the EC2 instance hosting your application. Now, let's move on to the next topic of this chapter, which is observability and monitoring.

Observability and Monitoring

Effective protection requires visibility, but too much information can make it hard to identify what matters. In this section, you'll explore tools that allow you to search for items of interest, tools to report, tools to remediate, and tools to automate. Gathering relevant logs and events, then surfacing and responding to what's important, is the art and science of observability.

When faced with a visibility question, consider three key points: what you need to see, how you'll find it, and what you plan to do with the information. Start by identifying what you're looking for—this could include traffic events, performance issues, or security incidents. Then, think about how you'll gather this information by selecting the right tools or services. Finally, determine what actions to take with the information—whether simply reporting findings or addressing any issues you discover.

By the end of this section, you'll know what visibility tools you have at your disposal and what they do. You can apply this information to deduce which tool or tools would be the best option to accomplish the task at hand.

VPC traffic mirroring is a powerful tool for troubleshooting and analysis, allowing you to capture and inspect network traffic within your VPC. In the next section, you'll explore how to use traffic mirroring to gain deeper insights into network behavior and identify potential issues.

VPC Traffic Mirroring

VPC Traffic Mirroring is an AWS feature that allows you to capture and inspect network traffic from Amazon EC2 instances within your VPC. By mirroring this traffic to security and monitoring appliances, you can perform deep packet inspection, threat analysis, and troubleshooting, ensuring greater visibility into your network's behavior and security posture. This is an important capability for gaining deep visibility of network traffic in your AWS environment.

VPC Traffic Mirroring use cases are as follows:

- **Deep packet inspection**: VPC Traffic Mirroring enables detailed analysis of network traffic by capturing packets for security tools to inspect. This allows you to detect malicious payloads and unauthorized data transfers, and ensure compliance by scrutinizing every packet.

- **Threat monitoring**: Traffic Mirroring helps in real-time threat detection by sending mirrored traffic to intrusion detection or prevention systems. It enables the identification of unusual traffic patterns, attacks, and vulnerabilities, allowing for quick mitigation.

- **Troubleshooting**: By capturing traffic from specific EC2 instances, Traffic Mirroring aids in diagnosing network issues such as connectivity problems, latency, or packet loss, helping you identify and resolve the root causes efficiently.

VPC Traffic Mirroring involves several key components:

- **Source**: The ENI of the EC2 instance you want to monitor
- **Filter**: A set of rules that specify the specific traffic you wish to capture and monitor
- **Target**: The destination for the mirrored traffic, which can be an ENI, an NLB, or a **Gateway Load Balancer (GWLB)**
- **Session**: The configuration that links the source, filter, and target, enabling the flow of mirrored packets according to the defined criteria

Traffic filters use 5-tuple ACL format, with individual filter lists for inbound and outbound traffic. Rules are processed in order until there is a match. In *Table 12.1*, SSH (port 22) from an administrator's computer and Microsoft Remote Desktop (3389) from the corporate network (10.0.0.0/8) are being filtered:

Rule Number	Action	Protocol	Destination Port Range	Destination Port Range	Source CIDR Block	Destination CIDR Block
100	Reject	TCP	-	22	1.2.3.4/32	0.0.0.0/0
200	Reject	TCP	-	3389	10.0.0.0/8	0.0.0.0/0
300	Accept	22	-	-	0.0.0.0/0	0.0.0.0/0

Table 12.1: Traffic filter ACL

Table 12.1 depicts a network ACL with three rules for managing traffic. Rule 100 rejects TCP traffic on port 22 from the source IP 1.2.3.4/32 to any destination. Rule 200 rejects TCP traffic on port 3389 from the source IP 10.0.0/8 to any destination. Rule 300 accepts traffic on port 22 from any source to any destination.

In the context of VPC Traffic Mirroring, it's essential to consider how packets are delivered to your monitoring appliance, especially when using load balancers as targets. NLBs or GWLBs can cause out-of-order packet delivery due to the way they distribute traffic, which may disrupt accurate analysis if your monitoring appliance doesn't support reordering. Only ENIs guarantee that mirrored packets will arrive in their original sequence, making them the preferred target type for precise, sequence-dependent analysis.

VPC Traffic Mirroring supports various source/target combinations, including the same VPC, VPCs within the same Region that are peered, and VPCs connected to a **transit gateway (TGW)**. The source and target can also be in different AWS accounts, as long as they are connected through peering or a TGW.

A closely related topic to traffic capture is **network virtual appliances (NVAs)**. NVAs can be used as mirroring targets, but they also have other uses.

Network Virtual Appliances

NVAs are EC2 instances that perform network-related functions. They can run third-party packet capture and analysis tools by making an ENI attached to the instance a Traffic Mirroring target. Another common observability use case for NVAs uses multiple instances running IPS software sandwiched between load balancers. *Figure 12.9* illustrates a common traffic inspection architecture using NVAs:

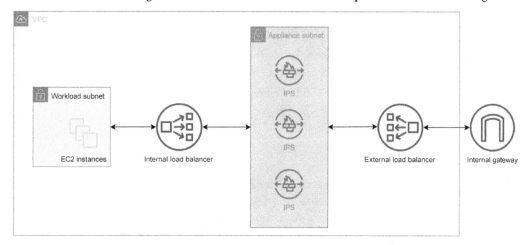

Figure 12.8: Load balancer sandwich

Figure 12.9 shows a network architecture within an AWS VPC environment. It includes a workload subnet with EC2 instances connected to an internal load balancer, which routes traffic to an appliance subnet containing multiple IPS appliances. The appliance subnet is also linked to an external load balancer, which connects to an internet gateway for external access. This setup enables security checks through IPS appliances in a scalable way as IPS appliances can be added or removed based on traffic load. It's important to remember the NVAs are not managed by AWS; the customer is responsible for supporting them.

Amazon CloudWatch

Amazon CloudWatch is a monitoring service in AWS that tracks performance metrics and logs for resources and applications. CloudWatch Metrics tracks and monitors your AWS resources and applications by collecting and visualizing performance data, such as CPU usage, memory, and network activity. CloudWatch Logs captures and stores log data from AWS resources, applications, and services, allowing real-time monitoring, querying, and analysis. The Unified CloudWatch Agent simplifies this process by collecting both metrics and logs from various sources, such as EC2 instances and on-premises servers, and sending them to CloudWatch for centralized monitoring and analysis.

Some examples of security-related items you can monitor with Cloudwatch are WAF, Shield, and VPC traffic, all of which are covered next.

Monitoring WAF

Monitoring AWS WAF with CloudWatch involves setting up separate metric namespaces for each web ACL and rule to gain detailed insights into the performance and effectiveness of your web application firewall. Key metrics to monitor include AllowedRequests, which tracks the number of requests that pass through the WAF rules; BlockedRequests, which logs requests that are blocked by the WAF; CountedRequests, which counts the requests that match the rules but are not blocked; and PassedRequests, which identifies requests that successfully bypass certain rules.

When monitoring WAF web ACLs associated with ELBs, the relevant metrics are available within the Region where the ELB is deployed. However, for WAF web ACLs associated with CloudFront distributions, all metrics are centralized and stored in the US-East-1 Region, regardless of where the requests originate. This is because CloudFront is configured and managed from US-East-1 on a global basis.

Monitoring Shield

AWS CloudWatch provides several key metrics for monitoring DDoS attacks. The DDosDetected metric indicates whether a DDoS attack has been detected, serving as an immediate alert to potential security threats. DDosAttackBitsPerSecond measures the volume of data being targeted at your resources during an attack, expressed in bits per second, which helps in understanding the scale of the attack. DDoSAttackPacketsPerSecond tracks the number of network packets being sent to your resources per second, offering insight into the intensity of the attack at the network level. Lastly, DDosAttackRequestsPerSecond monitors the number of requests aimed at your application per second, helping to gauge the pressure being applied to your web services during an attack. Together, these metrics provide a comprehensive view of the nature and severity of a DDoS attack, enabling better mitigation and response strategies.

Monitoring VPC Traffic Mirroring Metrics

VPC Traffic Mirroring metrics are reported by the monitored EC2 instances, providing insights into the mirrored traffic. Key metrics include NetworkMirrorIn and NetworkMirrorOut, which measure the volume of inbound and outbound mirrored traffic, respectively. NetworkPacketsMirrorIn and NetworkPacketsMirrorOut track the number of mirrored packets entering and leaving the instance. Additionally, NetworkSkipMirrorIn and NetworkSkipMirrorOut indicate the volume of traffic that was not mirrored, while NetworkPacketsSkipMirrorIn and NetworkPacketsSkipMirrorOut measure the number of packets that were skipped from mirroring. These metrics are essential for understanding the efficiency and coverage of your traffic mirroring setup.

Amazon EventBridge

Amazon EventBridge is a serverless event bus that connects applications using events from AWS services, your applications, and SaaS providers. It enables event-driven architectures, allowing you to automate workflows and respond in real time to changes.

Using Amazon EventBridge, you can automate the reporting of CloudWatch metrics and logs by setting up rules that trigger actions based on specific events or thresholds. For instance, when a particular metric reaches a defined threshold, EventBridge can automatically initiate notifications, trigger Lambda functions to process the data, or send the logs to other AWS services, such as S3 or SNS, for further analysis and reporting. This automation streamlines monitoring, enabling proactive management and rapid response to potential issues without manual intervention.

Amazon GuardDuty

Amazon GuardDuty is a threat detection service that continuously monitors your AWS environment for malicious activity and unauthorized behavior. It uses threat intelligence feeds and machine learning to identify threats and generate alerts. It has report-generation capabilities focused on AWS accounts and Organizations. GuardDuty is used to centralize findings and reporting services and provides actionable insights and alerts, helping you protect your AWS resources and maintain a secure cloud environment. Amazon GuardDuty can identify threats such as privilege escalation, exposed credentials, port or network scanning, and cryptocurrency mining activities.

A key point to remember about Amazon GuardDuty is that it serves as a detection and reporting tool, not a prevention or response solution. It alerts you to potential security issues, but it does not take direct action to resolve them.

Securing Traffic

In the context of the exam, traffic protection is synonymous with encryption. This includes transporting encryption methods such as **Transport Layer Security** (TLS) and utilizing various AWS infrastructure components, such as load balancers and CloudFront. A couple of smaller topics related to transport encryption on the exam are AWS managed databases, Amazon S3, and customer-implemented solutions on Amazon EC2. Those will be touched on under enforcing encryption on intra-AWS traffic, just as they pertain to the exam.

Part of securing traffic is the infrastructure required to generate keys and digital certificates to facilitate the use of encryption. This infrastructure includes **AWS Certificate Manager** (ACM), AWS CloudHSM, and **AWS Certificate Manager Private Certification Authority** (ACM PCA). These topics will be covered next.

Public Key Infrastructure

Whenever HTTPS or TLS encryption is being employed, you must supply a digital certificate to provide those keys. This section is a high-level overview of digital certificates and **public key infrastructure** (**PKI**) to provide some context for AWS offerings and options.

Just as a lock needs keys to function, cryptographic algorithms need keys to do their jobs. The challenge is how to distribute the keys to senders and receivers in a way that protects the information from being encrypted.

In the *DNSSEC* section of *Chapter 8, AWS Route 53: Advanced,* there is a discussion of public key cryptography and cryptographic concepts in general. The public keys are stored in DNS records, and the chain of trust for those records comes from the validation of the DNS servers up to the root domain, which in turn is secured through a key signing ceremony. The exact same technology is at play here, only you're using a different kind of infrastructure to create the chain of trust and store the keys. This infrastructure is called PKI, and the chain of trust uses the concept of **certificate authorities** (**CAs**). Instead of the keys being stored in DNS records, the keys are stored in what's called an X.509 certificate, more commonly known as a **digital certificate**.

Digital certificates contain various useful pieces of information, including details such as the certificate holder's identity, the issuing CA, the certificate's validity period, the public key, and the certificate's intended usage. The following includes commonly referenced attributes in a certificate:

- Information about the entity the certificate has been issued to. This is called the **subject**.
- Information about the issuing CA.
- Digital signature from the CA.
- The public key from the subject's public/private key pair.
- The cryptographic suites supported by the certificate.
- Validity and revocation information.
- Serial number of the certificate.
- Extensions in the certificate. Common extensions include:
  - **Authority information access (AIA)**
  - **Certificate revocation list (CRL) distribution points** (also called **CDP**)
  - What purposes the certificate can be used for (key usage)
  - **Subject alternative name (SAN)**

This is all to say that there's a lot of metadata in certificates to assure the recipient that the presenter of the encryption keys is who they say they are, and that the keys are valid. The entities that validate certificates and create the chain of trust are called CAs. There are three general categories of CA: public, managed, and private.

A public CA provides certificates on a global level to the public. They can be expensive and labor-intensive to work with in terms of issuing and usage, but they're the most widely trusted. Many domain registrars also operate public certification authorities, for example, Verisign and GoDaddy.

A managed CA is a service that features ease of use but comes with limitations on that usage. A managed CA service is typically a public CA who is also a registrar, providing additional hosted services that integrate with their CA and domain registration services. AWS is a prime example of this type of organization. If you register your domain with AWS, you can leverage ACM to automatically create certificates for ACM-integrated services such as Elastic Load Balancer and CloudFront. However, a service like ACM has limitations on the types of certificates it will issue. For example, you can't use ACM to create user or device certificates. For that type of usage, you need a private CA.

A private or internal CA is used to provide certificates for internal business infrastructure. This type has the most amount of flexibility and control, but is only trusted at the organization level, and can't be used on the internet. Private CAs provide strong proof of identity for internal resources. For example, if Trailcats had an application that held proprietary company secrets, the business could issue certificates and only allow users who can prove their identity with a certificate to log on to the application.

Now that you've had a brief overview of PKIs, it's time to discuss their relevance to AWS. AWS provides three choices for customers who want to use certificates: IAM, ACM, and AWS Private CA.

AWS IAM can be used as a certificate manager. This is typically used in Regions where AWS Certificate Manager is not available. IAM has the capacity to store certificates for use, but it is not a CA. That is, it cannot be used to sign and validate certificates. There is no interface in the console to import certificates; it must be done through API or the CLI.

ACM is best suited for securing websites and applications using TLS. ACM certificates are compatible with services such as ELB and Amazon CloudFront. They offer seamless integration and automatic renewal, especially when using Route 53 to host your domain. Like IAM, ACM also supports importing certificates that were generated outside of AWS. ACM is a free service.

For ACM to provision public certificates, you must prove domain ownership, which can be done in one of two ways: a DNS text record or via email. Certificates are valid for 13 months and support automated renewal. ACM provisioned certificates can only be used with ACM integrated services, as the private keys are not exportable. ACM is a regional service, which means you must provision the certificates in the Region they will be used in.

An important thing to keep in mind about ACM is that all CloudFront certificates are managed from us-east-1 and for load balancers, the certificate must be issued in the Region where the load balancer is deployed.

The **AWS Private Certificate Authority** service is for businesses that want to set up a PKI within AWS for private use. With AWS Private CA, you can create your own CA and issue certificates to authenticate users, computers, applications, services, servers, and other devices. The private certificate has a monthly charge ($400.00/month per CA at the time of writing), and there is a per-certificate fee.

There's a final piece of PKI that's relevant to the exam, and that is the secure storage of private keys. If you're using a service such as AWS Private CA, the security of your deployment is dependent on protecting the private key of the root CA. The technology that performs this function is called a **hardware security module (HSM)**. As you may have already guessed, Amazon has a service for that.

AWS CloudHSM is a managed service that offers highly available HSM clusters within AWS. ENIs are deployed to your VPC, and this is how you access the cluster. An HSM is a physical computing device designed to protect and manage secrets, especially digital keys. It performs encryption and decryption for digital signatures, strong authentication, and various other cryptographic functions.

CloudHSM is the only AWS service that provides FIPS 140-2 level 3-compatible hardware in AWS. Access to the HSM is achieved via industry-standard API calls and libraries, which is different than the typical AWS-managed services.

The CloudHSM cluster is hosted in an AWS-owned account and is not part of your infrastructure. All of the high availability aspects of the cluster, including load balancing and data synchronization, are handled by AWS.

CloudHSM use cases that are relevant to networking include offloading the cryptographic computation of TLS, storing Oracle **Transport Data Encryption (TDE)** and its master encryption key, and storing your CA private key.

Protecting Traffic

In this section, the focus will be on securing both external and internal traffic flows within your AWS environment. You will explore strategies to protect traffic originating from outside AWS, such as requests from the internet or other external networks, as it makes its way to your AWS-hosted services. Additionally, you will address the security measures needed to safeguard traffic that originates and terminates within AWS, ensuring that communication between your internal resources, such as EC2 instances, databases, and other AWS services, is protected against potential threats.

Enforcing Encryption on Traffic Originating Outside AWS

CloudFront, by default, will accept unencrypted traffic. You can change this through the Viewer Protocol Policy. Your options are to redirect to HTTPs or set to HTTPs only. S3 by default will accept unencrypted traffic. You can change this by setting a bucket policy condition that requires HTTPS. Here's an example of what that would look like:

```
{
      "Sid": "AllowSSLRequestsOnly",
      "Action": "s3:*",
      "Effect": "Deny",
      "Resource": [
        "arn:aws:s3:::DOC-EXAMPLE-BUCKET",
        "arn:aws:s3:::DOC-EXAMPLE-BUCKET/*"
      ],
      "Condition": {
        "Bool": {
          "aws:SecureTransport": "false"
        }
      },
      "Principal": "*"
}
```

ELB can be configured to enforce TLS traffic through the listener configuration. The options available depend on the type of load balancer. This section is a summary of your options for implementing TLS on ELBs. *Chapter 9, AWS Elastic Load Balancing*, goes into the details of TLS security policies, and it would be helpful to review that as well.

Classic Load Balancers (**CLBs**) support SSL, which is an old protocol that's been deprecated and shouldn't be used, but it may come up in the exam. The other thing to know about CLBs is that they only support one certificate per load balancer. Both ALBs and NLBs support multiple certificates per listener. The SNI extension allows the selection of a matching certificate, and if no match is found, the default certificate will be used.

Table 12.2 summarizes protocol and certificate support by load balancer type:

Type	Supported protocols	Multiple certificates	SNI supported
Classic Load Balancer	HTTPS or SSL	No	No
Application Load Balancer	HTTPS	Yes	Yes
Network Load Balancer	TLS	Yes	Yes

Table 12.2: Load balancer transport encryption support

Table 12.2 compares load balancer types in terms of supported protocols, support for multiple certificates, and SNI support. The CLB supports HTTPS or SSL but does not support multiple certificates or SNI. The ALB supports HTTPS, multiple certificates, and SNI. The NLB supports TLS, multiple certificates, and SNI.

Enforcing Encryption on Intra-AWS Traffic

Your network traffic in AWS is using shared infrastructure. Your organization may need to ensure that traffic in AWS is secured. In this section, you're going to learn about securing traffic for EC2 and container instances and database services to meet this requirement.

If you need to secure network connections between your instances, you'll need to implement instance-level encryption. This must be configured and managed by you, as this falls under the customer in the shared responsibility model.

For storage, Amazon storage services offer built-in encryption to secure data during transmission. For Amazon **Elastic Block Store (EBS)**, traffic between instances and EBS volumes is automatically encrypted by AWS, ensuring that data remains secure. Amazon **Elastic File System (EFS)** can be secured using TLS on a per-connection basis, providing encryption for data as it's accessed. Similarly, Amazon FSx encrypts traffic when file shares are accessed using SMB3 or Samba 4.2 or newer, adding a layer of protection for data transfers.

For databases, Amazon RDS provides built-in mechanisms to secure data through encryption. When an RDS database instance is provisioned, AWS automatically creates and installs a TLS certificate to protect data in transit. However, under the shared responsibility model, it is the user's responsibility to configure and enforce encryption as needed, and the exact methods for doing so can vary depending on the database engine and version. Additionally, any clients accessing RDS databases must have AWS root certificates installed in their trusted certificate store to ensure secure communication with the database.

Managing and Auditing Security Configurations

Managing and auditing AWS security configurations is crucial for maintaining a secure cloud environment. This process involves using specialized tools to ensure that your AWS resources are configured correctly, compliant with security best practices, and protected against potential vulnerabilities.

AWS Firewall Manager simplifies the management of firewall rules across multiple accounts and resources, ensuring consistent security policies across your organization. **Amazon Inspector** is a vulnerability management service that automatically assesses your AWS workloads for security vulnerabilities and deviations from best practices, providing actionable insights to mitigate risks. **AWS Trusted Advisor** provides real-time guidance to help optimize your AWS environment, including security recommendations that help you identify potential issues and ensure your resources are configured securely.

Together, these tools enable comprehensive management and auditing of your AWS security configurations, helping to safeguard your cloud infrastructure.

AWS Firewall Manager is a security management service where you can centrally configure and manage your firewall policies across accounts and applications in your organization. AWS Firewall Manager is an organizational-level tool rather than an account-level management tool. Before you can start using it, you must onboard with the AWS Organization, enable AWS Config, enable AWS Resource Manager, and designate an account as the Firewall Manager Administrator. AWS Firewall Manager can provide central management for the following:

- AWS WAF
- AWS Shield Advanced
- VPC Security Groups
- AWS Network Firewall
- Route 53 Resolver DNS firewall

In addition to centrally managing these services, AWS Firewall Manager sends findings to AWS Security Hub. AWS Firewall Manager allows you to create policies and either enforce those policies on AWS accounts (auto-remediation) or just provide visibility (notification of non-compliance).

Amazon Inspector is an automated vulnerability scanning service. Delegated Administrator (an AWS account that is authorized to manage Amazon Inspector settings and operations on behalf of an entire organization) allows centralized findings and reports. Amazon Inspector has two types of scans (also called assessments): network and host.

The network assessment is agentless and scans your AWS network for open ports and services that are accessible from outside your VPC. This assessment helps you identify network configurations that may be too permissive, such as misconfigured firewalls or security groups. Inspector scans EC2 and Lambda instances for reachability weekly by default, but it's configurable to a minimum interval of daily.

The host assessment performs a scan of the operating environment on the hosts using the AWS Systems Manager agent feature. It scans instances for deviations from best practices, checks for known vulnerabilities, and evaluates software for security risks. The assessment provides detailed findings, including severity levels and remediation recommendations, helping you strengthen the security posture of your instances by addressing identified risks.

AWS Trusted Advisor

AWS Trusted Advisor is a tool that helps you optimize your AWS setup by providing real-time recommendations to improve performance, security, cost, and reliability. It checks your account against best practices and suggests ways to save money, increase efficiency, and strengthen security. For the purpose of the exam, only the security checks will be discussed.

AWS Trusted Advisor provides two tiers of service based on your support level. All accounts receive the basic tier, which includes five free security checks. For accounts with Business or Enterprise support, Trusted Advisor provides an expanded set of 30 security checks. The five free checks include the following:

- **Amazon EBS public snapshots**: Checks the permission settings for your Amazon EBS volume snapshots and alerts you if any snapshots are publicly accessible

- **Amazon RDS public snapshots**: Alerts you if you have Amazon RDS snapshots that are publicly accessible, which could lead to unauthorized data access

- **Amazon S3 bucket permissions**: Identifies any publicly accessible Amazon S3 buckets that could expose sensitive data

- **MFA on root account**: Ensures **multi-factor authentication** (**MFA**) is enabled on your AWS root account for added protection

- **Security groups**: Checks security groups for rules that allow unrestricted access to specific ports

The AWS Trusted Advisor security checks identify potential security risks in your AWS environment and provide recommendations to enhance protection. They help secure root access, prevent public EBS snapshots, restrict public S3 bucket access, prevent unrestricted network access, and prevent public RDS snapshots, ultimately strengthening the security of your AWS resources.

Summary

In this chapter, you learned how the layered approach to defense applies to AWS, what services exist in each layer, and how to control traffic at each level. You learned what tools you have at your disposal to gain visibility into your environment, and how you can use that information. You learned how to secure traffic coming in from outside of AWS as well as intra-AWS traffic. Lastly, you learned what tools you have available to perform automated auditing, compliance checks, and automated security assessments. These skills are crucial in today's cloud-focused workplace, where security and compliance are paramount. By mastering these areas, you can proactively manage risks, streamline compliance efforts, and keep your environment secure against threats. This expertise not only strengthens your role in protecting your organization but also makes you an asset, positioning you for career growth in a field that values cloud security skills.

Exam Readiness Drill – Chapter Review Questions

Apart from mastering key concepts, strong test-taking skills under time pressure are essential for acing your certification exam. That's why developing these abilities early in your learning journey is critical.

Exam readiness drills, using the free online practice resources provided with this book, help you progressively improve your time management and test-taking skills while reinforcing the key concepts you've learned.

HOW TO GET STARTED

- Open the link or scan the QR code at the bottom of this page

- If you have unlocked the practice resources already, log in to your registered account. If you haven't, follow the instructions in *Chapter 16* and come back to this page.

- Once you log in, click the START button to start a quiz

- We recommend attempting a quiz multiple times till you're able to answer most of the questions correctly and well within the time limit.

- You can use the following practice template to help you plan your attempts:

Working On Accuracy		
Attempt	Target	Time Limit
Attempt 1	40% or more	Till the timer runs out
Attempt 2	60% or more	Till the timer runs out
Attempt 3	75% or more	Till the timer runs out
Working On Timing		
Attempt 4	75% or more	1 minute before time limit
Attempt 5	75% or more	2 minutes before time limit
Attempt 6	75% or more	3 minutes before time limit

The above drill is just an example. Design your drills based on your own goals and make the most out of the online quizzes accompanying this book.

First time accessing the online resources? 🔒

You'll need to unlock them through a one-time process. **Head to** *Chapter 16* **for instructions**.

Open Quiz `https://packt.link/ansc01ch12` OR scan this QR code →	

13
Infrastructure as Code

The concept of **infrastructure as code** (**IaC**) involves fundamentally changing how enterprises think about building, maintaining, and replacing the services that underpin their critical applications. While infrastructure has traditionally been seen as an immovable, unyielding, and static fixture, the IaC approach makes infrastructure more ephemeral and malleable, able to be created, modified, and removed using code. This chapter focuses on the framework for delivering IaC using DevOps and, specifically, how AWS supports the IaC approach with its premier service, CloudFormation. CloudFormation is the AWS IaC deployment toolkit that allows the deployment of scalable, repeatable network architecture. Finally, you will learn about the AWS **Cloud Development Kit** (**CDK**), a different AWS tool that also uses IaC to provision cloud resources.

In this chapter, you will learn about the following topics:

- DevOps
- Infrastructure as code
- AWS CloudFormation
- AWS Cloud Development Kit

By the end of this chapter, you should understand the framework of DevOps and IaC, as well as be able to use the AWS tools that support this approach.

DevOps

DevOps is a portmanteau of *development* and *operations*. It is a collaborative approach that integrates software development and IT operations to automate and streamline the software delivery process. It utilizes **continuous integration/ continuous delivery** (**CI/CD**) and IaC to achieve faster, more reliable deployments and improved team collaboration. IaC is an executable approach to managing infrastructure that DevOps makes possible via its approach to software management. As IaC fits within this larger context, it's useful to understand what DevOps is and what it evolved from. This understanding will make IaC more comprehensible and intuitive.

Prior to the invention of DevOps, creating, maintaining, and updating infrastructure to support applications was largely a manual process, as illustrated by the following example taken from Trailcats' history:

- **Planning and deployment were manual and slow**: Trailcats decided to fund the development of a new application. Through meetings with the developers, business owners, and other stakeholders, a specification was created. Part of the specification included infrastructure requirements such as network, servers, and firewalls. This specification was converted into a bill of materials, which was used to order everything. Once all of the materials came in, subject matter experts built and configured their pieces, and the infrastructure was wired together (integration). The application software was installed on top of the infrastructure, and the application and supporting infrastructure were placed into production.

- **Updates were full of friction and risk**: When a developer created updates to an application, they gave the code to the operations team, often as a ZIP file sent via email. The operations team was then responsible for figuring out whether any changes needed to be made to the supporting infrastructure in addition to implementing the changes, typically during a maintenance window. In this type of environment, the little testing that occurred was done in isolation or on a test environment that drifted significantly from production. This created a lot of risk. Due to this, the organization became averse to making changes. The change queue tended to grow, and developers ended up with batches of changes, which only increased the risk further.

- **Production drifts from specification**: Over time, the accumulation of changes and configuration drift led to what is known as technical debt. The system became brittle and ever more removed from the documented specification. As people came and went, undocumented changes were stacked on top of other undocumented changes. Eventually, a tipping point was reached, at which point it became almost impossible to make changes at any kind of fundamental level. This had a serious business impact as it affected the ability of the business to make improvements or updates to its offerings to reflect opportunities or risks in the marketplace.

Figure 13.1 depicts a situation where code updates are hand-deployed by operations staff:

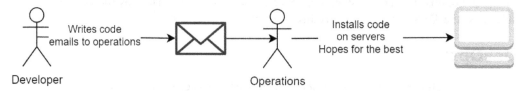

Figure 13.1: Manual software deployment

The preceding figure shows that the deployment of new software is very slow and manual, requiring multiple teams to be involved in the life cycle of software management.

The Three Pillars of DevOps

DevOps bridges the divide between development and operations and unites them into one seamless process. It emphasizes automation, CI/CD, and a culture of shared responsibility among different teams. DevOps reduces silos, accelerates release cycles, and enhances software stability and reliability. This is done through three pillars.

Collaboration

Collaboration is based on the concept of cross-functional work being shared between silos, with each team providing their specific expertise for a stronger outcome.

At Trailcats, the development team worked on building a new feature for their e-commerce website, while the operations team was responsible for managing the servers and infrastructure. Because these teams worked in silos, the developers were unaware of the server limitations and deployed a resource-intensive feature without consulting the operations team. As a result, the website experienced performance issues during peak traffic times. The lack of communication between the teams led to customer dissatisfaction and lost sales, highlighting how silos can cause significant communication problems and hinder the overall success of the business.

The principles contained in the DevOps collaboration pillar address this problem by fostering a culture where the development and operations teams work closely together from the start. Instead of working in isolation, the developers and operations team would collaborate during the planning phase of the new feature. They would discuss the feature's requirements, assess the server capacity, and identify potential challenges together. This open communication ensures that the development team understands the infrastructure constraints and the operations team is aware of upcoming changes. As a result, they can jointly find solutions, such as optimizing the feature for performance or scaling the infrastructure to handle the new demands. This collaborative approach prevents the issues caused by silos, leading to improved website performance and a better overall customer experience.

Processes

Processes refers to the practice of establishing a structured workflow for producing, maintaining, and retiring software. This workflow should be well understood by all stakeholders and have a clear delineation of responsibilities and flow from one step to another.

At Trailcats, the development team frequently pushed new software features to the e-commerce website without a structured process for testing and deployment. Because there was no standardized procedure, developers sometimes skipped thorough testing to meet deadlines, introducing bugs into the live environment. One day, a new feature designed to streamline the checkout process was deployed without proper testing, causing the checkout system to crash for several hours. This frustrated customers and resulted in a significant loss of sales. The lack of formal processes for testing and deployment at Trailcats led to inconsistent quality in software releases, causing disruption and harming the business's reputation.

The principles found within the processes pillar of DevOps solved this problem by introducing standardized, repeatable procedures for software development, testing, and deployment. With clear processes in place, the development team implemented automated testing pipelines that rigorously checked each new feature before it was deployed. A structured deployment process was also established, including staging environments where new features could be tested in conditions like the live environment. These processes ensured that every feature was thoroughly tested and properly deployed, reducing the risk of bugs and system crashes. As a result, Trailcats improved the reliability and quality of their software, minimizing disruption and enhancing customer satisfaction.

Tools

At Trailcats, the development and operations teams relied heavily on manual processes for building, testing, and deploying new software features. Each time a new feature was ready, developers manually compiled the code, and the operations team manually configured the servers and deployed the updates. This process was time-consuming and prone to human error. One day, a critical feature was deployed with a configuration mistake that went unnoticed during the manual setup, causing the website to go offline for several hours. The lack of automation tooling led to an increased risk of errors and ultimately resulted in significant downtime, negatively impacting customer experience and sales at Trailcats.

The principles of the tooling pillar of DevOps solved this problem at Trailcats by introducing automation tools that streamlined and standardized the software delivery process. Developers now used automated build tools to compile code consistently, while CI/CD pipelines automatically tested and deployed new features to the production environment. This automation reduced the risk of human error and ensured that configurations were correct and consistent across deployments. As a result, the process was faster, more reliable, and less prone to mistakes. This allowed Trailcats to deliver new features with fewer disruptions, ultimately improving customer satisfaction.

The next section sees these principles put into practice with the concept of IaC.

Infrastructure as Code

The goal of IaC is to treat the infrastructure layer supporting applications in the same way that you would treat the application code itself. This allows infrastructure to be more flexible in terms of deployment and operation as well as life cycle management. Integrating IaC into DevOps practices allows infrastructure to not only be created as code but also be managed as part of a code life cycle. Traditionally, infrastructure was highly static; changing anything required a lengthy change process because of the potential impact on the applications that the infrastructure supported.

IaC uses development tools such as **integrated development environments (IDEs)** to create infrastructure in the same way a developer would code an application. Typically, it starts with a declarative specification document written in a machine-readable format. A provisioning tool reads the specification and then provisions infrastructure based on that document. Typically, the documents live in a version-controlled repository. This means whenever the specification is updated, it automatically gets a new version number, and if a change to the specification causes a failure, it's easy to roll back the change and investigate what went wrong.

Because you will be deploying many changes to this infrastructure over time, it's important to be able to test those changes and have some assurance that they won't break the application. Additionally, you may want to perform security checks such as dynamic and static code testing and linting. This process of automating everything (including testing from the initial change all the way to the deployed running infrastructure) is called a **CI/CD pipeline**.

In a CI/CD pipeline, a process to test that code is kicked off when you commit changes to the repository. If the changes pass all tests, they can be deployed automatically or manually depending on the requirements of the business.

Another benefit of using a declarative document combined with version control and automation is that the specification itself is the source of truth. Version control is how you maintain a software life cycle over time. Each version should include changes that are specific to that version. Combine this with automation, the automated workflow of managing versions that involves tools built for that purpose. Over time, you can detect whether a production configuration has drifted from the specification and take corrective action.

Immutability

IaC-based deployments should be immutable, meaning that they are not different or special per deployment. If there's a problem with the deployment, you do not spend time troubleshooting and trying to repair the unhealthy element. Instead, you simply redeploy. If redeploying doesn't resolve the problem, then you work back through the testing and the specification until you find the source of the problem. You address it there and then you redeploy. A popular term for this is "cattle not pets," meaning you should treat this process as if the infrastructure has no particular or sentimental value. This saves a tremendous amount of time and effort when it comes to troubleshooting and fixing problems.

Figure 13.2 depicts a simplified version of a CI/CD pipeline where automated processes test the code, and the same code base can be deployed across multiple environments for development and testing purposes:

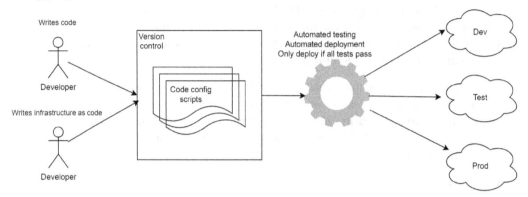

Figure 13.2: Simplified CI/CD pipeline

The value of a CI/CD pipeline is that it is structured and repeatable without needing input (which can lead to possible mistakes) from a human.

> **Note**
>
> DevOps and IaC are not about cost savings. Building a DevOps practice takes a lot of work upfront to realize its benefits. What DevOps provides is speed, consistency, and safety. Again, it is not used to save money. This is an important point to remember when having conversations about why you would want to implement DevOps or IaC.

In the next section, you will be introduced to the preferred AWS tool for IaC: CloudFormation.

AWS CloudFormation

CloudFormation is the AWS IaC provisioning tool. It uses a specification document called a template, which can be written in YAML or JSON. The template describes all the resources you want to provision and defines relationships between those resources. The template is uploaded to Amazon, where CloudFormation converts it into running infrastructure called a stack.

CloudFormation supports some programmatic constructs, such as parameters and intrinsic functions, to make life easier and to improve the reusability of your templates.

One of the key features of CloudFormation is rollback on failure. If an element in the deployment fails, CloudFormation, by default, will automatically roll back the entire deployment. There are ways to override this behavior through policies. The main thing to appreciate is that this is a huge advantage over most other automation tools, such as Terraform and Pulumi, where if there's a failure in the deployment, you may have to spend a significant amount of time cleaning up the mess by hand.

Once you have your stack up and running, you have two different options for making changes to your stack. The first is a stack update and the second is change sets.

All infrastructure has a life cycle, and this is another area where CloudFormation can help simplify matters. If you no longer need a stack, you can quickly delete all its resources by simply deleting the stack. CloudFormation knows how to delete all the resources in the correct order to prevent dependency errors. If you've ever had to hand-delete resources in a complex deployment, you will appreciate how convenient this is.

CloudFormation also has a built-in feature for detecting and logging configuration drift, and it integrates directly with version control repositories.

The main elements of CloudFormation are as follows:

- **Templates**: Files that define resources for CloudFormation to create. The files can be formatted in YAML or JSON. Think of them as blueprints.

- **Stacks**: The resources AWS created from the template. A stack is a set of resources created from a CloudFormation template. Its purpose is to allow the management of all the resources as a single unit. Where the template is the blueprint, the stack is how you manage the resources created from the blueprint.

- **Change sets**: A type of template for describing changes to a stack. CloudFormation uses the change set to analyze the proposed changes and give you feedback on their outcome. Thus, you can use the change set process to decide whether to make the change. It also creates a record of the changes that have been made to that stack.

Figure 13.3 depicts the role CloudFormation plays in AWS infrastructure automation:

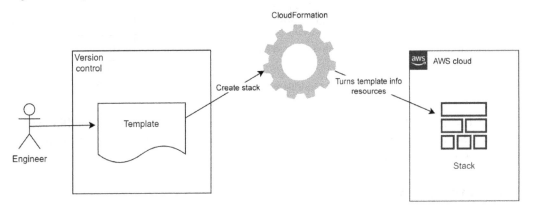

Figure 13.3: CloudFormation

In *Figure 13.3*, the engineer uploads a CloudFormation template to a version control repository and issues a create stack command. CloudFormation reads the template, and creates the resources specified by the template, producing a stack.

As previously outlined, a template is a specification file that can be composed in either YAML or JSON. The functionality of templates is divided into **sections**. The only mandatory section is resources. However, in most cases, numerous other sections will be employed.

Table 13.1 lists all the possible sections in a CloudFormation template and what they do:

Section	Description	Frequently used?
AWSTemplateFormatVersion	Description of the template format.	Yes
Description	Description of the template.	Yes
Resources	Describes resources to be created by CloudFormation.	Yes
Parameters	Values passed to the template at runtime. Can be used by the Resources and Outputs sections.	Yes
Outputs	Values returned when viewing stack properties.	Yes
Metadata	Additional information in machine-readable form.	No
Rules	Validates parameters.	No
Mappings	Key/value mappings for conditional parameter values.	No
Conditions	Similar to mappings only for resource creation.	No
Transform	Used for serverless applications.	No

Table 13.1: CloudFormation template sections

Table 13.1 maps out the sections and their relative frequency of appearance in CloudFormation templates. Something being frequently used means you're likely to see those sections in almost every template, as in the case of the Parameters section.

Template Authoring

Since the exam requires you to be proficient in interpreting CloudFormation templates, hands-on practice in creating them is essential. In the following sections, you will start with the simplest possible template and modify it, learning how CloudFormation's core features work in the process. If you work through the code examples and explanations, you will have a solid foundation for the real world and the exam.

The example code in this section can be found on GitHub at `https://packt.link/gZ5Io`.

For reference, the completed code examples and AWS CLI commands used in the exercises are available in the preceding repository.

The hands-on exercises make use of the AWS CLI, so it is necessary to have that installed. For instructions, refer to the AWS CLI user guide: `https://docs.aws.amazon.com/cli/latest/userguide/getting-started-install.html`.

The built-in editor in the CloudFormation console is useful for inspecting existing templates, but for authoring, you'll most likely find it lacking. If you're going to write CloudFormation templates, you're going to want a code editor.

Using a code editor can help you in several ways:

- Color coding and formatting
- Real-time syntax checking
- Auto-completion/tab completion
- Inline documentation

In the author's opinion, composing templates becomes a lot more efficient and pleasant with an editor such as **Visual Studio Code (VSCode)**.

> **Note**
> You can get more information about VSCode at `https://code.visualstudio.com`.

Let's start with a minimum viable configuration and add some elements one at a time. This is a useful pattern for creating IaC. As a general principle, small, incremental changes are easier to understand and troubleshoot. Try to make a habit of adding one small piece at a time. In the series of exercises in this chapter, you'll create a minimum viable template, revise the template use parameters, and then finally, add intrinsic functions to create a fully reusable CloudFormaton template.

Exercise 13.1: Basic Template

A minimum viable template requires that at least one resource is defined. Resources are the only mandatory element that must be defined in a template. Everything else is optional. In this exercise, you will create a VPC:

1. Enter the following code into your code editor:

    ```yaml
    Resources:
      myvpc:
        Type: AWS::EC2::VPC
        Properties:
          CidrBlock: 10.16.0.0/16
    ```

2. Save the file as basicvpc.yaml.

3. From a terminal, issue the following AWS CLI command:

    ```
    aws cloudformation create-stack --stack-name basicvpc
    --template-body file://basicvpc.YAML
    ```

4. If everything is in order, the command will return an object ID, which will look something like this:

    ```
    {
        "StackId": "arn:aws:cloudformation:us-east-
    1:361037072889:stack/basicvpc/534b6c50-5829-11ef-8cb3-
    0affdcf9ad87"
    }
    ```

5. Press q to exit the command output display.

6. To view the newly created stack, use the CloudFormation console in AWS, or issue the following AWS CLI command:

    ```
    aws cloudformation describe-stacks --stack-name basicvpc
    --output table
    ```

7. This should return information about your stack. --output table produces a more human-readable format. Feel free to browse the AWS console and inspect your newly created VPC.

Figure 13.4 is an example of what the output should look like:

```
|                                         DescribeStacks                                                          |
+------------------------------+---------------------------------------------------------------------------------+
|                                            Stacks                                                               |
+------------------------------+---------------------------------------------------------------------------------+
|  CreationTime                | 2024-08-12T00:34:39.649000+00:00                                                |
|  DisableRollback             | False                                                                           |
|  EnableTerminationProtection | False                                                                           |
|  StackId                     | arn:aws:cloudformation:us-east-1:361037072889:stack/basicvpc/a8df5870-5842-11ef-bab4-0e73da89f1cd |
|  StackName                   | basicvpc                                                                        |
|  StackStatus                 | CREATE_COMPLETE                                                                 |
+------------------------------+---------------------------------------------------------------------------------+
|                                         DriftInformation                                                        |
+------------------------------+---------------------------------------------------------------------------------+
|  StackDriftStatus            |                          NOT_CHECKED                                            |
+------------------------------+---------------------------------------------------------------------------------+
```

Figure 13.4: describe-stacks output in table form

8. Press q to exit the command output display.

9. Add the following information to the top of your YAML file:

```
AWSTemplateFormatVersion: "2010-09-09"
```

10. The template version is not required, but if AWS ever created a new version of the template format and you didn't have this field, this could potentially break your templates in the future. At the time of writing, there's only one template version: 2010-09-09.

11. Add the following to the YAML file in the Resources section to add a subnet to the template and attach the subnet to the VPC:

```
mysubnet:
    Type: AWS::EC2::Subnet
    Properties:
        VpcId:
            !Ref myvpc
        CidrBlock: 10.16.0.0/24
```

12. YAML is space- and tab-sensitive. The mysubnet resource needs to line up with the myvpc resource. This is what the file should look like in its entirety:

```
AWSTemplateFormatVersion: "2010-09-09"
Resources:
  myvpc:
    Type: AWS::EC2::VPC
    Properties:
      CidrBlock: 10.16.0.0/16
  mysubnet:
    Type: AWS::EC2::Subnet
    DependsOn: myvpc
```

```
Properties:
  VpcId:
    !Ref myvpc
  CidrBlock: 10.16.0.0/24
```

13. Save your file as `basicvpc.yaml` and run the following command:

```
aws cloudformation update-stack --stack-name basicvpc
--template-body file://basicvpc.YAML
```

14. If the command is successful, it will return an object ID like the one in *Step 4*. Press q to exit the command output display.

15. Enter the following AWS CLI command to the resources associated with the stack:

```
AWS cloudformation describe-stack-resources --stack-name
basicvpc -output table
```

16. This should return a list of two resources and their details: the VPC and the subnet. The format will look similar to the output of *Step 7*. Press q to exit the command output display.

17. Use the following command to delete the stack:

```
awd cloudformation delete-stack --stack-name basicvpc
```

Deleting the stack will delete the stack's resources in the correct order.

Note

You must delete the stack before continuing to *Exercise 13.2*.

You may have noticed that the `!Ref myvpc` property was used in the subnet resource definition. `Ref` is known as an intrinsic function. Intrinsic functions are one of the features that give CloudFormation its ease of use along with its power. `Ref` returns the value of the specified parameter or resource, which in this case is the VPC ID of `myvpc`.

In this exercise, you learned how to compose and execute a CloudFormation template to create resources. You used the `Ref` intrinsic function to reference one resource in another resource, in this case associating a subnet with a particular VPC.

Note

Investing a little bit of time into learning the CLI commands makes working with CloudFormation much faster than using the console. For more information, check out the AWS CloudFormation CLI reference.

The next section deals with resource attributes within CloudFormation.

Resource Attributes

Resource attributes can be added to resources to control behaviors and relationships. These attributes are important for controlling the interaction between resources and what will happen when actions are taken. Understanding the resource attribute options and what each one does is important.

Table 13.2 contains some frequently used resource attributes that you should know:

Attribute name	Purpose
DependsOn	Define explicit resource dependencies.
Metadata	Embed structured data within the resource.
CreationPolicy	Wait on resource creation (for example, install and configure an app on an EC2 instance before continuing).
DeletionPolicy	Preserve or back up a resource when a stack is deleted.
UpdatePolicy	Specify how CloudFormation handles updates.
UpdateReplacePolicy	Preserve or back up a resource when a stack update results in the resource being replaced.

Table 13.2: Frequently used resource attributes

Among the resource attributes in the preceding table, DependsOn is particularly important because it allows you to be specific about the order of resource creation, avoiding any dependency problems.

DependsOn is a resource attribute that allows you to be granular in the order of resource creation. It's important to appreciate that CloudFormation processes the creation of resources in parallel. Because some resources depend on others, CloudFormation needs a way to know about those dependencies. There are two ways CloudFormation does this:

- **Implicit dependencies** are created when using the !Ref, !GetAtt, or !Sub intrinsic functions, which are used to tie one resource to another. For example, when you use !Ref to refer to a VPC in a subnet resource, CloudFormation will wait until the VPC resource is available before it attempts to create the subnet because the subnet can only exist within the VPC. So, AWS understands implicitly that the subnet cannot be created before the VPC.

- DependsOn is used in cases where implicit dependencies aren't in place, and you need to guarantee that one resource is created before another. Taking the subnet resource as an example, it would look something like this:

```
AWSTemplateFormatVersion: "2010-09-09"
Resources:
  myvpc:
    Type: AWS::EC2::VPC
    Properties:
      CidrBlock: 10.16.0.0/16

  mysubnet:
    Type: AWS::EC2::Subnet
    DependsOn: myvpc
    Properties:
      VpcId: !ref myvpc
```

Normally, you use DependsOn when using an intrinsic function such as Fn::Ref isn't an option; this example is just a convenient way to illustrate how it works by referencing code you've already seen. Many templates will use this attribute, but you should focus on making templates more reusable to cut down the amount of work necessary to create them.

By now, it should be easy to see the power of CloudFormation. You can create arbitrarily complex designs and replicate them over and over simply by reusing templates that you've created. However, to maximize this reusability, you need to make a couple of changes in your template authoring strategy.

The first and biggest change is to parameterize the template. This simply means replacing hardcoded values with parameters. Hardcoding values requires you to edit the template every time you want to use it for something else. The second change is to leverage intrinsic functions to generate values for you at runtime dynamically. This often involves a little bit of logic and nested functions, but the results usually make the additional effort worthwhile.

Parameterization

Let's say you have a template that creates a VPC with four subnets. When a new application instance is needed, you want to reuse the template. In this case, you would need to hand-edit the template each time, adding a new CIDR block and changing the subnets, keeping track of what subnets and CIDR blocks you've used. A better approach is to use parameters. Parameters are used as placeholders, allowing the actual values to be assigned outside of the resource definition. This reduces the possibility of errors and opens the door to assigning values dynamically.

Default values for parameters can be assigned within the template under the parameters block, but those values can also be overridden from external sources. Since this is a networking book and not a DevOps book, you will stick to assigning your parameters in the template. In real-world usage and deployments, parameters are often stored in places such as the AWS Systems Manager Parameter Store, or passing them in as command-line arguments.

> **Note**
>
> The 12-factor app is a set of architectural pillars for designing a good software-as-a-service application. Importantly, it embodies key DevOps principles in a concise, easy-to-understand format. Pillar 3 states there should be a strict separation of configuration from code. This is why you don't want to store runtime values (such as VPC CIDR blocks) in your template. For more information, check out 12factor.net.

Parameters are an important part of understanding CloudFormation templates, as they help make templates reusable. The less that must be hardcoded in place, the more reusable a template becomes. Working with parameters will also help extend your knowledge of how they may appear on the exam in scenarios.

Exercise 13.2: Parameterizing Templates

In this brief exercise, you will update the CloudFormation template to use a parameter to specify the CIDR block in the VPC resource definition:

1. Copy the basicvpc.yaml file from *Exercise 13.1* and save it as vpc-parameterized.yaml.

2. Add the following block before the Resources block:

    ```
    Parameters:
      VPCParam:
        Type: String
        Default: 10.16.0.0/16
      SubnetParam:
        Type: String
        Default: 10.16.0.0/24
    ```

3. In your VPC resource, change the CidrBlock property to reference VPCParam:

    ```
    CidrBlock:
        !Ref VPCParam
    ```

4. In your subnet resource, change the `CidrBlock` property to reference `SubnetParam`:

```
CidrBlock:
        !Ref SubnetParam
```

5. Now, test it by running the following command:

```
aws cloudformation create-stack --stack-name vpc-param
--template-body file://vpc-paramaterized.yaml
```

6. If you get an object ID back, the command was successful. Use the CloudFormation console in AWS to inspect your stack.

7. When finished, delete the stack using the following command:

```
aws cloudformation delete-stack --stack-name vpc-param
```

In this exercise, you replaced a hardcoded value with a parameter. The key benefits of using parameters are you can reuse the same block of code without modification, and you can populate the specific details (the parameters) at runtime. This is a key building block of automation.

It may occur to you at this point that the subnet hasn't been parametrized. The reason is you're going to use Fn::Cidr to allocate those subnets. This represents the second major change for reuse-friendly template authoring.

Fn::Cidr takes three inputs: the CIDR block, the number of subnets to generate, and the number of host bits (called CIDR bits in the AWS documentation). If you wanted Fn::Cidr to create 4/23 subnets from 10.16.0.0/16, the YAML shorthand version of the Fn::Cidr call would look like this: !Cidr ["10.16.0.0", 4, 9].

Recall that an IPv4 address is 32 bits long. A subnet mask tells the computer which bits are the network identifier and which bits are the host identifier. Take the following example of 10.16.0.0/23:

Mask bits	11111111	11111111	11111110	00000000
Address in binary	00001010	00010000	00000000	00000000
Address in decimal	10	16	0	0

Table 13.3: Host bits example

Table 13.3 shows how to count the host bits when determining your CIDR. If you count from right to left in the mask row, the first 1 bit is in position 10, therefore there are 9 host bits. There's a small wrinkle: Fn:Cidr returns the subnets as a list. To use those subnets, you must select an individual item from the list. CloudFormation has a function for this.

Fn::Select returns a single object from a list by referencing an index value, starting at 0. Take the following list as an example:

```
[10.16.0.0, 10.16.1.0, 10.16.2.0, 10.16.3.0]
```

The first subnet in the list would be index 0, while the last would be index 3.

Exercise 13.3: Fn::Cidr

In this exercise, you will use multiple parameters to create a VPC instead of hardcoding the settings. Putting it all together requires the use of three functions: Fn::Ref, Fn::Select, and Fn::Cidr. You'll start by building on what you created in the last exercise:

1. Copy the completed vpc-param.yaml from the last exercise and name it vpc-fn-cidr.yaml.

2. Add the following parameters to the Parameter block:

```
NumSubnetsParam:
   Type: Number
   Default: 2
SubnetLenParam:
   Type: Number
   Default: 8
```

3. Modify the mysubnet block to look like the following:

```
mysubnet:
   Type: AWS::EC2::Subnet
   Properties:
      VpcId:
         Ref: myvpc
      CidrBlock:
         !Select [
            0,
            !Cidr [!Ref VPCParam, !Ref NumSubnetsParam, !Ref
SubnetLenParam]]
```

4. The tricky part here is the use of nested functions. Fn::Cidr runs, passing the list of subnets to Fn::Select, which then picks index 0, the first subnet in the list. The !Ref items refer to parameters that were defined at the top of the YAML file. Test by running the following command:

```
aws cloudformation create-stack --stack-name vpc-fn-cidr
--template-body file://vpc-fn-cidr.yaml
```

5. Use the CloudFormation console in AWS to inspect your stack. When you have finished, delete the stack using the following command:

```
aws cloudformation delete-stack --stack-name vpc-fn-cidr
```

At this point, the exercise is completed. You should now understand how to create a CloudFormation stack using parameters and intrinsic functions. While it's more work to set up a template with parameters at first, it makes for a highly reusable block of code. If you wanted to add more subnets, you could copy/paste the block, changing only the logical name of the subnet (e.g., call it mysubnet2), and the index value in the Select function (e.g., make it 1). The template as a whole is much more reusable as there are no hardcoded values within the template logic.

Example 13.1 Peering with a VPC in Another AWS Account

It's possible that you'll encounter a situation with cross-account VPC peering in the exam. You'll need to understand it well enough to be able to interpret the question and determine the correct answer. You can use this as an optional exercise if you wish. You'll need to work out a few things on your own, such as defining the necessary parameters and supplying the appropriate values. This example is intended to be a reference for what you might see on the exam.

Before attempting to replicate this example, you'll need the following:

*   A peer VPC ID

*   An accepter account that allows the peering, and a requester account that requests the peering

First, create the IAM cross-account access role. This allows CloudFormation to build the cross-account peering. The steps here are to help you understand that process:

1. Deploy a template that creates the VPC and the cross-account role in the accepter account:

```
Resources:
  peerRole:
      Type: 'AWS::IAM::Role'
      Properties:
        AssumeRolePolicyDocument:
          Statement:
            - Principal:
                AWS: !Ref PeerRequesterAccountIdParam
              Action:
                - 'sts:AssumeRole'
              Effect: Allow
        Path: /
        Policies:
          - PolicyName: root
            PolicyDocument:
```

```
Version: 2012-10-17
Statement:
  - Effect: Allow
    Action: 'ec2:AcceptVpcPeeringConnection'
    Resource: '*'
```

2. Deploy a template that creates a resource of type `vpcPeeringConnection` in the requester account:

```
Resources:
  vpc:
    Type: 'AWS::EC2::VPC'
    Properties:
      CidrBlock: 10.2.0.0/16
      EnableDnsSupport: false
      EnableDnsHostnames: false
      InstanceTenancy: default
  vpcPeeringConnection:
    Type: 'AWS::EC2::VPCPeeringConnection'
    Properties:
      VpcId: !Ref vpc
      PeerVpcId: !Ref PeerVPCId
      PeerOwnerId: !Ref PeerVPCAccountId
      PeerRoleArn: !Ref PeerRoleArn
```

If the deployment is successful, the peering will appear in the VPC console of both accounts under `Peering connections`. *Figure 13.5* shows where to find the peering connection that was created:

Figure 13.5: View the peering connection in the VPC console

In the preceding figure, you can see the VPC peering created in the AWS console. Moving on, the next section gives insight into what to do when CloudFormation deployments have errors.

Troubleshooting CloudFormation Stacks

When working with CloudFormation, you could potentially run into errors when creating, updating, or deleting a stack. You can look to get information about what went wrong in a couple of different places.

The first place to look is in the CloudFormation console. In the console, select your stack and then go to the `Events` tab and look for the error event of interest. The `Status reason` column often gives detailed and easy-to-interpret output to assist in investigating the error.

If the issues are related to EC2 instances, investigate the `cloud-init` and `cfn` logs. On a Linux instance, those will be located in `/var/log`. For Windows instances, the `cfn` log is located in `c:\cfn\log`, and the `cloud-init` log could be in one of three places: `%ProgramFiles%\Amazon\EC2ConfigService`, `%ProgramData%\Amazon\EC2-Windows\Launch\Logs`, or `%ProgramData%\Amazon\EC2Launch\log`.

AWS Cloud Development Kit

AWS CDK is an evolution of CloudFormation that allows programmers to take advantage of the power of automation within their programming language of choice, such as Golang, Python, or JavaScript. The CDK runtime performs synthesis, which executes the code and renders a CloudFormation template. CDK's key differentiator is the construct concept. Constructs are pre-written modular pieces of code that can be used to create and modify infrastructure. Constructs are like class objects in programming; they combine multiple functions in a way that simplifies their use.

There are three flavors of constructs you might use in your development code:

- **CloudFormation only (L1)**: CloudFormation-only (L1) constructs are a direct mapping to CloudFormation resources

- **Curated (L2)**: Curated (L2) constructs are a library of predefined resources with default settings prewired

- **Patterns (L3)**: Pattern (L3) constructs are multiple resources prewired to create an entire architectural pattern, with all the components prewired and just a small number of required parameters needed for deployment

The exam doesn't expect you to know how to use the CDK, but you will be expected to know what it is (enables programmers to drive CloudFormation using code) and why you would want to use it (leverage existing constructs and patterns to quickly create predefined resources and architectural patterns).

Summary

In this chapter, you explored the foundations of IaC and DevOps, focusing on AWS tools such as CloudFormation and AWS CDK. A key takeaway is understanding IaC as a dynamic approach to managing infrastructure through code, in contrast to traditional static setups. You learned that DevOps promotes collaboration between development and operations teams, automated workflows, and CI/CD to ensure smooth, reliable software releases.

You learned how to create AWS infrastructure through CloudFormation templates, parameterizing templates for flexibility, and using intrinsic functions such as `Ref`, `Cidr`, and `Select` to enable reusability. Additionally, you gained insight into automated infrastructure management with CI/CD pipelines, which learn to handle configuration drift, automate testing, and deploy updates seamlessly. AWS CDK allows developers to leverage existing constructs and patterns, simplifying cloud resource creation and improving efficiency.

After completing this chapter, you understand DevOps principles, are able to set up IaC with CloudFormation, and can use the AWS CLI to deploy, update, and troubleshoot resources effectively.

The next chapter will cover data analytics and optimization, in which you will learn about advanced topics for analyzing and troubleshooting as well as how to optimize your cloud network.

Exam Readiness Drill – Chapter Review Questions

Apart from mastering key concepts, strong test-taking skills under time pressure are essential for acing your certification exam. That's why developing these abilities early in your learning journey is critical.

Exam readiness drills, using the free online practice resources provided with this book, help you progressively improve your time management and test-taking skills while reinforcing the key concepts you've learned.

HOW TO GET STARTED

- Open the link or scan the QR code at the bottom of this page

- If you have unlocked the practice resources already, log in to your registered account. If you haven't, follow the instructions in *Chapter 16* and come back to this page.

- Once you log in, click the START button to start a quiz

- We recommend attempting a quiz multiple times till you're able to answer most of the questions correctly and well within the time limit.

- You can use the following practice template to help you plan your attempts:

Working On Accuracy		
Attempt	Target	Time Limit
Attempt 1	40% or more	Till the timer runs out
Attempt 2	60% or more	Till the timer runs out
Attempt 3	75% or more	Till the timer runs out
Working On Timing		
Attempt 4	75% or more	1 minute before time limit
Attempt 5	75% or more	2 minutes before time limit
Attempt 6	75% or more	3 minutes before time limit

The above drill is just an example. Design your drills based on your own goals and make the most out of the online quizzes accompanying this book.

First time accessing the online resources? 🔒

You'll need to unlock them through a one-time process. **Head to** *Chapter 16* **for instructions**.

Open Quiz	
https://packt.link/ansc01ch13	
OR scan this QR code →	

14

Data Analytics and Optimization

This chapter will focus on how to optimize and monitor network deployments in AWS based on business requirements. *Chapter 2, VPC Traffic and Performance Monitoring*, provided an overview of monitoring services and some basic monitoring options. This chapter will focus on how to optimize network performance and implement advanced monitoring capabilities. Data analytics and optimization are part of refining the basics introduced in *Chapter 3, Networking Across Multiple AWS Accounts*, to streamline the deployment of cloud networks, as well as improve outcomes by optimizing troubleshooting capabilities.

Here are the key topics covered in this chapter:

- Optimizing AWS networks via traffic reduction
- Choosing network services by the requirements
- Network topology mapping
- CloudWatch configuration
- Log ingestion and delivery
- Correlating and analyzing multiple data sources for network troubleshooting

By the end of this chapter, you will be able to optimize traffic within an AWS network, pick the best network services based on the business requirements, configure CloudWatch for advanced logging, and analyze and correlate the data for network troubleshooting.

Optimizing AWS Networks via Traffic Reduction

The word **optimization** is a broad term meaning "to make better and more efficient." In terms of networking, there are several ways to optimize a network, such as the following:

- Decrease the number of intermediate hops between two endpoints.

- Increase the speed and bandwidth of the network.

- Simplify the network complexity.

- Send as little traffic across the network as required to meet business objectives.

In previous chapters, the focus has been on the first three items in the list as they tend to have the most impact on network optimization. The last option in the list, however, involves minimizing the amount of traffic being sent across the network, which in turn minimizes data transfer costs and bandwidth charges.

Multicast in Five Minutes

Multicast is a networking technology almost as old as networking itself, built on a simple premise: what if I could send one packet and have every recipient get a copy of it without having to talk to each recipient individually? In this section, we will cover the basics of multicasting, enough for the exam, by looking at the problems it solves and how it solves them.

Standard communication between one client and one server is called uniflow, because there is only one path. There are times when a server might communicate with all devices on a local network, such as to send out an alert. This can be done by simply creating multiple one-to-one connections. This is called a broadcast.

Consider *Figure 14.1* as an example of how unicast traffic delivery works.

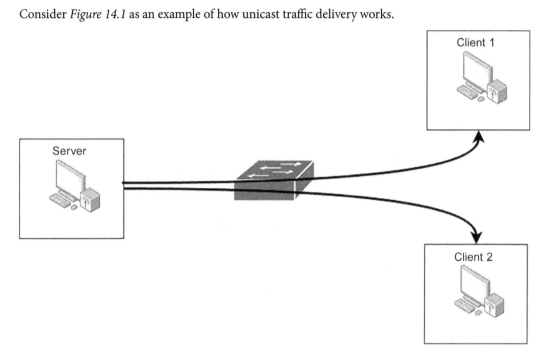

Figure 14.1: Unicast server flow

Figure 14.1 illustrates the normal server flow. If the data packet is small, for instance, a 1 MB file, that's hardly a problem even if the server has many clients. But what if the data is much larger, such as 1 GB? In the cloud, you pay per the amount of data moved, and this can get expensive, especially between AZs and AWS Regions. What if a server has 10,000 clients?

In the case of a managed AWS service, it is up to AWS to figure out how to scale the data transfers. But if you own and operate an EC2 instance, it is up to you to figure out the most efficient way of transmitting data. Services such as load balancers and auto-scaling groups help deal with increased demand and have the elasticity to contract when demand is lower; so, if fewer clients are on the network, the traffic will decrease. However, the number of connections might always be high, meaning that data streaming will remain expensive.

In a multicast model, this issue is solved by having a single flow from the server, with the network itself replicating data as needed only when a client requests that specific data.

To put it in more network-centric terms, the following is the same across all networking:

- Unicast flows are one to one between two endpoints.
- Broadcast flows are one-to-all flows within a network that require endpoints to expend processing power to determine whether traffic is even meant for them and discard it if not.
- Multicast strikes a balance between the two, allowing clients to signal their interest, which allows a one-to-many option that is efficient by its targeting and replication of traffic.

Multicast has its limitations, and not every type of application is meant for multicast traffic patterns. For example, multicast is a one-way flow from the server to the client. There is no transactional element to multicast traffic. Because of this, multicast does not support TCP flows, only UDP, which is transactionless. Think of multicast flows like tuning into a streaming media feed or a car radio type of consumption model. The server runs the application stream whether anyone is listening or not, but the interested receivers use their car radio dials to tune into the station's feed. IP television streaming is an example of multicast, as it will only transmit data if there are receivers that have signaled interest.

Here is an updated concept diagram using multicast:

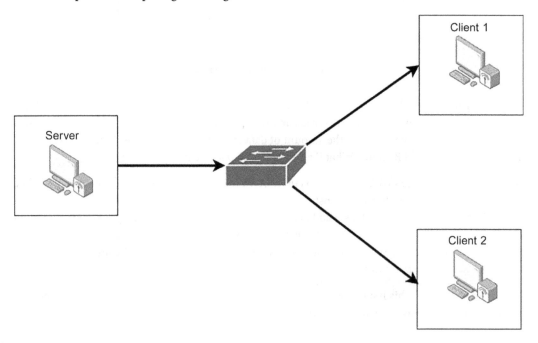

Figure 14.2: Multicast server flow

In *Figure 14.2*, the server sends only one stream of data, which is replicated by the network to the interested receivers. Multicast minimizes the load on the server, as it only needs to generate a single packet for any number of receivers. This also means you can design for resiliency only and not worry about the server load requiring scaling out or up. This could translate into cost savings if your application supports multicast flow.

Signaling interest in receiving a multicast group is done using the **Internet Group Management Protocol (IGMP)**. AWS only supports IGMP version 2 and only then with a **transit gateway (TGW)**. This limits the multicast application support and design patterns available to implement multicast in your cloud environment.

> **Exam Tip**
> Remember these requirements for the exam in case you run into a multicast question. You may be able to easily eliminate wrong answers that suggest a different IGMP version or use some solution other than TGW.

In AWS, multicast is tightly controlled and limited in its capacity compared to an on-premises multicast deployment. In a traditional multicast deployment, features such as a rendezvous point and multicast routing infrastructure using **Protocol Independent Multicast (PIM)** are necessary to carry multicast traffic. AWS takes care of all of this for you, requiring only that you define the following:

- **Multicast domain**: This is essentially a multicast segmentation solution. Having multiple multicast domains is optional but it provides an easy way to separate multicast sources and receivers.

- **Multicast group**: This identifies the "channel" for a source and receivers to use for a particular multicast flow. In AWS, this membership is at the ENI level attached to EC2 instances and not via the instance itself. This is an important distinction because an ENI can be migrated between instances.

- **IGMP**: As discussed earlier, the IGMP protocol is how multicast receivers signal their interest in receiving a multicast flow. This can be a dynamic join request sent by the instance itself (through the **elastic network interface (ENI)**) or it can be statically assigned to the ENI so that the ENI will always request that group. This is useful if the EC2 instance host OS itself is incapable of sending its own join request. Importantly, if the ENI is set statically, it cannot send its own traffic, only receive it.

- **Multicast source**: The multicast source is the originator of a multicast flow. For example, this could be a stock ticker program that ingests data from the stock market and then streams out the up-to-date stock information for others to consume on their workstations. In AWS, the source can simply send its multicast stream to a particular multicast group, and those receivers that use IGMP to request that multicast group should receive the stream.

- **Multicast receiver**: This is the ENI associated with a multicast group receiving traffic. As mentioned, if the static membership option is selected, this ENI can only receive traffic. If the host can use IGMP, the ENI could send traffic as well, becoming a multicast source as well as a receiver.

How multicast is used in AWS does not closely match the expectation of how multicast works in an on-premises environment. Subnets inside VPCs are added to multicast domains, and those domains govern what multicast streams can be received. Here is a basic diagram:

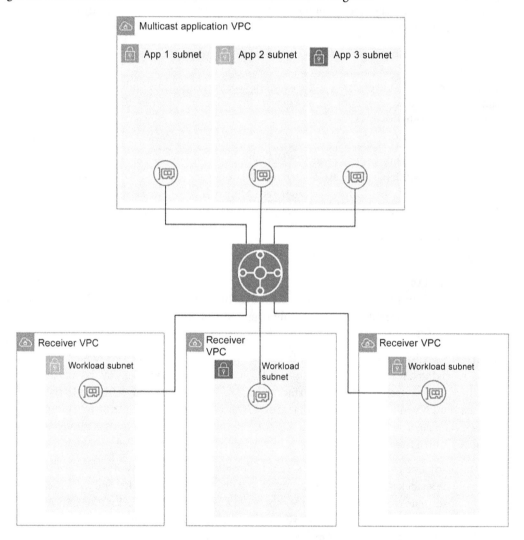

Figure 14.3: Multicast domains with TGW by subnet

Once subnets and TGW attachments are associated with a multicast domain, hosts in these subnets can be multicast senders or receivers dynamically, or only receivers statically. This is entirely determined by whether the ENI of the host is statically joined to the multicast group (thus making it receiver-only) or whether the ENI is dynamic and able to use IGMP from the host OS to drive the multicast group membership.

To understand how the membership is maintained, joined, or left, you need to understand that IGMP works on a timer and a query/join process. In the case of AWS, the TGW acts as a multicast router, issuing periodic queries into the subnets marked for the multicast domain to which it belongs. This query essentially asks, "Is there any host on this segment that wishes to receive a multicast stream to this group?" This is part of the optimization multicast offers to ensure that traffic is only sent where it is desired. The job of the host in a dynamic IGMP configuration is to answer this query in the affirmative, saying, "Yes, I want this stream," whenever asked, but also to issue its own join request when it first wishes to receive the stream, without waiting on a query. If a host wishes to leave a multicast group and uses the dynamic IGMP configuration, it issues a leave request instead, informing the TGW that it no longer wants to receive the multicast.

If a subnet has no hosts (ENIs) that want the multicast traffic, the TGW will stop delivering that traffic. This is the normal and preferred method that naturally optimizes traffic flow as hosts join and leave the stream. When an ENI is statically joined to a group, the ENI will always send a request to the group and answer queries and will never issue a leave request unless the ENI is reconfigured to drop the group.

This section will cover the limitations of multicast in AWS and attempt to provide context around them to make them easier for you to remember.

There are several architecture limitations when designing a multicast implementation within an AWS cloud network:

- A subnet can only belong to one multicast domain.
- Multicast routing is not supported over Direct Connect, S2S VPN, TGW peering attachments, or TGW Connect attachments.
- TGWs do not support fragmenting multicast packets.
- If a multicast source and receivers are in the same VPC, security group referencing cannot be used to accept traffic from the source security group. For example, if you would normally have a rule in a security group allowing traffic from any host belonging to another security group in the VPC, this is not sufficient to allow multicast traffic from that source; an explicit rule would be needed.
- Security groups for receivers must allow IGMP queries from the TGW; the source of these packets is `0.0.0.0/32` and the destination is `224t0.0.1/32`, with protocol 2. This is also true of any network ACLs that may exist on the subnets.

What this list details is that a multicast design differs from a unicast design. Because multicast is treated in a different way, there is less support over more primitive network constructs, and security must explicitly allow multicast traffic.

There are also technical limitations to consider when implementing multicast in an AWS cloud network:

- Current TGWs cannot be enabled for multicast; a new one must be created with multicast enabled.
- The AWS web UI has no way to manage multicast group membership; static membership must be managed via the AWS CLI, API, or dynamically by IGMP from the instance ENI.
- If using a non-Nitro instance, you must disable the source/destination check on the ENI.
- A non-Nitro instance cannot be a multicast sender.
- The TGW will automatically remove static multicast groups and sources for ENIs that are deleted by assuming a service-linked role to clean this up.

Since multicast is not an oft-requested feature, many of the technical limitations of the solution are because AWS simply has not implemented some of the features that are expected in unicast deployments, such as the lack of group membership support via the AWS console. Additionally, the more primitive network interfaces (non-Nitro) lack multicast feature parity.

Unlike multicast in an on-premises context, multicast in the cloud behaves differently because it is part of a larger tunnel-based architecture in the hypervisor. Here are some of the multicast behavior considerations with the TGW to be aware of:

- If a host sends 2-3 IGMP join messages and the TGW does not receive them, the host will not be added to the group. If this happens, IGMP must be re-triggered from the host side.
- If the TGW fails to receive a reply to an IGMP query three times from an ENI/host, that host is removed from the multicast group, but the TGW continues to send IGMP queries to it for 12 hours afterward. An IGMP leave message from the ENI/host cuts this off immediately.
- The TGW tracks the hosts that have joined a group, and if a TGW fails, the new TGW continues to send multicast data to those hosts for seven minutes after the last successful IGMP join message. New IGMP queries will continue to be sent for 12 hours or until the new TGW receives an IGMP leave message.

Ultimately, using multicast in AWS can involve a lot of unclear limitations and result in a deployment that is difficult to manage and troubleshoot. Care should be taken to evaluate whether the benefits of using multicast, such as minimized data transfer, are worth the increased complexity. Some applications are built only for multicast, so multicast may be required; in any case, it will be important to understand the way multicast is implemented in AWS to be successful.

> **Exam Tip**
>
> There is a lot here to memorize, so do not focus on the exact timers and instead focus on the behavior, especially caveats. For example, non-Nitro instances cannot be multicast senders. It is more likely that exam questions will attempt to test your ability to remember what is and is not possible rather than your memorization of timers.

CloudFront

CloudFront was covered in *Chapter 7, AWS Route 53: Basics*, but it is important to examine how a solution such as CloudFront helps to optimize traffic and reduce bandwidth utilization within an AWS network.

CloudFront is a content delivery network or CDN. By its nature, a CDN optimizes traffic in two major ways. A CDN is distributed across the globe so that end users can connect to the most geographically advantageous node for the content to download. This improves latency and reliability. CloudFront has both edge locations and regional edge caches. The regional caches serve as a backstop for content that is not being requested in large numbers but is still held in case it is requested. Edge locations are the closest geographical point of presence to users and are smaller and more distributed, and their cache is very ephemeral in comparison based on what is being requested. A CDN caches content so that repeated requests for content do not have to be pulled from the source, which might be geographically distant and unable to deal with a large scale of requests. This means that bandwidth through the AWS network is reduced, and repeated data transfers are not needed to serve content.

Because CloudFront handles caching in tiers like this, the total amount of bandwidth needed to serve end user requests is minimized, which lowers utilization and optimizes delivery.

CloudFront also offers some edge computing capability to customize end user requests and further optimize delivery with Lambda@Edge and CloudFront Functions.

The **CloudFront Functions** feature allows developers to adjust user-initiated requests using lightweight applications such as JavaScript. These functions can help with rewriting headers, evaluating headers, returning specific content, redirecting users to other URLs, and so on. These edge functions optimize bandwidth by requiring far less back and forth to a server located further within the AWS network that would perform the same work.

Lambda@Edge is a more specific serverless offering that allows you to use Lambda functions to do things such as check cookies, inspect headers, authorize users, and return specific content. This saves a ton of back-and-forth traffic to do this work from an EC2 instance or server in another geographical Region.

Choosing Network Services by Requirements

Throughout this certification guide, each chapter has endeavored to help you understand how to identify the correct services to use based on the business requirements. This section should be more of a recap than new information, conveniently collected and presented in a small, digestible format.

Generally, the more complex the requirements for workloads and services to connect to each other, the less likely you are to rely on cloud-native services alone. In addition, the more connectivity needs to scale, the more likely you are to utilize advanced AWS network services such as **Transit Gateway** and not rely on basic options such as VPC peering.

From a scale/complexity standpoint, the network services rank like this, in order of least complex and least scale to most complex and with the highest scale:

1. VPC peering
2. PrivateLink (customer)
3. Virtual private gateway (as with **Site-to-Site (S2S)** VPN and Direct Connect)
4. TGW (S2S VPN and Direct Connect gateway)
5. TGW Connect
6. TGW attachment
7. PrivateLink (provider)
8. **Network virtual appliance (NVA)**

Often at odds with other requirements, sometimes the need for high throughput is a critical requirement that needs special consideration. In the cloud, this is often not a challenge except when weighed against other requirements. For example, VPC peering can provide the highest bandwidth possible but also has the least ability to scale and routes must be statically managed. However, if connecting something within AWS to another location, such as a data center or another cloud, the bandwidth available is very low in comparison. This is expected as AWS has a robust and performant cloud fabric and backbone compared to extra cloud connectivity.

Here are the network services in AWS ranked from least bandwidth available to most bandwidth available. The list takes into consideration that VPC peering and PrivateLink bandwidth limits are generally determined by the underlying AWS infrastructure between the resources being connected, but should still be very high:

1. AWS Direct Connect (subrate, lower bandwidth via AWS Partner)
2. AWS S2S VPN with VGW
3. TGW (S2S VPN)
4. AWS Direct Connect (1 Gbps)

5. NVA

6. TGW Connect

7. AWS Direct Connect (10 Gbps)

8. TGW attachment

9. AWS PrivateLink

10. VPC peering

It is worth noting that several of the items in the list have variable speed (such as Direct Connect) and others are effectively the same speed, such as all flavors of S2S VPN. The list is close to an inverse of the complexity/scale list previously. This will not surprise network veterans; high bandwidth has always come at the cost of scale and complexity. The choices may seem obvious when only one requirement matters, or when the requirements complement each other (i.e., high bandwidth, low complexity), but what do we do when the requirement necessitates complexity and high bandwidth as well?

In these situations, it is best to verify with the business, developers, or both which requirements are critical for the application to function, and which requirements are more of a wish list. As an example, the need for high bandwidth is driven by application performance requirements, but not always. Often, the business sets forth "requirements" that are ideal but can be discarded if needed. Application requirements for functionality are the single largest contributor to the network requirements, and when there are no easy answers to meet everything that is needed, understanding the functionality of an application and its resiliency needs (based on business criticality) is how to make hard choices among competing priorities.

> **Exam Tip**
>
> You can expect questions that test your capability to make choices to meet minimally functional requirements in the AWS exam. For example, the questions often include several functionally correct choices, but the scenario will include language such as "cost-conscious," "resilient," and "minimal requirement" to test your understanding of the services and when to use them.

Network Topology Mapping

Network visualization has been a challenge since the early days of networking. Starting with paper napkins and working up to complex applications, there has always been a need to visually describe and reference network implementations. The cloud is no different in that respect. AWS has a network monitoring and visualization offering in TGW **Network Manager (NM)** (also referred to as AWS Network Manager). This was briefly introduced in *Chapter 2, VPC Traffic and Performance Monitoring*, but this section will cover how to build and register a TGW implementation to understand the deployed network topology.

Creating a Network Topology

Creating a network topology is free unless you also create a Cloud WAN core network or attach a TGW, which carries an hourly charge.

In AWS Network Manager, users can create a global network that is just a logical container for other resource objects. This is a top-level construct that has no functionality on its own but serves to collect and organize resources placed within it. This is also the level at which monitoring and visualization dashboards are created. When creating a global network, you only need to provide a name and description for it. Optionally, if your organization makes use of tags, you can also define those.

> **Note**
>
> Since AWS Cloud WAN is not on this version of the *Advanced Networking Specialty* exam, it is unlikely you will see any questions about the Cloud WAN elements of AWS Network Manager. However, be aware that in the real world, this is also how Cloud WAN is configured. When creating a global network, there is an option checked by default that also creates a core network, which is part of Cloud WAN. If you are practicing this lab for the exam, clear this checkbox and proceed to create a TGW-only global network.

In this example, you start by creating the global network for our example application, Trailcats:

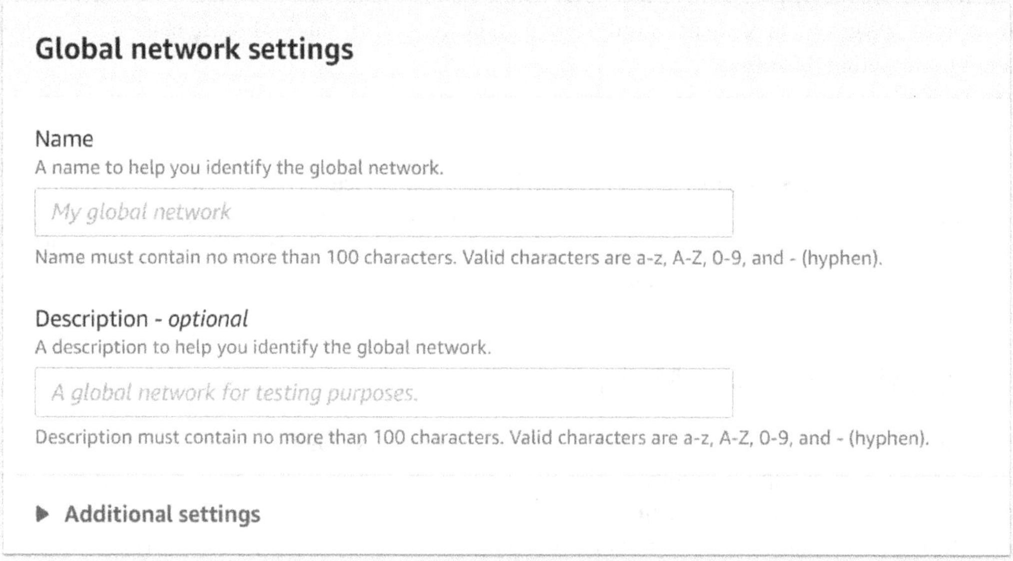

Global network settings

Name
A name to help you identify the global network.

    My global network

Name must contain no more than 100 characters. Valid characters are a-z, A-Z, 0-9, and - (hyphen).

Description - *optional*
A description to help you identify the global network.

    A global network for testing purposes.

Description must contain no more than 100 characters. Valid characters are a-z, A-Z, 0-9, and - (hyphen).

▶ **Additional settings**

Figure 14.4: Create a global network

Step 1: Provide Details

Provide the basic details of your global network, including a name, description, and any tags. Remember that a global network is a top-level container for the objects within it.

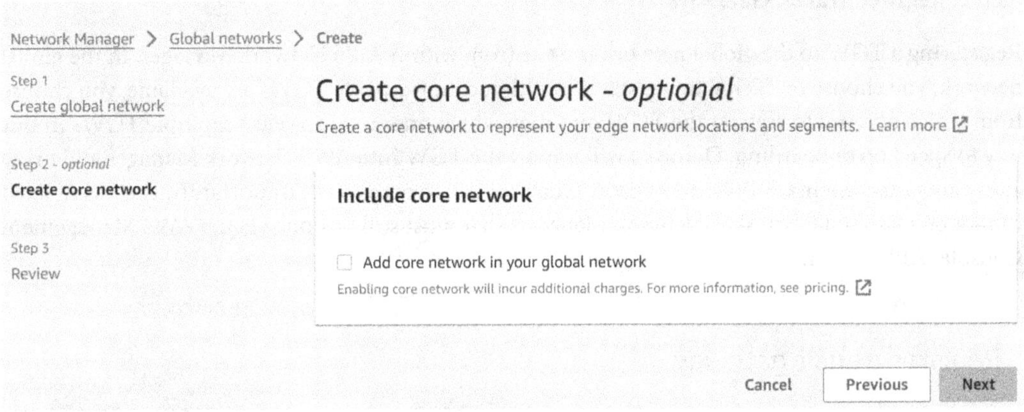

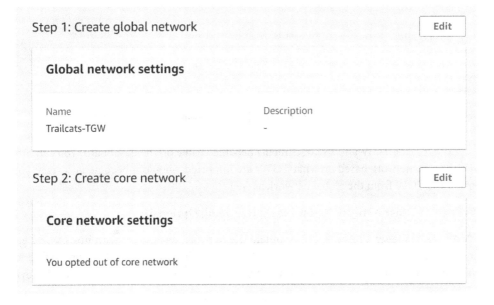

Figure 14.5: Create a global network

Be sure to leave Add core network in your global network unselected, as in *Figure 14.5*.

Figure 14.6: Create a global network

Complete the creation of the global network. Without any TGW instances or core network edges (CloudWAN constructs), this global network will be empty until you add the TGW and other logical objects such as sites and links.

Step 2: Register Transit Gateways

Registering a TGW to the global network is done from within AWS Network Manager. In the global network, you choose the TGW to register; if multi-account-capable TGWs are available, you choose from which account to register the TGW, and then which one. You can select multiple TGWs in this way to speed up onboarding. Doing so will import the TGW into AWS Network Manager and bring along any attachments, VPNs, and Direct Connect instances as well. Importantly, the TGW itself cannot be managed from AWS Network Manager; that must still be done via the AWS Management Console, API, or CLI.

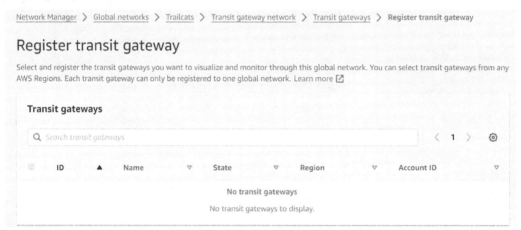

Figure 14.7: Network manager transit gateway registration menu

Once onboarded, the TGW and its attachments become visible on the dashboard. You can monitor and visualize the network based on what TGWs are imported. Should it become necessary, you can de-register the TGW from the global network as well.

Step 3: Register Other Network Resources (Extra Cloud, Optional)

AWS Network Manager allows you to populate the network dashboard with non-AWS network resources as well, to paint a better picture of a full topology. This involves creating (up to) four logical objects, which can be further associated with AWS resources to help logically define that deployment:

- **Sites**: A site is a logical representation of a physical location and includes physical details such as physical address, latitude, and longitude.

- **Devices**: Devices are logical representations of real equipment that connects to AWS, or they can logically represent **NVAs** and non-native resources. Details include vendor, model, serial number, and location type (on-premises, data center, cloud, etc.). If a physical device, this can also be given a physical address and latitude/longitude as well.

- **Links**: A link is a logical representation of an internet connection and must be associated with a site. The details of a link include the upload/download speed, provider, and connection type.

- **Connections**: This represents the connection between two devices. It includes associations with devices and can optionally reference a link object that has been created.

Remember that all of this is logical, for tracking and visualization purposes only. Creating these objects and associating them makes no change to the actual infrastructure.

Once these objects are created, they can be associated with any TGWs in use. For example, the TGW can be associated with a customer gateway via the TGW or `Devices` page in AWS NM, under `Global Networks`. This association is used to associate a link and device with an AWS S2S VPN landing on the TGW. The TGW can also be associated with a device and link via the TGW Connect peer, on either the TGW or `Devices` page under `Global Networks`. This helps to further visualize the deployment and visualize traffic flows.

The sample global network is now completed. If you are following the exercise, be sure to remove the TGW if not using it for other labs as there is an hourly charge associated with having one.

CloudWatch Configuration

To effectively optimize systems, you need to measure their performance, and as you have already learned, this is done with CloudWatch. Basic CloudWatch alerts and actions were discussed in *Chapter 2, VPC Traffic and Performance Monitoring*. In this section, more advanced CloudWatch features, such as custom metrics and automated alarms, will be detailed. CloudWatch offers a robust suite of monitoring, alerting, and reaction-based responses based on predefined thresholds.

Custom CloudWatch Metrics

The Advanced Network Specialty certification exam specifically mentions needing to be able to implement custom CloudWatch metrics. It does not, however, mention creating those custom metrics. This is an important distinction.

The details of basic metrics were discussed in *Chapter 2, VPC Traffic and Performance Monitoring* but custom metrics have several attributes that define what the metric collects and how they perform:

- **Resolution**: Resolution refers to the granularity of metric data. There are two custom metric options:

  - **Standard**: One-minute granularity and the default setting

  - **High-Res**: One-second granularity; can be viewed in 1-, 5-, 10-, 30-, and 60-second intervals

- **Dimensions**: Dimensions are the attributes of a metric indicating what the metric is and what data it has. The list of dimensions can be specified in a `put-metric-data` operation where you specify the name and value to push, or it can be specified in a `get-metric-data` or `put-metric-alarm` operation referencing the available dimensions. Often, you will specify dimensions for your custom metrics and then retrieve them for monitoring or use them for setting up alarms.

- **Single data points**: This custom metric simply publishes a log entry to CloudWatch specifying specific dimensions and data, one per entry.

- **Statistic sets**: This aggregation technique batches data by simply publishing statistical numbers to CloudWatch instead of raw data at a specified interval. Statistic sets make it harder to calculate a percentile metric; only in the following cases can percentiles be derived from them:

  - The sample count of the statistic is 1

  - The minimum and maximum values of the statistic are equal

- **Value zero**: If the metric you want to publish is only an occurrence or non-occurrence, you may elect to publish no data or publish a zero-value entry instead for the benefit of graphing, such as when the number of data point collections would be helpful to use whether or not there was data.

These custom attributes are used when an application or endpoint publishes its events to CloudWatch, and these events are pushed to CloudWatch from an external source using the AWS CLI or API.

Within CloudWatch itself, you can also create custom metrics to act based on what is ingested via a log group using `Create Metric Filter` under the `Actions` menu of the log group for which you wish to receive metrics. The actual creation of metric patterns is not part of the exam but you should understand that creating a filter pattern will allow you to receive logs based on that metric as well as to visualize and raise an alarm.

Here is an example of creating a custom metric:

1. Under `CloudWatch | Logs | Log Groups`, choose `Actions` and then `Create Metric Filter`.

2. Enter a filter pattern. As an example, regular expressions are commonly used, but you can also match on basic matching, JSON string matching, or combinations.

3. Enter a name for the metric filter, and under `Metric Details`, place the filter under a current namespace or create a new one. The namespace will be the top-level container for the custom metric when creating alarms and graphs.

4. `Metric Value` must also be selected. For example, if the goal is to increase a counter by 1 for every occurrence of a filter match, set that. Otherwise, you can use tokens such as `$size` where the value of that token is defined, and the filter match will increment a counter by that number.

5. After selecting `Create Metric Filter`, that custom metric is now available to be selected for use in the log groups in which it is defined for creating alarms and visualizations.

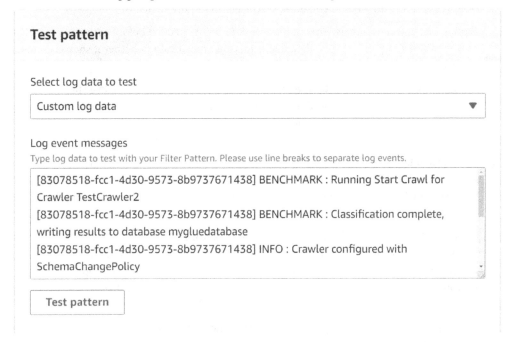

Test pattern

Select log data to test

Custom log data ▼

Log event messages
Type log data to test with your Filter Pattern. Please use line breaks to separate log events.

[83078518-fcc1-4d30-9573-8b9737671438] BENCHMARK : Running Start Crawl for Crawler TestCrawler2
[83078518-fcc1-4d30-9573-8b9737671438] BENCHMARK : Classification complete, writing results to database mygluedatabase
[83078518-fcc1-4d30-9573-8b9737671438] INFO : Crawler configured with SchemaChangePolicy

Test pattern

Figure 14.8: AWS CloudWatch custom metric creation

Creating a CloudWatch Alarm Using Custom Metrics

Assume that in the previous section, a custom filter metric called READY was created using a custom namespace called `Webserver`. An alarm can be created in CloudWatch to alert whenever the READY custom metric reaches or breaches a threshold.

This section walks through creating a custom CloudWatch alarm based on the previously created custom metric:

1. From the CloudWatch console, choose `Log Groups`|`[Log Group Name]`|`Metric Filters`.

2. Select the custom metric filter and then `Create Alarm`. This will automatically choose the correct metric information for the alarm.

3. On the `Alarm creation` screen, specify what statistic to use, such as `Sum` or `Average`, and over what period the condition should be above the data point before activating an alarm.

4. For Condition, select a static threshold. Normally, you would choose to activate an alarm if the value is greater than the threshold value, but you can also choose to do so if it is equal to or lower than the threshold depending on the alarm requirements.

5. Optionally, you can increase the number of data points needed before activating the alarm, as well as adjusting how to treat data that is missing; specifically, whether it should count as a breach, be ignored, or just be marked missing.

6. Next, configure the alarm state trigger. Remember that an Ok state is also a state that should be triggered when an alarm condition has ended and normal operation has returned. Multiple alarm state triggers can be built at once, so a return to normalcy will trigger an all-clear alarm, notifying the operations team that there is no longer a problem.

7. The normal behavior of an alarm is to notify a user of an issue through SNS by email or a group of emails. Within AWS, an SNS topic notification to such a group of users could be used to trigger further actions using a service such as EventBridge, but that is beyond the scope of this explanation.

8. Additionally, and especially with EC2 instances, automated actions can be taken when an alarm threshold causes an alarm to be activated. For example, an Auto Scaling group could be triggered to scale out or in a defined EC2 instance group. Although not applicable for most custom filters, for EC2-based metrics, the alarm can also trigger EC2-based automation such as terminating or rebooting the instance causing the alarm.

9. If using AWS Systems Manager (also not within the scope of the exam), an incident and ops event can be triggered.

10. Lastly, the alarm must be given a name.

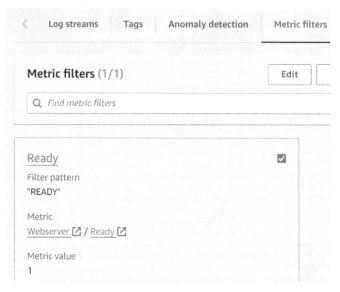

Figure 14.9: Create a custom alarm in CloudWatch

11. Click the `Create Alarm` button, which opens a new window.

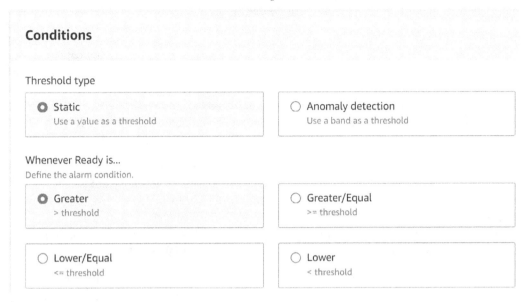

Figure 14.10: Create alarm in CloudWatch with conditions

12. Specify the alarm thresholds here and how long they should be in this state to trigger an alarm.

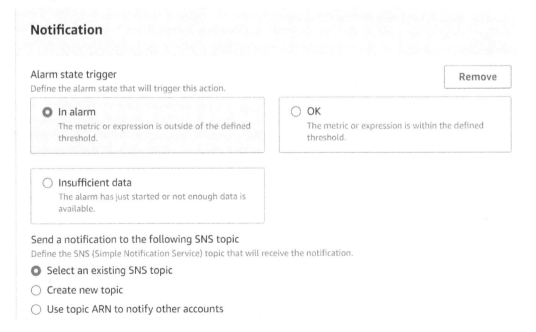

Figure 14.11: Create CloudWatch alarm notification

Specify the notification settings and whom to notify. It is recommended that you add a second notification action for when the alarm state is cleared as well.

When should you be using these custom metrics and EC2 metrics to trigger alarms as a cloud networking engineer? For custom metrics, it is usually good to use CloudWatch to pull relevant data from logs such as VPC Flow Logs and then create a filter looking for whatever you are interested in. This keeps you from having to run queries on demand; you can create the custom metric for the relevant log data and graph it or create an alarm. EC2 metrics are a clear method to monitor the health and throughput of NVAs. For example, if you have an NVA handling too much traffic and the CPU is spiking high enough that packets are being dropped, that is a critical problem to detect and solve quickly.

In general, custom metrics and automated alarms using CloudWatch should have a focus on early detection and notification. While there are automated actions (such as scaling out, scaling in, and rebooting) that can be taken, it is rarely advantageous with an NVA to do so. Directing traffic to or away from NVAs often requires changing route tables. If the route table points at a load balancer instead, this helps alleviate that problem, but more often, an NVA is not ready to forward traffic as soon as it is built and it may require configuration changes to make it operable.

As vendors support their NVAs in a cloud environment more and more, many vendors now support pulling a bootstrap configuration from an S3 bucket or passing in a minimal config via the user data at inception. A mature cloud services organization can more confidently automate actions with NVAs than a novice one; get to know your capabilities based on the vendors used, what methods they support, and your own skillset.

Log Ingestion and Delivery

Though different solutions and their logging capabilities are discussed in *Chapter 2, VPC Traffic and Performance Monitoring*, this section will focus on exporting those logs to another location for analysis by other solutions. CloudWatch Logs can be exported to an AWS S3 bucket, whether in the same AWS account or another one.

AWS Advanced Networking Specialty is not an exam that focuses on S3 in any capacity, but some knowledge of S3 as a solution is required to understand how to set up logs to export to it, and, later, how to use AWS-specific or third-party applications to consume the logs from S3.

Major restrictions for exporting from CloudWatch Logs to S3 include the following:

- S3 buckets can be encrypted with AES-256 or **server side encryption using Key Management Service (SSE-KMS)** encryption only. Dual-layer SSE KMS is not supported.

- Only CloudWatch log groups with the Standard class can be exported to S3. The Infrequent Access class is not supported. The Infrequent Access class is more for the analysis of logs that are not accessed often or only in an ad hoc manner.

The steps to create a log export are simple:

1. Create an S3 bucket for the CloudWatch log export.

2. Set access permissions on the bucket for IAM roles to be able to access them for writing or reading from the bucket. This is done by adding permissions to an S3 bucket policy or user identity system. In the bucket policy, specify the AWS source account number, from what Region the AWS logs will be coming, and the `S3:GetBucketAcl` and `S3:PutObject` permissions referencing the log groups and accounts.

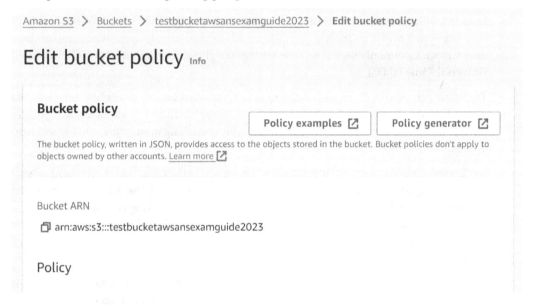

Figure 14.12: Edit an S3 bucket policy for CloudWatch log export

If using a bucket with SSE-KMS, extra policy actions need to be allowed so that exported data can be encrypted for the bucket.

3. The next step is to set up the log export in CloudWatch. In the CloudWatch console, go to the `Log Groups` section, select `Actions`, and then `Export Data to S3`.

4. Define the data export time range and, optionally, a prefix to assign to this log group export. This is for shared S3 buckets, so you can figure out which log files belong to which log group. For example, you might add a prefix of `nva-firewall` and the log file in S3 will start with that prefix.

5. Choose the S3 bucket to export to. Finally, select `Export`, and the logs will be sent to S3.

Of course, this only exports a point in time. Log exports are necessary to have any hope of correlating data or understanding points in time. It is recommended to try this on your own and become familiar with the process and options available in AWS.

Correlating and Analyzing Multiple Data Sources in Network Troubleshooting

Pulling together different log streams for analysis can be a daunting task. In network troubleshooting, there are an impressive amount of data sources that can provide insight into possible problems with AWS networking, but often these services exist on their own and without a unifying thread.

Here are some basic tips to enable unity in log collection so that correlation is possible:

- Ensure that log sources are all using the same date and time settings. It is impossible to correlate events across a variety of logs when each log carries a different timestamp or date/ time format. Commonly used date/time settings for global correlation rely on **Coordinated Universal Time (UTC)**.

- Try to use common prefixes and names to identify to what service or application a log data stream belongs. As discussed earlier, for example, CloudWatch allows setting custom prefixes for export to S3 so that log files will be prefixed with something meaningful, such as `webapp` or `rds`.

- To the extent possible, keep the logging clean. That is, log to the minimum requirement of what you are trying to capture. Depending on the service being logged, the default setting may be far too noisy and capture too much extraneous information, making correlation difficult or impossible. As an example, when setting up things such as VPC Flow Logs, log to the minimum amount required to capture what you think will be relevant. The relevance will change based on the problem; do not be afraid to change this up.

- Logging should include as many of the relevant fields from the service being logged as is prudent, but where possible should also include similar fields to the other data sources you may be correlating. For example, timestamps, ports, protocols, and other network-level details will be useful for the correlation process.

AWS also offers several services that can help with the correlation and analysis of logging data, such as CloudWatch Logs and AWS X-Ray. The largest is **OpenSearch**. Amazon OpenSearch Service is a log analytics engine that can intake metrics, logs, and other operational data and correlate the data through search functions as well as offering visualizations. Amazon OpenSearch Service Ingestion is a managed data collector that delivers this data to OpenSearch data domains directly. The basic format is that OpenSearch Ingestion is a data preparation service that can perform transformation and normalization on ingested log data before delivering it to OpenSearch Service for analysis. OpenSearch Ingestion is part of the larger OpenSearch Service, which can be a serverless, AWS-managed server or a client-managed server that provides log collection and correlation.

There are other ways to provide network troubleshooting data correlation as well. AWS offers other log correlation options and insights with Kinesis Data Streams and Kinesis Firehose, CloudWatch Log Insights, and CloudTrail Log Insights. There are also many third-party products that provide it as well.

The following AWS services have rich logging that can be used for log correlations when troubleshooting network issues:

- VPC Flow Logs (Including Transit Gateway)
- Network load balancers
- Web Application Firewall
- Firewall Manager
- CloudTrail
- EC2 instance logs
- Application Load Balancer (using access logs and X-Ray tracing for application activity)

This list is not exhaustive. Most of these services have been discussed elsewhere and can either log to CloudWatch or S3 directly. Depending on where the logging is sent, CloudWatch Logs Insights, Kinesis, or third-party services may be the best way to correlate the logs when investigating network problems.

For instance, imagine a network change that is automated by a code pipeline and utilizing an **infrastructure as code** (**IAC**) solution such as CloudTrail or Terraform. This change is examined, approved, and carried out by an automated process using a **continuous integration/continuous delivery** (**CI/CD**) pipeline during a maintenance window, resulting in changes to a VPC and some network ACLs.

A few days later, the developers complain that their application logs are filling with errors regarding being unable to connect to a certain database. The first instinct might be to simply roll back the change, but what if that change affects other applications or is needed for multiple reasons? Using a combination of VPC Flow Logs and AWS Firewall logs, to identify whether the traffic is being allowed or blocked, and CloudTrail logs indicates that the application traffic is being blocked, but not because of the automated change.

Rather, a junior cloud engineer added a new firewall rule as part of another unrelated change that had an unforeseen impact on the developer application due to the developers relying on a hardcoded IP somewhere they were unaware had been pushed to production. Sound implausible? Well, this is common when dealing with multiple teams making changes that affect each other's solutions. Without these log sources and the ability to correlate events, it would have been incredibly difficult to determine what happened and the order of operations that led to the incident.

Correlation among multiple data sources is among the most challenging tasks that a cloud network engineer will be called on to perform. The most important thing to do is ensure that multiple data sources have common criteria to aid in correlation, as well as to understand which tool provides the data that is most useful for the correlation activity.

Summary

This chapter covered the importance of optimizing traffic for AWS networks and the best ways to determine how to optimize it. Traffic reduction strategies such as multicast and CloudFront were discussed, as well as how to create and optimize logging to understand how to troubleshoot network problems. The concepts covered in this chapter may appear on the exam, especially in scenarios that ask you to pick the best option from multiple valid ones.

This chapter is a companion chapter to *Chapter 2, VPC Traffic and Performance Monitoring*. As such, the focus of this chapter was on expanding the details of that chapter and teaching you relevant skills needed to pass the AWS Advanced Networking Specialty exam.

Exam Readiness Drill – Chapter Review Questions

Apart from mastering key concepts, strong test-taking skills under time pressure are essential for acing your certification exam. That's why developing these abilities early in your learning journey is critical.

Exam readiness drills, using the free online practice resources provided with this book, help you progressively improve your time management and test-taking skills while reinforcing the key concepts you've learned.

HOW TO GET STARTED

- Open the link or scan the QR code at the bottom of this page

- If you have unlocked the practice resources already, log in to your registered account. If you haven't, follow the instructions in *Chapter 16* and come back to this page.

- Once you log in, click the START button to start a quiz

- We recommend attempting a quiz multiple times till you're able to answer most of the questions correctly and well within the time limit.

- You can use the following practice template to help you plan your attempts:

Working On Accuracy		
Attempt	Target	Time Limit
Attempt 1	40% or more	Till the timer runs out
Attempt 2	60% or more	Till the timer runs out
Attempt 3	75% or more	Till the timer runs out
Working On Timing		
Attempt 4	75% or more	1 minute before time limit
Attempt 5	75% or more	2 minutes before time limit
Attempt 6	75% or more	3 minutes before time limit

The above drill is just an example. Design your drills based on your own goals and make the most out of the online quizzes accompanying this book.

> **First time accessing the online resources?** 🔒
> You'll need to unlock them through a one-time process. **Head to** *Chapter 16* **for instructions**.

Open Quiz https://packt.link/ansc01ch14 OR scan this QR code →	

15
Conclusion

This chapter will focus on preparing for the AWS ANS-C01 exam itself. Since the rest of the book has focused on the content to pass this exam, it seems fitting that this last chapter focuses on the study strategy you should use to pass the exam, and how passing the exam will improve your career. Becoming certified can open new doors in your career progression as a cloud network engineer. Many recruiters are looking for AWS-certified individuals to fill job openings, and this can give you an edge in the hiring process or make some positions attainable that were previously unattainable.

Here's a brief rundown of the topics covered:

- Registering for the exam
- Reviewing for the exam
- Revising review based on results
- Passing the exam

This chapter focuses on the best method to consume the certification guide materials and how to build a strong path to certification. Building a study plan, familiarizing yourself with the types of questions you can expect to see on the exam, engaging with the study materials, and maximizing the time you can devote to studying the materials are the best steps leading toward exam success.

Registering for the Exam

Specialty exams usually exist outside the certification path, making it unclear at what point they can be scheduled. Many exams are part of a specific track with a clear line of progression, but because *ANS-C01* is a specialization exam, there is no technical requirement to take any prior AWS exam before scheduling this one. However, it is strongly recommended that you are familiar with the material introduced in the AWS SAA-C03 exam before tackling ANS-C01.

You can check the status of your current certifications on the AWS certification portal: `https://www.aws.training/certification`. If you have an active AWS Solutions Architect Associate certification, you can register for the *ANS-C01* exam using the same certification portal. The exam can be taken online with an online proctor or at a local testing center. In both cases, the exam is scheduled via Pearson Vue through the AWS certification portal.

The portal allows you to do the following:

- Track certification expiration dates and status
- Schedule certification exams
- See the history of past exams you have taken with AWS and their outcomes
- Track the usage of certification benefits such as future exam discounts
- Download digital badges for certifications you have earned

Registering for the Online Exam

If you have never taken an online proctored exam before, this section will cover what to expect when registering for the online exam.

To register for an online exam, you must agree to several Pearson policies, including the capture of photos of the space you will test in. Photo recognition software may be used to compare a photo of yourself to a photo ID, which you must supply as part of the test process. You will also need to confirm the language your proctor will use and what time zone, date, and local time plan to schedule the exam in. Lastly, ensure that you can start the check-in process 30 minutes before the scheduled time. If you start the check-in process on time and have technical issues, you may still be able to take the exam because a large window is reserved for the exam.

Registering for the Exam at a Test Center

If you have never taken an exam at a testing center before, this section will cover what to expect when registering for the exam.

To register for the exam to be taken at a testing center, you must agree to present a valid photo ID at the center on the day of the exam, as well as agree to standard Pearson testing policies. You will select the location, date, and time to take the exam based on available testing equipment. This is often done at college campuses that already have a computer lab, but it could also be a dedicated testing center. If the available date and time do not meet your needs, you may have to choose a different center. You should plan to arrive no later than 15 minutes before your test time to ensure check-in can be completed. Different testing centers may have their own policies regarding forfeiture should you miss your time slot, but generally, being over 15 minutes late will result in the refusal of admission.

Reviewing for the Exam

Once you have consumed the content in this book, it can serve as a quick reference guide as you prepare for your exam. The book is organized around the domains of the exam, and each chapter focuses on the knowledge of, and skills in, those domains. It can be used as a reference for specific topics you have now learned but need reinforcement to retain.

Don't overuse the chapter review questions. Reviewing them just after you first read the chapter is good for testing retention. For questions that you get wrong, do not focus solely on the right answer, but also on why it is the right answer. The ANS-C01 exam is filled with "pick the best answer from lists of valid answers" questions, so it is not enough to pick a correct answer; you must also understand why it was the correct one given the other options. The exam often presents multiple-choice questions that have answers that would provide a solution but are wrong due to some stated constraint or implied constraint based on business requirements. This is why it is imperative to understand why a solution is correct given multiple solutions; you need to be able to pick the best one in the circumstances.

This is not the only kind of question. There will be multiple scenarios and multiple response questions as well. In all cases, the scenario or question will give some qualifying information that should end up with only one complete solution.

The reason to review these chapter review questions as sparingly as possible is because your brain subconsciously pays more attention to memorization of the answer than the understanding of the answer. Over time, usually within two or three reviews, you will find that you know the correct answer without even finishing reading the question. This is a handy shortcut the brain uses to save energy, but it can give a false sense of understanding the material. For this reason, it is better to mix up the review activities, leaving space between the material and your brain to avoid building that false sense of understanding.

The best way to combat this unintended memorization is to switch up the materials often and spend time with each type. This gives your brain time to let go of some information and work on retaining it in a new way. The guide comes with a large amount of test preparation material besides the book itself: two mock exams, flashcards, and exam tips. Rotating regularly through the material will keep things fresh in your brain, allowing it to focus on learning material instead of answers.

Included with this book are two full mock exams that mimic the real exam as well. The same advice applies to these mock exams as the chapter questions: don't repeat them too often and focus on why an answer is correct instead of the correct answer itself.

Consult the Appendix

The very first suggestion upon completing this book is to give yourself a little brain break. Get a drink, watch a movie, or do something that resets your brain to a relaxed state. This is important to ensure that your brain recognizes that it's having a break between the initial study and review. Come back when you have some time and some more energy.

Once you are ready to start reviewing, consult the appendix of this book. The appendix covers the mapping of exam domains to specific chapters, which is a convenient way to optimize your review time. Based on your initial chapter review question performance, you may already have an idea of your weak points. The appendix allows you to refer directly to the section of the book that explains the topic.

Create a Review Plan

Using the appendix, mark down each exam objective, its reference chapter, and how weak you are in the topic. Use this as a launch point for building a review plan based on the time you can devote to review; if you can devote 30 minutes, make the plan small and bite-sized with frequent revisiting of the covered content to practice spaced repetition. If you can devote an hour, make the plan focus on only one or two things but in more depth.

For all review plans, you should be thinking about spaced repetition. Spaced repetition is a memory technique that involves being exposed to information at intervals to reinforce memory—more frequently when first being exposed to the material, then gradually less frequently (but still on a schedule) as the repeated exposure results in better retention.

A sample spaced-repetition-focused plan may look something like this:

Topic	Reference	Activity	Date/Time	Comfort Level
Load Balancing	*Chapter 9*	Flashcards	8/10/24 – 1 hr	Weak
Cross-Account VPC Sharing	*Chapter 3*	Chapter Review	8/12/24 – 1 hr	Weak
Load Balancing	*Chapter 9*	Flashcards	8/13/24 – 1 hr	

Table 15.1: Sample review schedule

A review plan keeps you on track and saves you from using up brain power deciding what to study during the time you devote to studying. Any review plan needs frequent adjustment based on the results of your last review session, which is why a review plan is better saved electronically, where it can be edited and referenced easily. Once you have added a topic and activity to the review plan, add further review using the spaced repetition method further on in the plan. Refrain from assigning a comfort level to future planned sessions covering the same topic; instead, fill this in based on the results of your previous topic review. This helps you see the results of your study and will also help you plan future sessions.

Revising the Review Plan

As you review the material, it is important to continually revisit the review plan and adjust the topics and activities. Some suggested activities to focus on with your review are as follows:

- **Chapter review**: Reread sections of a chapter that focus on areas you feel you have a weaker understanding of. Take compressed notes on just these topics for review when you have short periods of spare time, such as waiting for public transportation or between commercial breaks of a television show.

- **Flashcard review**: Review flashcards associated with topics where you feel your understanding is weak to improve retention. Make sure to use spaced repetition techniques to review multiple topics some time apart. While there are flashcards included with this book, feel free to create your own that covers specific weak points, which also improves memory.

- **End-of-chapter questions**: At the end of every chapter is a set of practice questions that focus on what was introduced in that chapter. These questions mimic the real AWS ANS-C01 exam. They should be reviewed regularly but not excessively; as mentioned earlier, your brain will trick you by learning the answers.

- **Mock exams**: This book comes with two mock exams. As mentioned, excessively completing mock exams has a detrimental effect on checking for actual exam readiness. For this reason, it's suggested to do the mock exam no more than once per week to lessen the effect of accidental memorization. The important thing is to use the mock exam to check exam readiness and not to learn content.

- **Other media**: This book alone cannot fully cover the material that AWS expects you to know for this exam, as it is a specialty exam that presupposes a lot of prior knowledge. The best way to shore up any gaps you have that are not covered by this book directly is to consult the AWS documentation for any topic you are expected to know already that was not covered here. Other than that, of course, the internet is full to bursting with helpful bloggers, YouTubers, and other content creators who have probably already created something to help you on your journey.

Tie topics to review activities where they make sense. Make sure you don't always use the same review techniques for the same topics, though. As part of your review plan, you should create a review log that you can refer to when planning new review sessions. This is not a separate artifact, but a persistent accounting of the review you have already accomplished. This makes revision simple.

For example, *Table 15.2* shows a sample review plan. Here is what a review log may look like:

Topic	Reference	Activity	Date/Time	Comfort Level	Notes
Load Balancing	*Chapter 9*	Flashcards	8/10/24 1 hr	Weak	Focused on GWLB
Cross-Account VPC Sharing	*Chapter 3*	Chapter Review	8/12/24 1 hr	Weak	Still not understanding shared VPC
Load Balancing	*Chapter 9*	Flashcards	8/13/24 1 hr	Average	Work on Network Load Balancer next time

Table 15.2: Sample review log

Using the notes, comfort level, and activity as a baseline, you can plan future review sessions to focus on other topics and activities.

Passing the Exam

On exam day, you may want to use the time before the exam to review any points you still feel unsure about, or you may want to relax and focus on being mentally ready for the exam. Both are valid options; you know yourself best. If you want to do a last-minute review, focus on short, quick checks of understanding paired with a simple review such as flashcards.

If you choose to relax and mentally prepare, it's important to keep track of time and be prepared. Make sure that everything is prepared for the exam before disconnecting from the exam for a while, so that you can go straight into the exam itself and not spend time panicking about preparations when the time comes.

The appendix covers the format of the exam and the types of questions to expect, and the practice questions in this book should have prepared you as well. The most important thing you can do now is to trust your knowledge, the process, and the journey that has got you to the exam date.

Summary

The AWS ANS-C01 exam will test not only your understanding of network engineering but also how AWS uses networking to deliver its solutions to empower businesses to be agile and effective. While networking is but a part of a business application strategy, it is arguably the most important. Networking underpins all application connectivity and functionality; its importance in accomplishing business objectives cannot be overstated. Passing this exam will mark you as an AWS-certified cloud network specialist, and hopefully, that will serve you well for the rest of your career.

It is the sincere hope of the team that created this certification guide that the book and its accompanying materials have delivered value and will help you pass this exam, but more importantly, that you have gained the skills and confidence to be an AWS network specialist.

16

Accessing the Online Practice Resources

Your copy of *AWS Certified Advanced Networking – Specialty (ANS-C01) Certification Guide* comes with free online practice resources. Use these to hone your exam readiness even further by attempting practice questions on the companion website. The website is user-friendly and can be accessed from mobile, desktop, and tablet devices. It also includes interactive timers for an exam-like experience.

How to Access These Materials

Here's how you can start accessing these resources depending on your source of purchase.

Purchased from Packt Store (packtpub.com)

If you've bought the book from the Packt store (`packtpub.com`) eBook or Print, head to `https://packt.link/ansc01unlock`. There, log in using the same Packt account you created or used to purchase the book.

Packt+ Subscription

If you're a *Packt+ subscriber*, you can head over to the same link (`https://packt.link/ansc01practice`), log in with your `Packt ID`, and start using the resources. You will have access to them as long as your subscription is active.

If you face any issues accessing your free resources, contact us at `customercare@packt.com`.

Purchased from Amazon and Other Sources

If you've purchased from sources other than the ones mentioned above (like *Amazon*), you'll need to unlock the resources first by entering your unique sign-up code provided in this section. **Unlocking takes less than 10 minutes, can be done from any device, and needs to be done only once**. Follow these five easy steps to complete the process:

STEP 1

Open the link `https://packt.link/ansc01unlock` OR scan the following **QR code** (*Figure 16.1*):

Figure 16.1: QR code for the page that lets you unlock this book's free online content

Either of those links will lead to the following page as shown in *Figure 16.2*:

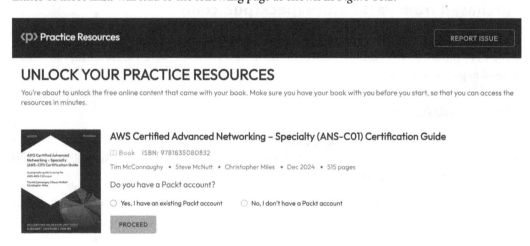

Figure 16.2: Unlock page for the online practice resources

STEP 2

If you already have a Packt account, select the option `Yes, I have an existing Packt account`. If not, select the option `No, I don't have a Packt account`.

If you don't have a Packt account, you'll be prompted to create a new account on the next page. It's free and only takes a minute to create.

Click `Proceed` after selecting one of those options.

STEP 3

After you've created your account or logged in to an existing one, you'll be directed to the following page as shown in *Figure 16.3*.

Make a note of your unique unlock code:

`CPP4153`

Type in or copy this code into the text box labeled 'Enter Unique Code':

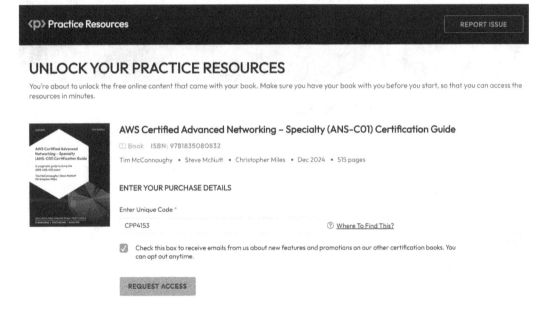

Figure 16.3: Enter your unique sign-up code to unlock the resources

> **Troubleshooting tip**
>
> After creating an account, if your connection drops off or you accidentally close the page, you can reopen the page shown in *Figure 16.2* and select `Yes, I have an existing account`. Then, sign in with the account you had created before you closed the page. You'll be redirected to the screen shown in *Figure 16.3*.

STEP 4

> **Note**
>
> You may choose to opt into emails regarding feature updates and offers on our other certification books. We don't spam, and it's easy to opt out at any time.

Click `Request Access`.

STEP 5

If the code you entered is correct, you'll see a button that says, `OPEN PRACTICE RESOURCES`, as shown in *Figure 16.4*:

PACKT PRACTICE RESOURCES

You've just unlocked the free online content that came with your book.

AWS Certified Advanced Networking – Specialty (ANS-C01) Certification Guide

Book   ISBN: 9781835080832

Tim McConnaughy • Steve McNutt • Christopher Miles • Dec 2024 • 515 pages

⊙ Unlock Successful
Click the following link to access your practice resources at any time.

Pro Tip: You can switch seamlessly between the ebook version of the book and the practice resources. You'll find the ebook version of this title in your <u>Owned Content</u>

OPEN PRACTICE RESOURCES ⬀

Figure 16.4: Page that shows up after a successful unlock

Click the `OPEN PRACTICE RESOURCES` link to start using your free online content. You'll be redirected to the Dashboard shown in *Figure 16.5*:

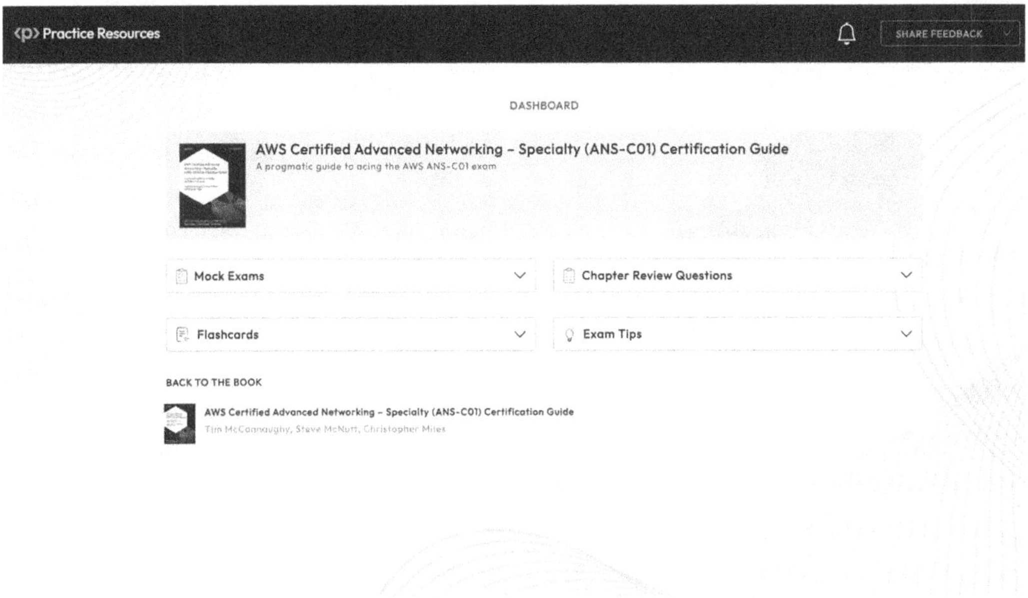

<p> Practice Resources   🔔   SHARE FEEDBACK ∨

DASHBOARD

AWS Certified Advanced Networking – Specialty (ANS-C01) Certification Guide
A pragmatic guide to acing the AWS ANS-C01 exam

| 📄 Mock Exams | ∨ | 📄 Chapter Review Questions | ∨ |
| 📄 Flashcards | ∨ | 💡 Exam Tips | ∨ |

BACK TO THE BOOK

AWS Certified Advanced Networking – Specialty (ANS-C01) Certification Guide
Tim McConnaughy, Steve McNutt, Christopher Miles

Figure 16.5: Dashboard page for AWS ANS-C01 practice resources

---

Bookmark this link

Now that you've unlocked the resources, you can come back to them anytime by visiting `https://packt.link/ansc01practice` or scanning the following QR code provided in *Figure 16.6*:

Figure 16.6: QR code to bookmark practice resources website

Troubleshooting Tips

If you're facing issues unlocking, here are three things you can do:

- Double-check your unique code. All unique codes in our books are case-sensitive and your code needs to match exactly as it is shown in *STEP 3*.

- If that doesn't work, use the `Report Issue` button located at the top-right corner of the page.

- If you're not able to open the unlock page at all, write to `customercare@packt.com` and mention the name of the book.

Share Feedback

If you find any issues with the platform, the book, or any of the practice materials, you can click the `Share Feedback` button from any page and reach out to us. If you have any suggestions for improvement, you can share those as well.

Back to the Book

To make switching between the book and practice resources easy, we've added a link that takes you back to the book (*Figure 16.7*). Click it to open your book in Packt's online reader. Your reading position is synced so you can jump right back to where you left off when you last opened the book.

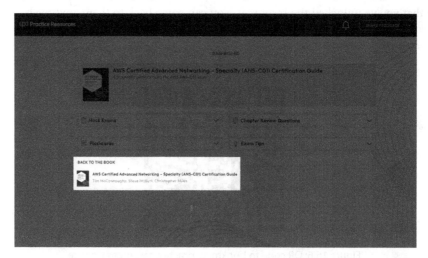

Figure 16.7: Dashboard page for AWS ANS-C01 practice resources

> **Note**
> Certain elements of the website might change over time and thus may end up looking different from how they are represented in the screenshots of this book.

Appendix 1
Network Fundamentals

Before diving into the specifics of AWS networking, it is important to understand some of the key fundamentals first. Some of you moving into the cloud engineering sector of **information technology** (**IT**) and, specifically, studying for this exam may not yet be seasoned experts in computer networking.

Computer networks have evolved a ton since the earliest conception of a packet-switched network in the 1960s. However, many of the fundamentals and core components have not changed much in the last 30 years. This chapter will cover many of those components, as they are crucial to understand when preparing for the *AWS Certified Advanced Networking – Specialty (ANS-C01)* exam.

The following are the learning objectives for this chapter on network fundamentals:

- The OSI model
- IP addressing
- **Classless Inter-Domain Routing (CIDR)**
- Basic subnetting
- IPv6

Since this chapter is predominantly focused on the fundamentals of networking, you may notice there will not be a strong emphasis on **Amazon Web Services** (**AWS**) or cloud networking, specifically. That is because the learning objectives listed previously require foundational knowledge. The first two chapters of this book are focused on laying a foundation prior to getting into AWS-specific networking constructs. The items covered in this chapter will allow you to understand the key structure and usage of **Internet Protocol** (**IP**) addressing, which is required to use all AWS network-based services covered on the *AWS Certified Advanced Networking – Specialty (ANS-C01)* exam.

The OSI Model

Several hardware and software protocols are used to facilitate and allow communication between systems across a network by dictating how data is transmitted, received, and processed. These protocols are standardized and organized into layers within the **Open Systems Interconnection (OSI)** model, developed by the International Organization for Standardization to provide an industry standard for understanding how these protocols interoperate. Refer to *Figure A1.1* to see how the OSI model is composed of seven layers.

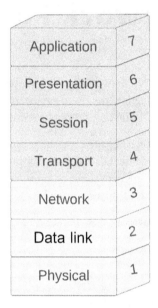

Figure A1.1: OSI model

As you can see in the preceding figure, each layer has a name that corresponds to the number. These numbers are often used as reference points when discussing the layers of the stack. Next, you will look at each of these layers in detail.

Seven Layers of the OSI Model

The OSI model consists of seven layers of abstraction that help visualize the process of encapsulation and decapsulation when two hosts need to communicate across a network. In *Table A1.1*, you will find details about each layer of the OSI model, including its function and common network protocols that are used at each of these layers.

Layers	Function	Protocols
Physical layer	Transmitting raw binary data over a physical medium, such as electrical signals or light pulses.	None; this layer uses physical signals instead of protocols.
Data link layer	Handles the framing of data for transmission purposes. Also responsible for error correction and access control to the physical medium. Commonly referred to as "Layer 2."	Ethernet (802.3) and Wi-Fi (802.11)
Network layer	Processing and routing data packets across separate networks. Commonly referred to as "Layer 3."	**Internet Protocol (IP), Border Gateway Protocol (BGP), Internet Control Message Protocol (ICMP), Open Shortest Path First (OSPF)**
Transport layer	Responsible for reliable end-to-end communications. Also handles data segmentation and reassembly.	**Transmission Control Protocol (TCP)** and **User Datagram Protocol (UDP)**
Session layer	Handles the establishment, management, and termination of sessions between two systems.	**Remote Procedure Call (RPC),** NetBIOS
Presentation layer	Ensuring capability between two systems by performing translation, compression, and even encryption.	No exclusive protocols associated with this OSI layer. There are many protocols at other layers that incorporate functionality at this layer.
Application layer	Provides a direct communication channel for user applications to interact with the network stack.	**Hypertext Transfer Protocol (HTTP** and **HTTPS), File Transfer Protocol (FTP** and **SFTP), Domain Name System (DNS), Simple Network Management Protocol (SNMP)**

Table A1.1: OSI layer details

From this table, you can see the functionality and protocols that operate at each layer of the OSI model. The protocols outlined at each of these layers are important to be familiar with so that you have context around their functionality, as it relates to networking.

Layers of Focus for Network Engineers

Looking at the complete OSI model, it's easy to grasp that this process can become very complicated, with many moving parts to consider. Understanding, implementing, and troubleshooting protocols at each layer of this stack can be a tall task for any IT organization.

Network engineers need to be skilled in five out of the seven layers of the OSI model. The primary focus of the network engineer will be from Layer 1 (Physical) up to Layer 4 (Transport). However, there are elements at Layer 7 (Application) that are very relevant as well, such as HTTP, SSH, FTP, and APIs, which network engineers may need to interact with and troubleshoot daily.

For network engineers transitioning to the cloud, there's even better news. Cloud networking offers additional layers of abstraction with the constructs that are provided in user space. With this in place, cloud network engineers rarely interact with Layer 1 (Physical) or Layer 2 (Data Link). Interaction with these layers typically only occurs when attempting to connect the cloud network to on-premises or third-party networks. This will be covered in later chapters.

Data Encapsulation

While the seven layers of the OSI model provide a great visual model for how data is transferred between two systems, it's not purely abstraction (i.e., obfuscation of the underlying components). The process of moving data through the layers of the stack and converting it into electricity or light pulses "on the wire" is referred to as data encapsulation. This process involves data being passed from each layer to the next in a "top-down" approach. This means data encapsulation starts at Layer 7 and moves down to Layer 1.

The process of encapsulation involves each OSI layer encapsulating the data within a header and a trailer as it moves down the stack. The header and trailer are essentially extra bits of data that can be used by the receiving device for several purposes, which you will read more about later in this chapter.

As far as encapsulation goes, an example would be when a web request is made, the Application layer will add an HTTP header and trailer that contains specific information related to caching, encoding, date, or even status. The header will be placed in **front** of the original application data, and a corresponding trailer will be placed **behind** the data. The data is now *encapsulated* with this application layer information and can be sent down to the next layer for further encapsulation. In this example, the original application data is considered the **payload**. When encapsulation occurs between the layers of the stack, that entire encapsulated unit becomes the payload for the next layer to process. Refer to *Figure A1.2* for a look at how application data is encapsulated.

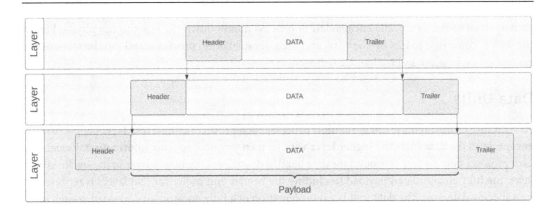

Figure A1.2: Data encapsulation payloads

As shown in the figure, as data moves down the stack of the OSI model, each layer's encapsulated data becomes the payload for the next layer.

In addition, *Figure A1.3* shows how this process moves down and up through the layers of the OSI model.

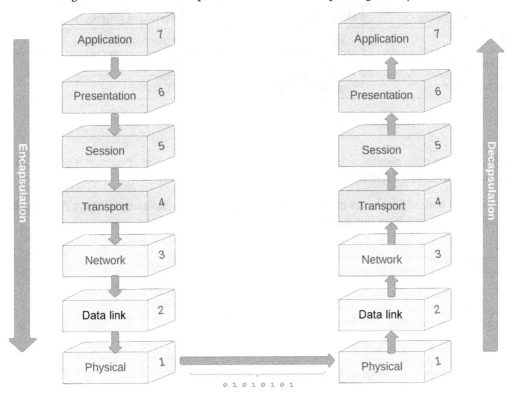

Figure A1.3: Encapsulation -> Decapsulation

As displayed in the figure, the encapsulation process moves down the stack to the physical layer, the data is converted to bits on the wire, and then decapsulation proceeds back up the stack as the encapsulation headers and trailers are removed.

Data Units

As data moves through the encapsulation process, each protocol at different layers appropriately encapsulates the data from the higher layer with its own protocol-specific information. Essentially, each protocol puts a wrapper around the data handed down from the higher layer in the stack. At each layer, the fully encapsulated payload (including the header and trailer for that layer) is represented by a specific data unit. The data unit is a representation of a fully encapsulated payload at each layer.

Each data unit contains a header and a trailer. The headers for each data unit contain structured information that instructs and/or assists the protocol stack with processing and/or forwarding the data elsewhere. The trailer typically does not contain any pertinent information about the payload but is used to do error detection and confirm the data unit is still fully intact once received.

Refer to *Table A1.2* for the classification of data units at Layers 2–4:

OSI layer: Name	Data unit	Standard protocols
2: Data Link	Frame	Ethernet, PPP, HDLC
3: Network	Packet	IP
4: Transport	Segment/Datagram	TCP/UDP

Table A1.2: OSI layers – Data unit

You will want to make note from the preceding table that protocols such as IP operate at Layer 3, while protocols such as TCP/UDP operate at Layer 4.

For example, look at the process of TCP traffic being encapsulated. Referring to *Figure A1.4*, there is data from the session layer that needs to be encapsulated.

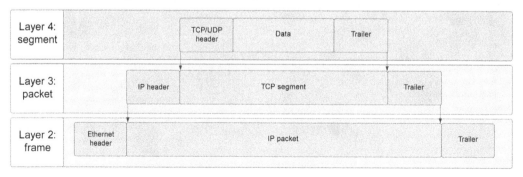

Figure A1.4: Data unit encapsulation

This application (which is shown simply as DATA) is TCP-based, so TCP will encapsulate the data with the appropriate TCP header information. This represents a Layer 4 **segment**. This TCP segment is passed to the network stack, which needs to be encapsulated in IP. Similarly, the TCP segment is then encapsulated with an IP header and forms a Layer 3 **packet**. As you can guess, the IP packet is then passed to the next layer and is encapsulated in an Ethernet header, forming a Layer 2 **frame**. This frame is then converted to bits of electricity on the wire.

> **Note**
>
> Layer 4 encapsulation uses two different data unit types. TCP encapsulated data is referred to as a segment, whereas UDP encapsulated data is a datagram.

IP Addressing

In the early days of the networking stack, several protocols were developed and used for specific purposes that had a wide range of adoption. There was not one protocol stack that was implemented at a universal level.

Networking protocol suites such as DECnet, AppleTalk, and Novell NetWare were common in enterprise networks, even sometimes coexisting on the same networks. They were often deployed for a specific use case and system proprietary. Over time, many of these protocols were either decommissioned or replaced with the TCP/IP stack. Being an open standard protocol, IP benefits from being well documented, publicly available, not owned by a single entity, and promoting interoperability. IP's adaptability played a major role in the protocol gaining dominance as the internet grew into what it is today. For example, unlike proprietary protocols such as DECnet and AppleTalk, IP's flexibility to support a wide range of hardware platforms and its ability to route data across diverse and interconnected networks allowed it to scale effortlessly as the internet expanded globally.

IP uses unique numerical identifiers, called IP addresses, to identify specific hosts or network interfaces connected to a computer network. When devices on a network want to communicate, they use these IP addresses to do so. They behave somewhat like a mailing address in that, as with a postal address on an envelope, a network-connected device uses the remote host's IP address as the destination when traffic is encapsulated at Layer 3 of the OSI model.

There are two versions of IP currently in use: **version 4 (IPv4)** and **version 6 (IPv6)**. The story of IP version 5 is an uneventful one and ended with a protocol that never became widely adopted. Today's internet still runs predominantly on IPv4, but the slow burn of conversion to IPv6 has gained momentum. IPv6 is ultimately a large redesign of the protocol with many enhancements and alterations to support better efficiency and scalability than IPv4, hence why it is not a simple task to migrate from IPv4 to IPv6. This book will not provide an extensive overview of IP versions, as they can be complex sets of protocols. However, this chapter will focus on the overall components of IP that could help someone new to networking prepare for the *AWS Certified Advanced Networking – Specialty (ANS-C01)* exam.

IPv4 Addressing Format

IPv4 addresses are written in dotted decimal format. This consists of 4 segments of 8 bits (1 byte), separated by decimals. Each IPv4 address, therefore, is 32 bits in length. These 8-bit segments are often referred to as **octets** and are separated by dots (periods) when written out.

Within each octet, the bits represent a binary number. In binary, the maximum value for 8 bits can be 255. This means the maximum value for a single octet is 255, so the full range for an IPv4 address is 0.0.0.0 through 255.255.255.255. That means a total of just over 4 billion addresses.

Refer to *Figure A1.5* to see the IPv4 addressing format.

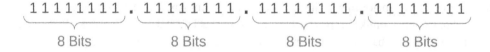

Figure A1.5: IPv4 address format

In *Figure A1.5*, you can see four groups of octets, which each contain eight binary bits. The value of 0.0.0.0 represents all the bits being set to 0, while 255.255.255.255 represents all the bits being set to 1. Next, you will learn more about how an octet is calculated and used within an IP address.

As noted previously, an IP address is a 32-bit value that is separated into 4 sections of 8 bits. The IP addresses are then written in dotted decimal format. This means each section is separated by a decimal point, or a dot, which gives the format you're all familiar with when you see an IP address.

With 8 bits, each octet has a possible value of 0 to 255, but how do you calculate that? Each bit is a binary value that is either 1 or 0. The easiest way to think of this is that 1 means the bit is set or **on**, while 0 means the bit is not set or **off**. The positioning of these bits plays a role in the corresponding value of the binary number that's represented. These bits are ordered from right to left, in positions ordered from 0 to 7 (totaling 8 bits). The value that corresponds to each bit position starts at 1 and increases by an exponent of 2. Refer to *Figure A1.6* to see these values.

Bit position		7	6	5	4	3	2	1	0
Value		128	64	32	16	8	4	2	1

Figure A1.6: Binary bit positions

You can see that the corresponding values tied to each bit position increase by an exponent of 2.

Exam Tip

You will need to memorize the powers of 2, all the way up to 256. It will make these mental exercises and subnetting much easier in the long run.

With these bit positions and their corresponding values, one can use simple math to calculate binary numbers that represent each value from 0 to 255. The easiest calculations occur when there is a single bit in the **on** position. See the examples in *Figure A1.7*.

00000001

Bit position	7	6	5	4	3	2	1	0	
Bit value	0	0	0	0	0	0	0	1	= 1
Value	128	64	32	16	8	4	2	1	

00000010

Bit position	7	6	5	4	3	2	1	0	
Bit value	0	0	0	0	0	0	1	0	= 2
Value	128	64	32	16	8	4	2	1	

Figure A1.7: Binary values – Single bit

As you can see in the preceding figure, when only a single bit is on, the value directly corresponds to the value for that bit position. So, a value of 00000001 corresponds to a value of 1, while 00000010 corresponds to a value of 2.

Things can get slightly more complex when there are multiple bits in the **on** position. Luckily, all that's required is some simple math to calculate these values. All the values for each **on** bit simply need to be added together to get the full value. Refer to *Figure A1.8* for an example.

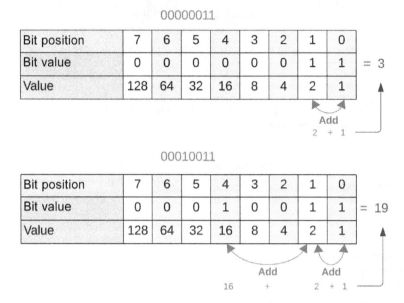

Figure A1.8: Binary values – Multiple bits

In this example, the values for 00000011 and 00010011 correspond to 3 and 19. These values are calculated by adding all the corresponding bits in the **on** position.

Using this logic of adding the values for the corresponding **on** bits, refer to *Figure A1.9* to see the maximum value for a single octet.

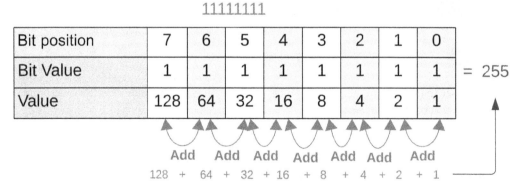

Figure A1.9: Binary values – Maximum bits

Adding each of these values together, you can see that the maximum value for a single octet is 255.

So far in this section, you have covered how to convert binary values into dotted decimal format, which is what an IP address comprises. This will be a key skill to have when evaluating IP addressing, allocating networks, and calculating what network an IP address belongs to. Next, you will look at different classification types of IP address ranges.

Private, Public, and Reserved IPv4 Addresses

Even having over 4 billion IPv4 addresses would still not suffice for the number of connected devices within the world. There are simply not enough addresses for the number of devices that would need an IP address to have connectivity. Because of this constraint, within the addressable IPv4 space, there are specific ranges that have been reserved for certain purposes and use cases. This classification allows for wider adoption and repeatable address space that could be used across different organizations or entities, while also allowing public address ranges to still be routable across the global internet. Now, take a closer look at each of these ranges.

Private IPv4 Address Range

This private IP range is reserved for addresses that are not **routable** on the global internet. This means these address ranges cannot be advertised and propagated over the public infrastructure that the internet runs on. Long story short, if the IP ranges are not routable, nothing on the public internet can talk directly to them because there would be no way to determine how to get to them.

This allows large ranges of IPs to be assigned to devices that do not need to be given a public, routable address. These ranges are nearly always used by any enterprise, organization, or other entity for addressing their internal systems. This can range from computers to servers to even heavy machinery that has network connectivity. For these devices to talk to the internet, a technology such as **network address translation** (**NAT**) will need to be used (which will be covered in *Appendix 3, VPC Networking Basics*). Refer to the following list for the reserved private IPv4 address ranges:

- **Class A**: 10.0.0.0 to 10.255.255.255 (10.0.0.0/8)
- **Class B**: 172.16.0.0 to 172.31.255.255 (172.16.0.0/12)
- **Class C**: 192.168.0.0 to 192.168.255.255 (192.168.0.0/16)

Any or all of these ranges can be used simultaneously for private networks within an organization. Each class has a different size of available addresses, with the Class A range having the most available addresses.

These address ranges are often referred to by the **Internet Engineering Task Force (IETF)** standard that defined them, RFC 1918. When the terms **RFC 1918 addresses** or **private IP address ranges** are used, the ranges listed previously are what is being referenced.

> **Note**
>
> Refer to the following URL for further details on the classification of RFC 1918 addressing: `https://www.ietf.org/rfc/rfc1918`.

Other Reserved IP Address Ranges

In addition to the private address ranges that can be used for internal resources, there are additional IP ranges that are reserved for specific use cases. Like the RFC 1918 range of addresses, these reserved ranges are also non-routable on the public internet. Also, much like the RFC 1918 range of addresses, these reserved addresses are based on well-known IETF standards. Virtually any appliance with a functioning network stack should adhere to these standards. Refer to *Table A1.3* for details of additional reserved IPv4 address ranges.

Name	Reserved IP address range	Purpose	Related RFC
Loopback address range	127.0.0.0 to 127.255.255.255 (127.0.0.0/8)	Reserved for assignment of loopback interfaces. Loopback interfaces are logical network interfaces. A well-known and widely deployed loopback address is 127.0.0.1 and is commonly referred to as "localhost."	RFC 5735
Link-local address range	169.254.0.0 to 169.254.255.255 (169.254.0.0/16)	Reserved for IPv4 addressing that has link-local scope. Not only are these addresses not routable over the public internet, they also don't even route traffic for this address range outside of the local link. These are commonly used for interconnections between network appliances where the link addresses do not need to be known by the traffic traversing them.	

Name	Reserved IP address range	Purpose	Related RFC
Multicast address range	224.0.0.0 to 239.255.255.255 (224.0.0.0/4)	Reserved for a series of multicast group address ranges.	RFC 5771
Future or experimental use range	240.0.0.0 to 255.255.255.254 (240.0.0.0/4)	Also known as the "Class E" range. This range is not currently in use and is reserved for future use.	RFC 5735
Broadcast address	255.255.255.255	Known as the "limited broadcast" address, which is limited to a single network or subnet.	RFC 919
Other notable reserved IPv4 ranges	0.0.0.0/8 to 0.0.0.0 to 0.255.255.255 100.64.0.0/10 to 100.64.0.0 to 100.127.255.255 198.18.0.0/15 to 198.18.0.0 to 198.19.255.255 192.0.0.0/24 to 192.0.0.0 to 192.0.0.255		

Table A1.3: Other reserved IPv4 ranges

While *Table A1.3* is not a full list of all the reserved IPv4 ranges, these are some of the most notable reserved ranges that a network engineer needs to be familiar with. Recognizing these address ranges can aid network engineers in their troubleshooting efforts, as they can draw correlations from analyzing network traffic. Using these ranges is also critical from a network design perspective. Consider the following example.

A network engineer or architect may use a range from the RFC 1918 address range to allocate network addresses for their entire enterprise network. In addition, they could use addresses from the link-local address range to allocate IP addresses for some interconnect links to third-party networks that do not need to be routable on their network. They could also allocate some addresses from the multicast address range for a media streaming application that broadcasts a digital video feed out to all their largest branch locations.

Public IP Addresses

If all the previously mentioned IPv4 ranges are reserved and not routable over the global internet, what addressing does the internet actually use? The answer is public IP addresses. Public IP addresses are **globally unique** addresses that are routable across the global internet. This means each public IPv4 address used ensures that traffic can be accurately routed only to the appropriate destination.

So, does that mean anyone can assign a public address to their machine and connect to the internet? Not exactly. There are quite a lot of rules about the delegation of public IPv4 addresses. The governing bodies responsible for the allocation of public IP addresses are called **regional internet registries (RIRs)**. Today, there are five organizations around the globe that serve as RIRs. These organizations are responsible for the allocation of public IP ranges to both **internet service providers (ISPs)** as well as other large entities. Here are the five RIRs that are in operation as of today:

- **AFRINIC: African Network Information Center** (Africa)
- **APNIC: Asia Pacific Network Information Centre** (Asia-Pacific region)
- **ARIN: American Registry for Internet Numbers** (North America)
- **LACNIC: Latin American and Caribbean Internet Address Registry** (Latin America and the Caribbean)
- **RIPE NCC: Réseaux IP Européens Network Coordination Centre** (Europe, the Middle East, and parts of Central Asia)

In addition to the allocation of addresses, an RIR also provides a searchable database of these addresses to allow users of the internet to look up who is allocated which addresses. Next, you will see how to use a common tool called **Whois** to do so.

Analyzing the Output of a Whois Lookup

The most common tool used to perform a lookup of RIR databases is called **Whois** (pronounced "who is"). A Whois lookup can provide specific registration and administrative information about a domain name or an IP address. An RIR will typically provide a web-based tool that users can refer to for specific lookups of their databases. For example, ARIN provides a site through which users can perform Whois lookups: `https://whois.arin.net/ui/`.

> Note
>
> There is also a commonly used built-in command-line tool on Unix-based (Linux and macOS) and Windows-based systems to perform these lookups as well. Both systems use a `whois` command to invoke the tool. More details about these tools can be found in the operating system-specific documentation.

As a network engineer, the top three use cases for performing a Whois lookup would be the following:

- **Domain ownership verification**: To identify the owner of a domain or IP address, especially in cases of troubleshooting, legal inquiries, or security investigations

- **Network abuse investigation**: To trace the source of malicious activity such as spam, denial-of-service attacks, or unauthorized access attempts by checking the registration details of suspicious domains or IP addresses

- **DNS or network configuration issues**: To confirm the domain's registrar information, expiration dates, and contact details for resolving DNS or configuration issues between organizations

The following is a list of common pieces of information you can receive when performing a Whois lookup:

- **Net Range or CIDR**: The range of IP addresses for the associated allocation

- **Net Name**: A name or identifier for the allocation

- **Organization**: The name of the organization that holds the address space

- **Administrative Contact**: Contact information for the person responsible for administrative aspects of the IP range

- **Technical Contact**: Contact information for the person responsible for technical aspects of the IP range

- **Country**: The country where the organization is based

- **Address**: The postal address of the organization or administrative/technical contact

- **Phone and Fax**: Contact phone numbers

- **Email**: Contact email addresses

- **ASN**: Associated **Autonomous System Number** (**ASN**) if relevant

- **Last Updated**: When the Whois record was last modified

- **Assignment Date**: When the IP range was assigned

As an example, say you run a Whois query against 8.8.8.8, which is a very well-known public DNS server offered by Google, as shown in *Figure A1.10*.

Network

Net Range	8.8.8.0 - 8.8.8.255
CIDR	8.8.8.0/24
Name	LVLT-GOGL-8-8-8
Handle	NET-8-8-8-0-1
Parent	LVLT-ORG-8-8 (NET-8-0-0-0-1)
Net Type	Reallocated
Origin AS	
Organization	Google LLC (GOGL)
Registration Date	2014-03-14
Last Updated	2014-03-14
Comments	
RESTful Link	https://whois.arin.net/rest/net/NET-8-8-8-0-1
See Also	Related organization's POC records.
See Also	Related delegations.

Point of Contact

Name	Google LLC
Handle	ZG39-ARIN
Company	Google LLC
Street	1600 Amphitheatre Parkway
City	Mountain View
State/Province	CA
Postal Code	94043
Country	US
Registration Date	2000-11-30
Last Updated	2022-11-10
Comments	
Phone	+1-650-253-0000 (Office)
Email	arin-contact@google.com
RESTful Link	https://whois.arin.net/rest/poc/ZG39-ARIN

Organization

Name	Google LLC
Handle	GOGL
Street	1600 Amphitheatre Parkway
City	Mountain View
State/Province	CA
Postal Code	94043
Country	US
Registration Date	2000-03-30
Last Updated	2019-10-31
Comments	Please note that the recommended way to file abuse complaints are located in the following links. To report abuse and illegal activity: https://www.google.com/contact/ For legal requests: http://support.google.com/legal Regards, The Google Team
RESTful Link	https://whois.arin.net/rest/org/GOGL

Point of Contact

Name	Abuse
Handle	ABUSE5250-ARIN
Company	Google Inc.
Street	1600 Amphitheatre Parkway
City	Mountain View
State/Province	CA
Postal Code	94043
Country	US
Registration Date	2015-11-06
Last Updated	2022-10-24
Comments	Please note that the recommended way to file abuse complaints are located in the following links. To report abuse and illegal activity: https://www.google.com/contact/ For legal requests: http://support.google.com/legal Regards, The Google Team
Phone	+1-650-253-0000 (Office)
Email	network-abuse@google.com
RESTful Link	https://whois.arin.net/rest/poc/ABUSE5250-ARIN

Function	Point of Contact
Admin	ZG39-ARIN (ZG39-ARIN)
Tech	ZG39-ARIN (ZG39-ARIN)
Abuse	ABUSE5250-ARIN (ABUSE5250-ARIN)

Figure A1.10: Whois example

You can see from the preceding figure several bits of information about the IP network, the organization, as well as points of contact for the organization. You'll see that the IP address 8.8.8.8 belongs to a CIDR of 8.8.8.0/24, which will be covered in the next section of this chapter. Additionally, you will also notice a point of contact listed for Abuse. This would be a contact to use in case you wanted to report some suspicious or malicious activity that was observed from this IP address or address range.

One important thing to remember is that all this information is **public**. That means anyone in the world who would like to look up this information will be able to see this level of detail that is published by the RIR.

Classless Inter-Domain Routing

You may have noticed in the previous section that the Whois results for the IP address 8.8.8.8 showed that the IP belongs to a CIDR of 8.8.8.0/24. If you're new to networking and the art of subnetting, then this may look quite odd at first. What does the /24 mean in that context? This section will explore CIDR (often pronounced *cider*) notation and how it is used in the world today. Before doing so, you will learn more about classful networking. While this type of network classification is not used in the current day, it is a key foundational element to understanding why things have evolved into their current form. That being said, the first subsection will cover classful networking, and then the next subsection will dive into CIDR.

Early Days: Classful Networking

In 1981, a rudimentary system was put in place to classify network ranges and boundaries by examining the high-order bits (first three bits) of an IP address. In layperson's terms, the first octet of an IP address could tell you where the range started and ended. The system followed a simple process of reserving bits within an IP address, which would let the user easily determine where an IP range started and ended.

> **Note**
>
> See RFC 791 for more information on the origin of these classful networks and their intended usage: https://datatracker.ietf.org/doc/html/rfc791.

IP networks were divided into five different class ranges:

- **Class A**: Large networks
- **Class B**: Medium-sized networks
- **Class C**: Small networks
- **Class D**: Multicast networks
- **Class E**: Experimental; reserved for future use

Since Class D and Class E were already covered in the previous section about reserved IP ranges, this section will focus on the Class A, B, and C ranges.

> **Note**
>
> These class ranges are not to be confused with the Class A, B, and C ranges that belong to private address ranges of RFC 1918. These are different classifications.

On the surface, this rudimentary system can seem a little difficult to comprehend. Once it is broken down into binary, you can see that it is relatively simple.

The entire network range that an IP address belongs to can be defined simply by looking at the first three bits of the IP address.

RFC 791 refers to these **high-order bits** to map each sequence into classes. Refer to *Figure A1.11* for a look at these high-order bits in the context of the octet.

High order bits

Bit position	7	6	5	4	3	2	1	0
Value	128	64	32	16	8	4	2	1

Figure A1.11: High-order bits

The binary values within these positions shown in the preceding figure let the operator know the entire range of the network the IP address belongs to. Each high-order bit combination is mapped to a specific class, as shown in *Table 1.4*.

High-Order Bits	Binary Value Range	Reserved Network (N)/Host (H) Format	Available Host Addresses	Class
0 _ _	0–127	N.H.H.H	16,777,214	A
1 0 _	128–191	N.N.H.H	65,534	B
1 1 0	192–223	N.N.N.H	254	C
1 1 1	224–255	N/A (reserved ranges)	N/A	D, E

Table A1.4: Classful network reservations

As displayed in the preceding table, each class has bits that are reserved for the network, while the remaining bits can be used for hosts. In short, the network bits are fixed and cannot be changed. With those bits being fixed, the rest of the bits can be used for host addressing. In addition, this means you can determine the full network range that the IP address belongs to just by the value in the first octet. These ranges are kept very simple by reserving entire octets for the network bits. Refer to *Figure A1.12* for a look at all the reserved bits based on each classful type.

Class A

Class B

Class C

Figure A1.12: Classful network reserved bits

As shown in the figure, each class has a reserved range of host addresses that can be consumed by hosts on the network. Next, you will look at some examples to illustrate this further. These will reference example IP addresses and what you can determine about the network they belong to by using classful networks.

Here are a few examples of how this can be applied to IP addresses:

- **Example A**: 124.64.208.10

    - High-order bits: 0 1 1 1 1 1 0 0

    - Class: A

    - Network bits: XXXXXXXX.XXXXXXXX.XXXXXXXX.XXXXXXXX

    - Host bits: XXXXXXXX.XXXXXXXX.XXXXXXXX.XXXXXXXX

    - First host address: 124.0.0.0

    - Last host address: 124.255.255.255

- **Example B**: 171.16.12.148

    - High-order bits: 1 0 1 0 1 0 1 1

    - Class: B

    - Network bits: XXXXXXXX.XXXXXXXX.XXXXXXXX.XXXXXXXX

    - Host bits: XXXXXXXX.XXXXXXXX.XXXXXXXX.XXXXXXXX

- First host address: `171.16.0.0`

- Last host address: `171.16.255.255`

- **Example C**: `192.148.1.216`

  - High-order bits: 1 1 0 0 0 0 0 0

  - Class: C

  - Network bits: XXXXXXXX.XXXXXXXX.XXXXXXXX.XXXXXXXX

  - Host bits: XXXXXXXX.XXXXXXXX.XXXXXXXX.XXXXXXXX

  - First host address: `192.148.1.0`

  - Last host address: `192.148.1.255`

As shown in the examples, using classful networks, it was easy to determine a lot of information about the entire network that an IP address belonged to, just by looking at the first three bits of the IP address.

Inefficiencies of Classful Networks in Network Planning

A common activity in which a network engineer plays a significant role is network planning. This could involve deploying a new branch location, data center, or even network expansion for an upcoming merger. During this planning process, the network engineer will need to allocate the appropriate network(s) to account for all the necessary host addresses to accommodate the systems that will be located there.

> **Note**
>
> The use of classful networks preceded the private address range of RFC 1918 by about 15 years.

Using the classful networking schema listed in the previous section, it's easy to see that there are many combinations available to accommodate the networks. Since this was before the adoption of RFC 1918 addressing, this would often involve the organization contacting the necessary ISP and getting an allocation of addressable networks to be assigned for different purposes. Network engineers using classful networks needed to be considerate of each possible number of host addresses per class:

- **Class A**: 16,777,214

- **Class B**: 65,534

- **Class C**: 254

With these numbers in mind, a network engineer could allocate a network per the number of required hosts. For example, a site with around 100 computers could warrant a single Class C network. Or a data center with about 300 servers that need to be on the same network? A Class B network would do the trick.

However, looking at the Class B reservation reveals a major problem with classful networking. The requirement was for 300 host addresses, but the reservation that gets handed out allows for over 65,000 addresses. That's a lot of wasted space. Luckily, a better way eventually came along.

Subnet Masks and CIDR Notation

Before covering what CIDR notation is, an important concept that needs to be understood is **subnet masks**.

Up to this point, you have specifically looked at IP addresses and their ranges. However, something that is always paired with an IP address is a subnet mask. Simply put, a subnet mask is a 32-bit value (same as the IP address) that indicates the network to which the IP address belongs. If you recall the detail about fixed bits (network) versus flexible bits (host) in the previous section, that's exactly what a subnet mask represents.

The format of a subnet mask is almost identical to an IP address, as they are also written in dotted decimal format. But how the bits are used slightly differs. Starting from the leftmost binary bit and moving toward the rightmost, any bits in the **on** (1) position symbolize that the corresponding bit in the IP address is reserved for the network portion. See the following example for the classful network ranges from RFC 791. It's very simple to calculate subnet masks for classful network ranges, as the entire octet is reserved for network host, as shown in *Figure A1.13*.

Figure A1.13: Classful subnet masks

As shown in the preceding figure, the reserved host ranges are always on the bit boundary, so the calculation of the subnet mask is very simple. Each octet is either fully reserved for the network or fully reserved for host addresses.

When you are configuring a network interface on an appliance, such as a router, you will always configure a subnet mask. See the following example for the interface configuration of a Cisco router. You can see the configuration of the IP address (host), immediately followed by the subnet mask:

```
interface GigabitEthernet1/2
  description LAN Interface
  ip address 192.168.16.25 255.255.255.0
```

CIDR Notation

CIDR was introduced by RFC 1518 and 1519 in 1993. The introduction of CIDR provided a simplistic way of representing subnet masks, using something called **CIDR notation**. As you can see in the preceding figure, continuously writing out two separate 32-bit values in dotted decimal format can become cumbersome. CIDR notation, sometimes referred to as slash notation, consists of trailing the IP address with a simple slash (/) and the number of reserved bits from the subnet mask. This means you could refer to the IP address and subnet mask from the Cisco interface example earlier like so: 192.168.16.25/24

> **Note**
> Some networking appliances will allow the configuration of interface IP addresses using slash notation. Refer to the vendor-specific documentation to determine whether this is supported.

Since an IP address is 32 bits in length, it is important to remember that CIDR notation refers to the total number of reserved network bits, not just within the explicit octet. This is why a subnet mask of 255.255.255.0 is referred to as a /24. Refer to *Figure A1.14* to see how the slash notation is derived from the full subnet mask.

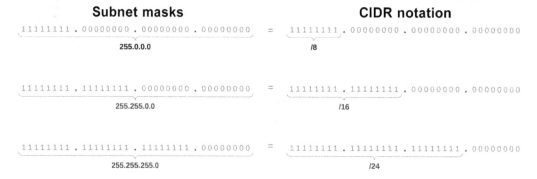

Figure A1.14: CIDR notation

From the figure, you can see that the slash value always relates to the number of network reserved bits within the IP address.

Variable Length Subnet Masks

One of the major shortcomings of classful networking was the inefficient consumption of IP addresses based on the network reservation. These network reservations (which correlate to a subnet mask) would reserve the entire octet, often resulting in allocations that were far too oversized and wasteful. Since there are eight bits within each octet, you can see how some flexibility in using those bits could be much more efficient.

CIDR introduced the concept of a **variable-length subnet mask** (**VLSM**). While adoption took several years, VLSM essentially marked the end of classful networking. With VLSM, as the name implies, the capability was introduced that allowed for the bit boundary (i.e., the delineation between network and host) to exist anywhere within the subnet mask. This allowed much more efficient use of IP address space when creating these subnetworks, or **subnets**.

With classful networking, the only options were based on all bit values for a particular octet being **on** or **off**. This means you could have a /8, /16, or /24. This meant you could set the boundary on any bit, allowing for subnets as large as a /1 or as small as a /32. Refer to *Figure A1.15* for some examples of VLSM.

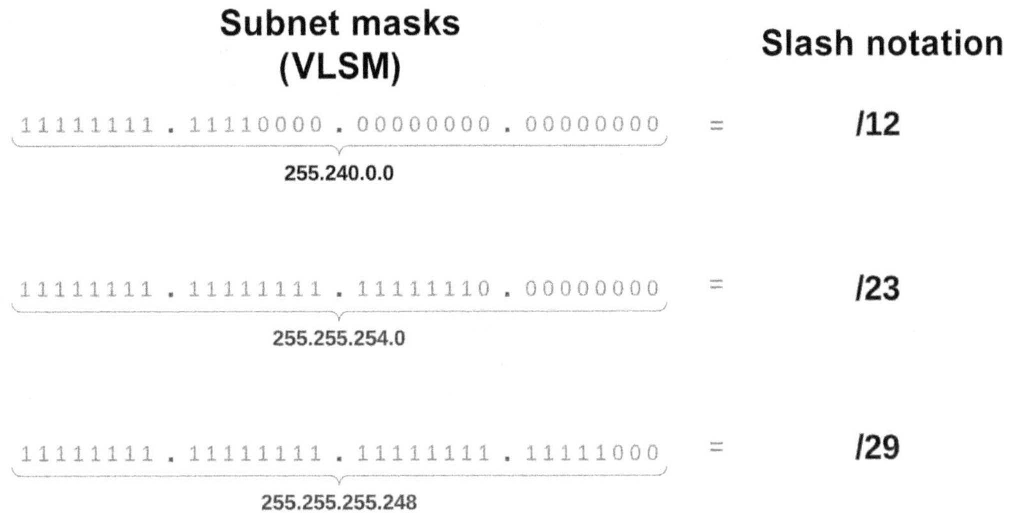

Subnet masks (VLSM)   Slash notation

| 11111111 . 11110000 . 00000000 . 00000000 | = | **/12** |

255.240.0.0

| 11111111 . 11111111 . 11111110 . 00000000 | = | **/23** |

255.255.254.0

| 11111111 . 11111111 . 11111111 . 11111000 | = | **/29** |

255.255.255.248

Figure A1.15: VLSMs

As you can see in the preceding figure, this adds a great deal of flexibility in the realm of reserving IP networks. This allows a specific number of bits within the IP address to be reserved rather than having to reserve the entire octet.

The next section will detail how network engineers work through the process of calculating and setting up these networks in an activity referred to as **subnetting**.

Basic Subnetting

Subnetting is a unique skill that all network professionals must have, whether they love it or hate it. Network engineers must have the ability to carve up large network ranges to allow for networks to be appropriately sized for the intended use case. This could be choosing the appropriate IP address for a data center, headquarters, or branch location. This section will dive into how this has been accomplished for the last 30 years of computer networking.

Since there a no classful associations in CIDR, this means the subnet mask is no longer determined by the value set within the IP address. As a network engineer, it's a common activity to allocate IP networks based on the requirements for the task at hand. For example, maybe someone on another team needs to spin up a new set of servers in the data center and they need to know what IP network they should belong to. In addition, the more common and important capability is for a network engineer to calculate what IP subnet an IP address belongs to by simply looking at the subnet mask. Whether allocating new networks in your **IP address management** (**IPAM**) system or calculating the IP network range while troubleshooting, these calculations are referred to as subnetting.

Reserved Addresses within IP Subnets

Before diving into the process of subnetting, it is important to understand which addresses within an IP subnet can be consumed by the hosts that will reside there. Within an IP subnet, certain addresses are reserved and cannot be assigned to individual devices or end hosts. At this point, you may be noticing that the subnet mask is already reserving bits for the network and now some of the addresses within that range are also reserved. However, it is not as complicated as it sounds.

Within an IP subnet, there are two addresses that are reserved: **network address** and **broadcast address**. The network address is the first address within the IP subnet and is used to identify the subnet itself. This address in binary is when all the available host bits are set to 0. The broadcast address, by contrast, is the last address within the IP subnet and is used to send **broadcast** traffic to all the hosts on that IP network. This can be used for discovery or network service announcements to all the hosts within that subnet. This address in binary is when all the available host bits are set to 1.

Refer to *Figure A1.16* to see a couple of examples of some IP subnets and the addresses reserved for the network address and broadcast address.

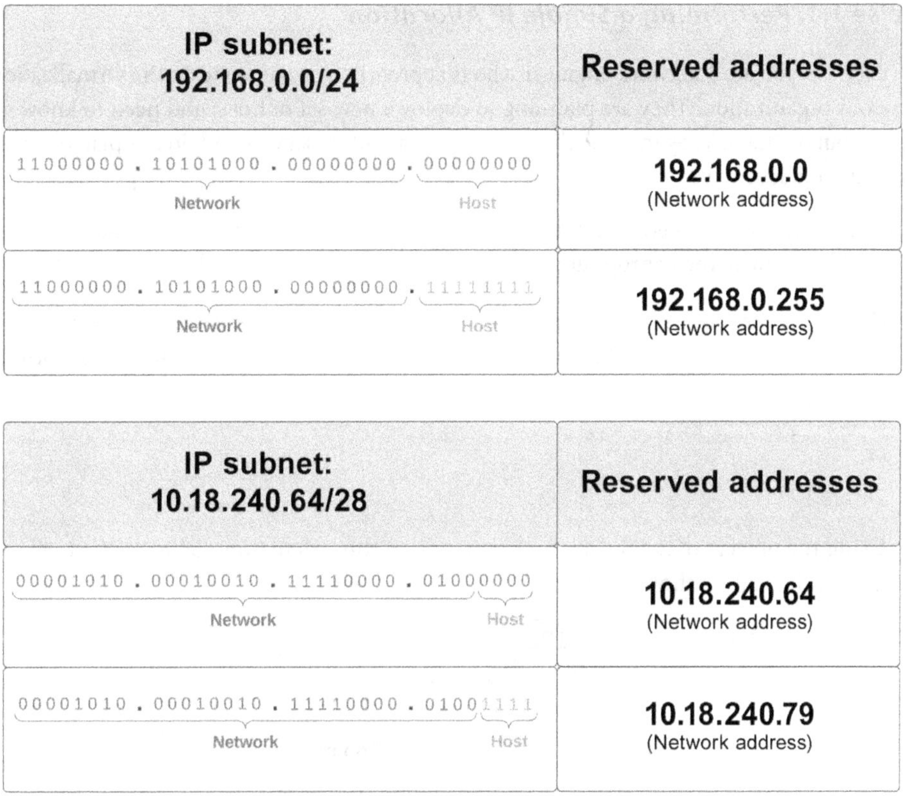

IP subnet: 192.168.0.0/24	Reserved addresses
11000000 . 10101000 . 00000000 . 00000000 Network Host	**192.168.0.0** (Network address)
11000000 . 10101000 . 00000000 . 11111111 Network Host	**192.168.0.255** (Network address)

IP subnet: 10.18.240.64/28	Reserved addresses
00001010 . 00010010 . 11110000 . 01000000 Network Host	**10.18.240.64** (Network address)
00001010 . 00010010 . 11110000 . 01001111 Network Host	**10.18.240.79** (Network address)

Figure A1.16: Reserved addresses in IP subnets

As noted before, the first usable address and last usable addresses within each subnet are reserved for network and broadcast.

Allocating IP Subnets

As mentioned before, network engineers need to be able to allocate subnets for usage across their organizations. They should be able to make calculations based on the number of host addresses required and allocate the appropriately sized network for use. This is where knowing the powers of 2 will come in very handy. Since each bit in the host portion of the IP address can be 0 or 1, each additional bit allocated to hosts doubles the number of available IP addresses (working from right to left). It is also important to remember that there are two reserved addresses in each IP subnet. Therefore, where H is equal to the number of hosts required, the formula to use is $2n \geq H + 2$.

To illustrate how this formula is effective in allocating networks, refer to the following exercises to see some real-world use cases.

Exercise 1.1: Performing a Simple IP Allocation

In this exercise, you are a network engineer who is approached by someone on the virtualization team within your organization. They are planning to deploy a new set of hosts and need to know what IP address to put on them. They indicate they will deploy about 50 servers and do not plan on expanding this capacity for several years.

For this exercise, you will need to take the preceding requirements and use the suggested formula of $2n \geq H + 2$ to calculate the appropriate subnet mask to use. You will use the following steps:

1. Given that the requirement is for 50 servers that means there need to be at least 50 different host addresses (i.e., host bit combinations). Now, put this through the following formula:

   $2n \geq 50 + 2$

   or

   $2n \geq 52$

2. Using the powers of 2, you can quickly perform this calculation. 64 is the first value that is greater than 52, as shown in *Figure A1.17*.

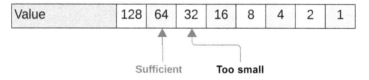

Figure A1.17: Subnetting – powers of 2

3. Since you know 26 is the answer, that tells you that you need at least 6 bits reserved for host addresses. Subtract that from the 32 bits that are available, and you arrive at 26. This means that our subnet mask in slash notation is a /26. This is represented in *Figure A1.18*.

Bit position	7	6	5	4	3	2	1	0	
Reservation	N	N	H	H	H	H	H	H	= /26
Value	128	64	32	16	8	4	2	1	(32 - 6)

Figure A1.18: Subnetting – Exercise 1 reservation

In this exercise, you calculated the appropriate subnet mask for a network with 50 hosts using the formula $2n \geq H + 2$. By determining that 6 bits are needed for the host portion, you arrived at a subnet mask of /26, which provides enough IP addresses for the required hosts while efficiently utilizing the available address space.

Exercise 1.2: Performing a Large IP Allocation

In this exercise, you are a network engineer who is tasked with assigning a new IP block for an acquisition that your company is going through. The company being acquired has around 4,000 endpoints, including laptops, servers, IP phones, and other devices. The company expects to expand the headcount for this company by around 300–400 people in the next two years. Based on this information, you want to allocate an IP subnet capable of having around 6,000 host addresses. Keep in mind that with VLSM, this large subnet can be carved up into smaller ones.

For this exercise, you will use the same formula as demonstrated in *Exercise 1.1*, but with a minor adjustment:

1. Since this subnet (or in this case, supernet) will be carved up for additional use, you don't need to necessarily anticipate two reserved addresses, as those reserved addresses will be consumed in all the smaller subnets carved out of this one. The formula will be as follows: $2n \geq 6,000$

2. Once again, you solve for n using the powers of 2 and determine that 13 is the value that you need, as shown in *Figure A1.19*.

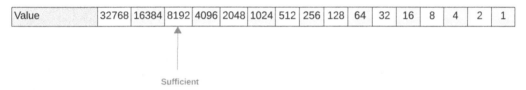

Value		32768	16384	8192	4096	2048	1024	512	256	128	64	32	16	8	4	2	1

Sufficient

Figure A1.19: Subnetting – powers of 2 continued

3. You know that means you need 13 bits reserved for host addresses. Subtract that from the entire 32 bits available, and you have 19. So */19* is our subnet mask, per *Figure A1.20*.

Bit position	15	14	13	12	11	10	9	8	7	6	5	4	3	2	1	0	
Reservation	N	N	N	H	H	H	H	H	H	H	H	H	H	H	H	H	= /19
Value	32768	16384	8192	4096	2048	1024	512	256	128	64	32	16	8	4	2	1	(32 - 13)

Figure A1.20: Subnetting – Exercise 2 reservation

Obviously, this calculation starts to become easier the more you do it. It's not expected that you'll be looking at a chart like the ones in *Figure A1.18* and *Figure A1.20* every time you need to make an IP subnet allocation. Instead, you should be able to perform this calculation mentally.

IP Subnet Identification

Calculating new IP subnet allocations for new deployments is a common activity for network engineers, but an even more common one is troubleshooting. When troubleshooting, a network engineer will often need to look at the host IP of a workstation or server to determine what IP subnet it resides in by the subnet mask assigned to the interface. This is a similar calculation to what's shown in *Exercise 1.1* and *Exercise 1.2* but differs slightly in that you ultimately must work your way backward. When you need to quickly identify the IP subnet that a host address belongs to, there are a few ways to do this, but this section will focus on using the **significant bit** method. The process involves looking at the subnet mask and performing a calculation based on the "significant bit." The significant bits are all the binary bit boundaries of the octets. Remember the classful subnet masks? These values represent the number of significant bits used for different subnet masks, with each step adding eight bits to the network portion:

- 8, or subnet mask 255.0.0.0

- 16, or subnet mask 255.255.0.0

- 24, or subnet mask 255.255.255.0

- 32, or subnet mask 255.255.255.255

You can see these are all multiples of 8, representing the full 8 bits of each octet. If you have a host address with a subnet mask of either /8, /16, or /24, determining those ranges is quite easy because each of the unreserved octets would have a range of 0 to 255. However, the subnet masks that are not on the octet boundary are the tricky ones. That's where the significant bit method comes in handy. Next, you will go through a demonstration of how to use this significant bit to calculate the IP subnet, then you will go through a couple of exercises demonstrating real-world scenarios.

Using the Significant Bit to Calculate the Subnet Address

To determine which subnet an IP address belongs to, you can follow a straightforward method using the significant bit. This process involves three key steps: subtracting the subnet mask, using powers of two, and calculating the range.

Step 1: Subtract the subnet mask from the next highest significant bit

Begin by identifying the next highest boundary bit from the subnet mask, which is usually expressed in slash notation (CIDR). Subtract the subnet mask value from this boundary bit to understand how many bits are reserved for the subnet.

Here are some examples:

- If the subnet mask is /18, the next highest boundary bit is 24

- If the subnet mask is /28, the next highest boundary bit is 32

This step reveals the number of reserved bits in the octet.

Step 2: Use powers of two

Once you have the difference between the next highest significant bit and the subnet mask, apply this value as an exponent to base 2. In other words, calculate $2^\wedge N$, where N is the value from *Step 1*. This result gives you the increment range of the octet.

Step 3: Calculate the range

With the value from *Step 2*, you can determine the ranges of the octet in the IP address. These ranges help you identify the subnet to which the given IP address belongs. Subnets will increase by the calculated value, and you can match the host IP address with its corresponding subnet.

By following these steps, you can quickly calculate the subnet for any given IP address, even in complex network configurations. Once you practice this method a few times, you'll be able to perform these calculations in seconds. Next, you will go through exercises of real-world scenarios in which you would perform this calculation as a network engineer.

Exercise 1.3: Subnet Calculation for a Corporate Network

Scenario: You are a network administrator responsible for managing the subnetting structure in a large corporate network. One of your tasks is to ensure that all devices are assigned to the correct subnet. A new employee in the IT department has been assigned an IP address of 10.208.14.48/26, and you need to confirm which subnet this IP address belongs to in order to configure the firewall rules accordingly.

Objective: Determine the subnet for the host IP address 10.208.14.48/26 so that you can properly assign it to the correct network segment and apply appropriate firewall policies

Step 1: Subtract the subnet mask from the next significant bit:

- Subnet mask: /26
- Next highest significant bit: 32
- Subtract 26 from 32 = 6

Step 2: Use powers of two:

- $2^\wedge 6 = 64$

Step 3: Calculate the range

With a range of 64, the subnets increase by 64. The possible subnets are as follows:

- 10.208.14.0/26
- 10.208.14.64/26

- `10.208.14.128/26`

- `10.208.14.192/26`

The IP address `10.208.14.48` belongs to the subnet `10.208.14.0/26`.

Exercise 1.4: IP Subnet Identification (Real-World Scenario)

Scenario: You are a network engineer, and an end user reports no connectivity to a system on the same subnet. The provided IP address of the user's laptop is `192.168.148.201` with a subnet mask of `255.255.255.128`. The machine they are trying to access has the IP address `192.168.148.64`.

Objective: Determine whether both IP addresses are on the same subnet.

Preliminary Step: Convert the subnet mask to slash notation:

- Subnet mask: `255.255.255.128`

- The first 3 octets are `255` (24 bits), and the last octet is `128` (1 bit)

- Therefore, the subnet mask is /25

Step 1: Subtract the subnet mask from the next significant bit:

- Subnet mask: /25

- Next highest significant bit: 32

- Subtract 25 from 32 = 7

Step 2: Use powers of two:

- $2^7 = 128$

Step 3: Calculate the range

Each subnet will increase by 128:

- `192.168.148.0/25`

- `192.168.148.128/25`

The IP address `192.168.148.201` belongs to the subnet `192.168.148.128/25`, while `192.168.148.64` belongs to `192.168.148.0/25`.

Conclusion

The two IP addresses are not in the same subnet.

In this section, you have explored the basics of subnetting, a critical skill for any network professional. You learned how to efficiently allocate IP addresses using CIDR notation, the powers of two, and how to identify reserved addresses within subnets. Real-world examples demonstrated how subnet calculations can be applied to network planning and troubleshooting. Whether you are assigning IP addresses to new servers or diagnosing network issues, mastering these techniques will help you manage network resources effectively.

As networks continue to expand, the limitations of IPv4 have become more pronounced, particularly in terms of address exhaustion. This leads to the need for a more scalable solution, which is where IPv6 comes into play. In the next section, you will explore the structure, benefits, and key features of IPv6, and how it addresses the shortcomings of IPv4. Understanding IPv6 will be essential as you work with modern networks that demand larger address spaces and more efficient routing.

IPv6

As mentioned at the beginning of the previous section, there is another version of IP that is in use today: IPv6. This follows many of the same principles as IPv4 with some added capabilities and enhancements. In this section, you will learn why IPv6 was developed to address the limitations of IPv4, particularly its address exhaustion. You will explore the key differences between IPv4 and IPv6, focusing on the expanded address space, the use of hexadecimal notation, and how IPv6 subnetting works. Additionally, you'll understand the concept of dual-stack networks, where IPv4 and IPv6 coexist, and the practical implications of transitioning to IPv6.

IPv6 was created to address the exhaustion of available addresses in IPv4. IPv4 offers over 4 billion unique addresses, which initially seemed sufficient to cover all potential use cases. However, researchers and engineers did not fully anticipate the significant expansion of the internet in the 1990s, nor the even greater growth in the years after 2000. The number of devices connected to the internet has skyrocketed. What once included primarily personal computers and servers now extends to devices such as manufacturing machinery, automobiles, and even refrigerators. Additionally, the widespread adoption of cell phones means nearly every person carries a device that requires an IP address. As a result, it's easy to see how 4 billion addresses are no longer enough.

When comparing IPv6 to IPv4, the obvious use case is to avoid address exhaustion. IPv6 expands the address space from 32 bits to 128 bits, allowing for approximately 340 undecillion (340 trillion trillion trillion) addresses. This vast address space ensures that the supply of available IP addresses will meet future demands for generations, preventing the exhaustion issues faced with IPv4. Additionally, IPv6 addresses are written in hexadecimal, making these large addresses more compact and readable.

In addition, the development of IPv6 provided an opportunity to enhance and optimize the IP protocol. IPv6 improves routing efficiency by reducing the size of routing tables and streamlining packet processing. It also includes built-in security features such as IPsec, which supports end-to-end encryption and authentication. Furthermore, IPv6 simplifies network configuration through features such as **Stateless Address Autoconfiguration (SLAAC)**. These improvements enhance the overall performance, security, and scalability of networks, making IPv6 more adaptable to modern and future networking needs.

IPv6 Addressing Format

IPv6 addresses have a total of 128 bits. Where IPv4 was separated into 4 octets of 8 bits each, IPv6 uses 8 **hextets**, which each hold 16 bits. Each hextet is separated by a colon. If you're doing the math, you can see that these addresses can get quite lengthy. Luckily, there are some mechanisms for shortening these addresses, which you will cover later in this section.

As mentioned before and shown in *Figure A1.21*, IPv6 addressing is also written in hexadecimal. This can look quite confusing at first when compared to IPv4, as now the values contain both numbers and letters. Now, you will have a quick look at calculating values in hexadecimal.

0000:0000:0000:0000:0000:0000:0000:0000

0000000000000000 : 0000000000000000 : 0000000000000000 : 0000000000000000 :

16 Bits   16 Bits   16 Bits   16 Bits

0000000000000000 : 0000000000000000 : 0000000000000000 : 0000000000000000

FFFF:FFFF:FFFF:FFFF:FFFF:FFFF:FFFF:FFFF

1111111111111111 : 1111111111111111 : 1111111111111111 : 1111111111111111 :

16 Bits   16 Bits   16 Bits   16 Bits

1111111111111111 : 1111111111111111 : 1111111111111111 : 1111111111111111

Figure A1.21: IPv6 address format

As shown in the figure, when all bits are set to 0, the address is 0000:0000:0000:0000:0000:0000:0000:0000. Alternatively, when all bits are set to 1, the address is FFFF:FFFF:FFFF:FFFF:FFFF:FFFF:FFFF:FFFF.

Hexadecimal Values in IPv6

In IPv4, a single numeric value (0–255) is used to represent the entire 8 bits of an octet. In IPv6, using hexadecimal, each character of the address is used to represent four bits. This is why each hextet contains 4 characters, as seen in *Figure A1.21*, totaling 16 bits. Since each character must represent a value of 0 to 15, hex is used to represent the additional values. Refer to *Figure A1.22* to see these hexadecimal values and how they are represented in binary.

Hexadecimal digit	Decimal equivalent	Binary representation
0	0	0000
1	1	0001
2	2	0010
3	3	0011
4	4	0100
5	5	0101
6	6	0110
7	7	0111
8	8	1000
9	9	1001
A	10	1010
B	11	1011
C	12	1100
D	13	1101
E	14	1110
F	15	1111

Figure A1.22: Hexadecimal values

As shown in the figure, not only are numeric values 0 through 9 present but also alphanumeric values for A through F.

Guided Example: Removing Leading Zeros in IPv6 Addresses

In this example, you will learn how to shorten an IPv6 address by removing unnecessary leading zeros and compressing consecutive zero blocks. This is an important technique for making IPv6 addresses easier to read and manage.

Step 1: Understand the Full IPv6 Address Format

A full IPv6 address is typically 128 bits long, represented by 8 hextets, each containing 16 bits. In hexadecimal format, this results in a total of 32 characters. However, writing out the full address isn't always necessary, as there are two shorthand rules that simplify it.

Step 2: Apply Rule 1 – Remove Leading Zeros in Each Hextet

The first rule for shortening IPv6 addresses allows you to omit leading zeros within each hextet. For example, let's take this full IPv6 address:

`2001:0db8:85a3:0000:0000:8a2e:0370:7334`

You can remove the leading zeros from each hextet:

`2001:db8:85a3:0:0:8a2e:370:7334`

Step 3: Apply Rule 2 – Replace Consecutive Zeros with a Double Colon (::)

The second rule allows you to compress consecutive blocks of zero-filled hextets into a double colon (`::`). This can only be done once within the address to avoid ambiguity. Continuing from the previous step, the address `2001:db8:85a3:0:0:8a2e:370:7334` has two consecutive hextets with all zeros. You can compress them using `::` as follows:

`2001:db8:85a3::8a2e:370:7334`

Step 4: Verify the Final Shorthand Address

By applying these two rules, you have successfully reduced the IPv6 address from `2001:0db8:85a3:0000:0000:8a2e:0370:7334` to the shorter, more manageable form, `2001:db8:85a3::8a2e:370:7334`.

Conclusion

Using these steps, you can significantly shorten IPv6 addresses, making them easier to work with. For addresses with more consecutive zero-filled hextets, such as `2001:0db8:0000:0000:0000:0000:0370:7334`, the shortened version becomes `2001:db8::370:7334`. Even an all-zero address can be reduced to just `::`. These shorthand rules will help you simplify IPv6 address representation while maintaining accuracy.

Subnetting in IPv6

Subnetting in IPv6 plays a crucial role in modern networking, allowing efficient allocation and management of the vast address space IPv6 provides. Unlike IPv4, where address exhaustion is a significant concern, IPv6 offers an expansive 128-bit address space, which greatly reduces the risk of running out of addresses. However, subnetting is still necessary to organize networks and ensure effective routing.

The process of subnetting in IPv6 closely mirrors that of IPv4 but with some simplifications. The large address space means that overallocation is no longer a concern, and network engineers can easily allocate subnets without worrying about running out of IP addresses. This eliminates many of the challenges that arose from the limitations of classful networking in IPv4. Thanks to the expanded range of addresses, the likelihood of address exhaustion in IPv6 is extremely low, ensuring scalability for future network growth.

Subnetting with a /64 Network

In IPv6, the general practice for subnet allocation is to assign a /64 subnet for each local network. This is considered standard because it provides a vast number of possible addresses for devices within a single subnet, ensuring that the network will not run out of addresses. While smaller subnets, such as /127 or /126, may be used for specific cases such as point-to-point links where only two host addresses are needed, the /64 subnet is optimal for most networks with end hosts.

At first glance, a /64 subnet may seem excessively large. With 64 bits allocated for host addresses, the total number of possible addresses in a /64 subnet is 2^{64}, or over 18 quintillion addresses. This enormous number underscores the scalability of IPv6, where even extremely large networks will have more than enough address space. In the context of the overall IPv6 address space, the use of /64 subnets still leaves a vast pool of available addresses, ensuring flexibility and growth for future network needs.

Procuring IPv6 Networks

Once you understand the basics of subnetting with IPv6, the next step is to look at how organizations can obtain publicly routable IPv6 address ranges. The process for procuring IPv6 addresses is similar to that used for IPv4. Organizations typically contact their local ISP or an RIR to request a block of IP addresses.

In IPv4, organizations need to carefully calculate and request an appropriate range of addresses, such as /21 or /24, depending on their specific needs. With IPv6, the most common allocation from ISPs or RIRs is /48. This allows organizations to create up to 65,536 subnets, each with a /64 address block, offering more than enough space for most use cases. Larger organizations with extensive requirements may request a bigger block, such as /44, provided they have sufficient justification.

The sheer size of the IPv6 address space means that organizations can allocate IP addresses with far less scrutiny compared to IPv4. The availability of vast address ranges simplifies the process, ensuring that networks have ample room for expansion without the concerns of running out of address space, which was a common issue in the IPv4 world.

Public versus Private Addressing in IPv6

IPv4 contains strict rules about what addresses are globally routable versus those that are not. For this reason, you'll never see a prefix for `10.0.0.0/8` advertised on the public internet. The vast majority of IPv6 addressing is, in fact, routable over the internet. IPv4 had shortcomings that eventually came to light, one of which was a limited address space available for end hosts. Thus, you had the introduction of RFC 1918 so that there was a large number of IP addresses that could be reused over and over by several organizations within their IP routing environments.

However, IPv6 does not have this problem. There are so many available addresses that you don't have to worry about this issue for the foreseeable future. As it currently stands, every device on Earth that needs an IP address could have a globally routable IPv6 address, and you would still have billions of addresses left over. This means every single host – every computer, every camera, every refrigerator, every sensor – can all have a globally routable address.

That all being said, IPv6 does have some reservations put in place for "private" addressing, but the use cases differ slightly from IPv4 in how they are used. Now, have a look into unique local addresses and link-local addresses, and when they would be used.

Unique Local Addresses

Unique local addresses (ULAs) were defined in RFC 4193 for local communication within a specific site or location and are not intended to be routed across the global internet. ULAs provide a stable, private address range that can be used within an organization's internal network, offering the benefit of avoiding address conflicts with global addresses. Unlike globally routable IPv6 addresses, ULAs are similar in purpose to private IPv4 addresses (e.g., `10.0.0.0/8`) but allow for larger address spaces and more efficient internal routing. For example, ULAs are useful for internal services such as management networks, where global connectivity is not required, ensuring that sensitive or administrative traffic remains isolated from the broader internet.

In terms of format, ULAs are defined as belonging to the prefix `fc00::/7`. However, as of the time of writing, only the range `fc00::/8` is actively used for ULA allocation. This gives administrators a broad, structured range of addresses that can be easily segmented within internal networks.

As for usage, ULAs are often employed in scenarios where global addressing is unnecessary or undesirable. They function similarly to the reserved RFC 1918 ranges in IPv4, providing a mechanism for internal communication without risking overlap with public addresses. This is particularly useful for administrative functions, such as managing internal infrastructure, configuring devices, or avoiding IP address conflicts. By using ULAs, organizations can maintain a clear separation between internal network traffic and external, globally routed traffic.

> **Note**
> Further reading on ULA addressing can be found in RFC 4193 at the following URL:
> `https://www.ietf.org/rfc/rfc4193`

Link-Local Addresses

Link-local addresses were defined in RFC 4291 as a type of private IP range within IPv6 that is restricted to a limited routing scope. These addresses are specifically used for communication within a single network segment, or **link**, and are not intended for routing outside that segment. Link-local addresses play a crucial role in internal networking tasks, such as discovering and interacting with nearby devices.

In terms of format, link-local addresses are defined as belonging to the `fe80::/10` prefix. This range ensures that the addresses are reserved exclusively for local communication within a given link and are easily identifiable within the broader IPv6 address space.

Link-local addresses are commonly used for tasks such as address resolution and neighbor discovery, which help network devices identify and communicate with each other on the same local network. For example, when a new device connects to a network, it uses a link-local address to announce its presence and discover other devices in the segment. These addresses are automatically generated by devices and require no manual configuration, making them essential for basic network functions and internal communication that does not require global routing.

> **Note**
>
> Further reading on link-local addressing can be found in RFC 4291 at the following URL:
>
> `https://www.ietf.org/rfc/rfc4291`

Usage of ULA and Link-Local Addressing

ULAs can be used to meet the network security compliance criteria of an organization. For instance, some organizations require that internal networks, such as those handling sensitive data or administrative tasks, use private address spaces to ensure that traffic remains isolated from the global internet. A practical example might be a financial institution's internal database network, where ULAs are assigned to ensure that sensitive customer data is protected from external exposure by keeping it within the organization's private network. This approach helps organizations enforce internal security policies while avoiding address conflicts with public IPs.

Link-local addressing, on the other hand, is automatically used by the IPv6 protocol stack for communication within a local network segment. For example, when a new device such as a printer is connected to a company's **local area network** (**LAN**), it automatically generates a link-local address without needing manual configuration. This address allows the printer to communicate with nearby devices such as computers and routers for functions such as network discovery and configuration, without ever needing to communicate outside the local network. This automatic generation and local-only scope make link-local addresses essential for seamless, internal network operations.

Transitioning to IPv6

The adoption of IPv6 across the internet has been a slow process. For decades, there has been a push to transition to IPv6 as IPv4 addresses have largely been exhausted or allocated. However, IPv4 still accounts for approximately 70% of internet routing tables, and most of the traffic on both the global internet and enterprise networks still relies heavily on IPv4. This raises the question: why does IPv4 continue to dominate, even though IPv6 offers clear advantages?

Migrating network infrastructure from IPv4 to IPv6 is not as simple as flipping a switch. It involves careful consideration of factors such as addressing, routing, and security, as well as support from vendors and service providers for IPv6. Additionally, the transition affects services such as NAT and DNS. While IPv6 offers benefits such as a larger address space and improved efficiency, the complexity of implementing these changes, coupled with the fact that it doesn't immediately enhance higher-level functions (e.g., applications), has led many network operators to deprioritize the transition.

A real-world example of IPv6 adoption can be seen in the case of Belgium, which has one of the highest IPv6 adoption rates globally. The transition was driven largely by ISPs such as Telenet and Proximus, who implemented IPv6 to future-proof their networks and handle increasing demand for IP addresses. However, the transition was not without challenges. Both providers had to manage dual-stack environments (running both IPv4 and IPv6), which added operational complexity and increased costs for maintaining two parallel systems. Additionally, ensuring compatibility with older equipment and services that still rely on IPv4 proved difficult, requiring extensive testing and phased implementation.

Despite these hurdles, Belgium's successful deployment of IPv6 demonstrates the long-term benefits of adopting the protocol. Organizations that manage to overcome these initial challenges stand to gain from the scalability and efficiency IPv6 offers, paving the way for a more sustainable internet infrastructure.

The process of transitioning to IPv6 introduces the concept of a **dual-stack** network. Dual-stack is a configuration where devices on a network, including servers, switches, and routers, run both IPv4 and IPv6. In this scenario, these devices can understand and/or process network traffic encapsulated in either IP version. This coexistence of IPs typically offers the first step in migration from IPv4 to IPv6. Dual-stack configuration does not force an "all or nothing" approach, as you can have a network that should support the same networking capabilities within both protocol stacks.

While dual-stack networks offer a practical solution for transitioning from IPv4 to IPv6, they also introduce significant complexities. Running both protocols in parallel requires additional resources, as network devices must be configured to handle both IPv4 and IPv6 traffic. This can lead to increased operational costs, more complicated network management, and the need for staff to be proficient in managing two protocol stacks. Additionally, ensuring compatibility between devices and applications that may rely on either IPv4 or IPv6 adds another layer of complexity.

These complexities influence several key decision points that network operators must carefully evaluate when designing and managing dual-stack networks. First, resource requirements must be considered, as selecting the right network appliances and vendors is critical for supporting both IPv4 and IPv6 traffic. Security is another concern, as traffic for both protocols must be managed separately, requiring different configurations and strategies to ensure they are both secure. When it comes to implementation, operators need to decide how to configure interfaces and routing protocols for both IPv4 and IPv6, ensuring that they function effectively in parallel. Interoperability is also a key factor – network operators must determine how IPv4 and IPv6 networks will communicate with each other when necessary, and how to handle the translation between protocols. Finally, ensuring internet connectivity for both protocol stacks presents another challenge, as operators need to configure their networks to support external communication via both IPv4 and IPv6.

For example, an organization transitioning to IPv6 while still supporting legacy IPv4 applications would need to deploy dual-stack routers and switches. These devices must be configured to route both IPv4 and IPv6 traffic, maintain separate security policies for each, and ensure proper translation mechanisms are in place for services that require communication between the two protocols. This requires careful planning and resource allocation to ensure seamless interoperability and secure, stable network operations.

Summary

In this chapter, you have explored essential network fundamentals, laying a strong foundation for understanding cloud networking and AWS-specific constructs. Key concepts covered include the OSI model, IP addressing, CIDR, basic subnetting, and an introduction to IPv6. You learned how data is transmitted across networks using the OSI model and the role of each layer in this process. The chapter also covered how IPv4 and IPv6 address structures work, and the importance of subnetting for network allocation and troubleshooting. Additionally, you practiced key skills such as calculating subnet masks and IP address ranges, which are crucial for efficiently managing network resources.

These concepts not only prepare you for the *AWS Certified Advanced Networking – Specialty (ANS-C01)* exam but also equip you with practical skills that are foundational to a career in cloud networking. A strong grasp of these fundamentals will enable you to navigate complex networking environments, allocate IP subnets effectively, and troubleshoot network issues – skills that are invaluable for any network or cloud engineer.

Appendix 2
IP Routing Fundamentals

This chapter will add to the foundation of networking fundamentals that you need to understand prior to tackling the *AWS Certified Advanced Networking - Specialty (ANS-C01)* exam. In the previous chapter, you explored concepts such as the OSI model and data encapsulation, as well as protocols such as IP. This chapter will focus on IP routing. Building connectivity over IP networks includes not only the sending and receiving of IP packets but also the advertisements about IP networks themselves.

This chapter covers the following main topics:

- Routing concepts
- Basic **Border Gateway Protocol (BGP)**

By the end of this chapter, you should have an understanding of how IP routing works at a basic level, what takes precedence when deciding the best path with routing, as well as BGP fundamentals, attributes, and path selection. This will aid you as you learn more about how AWS networking works.

Routing Concepts

When looking at the example of encapsulation and decapsulation in the previous chapter, the fundamental principles of communication using direct connectivity seem evident. The system initiating the communication encapsulates the data and puts the bits on the wire, and then the receiving system decapsulates what has been sent. But how does this function when the two end systems are not directly connected? How does this information get sent across a network that exists between the two? The process of sending data between networks is called routing, and it is used to perform packet forwarding.

Forwarding Data between Networks

As covered earlier, the data unit that represents the encapsulated data at layer 3 is called a packet. Network devices that receive packets from end hosts and forward them along to their destination are referred to as routers. Routers forward data in the following way:

1. Receive encapsulated data.

2. Decapsulate enough of the data to see the layer 3 information.

3. Make a routing decision and determine where the "next hop" destination is.

4. Re-encapsulate the data and forward the data to the next hop.

Figure A2.1 shows how traffic is routed over an IP network:

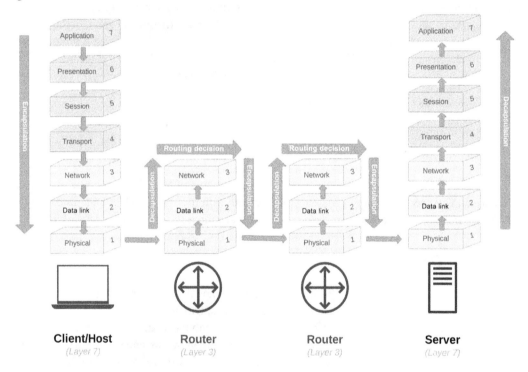

Figure A2.1: IP routing – client to server

In *Figure A2.1*, as traffic is moved from the application layer down to the physical layer, the payload of data is encapsulated. As traffic moves through the network, it is unnecessary to decapsulate the traffic beyond the network layer in most cases; rarely, a security appliance (not shown) could do a higher layer inspection for malware.

As IP packets are forwarded through a network, the information contained within the IP header is used to make the routing decision by the routing appliance.

Next, you will have a look at the structure of an IP header and how the data is arranged.

IP Headers

The IP header is a set of structured data that is appended to the beginning of a data packet to assist the receiving system with the successful delivery of the packet. Here are some key functions of the IP header:

- **Routing**: The IP header contains source and destination IP information that can be used to forward (or route) the IP packet to the proper destination.

- **Fragmentation and reassembly**: In IPv4, certain fields within the IP header are associated with fragmenting large IP packets. This involves breaking them out into smaller packets.

- **Error detection**: IPv4 includes a checksum that assists with validating the header's integrity.

- **Quality of service**: A set of fields is used to identify specifications on how the specific packet should be prioritized when forwarding.

- **Payload specification**: Identification of the type of protocol used in the "next header," aka layer 4. This could specify that the next header is either TCP or UDP.

The data within the IP header is structured in a specific format, as shown in *Figure A2.2*. This IP header structure is based on standards put forth by the **Internet Engineering Task Force** (**IETF**) to ensure that systems can communicate. As several header fields have been updated, removed, or added in the interest of improvement for IPv6, the structure of these IP header formats differ between IPv4 to IPv6, but there are common fields between the two. The standard **IPv4 header format** is 20 bytes in length and contains 13 separate fields. That may sound like a large number, but luckily, for the sake of IP routing, there are only a few fields that are significant: mainly the source and destination address. However, it is useful for any network engineer to know what these fields are and how they are used by the systems involved. Refer to *Table A2.1* for details about these fields.

Field No.	Field Name	Field Description	Field Size
1	Version	Specifies the **Internet Protocol** (**IP**) version.	4 bits
2	Header length	Identifies the length of the IP header. The minimum value is 5, which correlates to 20 bytes.	4 bits
3	**Type of service (TOS)**	Defines the **quality of service** (**QoS**) desired.	8 bits

Field No.	Field Name	Field Description	Field Size
4	Total length	Indicates the entire length of the packet (or datagram), including the header and data.	16 bits
5	Identification	Used for identification of data fragments for the purpose of reassembly.	16 bits
6	Flags	Set of control flags.	3 bits
7	Fragment offset	Indicates the location of where the fragment belongs within the entire datagram.	13 bits
8	**Time to live (TTL)**	The maximum time (or # of hops) that a datagram is permitted to be in transit.	8 bits
9	Protocol	Indicates the protocol used within the next layer of encapsulation (layer 4).	8 bits
10	Header checksum	Checksum used to verify that the content of the IP header is valid.	16 bits
11	Source address	Source IP address of the IP packet.	32 bits
12	Destination address	Destination IP address of the IP packet.	32 bits
13	Options and padding	Zero or more optional fields and/or padding to ensure the header is a standard length.	Variable
Total			**160 bits (20 bytes)**

Table A2.1: IPv4 header fields

The IP header fields consist of very specific information conforming to a standard, decided by the IETF. The IETF is the standards body that decides how all vendors and organizations must conform for interoperability. The IP header is a collection of these fields and provides details about every data payload. The arrangement of the fields is shown in *Figure A2.2*:

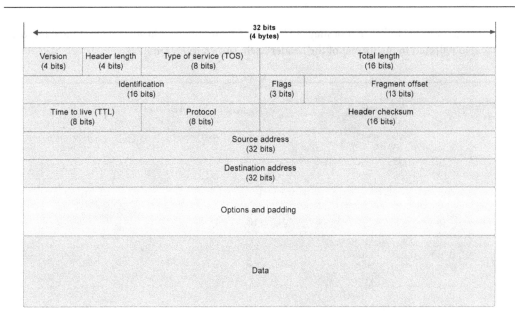

Figure A2.2: IPv4 header format

The **IPv6 header format** has a fixed length of 40 bytes, rather than a variable one. This aids in the processing of the packets and uniformity. Refer to *Table A2.2* for details of these fields:

Field no.	Field name	Field description	Field size
1	Version	Specifies the IP version.	4 bits
2	Traffic class	Used for designing **quality of service (QoS)**. Similar to the **Type of Service** field within the IPv4 header.	8 bits
3	Flow label	Used by the source to indicate packets in a sequence that belong to the same flow.	20 bits
4	Payload length	The length of the data following the IPv6 header.	16 bits
5	Next header	Identifies what type of header immediately follows the IPv6 header.	8 bits
6	Hop Limit	Similar to **time to live (TTL)** in the IPv4 header.	8 bits
7	Source address	Source IP address of the IP packet.	128 bits
8	Destination address	Destination IP address of the IP packet.	128 bits
Total			**320 bits (40 bytes)**

Table A2.2: IPv6 header fields

There are far fewer fields that may or may not be used, streamlining the header and making it uniform. This is also reflected in the format of the header itself, shown in *Figure A2.3*:

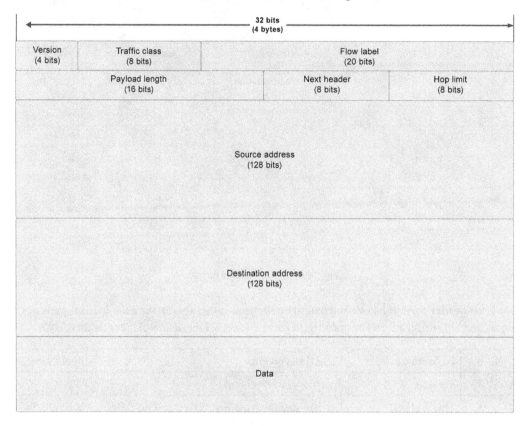

Figure A2.3: IPv6 header format

Routing Decisions and Route Tables

Network appliances (most often, routers) need to make a routing decision once they receive an IP packet to forward the packet along to the appropriate destination. When a router needs to make this decision, it consults its local route table.

A route table is essentially a pre-calculated list of all routing decisions that a router would need to make. These decisions being calculated ahead of time can assist with the efficiency of forwarding packets through the network, as they could be required to process hundreds of thousands of packets within seconds.

Every router maintains at least one primary route table for each version of IP that it can support. This means you will require a primary routing table for IPv4 and a separate table for IPv6. This primary table (for both protocol versions) is often referred to as the **routing information base** (**RIB**). The RIB is then populated with several entries known as routes that are determined by routing decisions for destination IP prefixes and/or subnets. This routing table acts as the brain of the routing engine.

As an example, if a router has been configured to forward data to the network $10.110.0.0/24$, the router will need that prefix to appear in the routing table, along with the next hop to forward the packet to. This next hop is often another router in the network path, or a network attached locally. When a data packet is received with a destination address of that network, the router determines how and where to forward that traffic based on the route table entry.

> **Note**
>
> Routers are often able to run multiple route tables. These are often referred to as **virtual routing and forwarding** (**VRF**) instances. However, the discussion of VRF is beyond the scope of this book.

Each route entry within the RIB is comprised of three major attributes, as shown in *Table A2.3*:

Attribute No.	Attribute	Description
1	Source	How the route was programmed into the route table (i.e., manually or dynamically).
2	IP prefix	The destination IP prefix/subnet intended to route to.
3	Next hop information	Details about the next hop device and how to reach it. This is often the combination of the next hop's IP address and the egress interface used to reach it.

Table A2.3: RIB route entry components

As always, the use of a visual component works much better to bring all these concepts together. *Figure A2.4* is an example topology of a client machine that needs to communicate with a server across an IPv4 network. The client is configured with an IP address of 192.168.10.1/30 and the server with an IP of 192.168.30.2. There are two routing appliances along the path that need to facilitate this connectivity, and their route tables have been configured to do so.

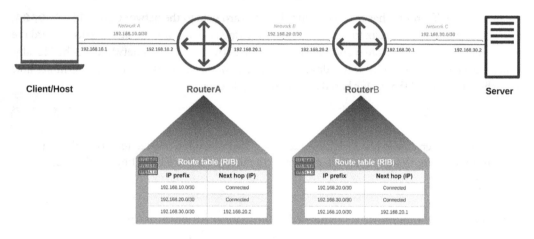

Figure A2.4: Route tables

Figure A2.4 shows the process of forwarding traffic between two servers and the route table lookups that determine the traffic path.

To see this process step by step, you can perform something called a **packet walk**. A packet walk is essentially a walk-through of each individual step that a packet takes to get from point A to point B. *Figure A2.5* shows the packet walk of the client communicating with the server:

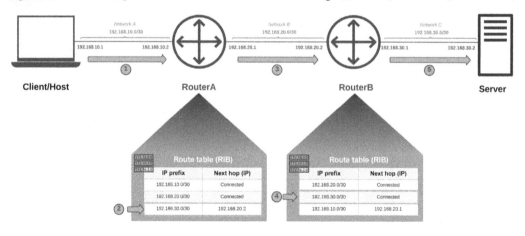

Figure A2.5: Route table packet walk

Here is a step-by-step discussion of the preceding figure:

1. The client machine (192.168.10.1) encapsulates the application data within an IP packet. The destination of the IP packet is set to the server (192.168.30.2). The client machine sends this packet to RouterA.

2. RouterA receives the IP packet on its interface configured with 192.168.10.2. RouterA inspects the IP packet and sees the destination is set to 192.168.30.2. RouterA then compares this IP to the prefixes listed in the route table and sees a match for the IP prefix: 192.168.30.0/30. The next hop for this prefix is set to RouterB's address of 192.168.20.2. RouterA has a directly connected interface on that network (192.168.20.1).

3. RouterA re-encapsulates the packet and forwards it to RouterB (192.168.20.2).

4. RouterB receives the IP packet on its interface configured with 192.168.20.2. Router B follows the exact same process that RouterA did in *Step 2*. The local routing table (RIB) is consulted and the route for the destination IP address is directly connected.

5. RouterB forwards the IP packet directly to the server.

When the steps are broken down into very basic components, there can be a lot that goes into the IP routing process across a network. Assuming that these two endpoints are not on the other side of the planet from one another, this process typically happens within a fraction of a second.

One important thing to note about destination IP information in the preceding packet walk is that the destination IP address does not change throughout the end-to-end routing process. Since the client that crafted the original IP packet sets the destination IP address, it is important that every device along the path makes a routing decision based on that information. This is also known as destination-based routing. Altering the destination IP information within a packet could drastically alter the true destination of the packet.

Details about when/why you would need to change IP information within the packet will be covered later within the *NAT* section.

Longest-Match Routing

As you can imagine, there are a lot of IP prefixes in use within the world today. With the vast amount of usable IPv4 and IPv6 space with the combination of **variable length subnet mask** (**VLSM**), there are potentially millions of IP prefixes. Route tables would simply scale if each router had to maintain a route table with every possible prefix entry listed as an individual route. This is why routers implement something called longest-match routing to make the process more efficient.

Back in the *CIDR Notation* section of *Appendix 1, Networking Fundamentals*, you learned about the concept of bits within the IP address being reserved for the network or subnet. Longest match routing builds upon this concept for the route table entries. Remember, the destination address within an IP packet is a **single IP address**, whereas a route entry within a route table is an **IP subnet**.

Longest-match routing means that the destination IP address of an IP packet is compared with the route entries of the RIB, and the route entry with the most bits in common with the IP address is selected as the best path by the router. In other words, the "longest match" (that is, the most specific match) when comparing bits from left to right wins. Consider an example. A router receives a packet with a destination IP address of 10.150.128.4 and needs to forward the packet. The RIB contains the following entries:

IP Subnet	Next Hop IP
10.0.0.0/8	192.168.10.1
10.150.0.0/16	192.168.20.1
10.150.128.0/15	192.168.30.1
10.150.128.0/30	192.168.40.1

Table A2.4: RIB example 1

At first glance, you'll notice that the destination IP address falls within the range of all the IP subnets listed. But since the routing logic is built upon the **longest match**, this means that all the IP subnets that match the destination (that is, the most specific route) will be chosen. The last IP subnet, 10.150.128.0/30, is the most specific as it has the most bits in common with the destination IP address. In turn, the router will forward the destination to the next hop of 192.168.40.1.

Consider an alternate situation in which the router has a slightly different list of route table entries:

IP Subnet	Next Hop IP
10.0.0.0/8	192.168.10.1

Table A2.5: RIB example 2

While the route entry is not overly specific and covers a very wide range of IP addresses, it is still a match. And since, in this RIB example, there are no more specific matches to contend with that match, the IP packet will be forwarded to 192.168.10.1.

Default Route

Up to this point, the routing decisions covered have been for **known** destinations. Regardless of whether that address contains a more specific prefix (/30) or something vaguer (/8), the destination is still known. But what about unknown destinations? Previously, in the *Public IP Address Ranges* section in *Appendix 1, Networking Fundamentals*, it was mentioned that several public IP registrars are responsible for managing and distributing billions of IP addresses across the world, between both IPv4 and IPv6. Putting all of these IP subnets into the route table of every router would simply not be manageable or scalable.

Rather, using the logic of longest match routing, you can introduce a single route entry that can account for all the **unknown** prefixes. The idea is that a route entry can be created that is the absolute least specific entry that it can be (all zeros) so that it can be used for any unknown destination. This is referred to as a **default route**. The default route can be thought of as a catch-all route and will generally appear as 0.0.0.0/0, though you may see this written with the full subnet mask as well, 0.0.0.0 0.0.0.0.

Figure A2.6 displays the same network topology as seen in *Figure A2.5*, but instead of a single destination, now there are two servers attached to RouterB. Each server has a public IP address assigned. Notice that instead of adding individual routes to the route table of RouterA, a default route has been configured instead.

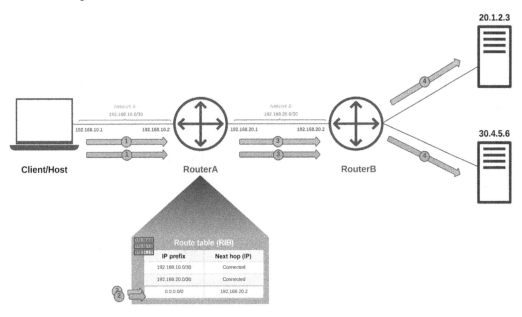

Figure A2.6: Default route

This sequence is similar to the previous packet walk, but instead of exact matches in the route table, the forwarding path is chosen by the default route, which is the longest match in the route table for this traffic. The use of a default route is very scalable.

A common example of using a default route within enterprise networks is when an **internet service provider** (**ISP**) is contracted to provide internet connectivity to a specific office, branch, or headquarters. The ISP will often terminate this connection on a router that they manage. The network engineering team for the enterprise configures their router with a single default route toward this ISP router for connectivity to all public IP subnets, as all traffic not destined for the local network must go through the same path to the ISP.

Default Gateway

In the previous routing examples, you've seen a client or host trying to reach a set of servers across an IP network. The client initiates the communication by sending IP packets to the first-hop router (RouterA). But how does this host know to do that?

If you consider your home network, there are probably several devices connected: laptops, game systems, smart TVs, and so on. You also likely have some type of router/gateway from your ISP that provides connectivity to the outside world over the internet. With this many devices connected to a single network, how does the client know which device could be responsible for routing packets to the outside world?

Devices such as laptops and servers are typically not responsible for routing or forwarding packets. In other words, they do not receive IP packets destined for other systems, make a routing decision, and forward those packets to another system. However, they do maintain a local route table that instructs the local network stack on how to process and/or generate IP packets. The route tables of end user devices are typically much smaller in size compared to the typical network router.

With that in mind, these devices can follow the same logic that was previously touched on with the concept of the default route. These systems can send traffic for *all* unknown prefixes to their **default gateway**. Since these types of nodes (end hosts) are meant to keep relatively simple IP route tables, the default gateway is typically used to reach *any* prefix that is not on the local network.

For example, say an end host is configured with an IP address of 192.168.10.100/24 and a default gateway of 192.168.10.1. If the end host were to send traffic to any IP address not within that local network of 192.168.10.0/24, then the traffic would be first sent to the default gateway.

Figure A2.7 shows the route table of a Windows PC, including the local and default routes the PC will use. It is common for the default gateway to not be manually configured but dynamically instead. This is accomplished through **Dynamic Host Configuration Protocol (DHCP)**. Routers often possess the capability to act as a DHCP server and assign details such as IP address, default gateway, and DNS servers to each end host dynamically.

```
C:\Users\[   ]>route print
===============================================================================
Interface List
 17...b4 2e 99 39 07 40 ......Intel(R) Ethernet Connection (7) I219-V
  7...02 13 ef 1b 01 06 ......Microsoft Wi-Fi Direct Virtual Adapter
 11...00 13 ef 1b 01 06 ......Microsoft Wi-Fi Direct Virtual Adapter #2
 10...00 50 56 c0 00 01 ......VMware Virtual Ethernet Adapter for VMnet1
  6...00 50 56 c0 00 08 ......VMware Virtual Ethernet Adapter for VMnet8
  8...00 13 ef 1b 01 06 ......Realtek 8812BU Wireless LAN 802.11ac USB NIC
  1...........................Software Loopback Interface 1
===============================================================================

IPv4 Route Table
===============================================================================
Active Routes:
Network Destination        Netmask          Gateway       Interface  Metric
          0.0.0.0          0.0.0.0      192.168.0.1    192.168.0.37      35
        127.0.0.0        255.0.0.0         On-link        127.0.0.1     331
        127.0.0.1  255.255.255.255         On-link        127.0.0.1     331
  127.255.255.255  255.255.255.255         On-link        127.0.0.1     331
      192.168.0.0    255.255.255.0         On-link    192.168.0.37     291
     192.168.0.37  255.255.255.255         On-link    192.168.0.37     291
```

Figure A2.7: Route table – Windows

Unlike *Figure A2.7*, in *Figure A2.8*, the route table looks very different for a macOS/Linux operating system. The default route is listed as `default` instead of the `0.0.0.0` entry.

```
[      ]@ubuntu:~$ netstat -r
Kernel IP routing table
Destination     Gateway         Genmask         Flags   MSS Window  irtt Iface
default         _gateway        0.0.0.0         UG        0 0           0 ens33
10.10.0.0       0.0.0.0         255.255.0.0     U         0 0           0 br-1bd7568ce6af
link-local      0.0.0.0         255.255.0.0     U         0 0           0 ens33
172.17.0.0      0.0.0.0         255.255.0.0     U         0 0           0 docker0
192.168.18.0    0.0.0.0         255.255.255.0   U         0 0           0 ens33
```

Figure A2.8: Default gateway – macOS and Linux

Understand that even endpoints use routing to send data packets, and they usually send them via the default route as very few other endpoints are accessible locally, if any.

Routing Types

The configuration and populating of IP route tables across all the network systems is critical for establishing connectivity for all the appropriate end systems. As you've seen, each router maintains a route table that is filled with individual route entries for a variety of destination prefixes. The configuration of these routes within these tables can be accomplished in a couple of different ways: **static** or **dynamic**.

Static routing is the simplest form of configuring routes within a route table. A static route entry is manually defined by the network operator directly within the configuration of the router's configuration. This could be using the router's vendor-specific **command-line interface** (**CLI**), programmatically, or even through an API call (depending on what the vendor supports). Nonetheless, this is a manual entry that will always be present unless it is removed from the configuration.

Take the following example of a static route configuration via the CLI of a Cisco router:

```
Router(config)# ip route 192.168.0.0 255.255.255.0 10.1.1.1
```

This configuration creates an IPv4 static route entry for `192.168.0.0/24` with a next-hop address of `10.1.1.1`.

Here is another sample configuration from a Cisco router CLI, but for an IPv6 static route:

```
Router(config)# ipv6 route 2001:DEAD:BEEF::/64 2001:1::1
```

This configuration creates an IPv6 static route entry for `2001:DEAD:BEEF::/64` with a next-hop address of `2001:1::1`.

Dynamic routing, by contrast, is a mechanism to populate route tables, but via a system of routers all sharing information. Large groups of routers can all be configured by network operators to run a specific **dynamic routing protocol**. When routers are all configured to run the same routing protocol, they can start to exchange information about their route tables with the routers that are adjacent to them.

Figure A2.9 illustrates this concept in a simple fashion:

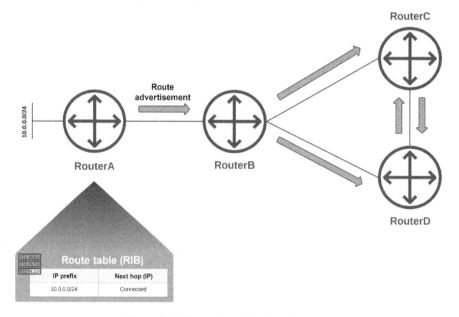

Figure A2.9: Dynamic route advertisement

All the routers in *Figure A2.9* are running a dynamic routing protocol, which allows them to exchange route information. RouterA knows about a locally connected network, `10.0.0.0/24`. RouterA generates a route advertisement using the protocol's specified format and sends that information to its **neighbor** or **peer**, RouterB. RouterB then processes the route information and does the same to advertise this route information to its neighbors, RouterC and RouterD. By the end of this process, all routers will know a specific network path that can be used to reach `10.0.0.0/24`.

Static Routing versus Dynamic Routing

Each of these routing mechanisms warrants some mental calculation regarding when you should use one over the other. Network engineering purists may say that dynamic routing is always the answer, but that can be purely subjective based on the situation at hand. As always, you need to consider the pros and cons of each solution. These are addressed next.

The advantages of static routing are as follows:

- **Simplicity**: The simplistic nature of static routing speaks for itself. If you desire a route to be present in the route table, you configure it to be there.

- **Predictability**: When routes are manually configured, it becomes simple to predict how all routing decisions will be made for those specified destinations.

- **Resource consumption**: Manual configuration of routes within the routing table does not require the router to utilize any other computational resources to perform routing. Once route tables are configured, routers often possess forwarding engines that can be used to offload the actual forwarding of packets. With static routing, there is less churn within that process.

- **Inherent network "security"**: The use of static routing can ensure in some cases that a neighboring router that has been compromised cannot inject malicious or faulty routes into that route table.

> **Note**
>
> The word *security* was placed inside quotation marks because this should not be relied on as a strong form of network security. Security is always relative. In the preceding situation, the router itself may be protected from a dynamic protocol injecting routes into its table. If the router itself is not properly secured from a management perspective, a bad actor could easily access the device and maliciously configure said static routes instead. This is why "security" should be considered a very caveated benefit to static routing.

The disadvantages of static routing are as follows:

- **Scalability**: Requiring manual intervention every time a route needs to be injected into a route table may work fine for a single router, but network engineers are often managing a fleet of routers across a wide-scale network.

- **Operational overhead**: Maintaining a large amount of route tables can become quite a tall task for operators. Not only for the sake of configuring routes but also for designing how an entire network should function from a routing perspective.

- **Lack of adaptability**: Static routes do not have the capabilities to react to a change in the environment. That is, if a link fails, your static route remains, potentially pointing that network traffic to a "black hole."

The advantages of dynamic routing are as follows:

- **Scalability**: When all the routers in a network are running a routing protocol, this enables the propagation of those routes to extend to the far reaches of the network without manual involvement. This can also allow the exchange of routing information with third-party networks in a dynamic nature.

- **Adaptability and fault tolerance**: Routing protocols are capable of making routing decisions based on the current state of the existing network. This can involve evaluating metrics of a route learned from multiple paths or even reacting to the failure of a network link that is currently in use by switching the routing over to a **failover** path.

- **Reduced operational overhead**: Routers running dynamic routing protocols can maintain the exchange of routing information alone, without an operator's involvement. The operator can simply configure the routing protocol process itself, networks to advertise, metrics tied to those networks, and so on, and that network information can be propagated automatically.

The disadvantages of dynamic routing are as follows:

- **Complexity**: Implementation of a dynamic routing protocol requires the network operators to possess a strong understanding of the protocol at hand. This can include how adjacencies are formed, how routing metrics are calculated, how to troubleshoot said protocol, and much more.

- **Security concerns**: The introduction of a dynamic routing protocol requires a strong layer of security over that protocol to prevent it from potentially being hijacked. Additionally, security needs to be instituted about what routing information can be exchanged within certain scenarios. That is, an ISP would not want one of their regional customers to start advertising them IP space that belongs to Google or Amazon and affect how routing is done to those companies.

- **Resource consumption**: Routing protocols require the system to consume additional resources to run the process and exchange routing information with other systems. Additionally, if those compute resources were to be constrained and/or crash, the routing process would also be affected.

The choice to use static or dynamic routing purely depends on the specific requirements of the implementation. Static is often used for smaller, unchanging use cases and sometimes "one-off" scenarios. Dynamic routing is better suited for large-scale networks that require scalability and failover capabilities. In enterprise networks, there is often a combination of both deployed simultaneously.

Dynamic Routing Protocols

As mentioned in the previous section, dynamic routing protocols allow for the exchange of routing information between routers on a network. All major routing protocols enable not only the exchange of routes but also have built-in components, logic, and algorithms that allow for calculating optimal routes. Here are some of the primary routing protocols used in enterprise networks today:

- **Border Gateway Protocol (BGP)**
- **Open Shortest Path First (OSPF)** protocol
- **Enhanced Interior Gateway Routing Protocol (EIGRP)**

You can picture a router logically separated into these two planes, each of which has its own roles and responsibilities for how the router can interact with the rest of the network. These are known as the **control plane** and the **data plane**, and their logical separation must be fully understood before these routing protocols may be explored further. The control plane is often considered the "brain" of the network appliance because it owns the decision-making process for how traffic will be routed. The control plane spans all processes related to routing protocols and how they operate. This involves not only route table management but also running the calculations that determine which routes are the best based on the current network topology. In addition, all the communication required for the routing protocols to exchange routing information falls under the control plane.

The key functions of the control plane include network topology discovery and management via routing protocols, route table management, best path route calculations per routing protocol, and processing any packets destined for the router's own IP addresses/CPU (this could be ICMP, DNS, or even HTTP, depending on what services the router is configured to run).

The **data plane**, often also referred to as the forwarding plane, is essentially the enforcement of the decisions made by the control plane. This encompasses the processing of data packets and forwarding them to a destination, performing an inspection on the packet, or potentially manipulating the packet in some form (i.e., network address translation, which will be covered later in this exam guide).

Routers typically have several physical interfaces that allow physical cabling (often copper or fiber optic) that allows for the ingress and egress of packets to/from the device. The simplest way to envision the data plane is the movement of packets from one of these interfaces to another.

The key functions of the data plane include processing and forwarding of ingress data IP packets, packet manipulation, inspection, filtering, and enforcement of QoS by managing queues that prioritize certain classes of traffic.

Figure A2.10 visualizes how the control plane and data plane work together:

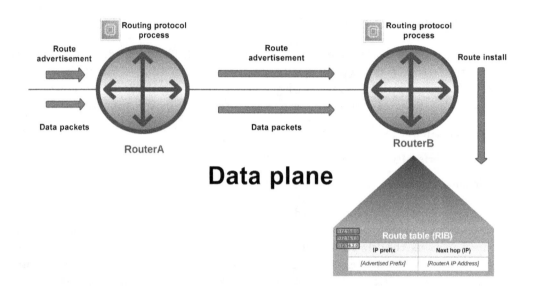

Figure A2.10: Control plane and data plane

In *Figure A2.10*, while the control and data planes remain logically separated within the router's internal processes, they are all still based on IP packets that use the same physical media as the data plane. This means that when the control plane protocols need to communicate, they encapsulate that communication data within IP packets and put them across the same physical cabling that data plane traffic is traversing. This is why the prioritization of specific traffic can become critical because, if data plane traffic is interfering with control plane communications, then you could have unfortunate consequences and/or outages for the data plane traffic. In network engineer speak, if you're performing any action that "impacts data plane traffic," then that means you're performing some kind of change/ update that will have an impact on application communications. This is why you will often see maintenance windows scheduled for after-hours to avoid this during peak business hours.

Dynamic Routing Protocol Functions

There are a handful of select routing protocols that have widespread adoption across the networks of the world today. They come in only a few different flavors as to how they function, but at a basic level, all routing protocols possess a certain subset of attributes that they all accomplish:

- Neighbor adjacencies (peering with other routers)
- Advertisement of routes

- Path selection

- Route table maintenance

- Loop prevention

- Dynamic updates (based on network topology changes)

- Load balancing

Each one of these functions provides critical input on how the routing protocol works. A more descriptive analysis of each will be provided in the following sections.

Neighbor Adjacencies

A neighbor adjacency is when two routers initiate and complete some initial communications with each other and agree to form a peering for them to further exchange routing information. This involves the two routers advertising details about the routing protocol process they are running to potential neighbors. If the router also receives these details from an adjacent router and the details meet the specific requirements of the protocol, then an adjacency will be formed.

Depending on the protocol, neighbor adjacencies may have the option to be discovered automatically or explicitly defined within the router's configuration. When using automatic discovery, the process is as follows:

1. The network operator applies a configuration to each router to start the routing protocol process. The configuration also includes what interfaces the routing protocol should start running the discovery process on (Eth0, aka Ethernet0, in the following example).

2. The router will begin sending discovery messages on this interface that include initial details about the routing process that it is running. This will often be done via multicast, which is a form of IP communication that allows devices to "listen" on a well-known IP address and all receive the same messages.

3. The candidate neighbor router will also advertise details about the routing protocol it is running. Note that this could be a specific message that is a "reply" to the discovery message from the first router or just an open discovery message like the one the first router sent, depending on the protocol.

4. Now that each router has seen there's another device on the same network running the same protocol, they will initiate forming an adjacency with that router. Depending on the protocol, this could be a single or multi-step process.

Figure A2.11 illustrates the process of automatic neighbor discovery:

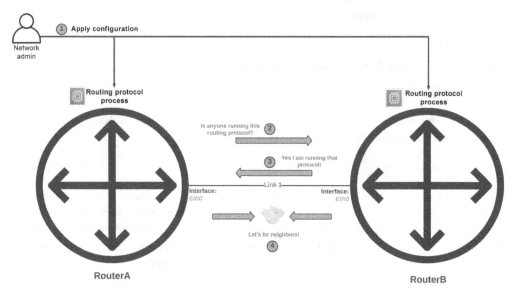

Figure A2.11: Automatic neighbor discovery

Automatic neighbor discovery is about being able to create a neighbor adjacency without explicit configuration and exchanging routes automatically.

Alternatively, with some routing protocols, they do not need to use a discovery process to find their neighbors. They allow network operators to explicitly define within the router's configuration what routers they should be forming a peering with, which is represented by the neighbor router's IP address. This can be viewed as a more secure measure in some scenarios to ensure that the router only forms adjacencies with neighboring routers that are approved by the network administrators.

This process looks very similar to the dynamic process, with some slight tweaks:

1. The network operator applies a configuration to each router to start the routing protocol process. The configuration also includes the target IP addresses of neighboring routers that the process should form adjacencies with.

2. The router will begin sending neighbor advertisement messages to the configured neighbor address, which includes the details about the routing protocol process. Since the traffic is for an explicit neighbor address, the traffic can be sent via unicast (i.e., directly to that IP address).

3. The candidate neighbor, which has also been configured to form an adjacency with the first router's IP address, responds to that message with details about the routing protocol process it is running.

4. The routers then begin the process of forming the adjacency, following the routing protocol-specific procedure.

Figure A2.12 explains the process visually:

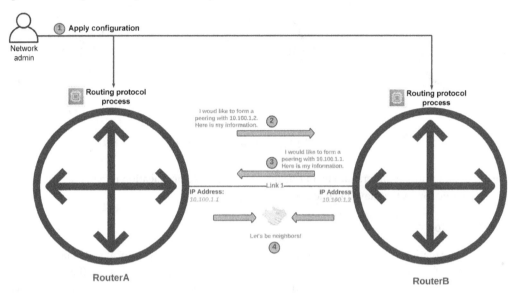

Figure A2.12: Explicit neighbor adjacency

Explicit configuration is usually needed when the connection between two routers does not support automatic discovery.

Advertisement of Routes

Route advertisement is the purpose of routing protocols, but the terminology for IP prefixes may differ between the protocols. For example, BGP uses the term **Network Layer Reachability Information** (**NLRI**), OSPF uses the term "link states," and EIGRP uses the terms "routes" and "topology entries."

However, regardless of the terminology, the target outcome remains the same: to advertise known network information to neighbor routers. This could include information about IP prefixes that are directly connected to that router or routes learned from other neighbors.

Figure A2.13 visualizes route propagation between routers:

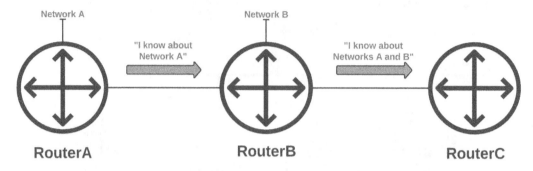

Figure A2.13: Route advertisements

The important thing to understand from this figure is that route propagation is generally inclusive of other dynamic routes that have been learned.

Path Selection

As route information is propagated throughout a system of routers dynamically, it is possible for a router to learn the same prefix from two different sources. Routing protocols all utilize their own unique algorithms and logic to calculate which is the most optimal path. *Figure A2.14* shows a network with routers connected together and multiple paths:

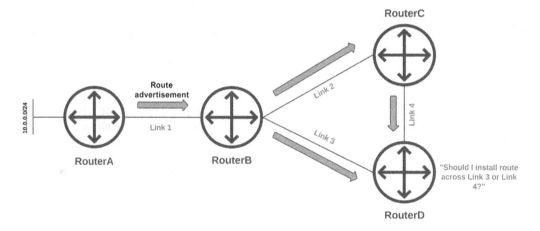

Figure A2.14: Path selection

From the perspective of RouterD in *Figure A2.14*, the prefix `10.0.0.0/24` has been learned via two paths:

- **Path 1**: Advertisement from RouterB (across Link 3)
- **Path 2**: Advertisement from RouterC (across Link 4)

When comparing the two paths to select the best one, RouterD must consider the distance from the IP prefix as follows:

- The path via RouterB contains two "hops" (RouterB -> RouterA -> Destination)
- The path via RouterC contains three "hops" (RouterC -> RouterB -> RouterA -> Destination)

If put simply in those terms, Path 1 may look preferable due to a lower hop count. However, if the bandwidth of each of these links is also considered, the best path might be different. Consider the following link speeds:

- **Link 1**: 10 Gbps
- **Link 2**: 10 Gbps
- **Link 3**: 200 Mbps
- **Link 4**: 10 Gbps

Now, things start to look a little different. While Path 1 (from RouterB, over Link 3) might be shorter, it also has a bottleneck of 200 Megabits per second, whereas Path 2 (from RouterC, over Link 4) has an end-to-end 10 Gigabits per second path.

Routing protocols each possess their own algorithm for calculating the best path between Path 1 and Path 2. Some routing protocols may only look at the hop count associated with each prefix, whereas others might also consider the link speeds across the path. This requires the protocol to not only advertise the IP prefix itself but also that that route advertisement contains additional data about the prefix, such as a simple **metric** or even a list of **attributes**.

Route Table Maintenance

Advertising routes and path selection are critical functions for routing protocols but the installation of the routes into the route table is what completes the process. Once the path selection calculation has run and the best route has been chosen, it is now a candidate for installation into the route table.

In *Figure A2.15*, RouterB is running both OSPF with RouterA and BGP with RouterC. RouterB has learned the prefix 10.0.0.0/8 from both protocols:

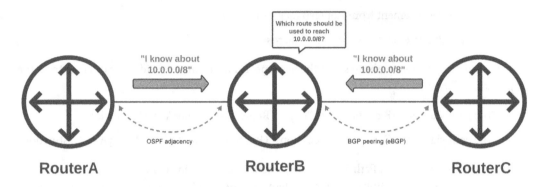

Figure A2.15: Route advertisement from two protocols

Administrative distance (**AD**) is essentially a scale of priority and/or trustworthiness of routes that can be installed into a routing table. As shown in *Figure A2.15*, this trust rating is a numerical value between 0 and 255, with certain values associated with each routing protocol. If multiple routes exist for the same prefix from different protocols (including static or connected routes), then the protocol with the lower AD value is chosen to be inserted into the routing table.

The AD values for each routing protocol do slightly differ depending on the vendor. The following is an example of the AD values for a Cisco router:

Protocol	AD
Directly Connected Network	0
Static Route	1
EIGRP Summary Route	5
External BGP (eBGP)	20
Internal EIGRP	90
OSPF	110
IS-IS	115
RIP	120
External EIGRP	170
Internal BGP (iBGP)	200

Table A2.1: Route protocol AD values

As you can see, sometimes, even the type of route from each protocol may have a different AD. For instance, an eBGP route has an AD of 20, while iBGP is 200. You will discover more about how BGP works in the next section of this chapter.

Now, refer again to *Figure A2.15*. Since the route for 10.0.0.0/8 was learned from both OSPF and BGP (eBGP, specifically), then the eBGP route would win that tiebreaker because the AD of 20 for eBGP is lower than OSPF's at 110, and the BGP route would be installed into the route table (RIB).

Loop Prevention

There is more to path selection than determining which route is most efficient and/or optimal. The calculation must also confirm that the path is free of **routing loops**, as these can be crippling events for a network. A routing loop is essentially when a set of routers has an incorrect next hop address for a route and ends up sending packets for that destination back to a router that has already processed the packet, resulting in a loop. See the following figure for a visual example of a routing loop:

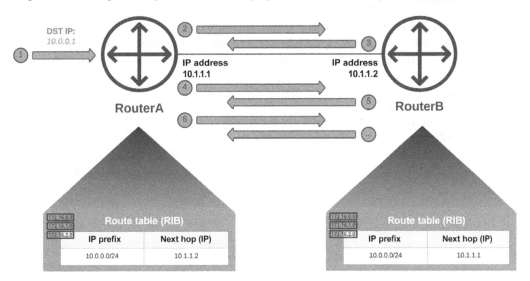

Figure A2.16: Routing loop

In the preceding example, both **RouterA** and **RouterB** have installed routes for the same prefix (10.0.0.0/24) that point to each other over the link connecting them. First, (1) a data packet arrives at **RouterA** with a destination address of 10.0.0.1, and based on **RouterA's** RIB, (2) the packet is forwarded to **RouterB**. Since **RouterB's** RIB has **RouterA** as the next hop for that prefix, (3) **RouterB** sends the packet right back to **RouterA**. This continues in a loop pattern (4 and beyond).

If you recall in the *IP Headers* section, there is a **time to live** (TTL) field within each header that starts at **255** when the original IP header is crafted. As a router processes a packet for IP routing, the value in this field is decreased by 1. When the TTL reaches **0**, the packet is dropped. This stops looped traffic from going on "forever" as each packet is only permitted to traverse 255 "hops." However, if you have a routing loop that occurs between routers that are processing 10 or even 100 Gigabits of data per second, you can see how this could quickly cripple a network with packets that will be dropped and connectivity for these destination prefixes will be broken altogether.

Routing protocols must be equipped with certain logic in their calculation algorithms to ensure that the paths that are installed are free of routing loops. Each major routing protocol does this in a specific fashion. For example, EIGRP does this by looking at metric data that is tied to each IP prefix and the adjacency, while OSPF relies on each router having raw data about the entire topology to ensure all routers make the same decisions.

Dynamic Updates

After establishing the best path and installing a route into the routing table, there must be timely updates to the dynamic route exchange, especially when changes occur. The protocol needs a mechanism to perform new calculations when the state of the network has changed. Not only does this include recalculating the best path but also notifying potential neighbors that you may no longer be their best route to reach that destination.

In *Figure A2.17*, RouterA has two possible paths available to reach 10.0.0.0/24 via Link 1 and Link 2. The routing protocol was able to calculate that the top path over Link 1 via RouterB is the most optimal. That route is then installed into the RIB of RouterA. RouterB has also installed a route to this destination via Link 3 to RouterD. This is the steady state of the network, and all is well.

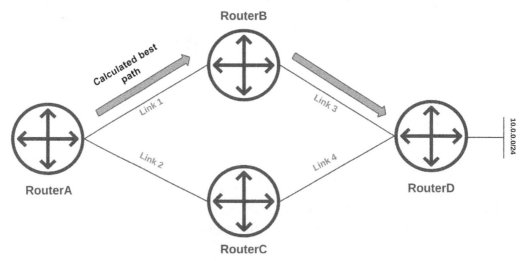

Figure A2.17: Dynamic updates – original path

Figure A2.18 illustrates a link failure on Link 3. This could simply mean that the cable was damaged or an interim device lost power. Nonetheless, the link is now down. RouterB's route to `10.0.0.0/24` is now invalid because that next hop is no longer reachable. If RouterA were to forward traffic to RouterB that is destined for `10.0.0.0/24`, then it would potentially dropped or black-holed.

Each routing protocol has a mechanism for RouterB to essentially notify RouterA that the routing information that was previously advertised for `10.0.0.0/24` is now invalid. This can be advertising the prefix again with an "infinite" metric or even a specific route withdrawal message.

When RouterB sends this route withdrawal to RouterA, the original route will be deleted and the process of calculating the best path will start again. This could include RouterA asking RouterC for an update as well. Consider *Figure A2.18* to see a link failure on Link 3:

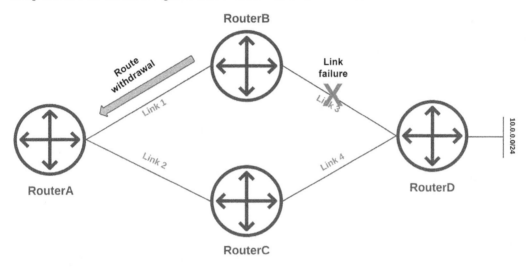

Figure A2.18: Dynamic updates – link failure

As shown in the preceding figure, the link failure results in immediate triggered updates to neighbors about the dropped route.

In *Figure A2.19*, once RouterA has updated messaging about the destination, it will rerun the best path calculation for 10.0.0.0/24 and install the path over Link 2 via RouterC:

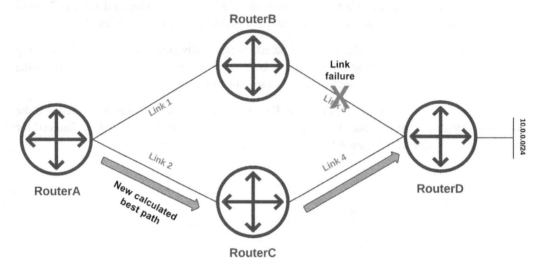

Figure A2.19: Dynamic updates – New path

This example looks at an individual link failure, but a route update and path recalculation could also occur if a better advertisement were to come from another path or if the prefix attributes change.

Load Balancing

To this point, all examples you've seen have been about routers calculating a single best path by evaluating the metrics from all the available options and selecting one. By default, routers typically only install one path. Routers can potentially use multiple paths, assuming the routing protocol supports choosing multiple paths at once. This is possible via something called **load balancing**.

Routing protocols typically provide configuration options that allow for load balancing. Load balancing allows for multiple entries for the same prefix to be installed into the router's RIB, which means traffic forwarded from the router will be balanced over multiple paths. This optimizes traffic distribution. The router itself is also configured with a specific algorithm that is used to properly distribute traffic across these multiple route entries.

In *Figure A2.20*, RouterA has received a route for `10.0.0.0/24` from both RouterB and RouterC. RouterB is advertising with a metric of "A" and RouterC using a metric of "B." Each protocol uses specific metrics/attributes, but for the sake of this example, envision that they are the exact same value.

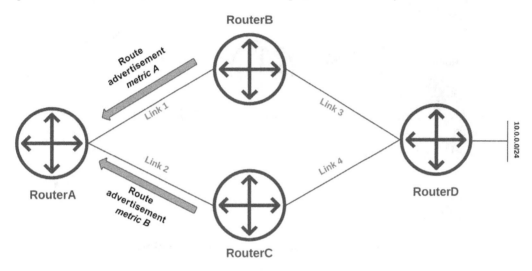

Figure A2.20: Load balancing – advertisements

Using a Cisco router as an example, the RIB will often have labels on the extra routes to indicate they are using **equal-cost multi-path** (**ECMP**) routing. Within BGP, for example, this is enabled using something called multipath, hence the "m" that you see on the second installed route.

In *Figure A2.21*, with the routing protocol configured to select multiple best paths, both advertisements can be installed into the RIB. RouterA will use both paths to distribute traffic destined for 10.0.0.0/24.

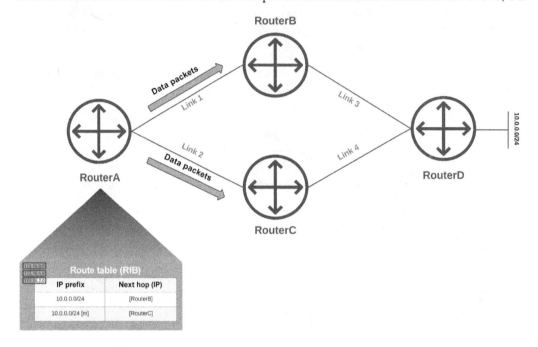

Figure A2.21: Load balancing – distribution of packets

Using multiple data paths improves resiliency and shares the load of traffic, effectively increasing the available bandwidth for sending data.

Basic BGP

BGP is widely considered "the protocol of the internet." This is because BGP is responsible for the advertisement of over one million prefixes, including IPv4 and IPv6, on the public internet. BGP is not only a very extensible protocol but users also maintain a large amount of control over how prefixes are shared, which caters well to interconnecting third-party networks to each other. After all, the internet is just the largest collection of third-party networks communicating with each other.

BGP is not just limited to the internet backbone; it is also heavily used in enterprise networks. BGP may be used to connect to an internet provider, build enterprise backbones, or connect to third-party networks for **business-to-business (B2B)** connectivity.

> **Note**
>
> In the world of cloud networking, the only routing protocol in use for consumers of the cloud is BGP. Enterprise networks can run several routing protocols, but the only option for dynamic routing within the cloud as of today is based on BGP. This is why this chapter has a strong emphasis on BGP.

In the following sections, you'll dive into the basics of BGP and how the protocol functions. Keep in mind the fundamental functions of a routing protocol that were mentioned in the previous section, as BGP implements that functionality as well. BGP is a very complex protocol and becoming an expert on BGP requires exploring additional textbooks, whitepapers, RFCs, and so on beyond the scope of this book. This chapter will focus on the basics that a network operator would need to describe the logic and theory behind BGP.

BGP Autonomous System

A common trait of routing protocols is the use of something called an autonomous system, and BGP is no exception. An **Autonomous System** (**AS**) is essentially a collection of routers all running the same routing protocol process and sharing IP prefixes together. Routers all running the same AS generally all fall under the same jurisdiction or administrative entity from a management perspective as well. The internet, for example, is a large collection of ASs exchanging routing information with each other. You can reach destinations such as Amazon, Google, Facebook, and so on because your ISP either exchanges route information with those entities directly or with another carrier that does.

An AS is often defined by a numerical value. Initially, BGP **AS Numbers** (**ASNs**) were defined with a 16-bit value (2 bytes). This allowed for 65,536 unique BGP ASNs (with 0 and 65,536 being reserved). With the expansion and growth of the internet, BGP was expanded to support 32 bits (4 bytes), which allows for up to 4,294,967,295 unique ASNs.

It is less common to see 4-byte ASNs in use within enterprise networks, but it is possible. This is because enterprises make use of a reserved private AS range that typically addresses their needs.

Public versus Private ASNs

As with public vs private IP addresses, ASNs can be used privately within a controlled network or used on the public internet. The previous section on private, public, and reserved IPv4 addresses mentioned the use of **regional internet registries** (**RIRs**). These same RIRs used to allocate public IP addresses can also be used to allocate public BGP ASNs. **Public ASNs** are meant for use within networks that connect directly to the global internet. Public ASNs are responsible for directly injecting public IP prefixes into global routing tables. These ASNs are globally unique.

By contrast, **private ASNs** are meant for use in private networks that do not require exchanging routing information directly with the global internet. The private ASN range is as follows:

- **16-bit ASN**: 64512 through 65534

- **32-bit ASN**: 4200000000 through 4294967294

While private ASNs do not exchange information directly with the global internet, this does not mean they cannot have an ISP do this on their behalf.

For example, the ISP may have a larger public IP prefix that it allocates to its customers, such as `1.2.0.0/16`. The ISP can reserve a portion of that larger block for its customers using VLSM. In this scenario, let's say the ISP reserves `1.2.3.0/24` for the end customer. The customer could then peer BGP with ISP using its private ASN and advertise this network. The ISP advertises the larger `/16` out to the global internet without the need to advertise each customer's specific allocation separately. This is referred to as **route aggregation**. ASNs are also used to determine whether a BGP neighbor is using internal or external BGP.

iBGP versus eBGP

BGP functions in two different operational modes: **iBGP** and **eBGP**. Each mode has its own functional components and characteristics. iBGP is used when exchanging routing information via BGP between two systems within the same AS. This allows all the routers within an AS to have a single, unified view of all external routing information learned from outside the AS.

The major characteristics of iBGP are as follows:

- iBGP peers do not require being directly connected to their peered neighbors.

- iBGP requires a full mesh of connectivity. This means all iBGP routers need to be peered together. That is, if there are five routers within a BGP AS, they all need direct peering to the other four routers. This can be overcome with the use of something called a route reflector, but this is outside the scope of this book.

- Within an iBGP mesh, a router will not advertise routes learned from one iBGP neighbor to another iBGP neighbor. This aids with loop prevention within iBGP.

- iBGP does not modify the attribute for "AS Path" when sending prefixes to other iBGP neighbors. This attribute is covered in the next section titled *BGP Attributes*.

- Since iBGP does not require devices to be directly connected, an underlying routing protocol is typically running to allow for iBGP peers to communicate with each other. See the following figure to visualize this concept. RouterA learns the address of RouterC via the OSPF adjacency, and vice versa. Using that connectivity via RouterB, an iBGP peering is formed between the two routers.

- *Figure A2.22* shows how an underlying routing protocol is used to establish reachability for two iBGP neighbors:

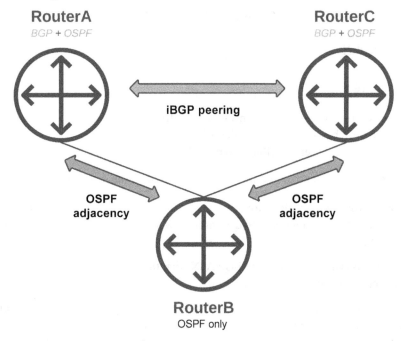

Figure A2.22: iBGP peering

As shown in the preceding figure, iBGP neighbors do not have to be directly connected, unlike most other routing protocols.

eBGP is used when exchanging routing information *between* BGP ASNs. This is the primary operational mode used when exchanging routing information across the global internet. eBGP is used any time two different ASNs are peered. This can be between two separate organizations exchanging public prefixes, or even between two private ASNs within your own organization exchanging private prefixes (if you were to choose BGP as your routing protocol).

The major characteristics of eBGP are as follows:

- eBGP peers are often directly connected. An example would be a direct peering with your ISP from your network's edge router. Since peers are directly connected, this can remove the need for another routing protocol underneath.

- eBGP can be used to advertise prefixes learned from within your own AS as well as learned via another AS (via other eBGP neighbors).

- eBGP does alter the `AS Path` attribute when advertising prefixes to neighbors. More in the next section titled *BGP Attributes*.

- eBGP "multihop" can be used to peer BGP between ASNs when the devices are not directly connected. This may/may not reintroduce the need for an underlying routing protocol in addition to BGP.

BGP Attributes

As stated before, routing protocols perform a series of calculations and evaluations to select the "best" route for a destination based on a set of criteria. This could be a simple metric value or something more complex. BGP uses several attributes to make this decision.

Every IP prefix advertised via BGP has the following attributes attached to it:

- **Next Hop**: The next hop IP address that traffic should be forwarded to in order to reach the destination prefix. Its function is to determine how forwarding should be implemented within the route table.

- **AS Path**: This is a full list of all the BGP ASNs that the IP prefix has traversed and is used for path selection and loop prevention.

- **Origin**: This refers to the origin of the IP prefix and could be **Interior Gateway Protocol (IGP)**, such as OSPF or EIGRP, **Exterior Gateway Protocol** (EGP), which is extremely rare, or **Incomplete** (origin unknown). It is used as a tiebreaker in the BGP path selection process.

- **Multi-Exit Discriminator** (MED): MED can influence incoming traffic from other ASs by signaling to any external neighbors in another AS which path into the advertising AS should be preferred. This is commonly used when there are multiple peering points between two separate ASs. This term is often referred to as "multi-homed." A lower MED value is preferred over a higher one.

- **Local Preference**: This is a numeric value that signals a preference for how traffic should exit an entire AS for the destination prefix, thereby influencing how outbound traffic flows from a particular AS. This value is typically replicated to all routers within the same AS. A higher local preference value is more preferred than a lower one.

- **Community**: This is a "label" that can be attached to a specific prefix advertised by BGP. Communities only serve to "tag" or identify routes in some way. Routers must be configured to enforce behavior based on these community values. This could be setting higher local preference values on routes with a certain community value, or even filtering them out altogether.

There are several other BGP attributes that are also associated with each BGP prefix that are not relevant to the exam, and so will not be covered here.

> **Note**
>
> If you wish to learn more about BGP and its attributes, extensibility, and features, then consider looking into the following RFCs:
>
> `https://www.ietf.org/rfc/rfc4271`
>
> `https://www.ietf.org/rfc/rfc4456`
>
> `https://www.ietf.org/rfc/rfc4760`

Path Selection

The process BGP follows when evaluating the attributes tied to one or more copies of a BGP prefix to determine which is the best is referred to as **path selection**. If a router were to receive multiple advertisements for the same prefix, path selection is how the router would choose which one is the best to be a candidate for installation into the RIB. This process is well-defined to remain predictable and consistent, through a series of steps laid out in the following order.

For each step, the process of evaluating a prefix to determine the best path will compare the attributes in this order; in the case of a tie, the next attribute in the list will be evaluated until the best path is chosen.

As an example, imagine there are two possible BGP paths for the prefix 10.10.10.0/24, and each has no configured weight, local preference, or the same AS Path, but one was learned over an eBGP neighbor while the other was learned from an iBGP neighbor. Since all path attributes match until reaching *Step 7*, best path evaluation must continue until that step, at which time, the tie is broken and the path learned from the eBGP neighbor is preferred.

1. **Highest Weight:** Weight is a Cisco proprietary piece of path selection and only locally significant (i.e., this value for "weight" is not advertised to other neighbors). While this is a vendor-proprietary method of path selection, it is worth mentioning due to the wide adoption of Cisco appliances within IP-routed networks.

2. **Highest Local Preference**: Local Preference (**LOCAL_PREF**) is an attribute advertised throughout an AS and is used to identify the preferred path to exit an AS to reach a particular prefix. LOCAL_PREF is a 32-bit value and the default value for most vendors is 100.

3. **Self-Originated**: Any prefix originated by the local router performing the path selection process will be preferred over an external one. The local preference and weight values must be the same.

4. **Shortest AS Path Length**: The BGP attribute for AS Path contains a sequential list of all the BGP ASNs that the prefix has passed through prior to being received. A prefix with a shorter list of ASNs will be preferred. This is often the default method of path selection if no other attributes are modified.

5. **Lowest Origin Type:** The term "lowest" is a little misleading here, but the three origin types are organized from lowest to highest: **IGP -> EGP -> Incomplete**. By this logic, **IGP** is the most preferred and **Incomplete** is the least preferred.

6. **Lowest MED:** MED is a 32-bit value, so the numeric value can be quite large. A lower MED value is preferred over a higher one.

7. **eBGP over iBGP:** A route learned via an eBGP neighbor will be preferred over an iBGP neighbor. The logic is that a route learned from outside your local AS would be a more direct path to the destination than something internal to your AS.

8. **IGP Metric:** When routes are redistributed from another internal gateway protocol (including static routes), the metric tied to that route is inherited and included in the route advertisement within BGP. With that, a lower IGP metric value will be preferred in BGP path selection.

9. **Lowest Router ID:** Within the BGP routing process (as well as other routing protocols) on any router, there is a requirement for a router ID to be configured for that process. The router ID is an identifier that is used in many operations tied to BGP, especially when advertising routes. The router ID is a 32-bit value and written in dotted decimal format, similar to an IP address. BGP path selection may use the "lowest" router ID as a tiebreaker.

10. **Lowest IP Address:** If all else fails and all the previous attributes are the same between multiple prefixes, the lower IP address of the advertising routers will be selected as the best path.

Loop Prevention

As noted in the previous *Dynamic Routing* section, all routing protocols need a mechanism to do loop prevention. BGP loop prevention slightly differs between iBGP and eBGP, but for the sake of cloud networking within AWS, you will focus primarily on eBGP. Loop prevention within eBGP is purely based on the AS Path attribute.

The AS Path attribute attached to a network prefix is shown as a list of BGP ASNs in the details of the prefix itself in a routing table. As BGP routers advertise network prefixes to other eBGP neighbors, they append their own BGP ASN to the list. This means each network prefix should have a comprehensive list of all the ASNs it has passed through along the path. In the following figure, you can see RouterA and RouterB adding their own respective BGP ASN to the AS Path as the network is propagated to the next neighbor.

The logic for preventing loops within eBGP is as follows: an eBGP router should not accept any prefixes that contain its own ASN within the AS Path attribute. If a prefix already originated or passed through the local router's ASN, then routing to an external neighbor could end up causing a loop. So, by that logic, the router will drop or not accept any inbound route advertisements that fit these criteria.

Refer to *Figure A2.23*. A prefix was advertised through RouterB via BGP and was sent along to RouterC. RouterC then sent the prefix to RouterD, which, in turn, sent the prefix to its neighbor, RouterB.

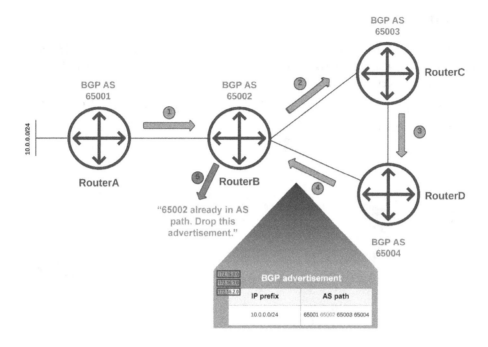

Figure A2.23: BGP loop prevention

Because RouterB saw its own ASN in the path, it dropped the prefix as a loop.

Keep in mind that during BGP path selection, AS Path Length is often the default method of determining the best path. This is assuming that Local Preference values are using the default value. In the preceding figure, RouterD would likely have received the same prefix (10.0.0.0/24) from both RouterB and RouterC. The advertisement from RouterB would have an AS Path length of 2, whereas from RouterC, it would have 3. Hence, RouterD would choose the path through RouterB.

Assume you are a network engineer putting together your routing design for an enterprise network and have chosen BGP to connect your locations together. Because of the method by which eBGP accomplishes loop prevention, this requires proper planning to ensure you do not have ASN overlaps as this can cause issues with route propagation, such as ensuring that you do not reuse BGP ASN across the network.

Summary

This chapter reviewed IP fundamentals, including the way routing decisions are made, what routing protocols are, and how they work statically and dynamically. It also covered BGP, which is the routing protocol you can expect to use within the AWS cloud.

In this certification guide, you'll see that these fundamental concepts are the bedrock upon which AWS networking is built.

Appendix 3
VPC Networking Basics

Networking in the public cloud is quite different from traditional, on-premises computer networking. While many of the familiar networking concepts and protocols still exist, the way you interact with them differs greatly. In addition, the exposure of the configurable elements is purely defined by the **cloud service provider** (**CSP**). This chapter is going to dive into the basics of the biggest fundamental piece of AWS networking: Amazon **Virtual Private Cloud** (**VPC**).

An Amazon VPC is composed of several networking elements that focus on not only connectivity but security as well. The *AWS Certified Advanced Networking - Specialty (ANS-C01)* exam is geared toward individuals who will be specialists within AWS networking, and the VPC is the foundation. Many AWS services rely purely on a VPC to provide connectivity not only within the AWS cloud but in hybrid environments as well.

In this chapter, you will cover the following topics:

- VPC basics

- IP addressing in VPCs

- Subnets in VPCs

- Route tables

By the end of this chapter, you should understand the theory behind Amazon VPC and the primary constructs that are used to build functionality and connectivity for the VPC.

VPC Basics

An Amazon VPC is the method by which networks are built within AWS. An Amazon VPC is comparable to an isolated network that you would deploy within your own data center. Within this virtual network, you can deploy other AWS resources, such as EC2 instances. By default, a VPC is isolated from any other VPC, unless you explicitly configure connectivity between them. Constructs such as **endpoints** and **gateways** can also be deployed within a VPC to provide connectivity to the outside world (internet) or other AWS services.

In this section, you will delve into the fundamental components and features that make up an Amazon VPC. You'll begin by exploring the core elements such as VPCs, VPC CIDRs, subnets, route tables, and gateways and endpoints. Understanding these components will establish a baseline for building and managing connectivity within the AWS cloud. You will then examine the differences between default and non-default VPCs to help you decide which is more suitable for your networking needs.

VPC Components

The first component is the overall VPC itself. An Amazon VPC is a **regional** construct, meaning its boundaries cannot extend beyond a single AWS Region. The VPC expands across all the **Availability Zones (AZs)** that exist within that region. Refer to *Figure A3.1* for a visual representation of a simple VPC. The VPC spans three AZs, but this number can be higher in AWS Regions that support more than three AZs.

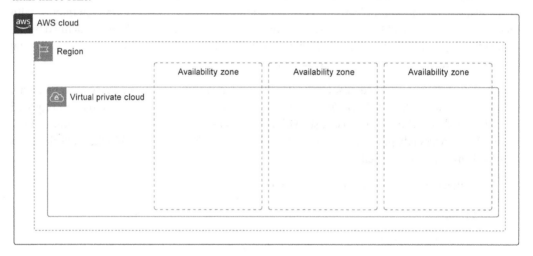

Figure A3.1: Amazon VPC

As mentioned, in the preceding figure, you can see a VPC that is created within an AWS Region and spans three AZs within that region. Assume this is a newly created VPC for Trailcats. This VPC can now be used to house components such as subnets, route tables, EC2 instances, and more.

> **Note**
> At the time of writing, most AWS Regions support three AZs. There are some regions that support more. For example, `us-east-1` supports up to six.

VPC CIDR

Additionally, VPCs also require a **Classless Inter-Domain Routing (CIDR)** to be defined. The CIDR range represents the overall range of IP addresses that can be used within the VPC. This is sometimes referred to as a **supernet** in common networking terminology. VPCs can have both IPv4- and IPv6-based CIDR blocks.

> **Note**
>
> At the time of this writing, VPCs must have at least one IPv4 CIDR block defined. AWS does not currently allow IPv6-only VPCs.

A VPC must have at least one CIDR block defined. A VPC can also have secondary CIDR blocks attached to increase the IP space usable within the VPC. The default service quota for the number of both IPv4 or IPv6 CIDR blocks associated with a VPC is 5, but this is increasable up to 50.

The VPC CIDR blocks are used or essentially *carved up* for the purpose of creating **subnets** within the VPC.

Subnets

Subnets are used to build specific networks within the range of the VPC. Within AWS, subnets are bound to a single AZ. VPC subnets can also be classified as either **public subnets** or **private subnets**. The definition of public versus private differs slightly from what you may already know as public and private IP addresses. This will be discussed further in the next section, *IP Addressing in VPCs*.

Expanding on the previous figure, *Figure A3.2* contains public and private subnets deployed into each of the AZs:

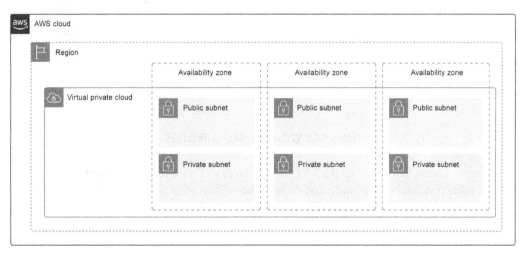

Figure A3.2: VPC subnets

The VPC represented in *Figure A3.2* now has adequately deployed private and public subnets that can be used to house components such as EC2 instances.

Route Tables

As with all IP-based systems, there needs to be a way to direct the network traffic to the appropriate destination. **Route tables** control the routing of network traffic within and out of a VPC. The route tables created within the VPC configure the **VPC router**. The VPC router is essentially a virtual router that is created by default within every VPC. You can think of the VPC router as a routing appliance with a physical interface attached to every single subnet that exists within the VPC. This router can also act as the **default gateway** within each subnet. More on this will be discussed in the next section, *IP Addressing in VPCs*.

A route table is defined within a VPC and cannot span multiple VPCs. Within the VPC, the route table can be associated with one or more subnets. Route tables consist of specific route entries and a corresponding next-hop address (or **target**) for each entry. The next hop for a route entry can vary from the local VPC, several AWS services (such as Internet Gateway or **Transit Gateway** (**TGW**)), or even the IP address of another EC2 instance.

Refer to *Figure A3.3* for a visual representation of the VPC router and its associated route tables:

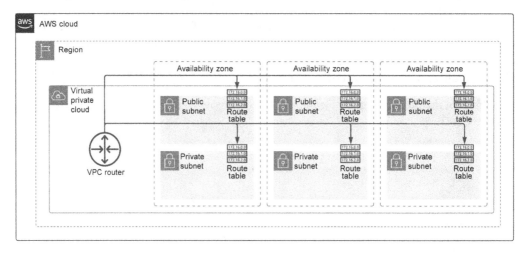

Figure A3.3: VPC router with route tables

For example, assume that *Figure A3.3* is a VPC that belongs to Trailcats. The VPC has been configured with six subnets in total (three public and three private). Then, there is a route table associated with each of those subnets, which corresponds with the VPC router.

Gateways and Endpoints

As mentioned in the previous *Route Tables* section, other constructs that can assist with connecting a VPC to other networks are VPC gateways and endpoints. Gateways typically have specific use cases for accessing or performing a function, such as connectivity to the internet or access to another private network over a VPN. More detail on the various gateway types will be covered in *Internet, NAT, and Egress-Only Gateways*.

VPC endpoints typically facilitate access to another AWS service, such as Amazon S3, DynamoDB, or services powered by AWS PrivateLink. VPC endpoints can be deployed as either gateway endpoints or interface endpoints. More details on interface endpoints will be covered in *Chapter 3, Networking Across Multiple AWS Accounts*.

Figure A3.4 shows different types of VPC gateways and endpoints deployed within a VPC. Traffic is typically directed to these gateways and endpoints via the use of the route tables associated with each subnet.

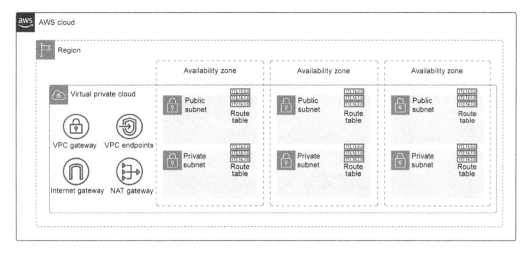

Figure A3.4: VPC gateways and endpoints

You can now see from this figure a more complete composition of a VPC for Trailcats. This VPC has the basic networking constructs configured, such as subnets and route tables, as well as other services such as endpoints and gateways. This allows Trailcats to build network connectivity from this VPC.

Default and Non-Default VPCs

When a new AWS account is created, a **default VPC** is deployed into each region that the account is opted into. The default VPC is meant to allow easy deployment of resources in each region. AWS users can deploy a new account and immediately start running workloads (such as EC2) in any region. Default VPCs are all deployed with a common set of components:

- VPC CIDR range: `172.31.0.0/16`

- A default subnet in each AZ (public)

- Internet gateway

- Main route table (associated with all subnets)

In *Figure A3.5*, you'll see an example of default VPCs deployed across two separate AWS Regions:

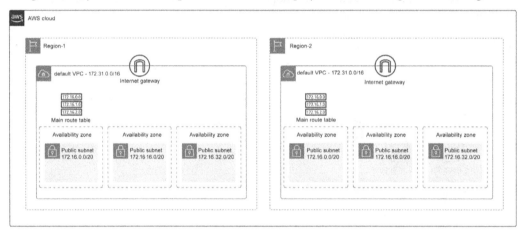

Figure A3.5: Default VPCs

Assume this scenario was within an AWS account for Trailcats. By default, these VPCs would exist in all AWS Regions that Trailcats opted into when the account was created. As mentioned before with default VPCs, the default VPC in each region uses the same IPv4 CIDR and has public subnets configured in all available AZs for that region. This could be a quick way for Trailcats to immediately deploy resources in a region, but there are many considerations about default VPCs they need to keep in mind. You will touch on those next.

There are some limitations with using default VPCs, one of which is that all default VPCs are deployed using the same CIDR range of 172.31.0.0/16. This CIDR range cannot be changed or removed, which can complicate connectivity due to overlapping IP ranges. Due to this limitation, it is generally recommended that anyone planning a network within AWS should use **non-default VPCs**. A non-default VPC is essentially any user-created VPC within an AWS account. Many organizations use only non-default VPCs in their environments because this is the recommended practice from AWS and allows for the most flexibility and interoperability. These VPCs follow a "build your own" approach where all the desired components, including CIDR ranges, subnets, and gateways, must be user-defined.

It is possible to delete the default VPC that is deployed into each region. It is also possible to create a default VPC within a region if the previous one has been deleted or an AWS Organizations policy is configured to not create one.

By now, you have gained an understanding of the key components and features of an Amazon VPC, including VPCs themselves, VPC CIDRs, subnets, route tables, and gateways and endpoints. You have also explored the differences between default and non-default VPCs, along with their respective advantages and limitations. This knowledge establishes a solid foundation for building and managing network connectivity within the AWS cloud.

Now that you have a solid grasp of VPC creation, it's time to dive deeper into the specifics of IP addressing within VPCs. In the next section, you will explore how to allocate and manage both IPv4 and IPv6 CIDR blocks in your VPCs. This knowledge is crucial for designing scalable and flexible network architectures within AWS. You will learn about the rules and best practices for assigning IP address ranges and how to add or modify CIDR blocks to meet your networking needs.

Exercise: How to Create an Amazon VPC

In this exercise, you will learn how to create an Amazon VPC using both the AWS Management Console and the AWS **Command Line Interface** (**CLI**). This hands-on activity will help you apply the theoretical concepts covered in the previous section. By the end of this exercise, you will be able to create a new VPC with a specified CIDR block, configure essential settings such as subnets and tags, and verify the creation of the VPC and its components.

Before beginning this exercise, ensure you have completed the following:

- Ensure you have access to an AWS account with sufficient permissions to create VPCs

- Familiarize yourself with the basic VPC concepts covered earlier

In the first part of this exercise, you will create a VPC using the AWS Management Console:

1. Sign in to the AWS Management Console.

2. Navigate to VPC dashboard by selecting VPC from the Services menu.

3. Click on the Create VPC button located in the top-right corner of VPC dashboard, as shown in *Figure A3.6*:

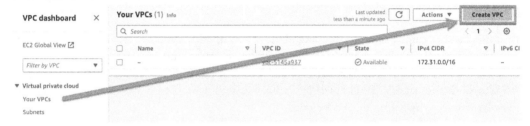

Figure A3.6: Create VPC

4. Configure the following VPC settings:

- Name tag: Enter a descriptive name for your VPC, such as MyFirstVPC.

- CIDR block: Enter the CIDR block for the VPC, for example, 10.0.0.0/16.

- Tenancy: Choose the tenancy option. Select Default unless you have specific requirements for a Dedicated tenancy.

A view of these options is shown in *Figure A3.7*:

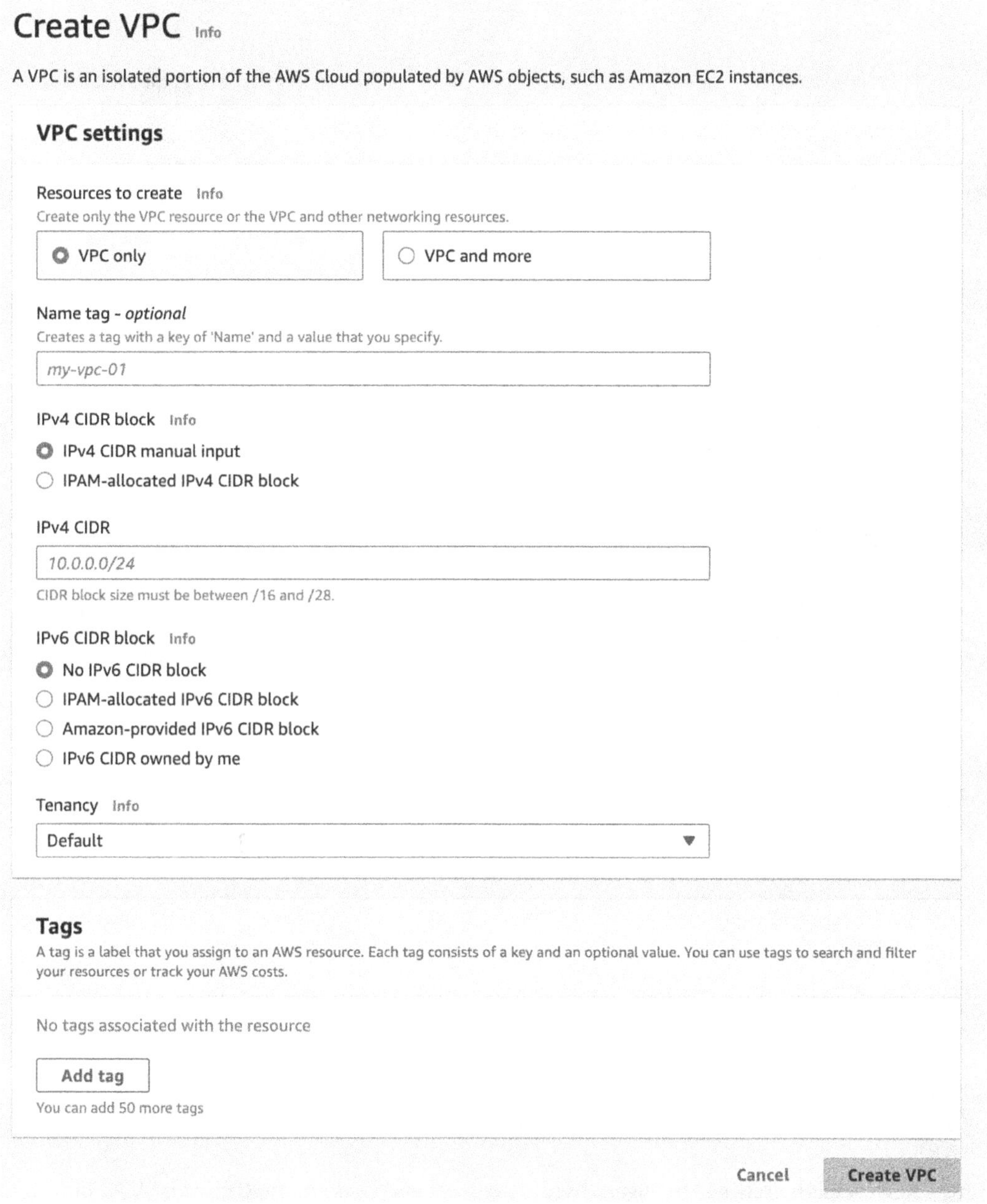

Create VPC Info

A VPC is an isolated portion of the AWS Cloud populated by AWS objects, such as Amazon EC2 instances.

VPC settings

Resources to create   Info
Create only the VPC resource or the VPC and other networking resources.

- ○ VPC only
- ○ VPC and more

Name tag - optional
Creates a tag with a key of 'Name' and a value that you specify.

| my-vpc-01 |

IPv4 CIDR block   Info

- ○ IPv4 CIDR manual input
- ○ IPAM-allocated IPv4 CIDR block

IPv4 CIDR

| 10.0.0.0/24 |

CIDR block size must be between /16 and /28.

IPv6 CIDR block   Info

- ○ No IPv6 CIDR block
- ○ IPAM-allocated IPv6 CIDR block
- ○ Amazon-provided IPv6 CIDR block
- ○ IPv6 CIDR owned by me

Tenancy   Info

| Default   ▼ |

Tags

A tag is a label that you assign to an AWS resource. Each tag consists of a key and an optional value. You can use tags to search and filter your resources or track your AWS costs.

No tags associated with the resource

| Add tag |

You can add 50 more tags

Cancel   **Create VPC**

Figure A3.7: Create VPC details

5. Click on VPC and more to configure additional settings (options):

    - **Subnets:** Add subnets by specifying the subnet name, CIDR block, and associated AZ

    - **NAT gateways:** Optionally, create a NAT gateway for outbound internet access for private subnets

    A detailed view of these options is shown in *Figure A3.8*:

Figure A3.8: Create VPC and more

6. Review all the settings to ensure they are correct and click on the Create VPC button to create your VPC.

7. Once the VPC has been created, it will appear in the VPC dashboard. Verify that the VPC has the correct CIDR block and any associated subnets, as shown in *Figure A3.9*:

Figure A3.9: VPC list view

> **Note**
>
> The AWS CLI is a commonly used text-based tool to deploy AWS resources instead of the **graphical user interface (GUI)** of the AWS console. Details about the AWS CLI are outside the scope of this certification guide, but further details on the AWS CLI can be found at the following link: `https://docs.aws.amazon.com/cli/latest/userguide/cli-chap-welcome.html`.

In the second part of the exercise, you will create a VPC using the AWS CLI:

1. Open `Terminal` or `Command Prompt` and ensure that the AWS CLI is installed and configured with appropriate credentials.

2. Enter the following command to create a new VPC with the specified CIDR block:

    ```
    aws ec2 create-vpc --cidr-block 10.0.0.0/16
    ```

    Note the `VpcId` value returned by the command, as you will need it for subsequent steps.

3. Use the following command to add a name tag to your VPC:

    ```
    aws ec2 create-tags --resources <VpcId> --tags \
    Key=Name,Value=MyFirstVPC
    ```

    Replace <VpcId> with the ID of your VPC from the previous step.

4. Run the following command to list your VPCs and verify that your VPC has been created successfully:

    ```
    aws ec2 describe-vpcs --vpc-ids <VpcId>
    ```

    Check the output to ensure that the VPC has the correct CIDR block and tags.

In this exercise, you have successfully created an Amazon VPC using both the AWS Management Console and the AWS CLI. You have also learned how to configure basic settings such as CIDR blocks and tags. This foundational knowledge will help you as you move forward to more advanced VPC configurations in the following sections.

Exercise: How to Create a Default VPC

In this exercise, you will learn how to create a default VPC using both the AWS Management Console and the AWS CLI. A default VPC comes preconfigured with subnets, route tables, and an internet gateway, making it easier to launch and manage resources. By the end of this exercise, you will be able to create a default VPC in a selected region using both the AWS Console and AWS CLI and understand the configuration and components of a default VPC.

Before beginning this exercise, ensure you have completed the following:

- Ensure your AWS account has the necessary permissions to create and manage VPCs.
- The selected AWS Region should not already have a default VPC. If it does, you will need to delete the existing default VPC before proceeding.

In the first part of the exercise, you will create a default VPC using the AWS Management Console:

1. Sign in to the AWS Management Console and navigate to the VPC dashboard by selecting VPC from the Services menu.
2. In the VPC dashboard, verify that the selected region does not already have a default VPC. You can see this under the Your VPCs section.
3. If a default VPC exists, delete it before proceeding.
4. Click on the Actions button at the top of the VPC dashboard and select Create default VPC from the drop-down menu.
5. A confirmation dialog will appear. Click on Create default VPC to confirm.

These options are shown in *Figure A3.10*:

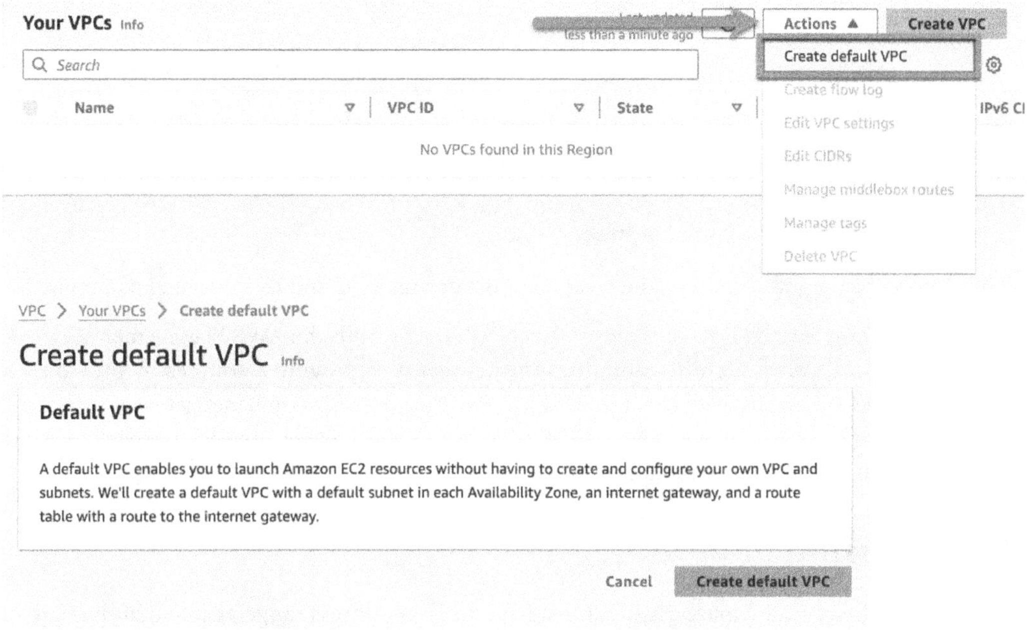

Your VPCs Info

less than a minute ago

Actions ▲   Create VPC

Create default VPC

Create flow log

Name	▽	VPC ID	▽	State	▽	IPv6 CI

Edit VPC settings

Edit CIDRs

No VPCs found in this Region

Manage middlebox routes

Manage tags

Delete VPC

VPC > Your VPCs > Create default VPC

Create default VPC   Info

Default VPC

A default VPC enables you to launch Amazon EC2 resources without having to create and configure your own VPC and subnets. We'll create a default VPC with a default subnet in each Availability Zone, an internet gateway, and a route table with a route to the internet gateway.

Cancel   **Create default VPC**

Figure A3.10: Create default VPC

Once the default VPC is created, you will see a success message. The new default VPC and its components, such as subnets and route tables, will be visible in the VPC dashboard.

In the second part of the exercise, you will create a default VPC using the AWS CLI:

1. Open your `Terminal` or `Command Prompt` and ensure that the AWS CLI is installed and configured with appropriate credentials.

2. Use the following command to list the VPCs in the selected region and verify whether a default VPC exists:

   ```
   aws ec2 describe-vpcs --filters
   "Name=isDefault,Values=true"
   ```

3. If a default VPC exists, delete it using the `delete-vpc` command and the corresponding VPC ID.

4. Enter the following command to create a default VPC:

```
aws ec2 create-default-vpc
```

The command will create a new default VPC along with the default subnets, route tables, and an internet gateway.

5. Run the following command to verify that the default VPC has been created successfully:

```
aws ec2 describe-vpcs --filters
"Name=isDefault,Values=true"
```

Check the output to confirm the creation of the default VPC and its associated components.

In this section, you learned how to create a default VPC using both the AWS Management Console and the AWS CLI. A default VPC simplifies the process of deploying resources by providing a preconfigured network environment that includes public subnets, an internet gateway, and route tables. Understanding the creation and configuration of a default VPC is essential for quick resource deployment and effective network management.

IP Addressing in VPCs

As mentioned before, all Amazon VPCs must have an IPv4 address range assigned at the time of creation. This IPv4 range is represented using **Classless Inter-Domain Routing (CIDR)** notation. You have the option to associate additional IPv4 CIDR blocks as well as IPv6 CIDR blocks with the VPC as well.

In the following sections, you will delve deeper into IP addressing within VPCs, exploring the specifics of IPv4 and IPv6 CIDR blocks. You will learn about the permissible address ranges, best practices for selecting IP ranges, and how to associate additional CIDR blocks with your VPC. This will include understanding the requirements, restrictions, and methods for expanding your VPC's IP address space. By grasping these concepts, you will be better equipped to design scalable and efficient network architectures on AWS.

IPv4 CIDR Blocks

An IPv4 CIDR block assigned to an Amazon VPC must be between /16 and /28. This means the number of IP addresses within the CIDR block can range from 65,536 down to 16 addresses. While additional capacity can be added by associating additional IPv4 CIDR blocks, it is important from a planning perspective to allocate an adequate CIDR range at the time of VPC creation.

The available IPv4 ranges that can be assigned to a VPC can technically be private or public. However, it is recommended in nearly every case that IPv4 VPC CIDR assignments fall within the range of RFC 1918 addressing. As a reminder, these ranges are listed in *Table A3.1*:

RFC 1918 Range	Available IP Range
10.0.0.0/8	10.0.0.0 – 10.255.255.255
172.16.0.0/12	172.16.0.0 – 172.31.255.255
192.168.0.0/16	192.168.0.0 – 192.168.255.255

Table A3.1: RFC 1918 address range

The IP ranges referenced in *Table A3.1* should simply be a review. If you are unfamiliar with the RFC 1918 range of IP addresses, please refer to *Appendix 1*, *Network Fundamentals*.

> **Note**
>
> While the address range is available for assignment to a VPC, it is generally encouraged to not use the 172.17.0.0/16 range for non-default VPCs. This is due to some existing AWS services that use this range and can cause overlapping IP conflicts.

By default, an Amazon VPC can support up to five assigned IPv4 CIDR blocks. This number is adjustable up to 50 via a service quota increase. When you successfully associate a new CIDR block with a VPC, this range becomes available for assignment to newly created subnets across all AZs within the VPC. However, there are several restrictions in place for additional IPv4 CIDR block associations. Some of the important restrictions are listed here:

- The permitted block size remains between a /16 and /28 subnet mask

- The newly associated CIDR block cannot overlap with any existing CIDR blocks associated with the VPC

- You cannot increase or decrease the size of an existing VPC CIDR block

- Additional CIDR blocks that have been associated with a VPC can also be disassociated and removed from the VPC if they are not in use

- The primary IPv4 CIDR block associated with a VPC cannot be removed

- When adding or removing a CIDR block from a VPC, you may see the status of the CIDR block go through the following states: `associating` | `associated` | `disassociating` | `disassociated` | `failing` | `failed`

> **Note**
>
> It is highly recommended to become familiar with the restrictions mentioned in the above list prior to sitting the *ANS-C01* exam. There are several caveats and unique restrictions that can trip you up on overly detailed exam questions.

In addition to the preceding restrictions, there are also several restrictions regarding what IP address ranges can be additionally associated with a VPC. These restrictions often relate to the existing IP address ranges already associated with the VPC.

> **Note**
>
> For a full list of all the rules applicable to assigning VPC CIDR blocks to a VPC, see the following link: `https://docs.aws.amazon.com/vpc/latest/userguide/vpc-cidr-blocks.html`.
>
> See the following link for a full list of restricted and permitted associations based on each address range: `https://docs.aws.amazon.com/vpc/latest/userguide/vpc-cidr-blocks.html#add-cidr-block-restrictions`.

IPv6 CIDR Blocks

All Amazon VPCs must have an IPv4 CIDR block attached, but IPv6 capability can be added to the same VPC by associating an IPv6 CIDR block. These IPv6 CIDR blocks must be between a /44 and /60 subnet mask. One small caveat is that while the subnet mask must be between /44 and /60, it also must be in increments of 4. For example, an associated IPv6 VPC CIDR block could be /44, /48, /52, /56, or /60. This allows AWS operators some flexibility in creating a hierarchical structure of IPv6 addressing.

When assigning an IPv6 CIDR block to your VPC, you are given three options:

- `IPAM-allocated IPv6 CIDR block`
- `Amazon-provided IPv6 CIDR block`
- `IPv6 CIDR owned by me`

An IPv6 CIDR block can be allocated via the Amazon VPC **IP Address Manager** (**IPAM**). This service is outside the scope of this exam, but you can read more about AWS IPAM at the following link: https://docs.aws.amazon.com/vpc/latest/ipam/what-it-is-ipam.html

The option using an AWS-provided IPv6 CIDR block allows AWS operators to assign IPv6 addresses to a VPC that is owned and managed by Amazon. The user will be required to specify a **network border group** when allocating these addresses. The network border group aligns with a specific group of AZs where Amazon will advertise these addresses to the public internet. Therefore, these border groups typically align with all the AZs within that particular AWS Region. A snapshot of this process via the AWS console is shown in *Figure A3.11*:

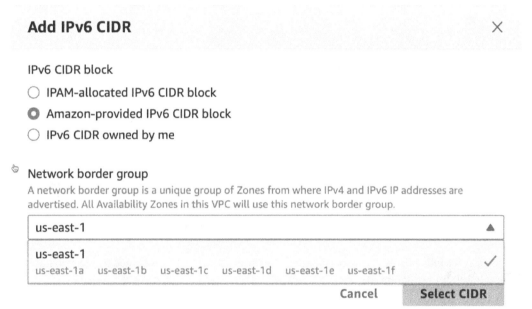

Figure A3.11: Amazon-provided IPv6 CIDR block association

As you can see in *Figure A3.11*, once selecting the option for Amazon-provided IPv6 CIDR block, you can select the appropriate network border group.

Lastly, AWS provides the ability for users to procure their own blocks of IPv6 addresses from the appropriate regulating body and advertise them through Amazon to the public internet. This is referred to as **bring your own IP** (**BYOIP**) in AWS. The BYOIP addresses can then be used within the configuration of Amazon VPCs.

BYOIP address ranges can be assigned to an Amazon VPC within ranges of /44 and /60 in increments of 4. The total available address range depends on the available BYOIP pools.

When selecting **IPv6 CIDR owned by me**, the operator will be presented with both an overall BYOIP pool to choose from, as well as a list of pool CIDRs. An example is shown in *Figure A3.12*:

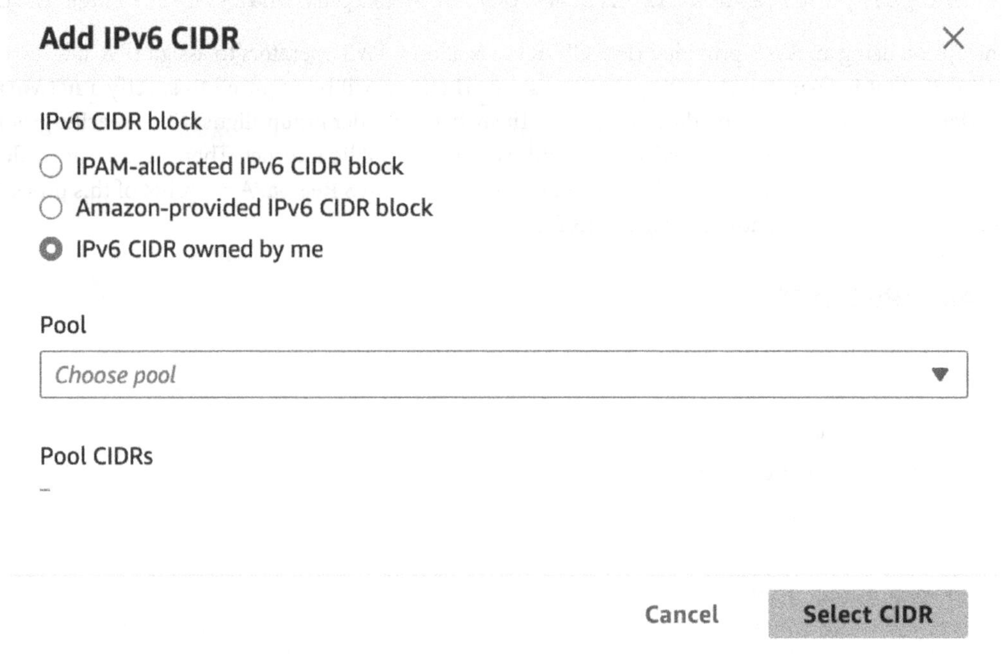

Add IPv6 CIDR   ✕

IPv6 CIDR block

○ IPAM-allocated IPv6 CIDR block
○ Amazon-provided IPv6 CIDR block
◉ IPv6 CIDR owned by me

Pool

Choose pool ▼

Pool CIDRs

—

Cancel   **Select CIDR**

Figure A3.12: BYOIP IPv6 CIDR block association

As you can see in *Figure A3.12*, once you select the option for IPv6 CIDR owned by me, you have the option to select the appropriate pool.

> **Note**
> The process of bringing your own IPv6 addresses to AWS requires several steps and verifications. The full process can be found at the following link: https://docs.aws.amazon.com/ AWSEC2/latest/UserGuide/ec2-byoip.html.

Exercise: Adding IPv4 or IPv6 CIDR Blocks to a VPC

In this exercise, you will learn how to add additional IPv4 and IPv6 CIDR blocks to an existing VPC using both the AWS Management Console and the AWS CLI. This will help you expand the IP address space within your VPC, allowing for more subnets and resources to be created within your network. By the end of this exercise, you will be able to add secondary IPv4 and IPv6 CIDR blocks to an existing VPC using the AWS console, use the AWS CLI to associate additional CIDR blocks with a VPC, and understand the constraints and limitations associated with adding CIDR blocks to a VPC.

The prerequisites are an existing VPC in your AWS account to which additional CIDR blocks will be added. You also need access to the AWS Management Console and/or AWS CLI with sufficient permissions to modify VPCs.

In the first part of this exercise, you will add IPv4 or IPv6 CIDR blocks using the AWS Management Console. The steps are as follows:

1. Sign in to the AWS Management Console and navigate to the VPC dashboard by selecting VPC from the `Services` menu.

2. From the list of VPCs, select the VPC to which you want to add a new CIDR block.

3. Click on the `Actions` button and select `Edit CIDRs` from the drop-down menu, as shown in *Figure A3.13*:

Figure A3.13: VPC Edit CIDRs

4. Click on `Add new IPv4 CIDR`.

5. Enter the desired IPv4 CIDR block (e.g., `10.2.0.0/16`) and click `Save`.

6. Verify that the new IPv4 CIDR block has been successfully associated with your VPC.

7. Click on `Add new IPv6 CIDR`.

8. Choose one of the following options for the IPv6 CIDR block:

    • `Amazon-provided IPv6 CIDR block`: Select the appropriate network border group

    • `IPAM-allocated IPv6 CIDR block`: Choose from IPAM pools if available

    • `IPv6 CIDR owned by me`: Select from your BYOIP pools

9.   Click Save to associate the IPv6 CIDR block with your VPC.

Snapshots of the AWS console options for *steps 3* and *4* can be seen in *Figure A3.14*:

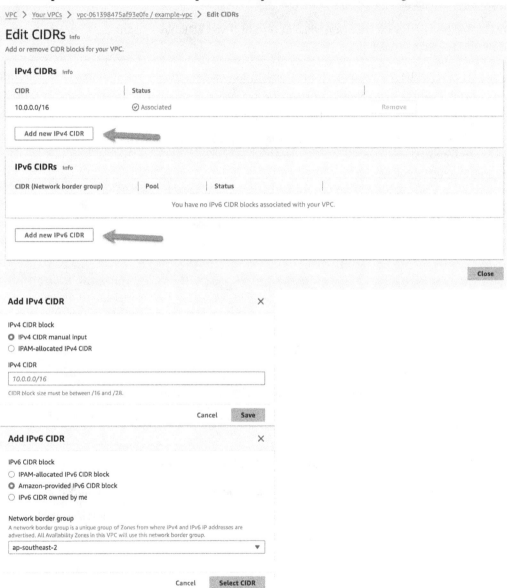

Figure A3.14: VPC Edit CIDRs details

10. Return to the VPC dashboard and confirm that the new IPv4 and/or IPv6 CIDR blocks are listed under the selected VPC.

In the second part of this exercise, you will add IPv4 or IPv6 CIDR blocks using the AWS CLI:

> **Note**
>
> In the following steps, replace `vpc-12345678` with your VPC ID and `10.2.0.0/16` with the desired CIDR block.

1. Open `Terminal` or `Command Prompt` and ensure that the AWS CLI is installed and configured with the necessary permissions.

2. Run the following command to add an IPv4 CIDR block to your VPC:

   ```
   aws ec2 associate-vpc-cidr-block --vpc-id vpc-12345678
   --cidr-block 10.2.0.0/16
   ```

   Check the output to confirm that the IPv4 CIDR block has been associated successfully

3. Run the following command to add an Amazon-provided IPv6 CIDR block:

   ```
   aws ec2 associate-vpc-cidr-block --vpc-id vpc-12345678
   --amazon-provided-ipv6-cidr-block
   ```

4. Verify that the IPv6 CIDR block has been associated successfully by listing the VPC details:

   ```
   aws ec2 describe-vpcs --vpc-id vpc-12345678
   ```

5. Run the following command to ensure that the new CIDR blocks are associated with your VPC:

   ```
   aws ec2 describe-vpcs --vpc-id vpc-12345678
   ```

   Confirm that both the IPv4 and IPv6 CIDR blocks are present.

In this exercise, you successfully added both IPv4 and IPv6 CIDR blocks to an existing VPC using the AWS Management Console and the AWS CLI. You have learned how to expand your VPC's IP address space, which is crucial for accommodating additional subnets and resources. Understanding how to effectively manage CIDR blocks is essential for maintaining a scalable and organized network within AWS.

Next, you will explore how to configure and manage subnets within your VPC, allowing you to segment your network and implement fine-grained control over resource placement and security.

Subnets in VPCs

Once you have deployed an Amazon VPC, subnets are how you divide the overall VPC CIDR into individual networks where resources will reside. As previously mentioned, a subnet within a VPC is bound to a **single AZ**. Since subnets cannot span AZs, this will have an impact on how you properly design and plan the deployment of subnets within your VPC. It is also important to know the classification of subnets that can belong to a VPC.

IP Ranges in Subnets

When creating Subnets within a VPC, there are a few options to choose from depending on the configuration of the VPC. The IP classifications that can be configured in a subnet are as follows:

- **IPv4 only**: This is the most common type of subnet deployment. Resources residing in an IPv4-only subnet may only have IPv4 addressing and communicate only using IPv4.

- **Dual stack**: Dual stack subnets can support both IPv4 and IPv6 addressing. Deployment of a dual-stack subnet requires that the VPC it belongs to has both IPv4 and IPv6 CIDRs associated. Resources residing within these subnets can communicate using either IPv4 or IPv6.

- **IPv6 only**: While VPCs require an IPv4 CIDR to be associated, you have the ability to create subnets within a VPC that only use IPv6. As with dual-stack subnets, deployment of an IPv6-only subnet requires that the VPC have an IPv6 CIDR associated. Resources within these subnets can communicate using IPv6 only.

Subnet Sizing

Subnets are deployed within a VPC using CIDR notation. Within each subnet, AWS reserves a small subset of the IP addressing for AWS components that reside within each subnet. Across both IPv4- and IPv6-based subnets, AWS reserves **five IP addresses** within each subnet. The reserved AWS addresses are always predictable and follow a common pattern. The following sections will provide details on the subnet sizing and reserved addresses for both IPv4 and IPv6 subnets.

Subnets using IPv4 addressing can support subnet masks as large as /16 or as small as /28. The subnet deployed must be smaller than the VPC CIDR associated with the VPC. For example, if the associated IPv4 range for a VPC is 10.100.0.0/24, you could not create a subnet that uses 10.100.0.0/16. As mentioned, AWS reserves a total of five IP addresses within a given subnet.

Within an IP subnet, the **network** and **broadcast** addresses are both reserved. Amazon VPCs do not support the use of the broadcast address, which is why this address is reserved. In addition, AWS reserves the first three usable IP addresses within the subnet for service components that function within each subnet. Refer to the following example for details on these reservations.

Trailcats has created a VPC subnet with the following IPv4 CIDR: `10.100.0.0/24`. The following addresses would be reserved:

- `10.100.0.0`: The network address for the subnet.
- `10.100.0.1`: Assigned to the VPC router for the overall VPC. This provides a default gateway function for the subnet.
- `10.100.0.2`: Assigned as a DNS server for the VPC subnet. This is also referred to as the Route 53 Resolver or Amazon DNS server.
- `10.100.0.3`: Reserved by AWS for future use.
- `10.100.0.255`: The broadcast address for the subnet.

Following this pattern, Trailcats can easily calculate both the VPC router and Route 53 Resolver addresses of any subnet. For the VPC router address, simply calculate the network address + 1. For the Route 53 Resolver, calculate the network address + 2. More details on the Route 53 Resolver will be covered in *Chapter 7, AWS Route 53: Basics*.

Subnets using IPv6 addressing can support subnet masks as large as `/44` or as small as `/64` but in **increments of 4**. The subnet deployed must be smaller than the VPC CIDR associated with the VPC, similar to IPv4. AWS also reserves the same set of IP addresses for service components within the VPC. However, you'll notice there is no specified address for the Amazon DNS server. With IPv6, the Route 53 Resolver has a dedicated address that will be discussed in *Chapter 7, AWS Route 53: Basics*. Refer to the following example for AWS reserved addressing with IPv6 subnets.

Trailcats has created a VPC subnet with the following IPv6 CIDR: `2001:0db8:85a3:0000::/64`. The following addresses would be reserved:

- `2001:0db8:85a3:0000::`
- `2001:0db8:85a3:0000::1`: Assigned to the VPC router for the overall VPC
- `2001:0db8:85a3:0000::2`
- `2001:0db8:85a3:0000::3`
- `2001:0db8:85a3:0000:ffff:ffff:ffff:ffff`

In addition to the preceding addresses, there are a couple more addresses reserved for use by the VPC router:

- A link-local address within the `fe80::/10` range, which is generated using EUI-64
- The link-local address: `fe80:ec2::1`

Public versus Private Subnets

Within Amazon VPCs, there are both **public** subnets and **private** subnets. You may think that it's as simple as public subnets having public IP addresses and private subnets having private IP addresses. However, it's not that simple. Technically speaking, a subnet is deemed private or public based on the contents of the route table that is assigned to it.

All subnets deployed must have a route table attached. A subnet with a route table that has a route entry pointed toward an internet gateway (likely a default route, 0.0.0.0/0) is classified as a **public** subnet. A **private** subnet can either have no default route at all, or it can use something such as a NAT gateway to reach external resources. The functionality of these gateways will be discussed later in this chapter under the *Internet, NAT, and Egress-Only Gateways* section.

Public subnets are to be used when resources need direct access to the internet. This could be as simple as an EC2 instance that hosts an internet-facing service, or even an AWS Application Load Balancer that is front-ending a web application. Private subnets typically do not host any external facing service. Workloads in private subnets can be allowed internet access by using a NAT gateway. This is common practice to allow outbound internet communications for things such as OS system updates or even external API calls.

> **Note**
> It is important to ensure that EC2 instances deployed directly in public subnets are properly secured from external threats. This could be a combination of hardening the system of the workload itself and configuring things such as AWS security groups and **network access control lists (NACLs)**.

Refer to *Figure A3.15* for an example of public and private subnets with their assigned route tables.

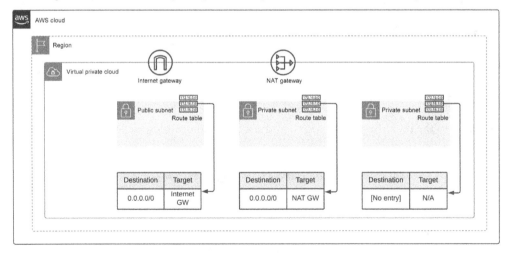

Figure A3.15: Public and private subnets

In *Figure A3.15*, assume Trailcats has an Amazon VPC with private and public subnets. The private subnets need to be allowed access to the internet. In turn, the private subnets have been configured with a default route toward the local NAT gateway within the VPC for external access.

Best Practices for Subnet Size and Segmentation

This section will focus on some best practices you should keep in mind when planning and deploying subnets within your Amazon VPCs. These best practices can assist in ensuring optimal subnet performance, appropriate sizing, and adequate segmentation.

Distribute Subnets across Multiple AZs

Subnets for your applications should be deployed across multiple AZs to ensure proper availability and fault tolerance. This can help prevent negative impacts on your application(s) in the case of a single AZ failure within AWS.

Plan for Growth

The requirements of your application today may not be the same as what they will be in the future. While it's impossible to plan for every possible scenario, you should always plan for growth by selecting appropriate subnet sizes for all components of your applications. Since the size of a subnet cannot be expanded or contracted without deleting and recreating it—which can lead to downtime and configuration challenges—it's important to choose a subnet size that accommodates future expansion at the time of creation.

Consider the number of usable IP addresses within each subnet size to ensure your network can handle increased demand. For example, a /28 subnet provides 16 IP addresses, but only 11 are usable due to AWS reserving the first 4 and the last IP address in each subnet. If your current application requires 10 IP addresses, a /28 subnet might seem sufficient, but it leaves little room for growth. Opting for a /26 subnet, which offers 59 usable IP addresses, allows your application to scale without the need for network reconfiguration. Similarly, larger subnets such as /24 provide 251 usable IP addresses, accommodating substantial growth and high levels of scalability.

By planning for growth and selecting subnets with adequate IP address space, you can avoid the complexities associated with resizing subnets later. This proactive approach ensures that your network infrastructure remains flexible and scalable, aligning with best practices for AWS networking.

Segmentation by Function

Subnets should be defined by function so that resources can be separated from a network perspective by their desired role. For example, if you consider the common "three-tier app," there is typically a web tier, an app tier, and a database tier. All these functions should be split out into dedicated subnets. This can assist cloud network operators with controlling routing and security between subnets.

In *Figure A3.16*, you can see a three-tier application separated into subnets by function. Each tier has subnets aligned to each tier of the application.

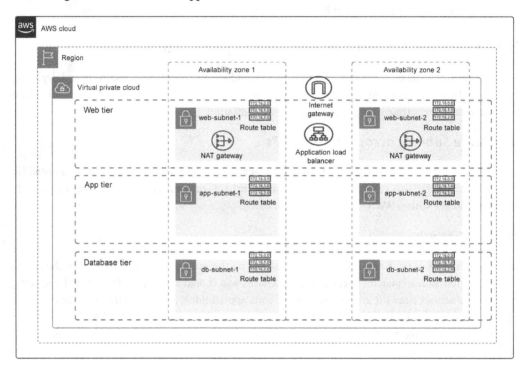

Figure A3.16: Subnet segmentation by function

Within *Figure A3.16*, you can see that subnets are separated based on function. This is a common architecture deployed within AWS environments where resources can then be placed into the appropriate subnet by their function.

For example, Trailcats could have public-facing web applications within this VPC. The web servers will reside in the web subnets, app servers in the app subnets, and database servers within the database subnets.

Next, you will learn how to configure route tables to control traffic flow within your network.

Exercise: Creating VPC Subnets

In this exercise, you will learn how to create subnets within a VPC using both the AWS Management Console and the AWS CLI. Subnets help you logically divide your VPC and manage your resources effectively. By the end of this exercise, you will be able to create subnets in a specified VPC using both the AWS Console and AWS CLI and configure basic settings such as AZs and CIDR blocks.

Before beginning this exercise, ensure you have the following:

- An existing VPC in your AWS account
- Access to the AWS Console and/or AWS CLI with the necessary permissions

In the first part of this exercise, you will create VPC subnets using the AWS Management Console:

1. Sign in to the AWS Management Console, navigate to the VPC dashboard, and select `Subnets`.
2. Click on `Create Subnet`, as shown in *Figure A3.17*:

Figure A3.17: VPC subnet creation

3. Configure the following in the sections shown within *Figure A3.18*:

- **VPC ID**: Select your VPC

- **Subnet name**: Enter a name (e.g., `PublicSubnet`)

- **AZ**: Choose an AZ (e.g., `us-east-1a`)

- **CIDR block**: Enter a CIDR block (e.g., `10.0.1.0/24`)

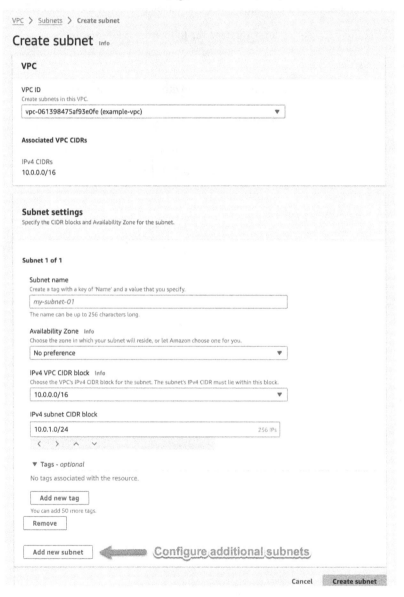

Figure A3.18: Subnet creation details

4. Click `Create subnet` to complete the process. Check the `Subnets` section to verify that the new subnet has been created successfully.

In the second part of this exercise, you will create VPC subnets using the AWS CLI:

1. Open `Terminal` or `Command Prompt` and ensure the AWS CLI is configured properly.

2. Run the following command:

```
aws ec2 create-subnet --vpc-id vpc-12345678 --cidr-block
10.0.2.0/24 --availability-zone us-east-1a
```

Note the `SubnetId` value from the command output.

3. Assign a name to the subnet:

```
aws ec2 create-tags --resources subnet-12345678 --tags
Key=Name,Value=PublicSubnet
```

You have successfully created VPC subnets using both the AWS Console and CLI. This is a key step in organizing and managing your resources within the VPC.

Route Tables

As mentioned in the *VPC Basics* section, route tables are utilized to direct traffic in and potentially out of a VPC using the VPC router. If you are familiar with the concept of **virtual routing and forwarding (VRF)** instances, route tables function in a comparable way within the VPC router. Refer to the following logic.

It is important to remember that the VPC router essentially has an interface within every single subnet that uses the IP address of the network address + 1. Also, each one of these subnets must be associated with a route table. You can visualize it as if all of the subnets associated with the same route table belong to the same VRF instance, that is, they all share the same routing rules when traffic is sent to the default gateway, which is commonly the VPC router.

However, there is one caveat with this analogy. Associating subnets with separate route tables does not inherently imply any segmentation inside the VPC. Segmentation is commonly addressed in other ways, such as security groups and NACLs, covered later in the *Network Security in Amazon VPC* section.

Refer to *Figure A3.19* for a visual representation of subnets being associated with separate route tables. You will observe that multiple subnets are associated with a single route table in some cases.

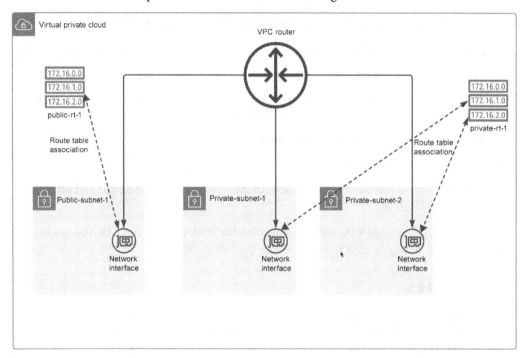

Figure A3.19: Subnet route table association

Assume *Figure A3.19* represents the Trailcats environment. There is a public subnet within the VPC associated with the `public-rt-1` route table, while the two private subnets are associated with the `private-rt-1` route table. This means that traffic sent to the VPC router (default gateway) within `private-subnet-1` and `private-subnet-2` will use the route entries of `private-rt-1`, while traffic sent to the VPC router in `public-subnet-1` will use the route entries within `public-rt-1`. In this example, the public route table could have a `0.0.0.0/0` route toward an internet gateway to get out to the internet, while the private route table may not have a `0.0.0.0/0` route at all, hence separating the function of each subnet.

Route Table Function

Route tables are built by **route** entries. Each individual route entry consists of a **destination** prefix and a **target** (next-hop). Details on the status and source (propagated or not) are also readable. Descriptions of each of these route entry components are shown in *Table A3.2*:

Route Entry Component	Description
Destination	The destination IP prefix for the route entry.
Target	The next-hop address or AWS component to route the traffic to. This could range from things such as AWS Gateway services, VPC peering connections, or even another EC2 instance.
Status	The status of the route entry, as defined by the availability of the next hop.
	For example, if a route entry was created with a target of an internet gateway and that internet gateway was then deleted, the status will change from `Active` to `Blackhole`. This is because the target no longer exists.
Propagated	Defines the source of the route. It is possible to propagate learned BGP routes into route tables in certain cases. Otherwise, route entries will show as `No` in this field to signify they are static entries.

Table A3.2: Route entry components

The table highlights essential components of AWS VPC route entries, including the destination prefix, target for routing traffic, status, and propagation details. Understanding these elements is crucial for effectively managing and troubleshooting VPC route tables, as they define how traffic is directed within and outside the VPC. This foundational knowledge prepares you for more advanced topics, such as configuring custom routes to control network traffic flow.

As mentioned, the target of a route entry can be one of several values, most of which are other AWS services. Being familiar with these naming conventions will help you identify the types of target entries that are present in a route table when looking at one in the AWS console. Refer to *Table A3.3* for a reference of the most common route table target entries and the naming convention associated with each.

Target	Target naming convention
Local	`Local`
Internet gateway	`igw-xxxxxxxxxx`
NAT gateway	`nat-xxxxxxxxxx`
VPC peering connection	`pcx-xxxxxxxxxx`
Virtual private gateway	`vgw-xxxxxxxxxx`
Egress-only internet gateway	`eigw-xxxxxxxxxx`
EC2 instance	`i-xxxxxxxxxx`
Elastic network interface	`eni-xxxxxxxxxx`
VPC endpoint	`vpce-xxxxxxxxxx`
AWS transit gateway	`tgw-xxxxxxxxxx`

Table A3.3: Route table target options

The preceding table outlines the naming conventions for different targets used in AWS VPC route tables. Each target type, such as internet gateway, NAT gateway, or transit gateway, follows a specific identifier format, which helps in distinguishing them within route table entries. Familiarity with these conventions is essential for correctly configuring and managing route tables, ensuring that traffic is directed to the appropriate AWS resources and services. This understanding is crucial for advanced routing configurations and troubleshooting within the VPC.

Refer to *Figure A3.20* for how route tables are displayed within the AWS console.

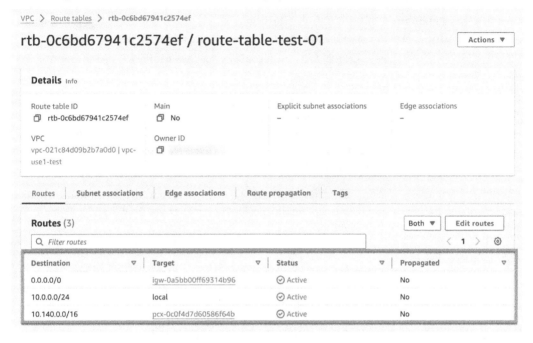

Figure A3.20: Route table in the AWS console

In this example, you will see three route entries, all with different target types.

In *Figure A3.21*, you can see what happens when the target is deleted and the route entry status is updated to Blackhole.

Figure A3.21: Route table Blackhole entry

In this example, a VPC peering connection was deleted and the existing route entry across that peering changed status from Active to Blackhole. VPC peering connections are covered in *Chapter 1, Advanced VPC Networking*.

Main Route Table versus Custom Route Tables

Each VPC has a **main route table** by default, which automatically controls the routing for all subnets not explicitly associated with any other route table. **Custom route tables** are user-created route tables that allow you to define specific routing rules for one or more subnets, providing greater control over network traffic flow.

For instance, imagine you have a VPC with both public and private subnets. The main route table could direct internet-bound traffic through an internet gateway, making any associated subnets public. To keep certain subnets private, you can create a custom route table without a route to the internet gateway and associate it with those subnets. This setup ensures that sensitive resources remain inaccessible from the internet while still communicating within the VPC.

Understanding the roles of main and custom route tables is essential for effective network segmentation and security. In the next section, we'll explore the specifics of the main route table and how it impacts subnet associations within your VPC.

Main Route Table

Since all subnets within a VPC must be associated with a route table, this is the role of the main route table. All newly created subnets will be implicitly associated with the VPC's main route table.

It is important to consider the following rules that apply to the main route table:

- You **can** add, remove, and modify routes in the main route table
- You **cannot** delete the main route table
- You can replace the main route table association with a subnet by associating a custom route table with the subnet
- You can explicitly associate a subnet with the main route table, even if it is already implicitly associated

Within the AWS console, you can view which route table has the role of the main route table by referring to the `Main` column or the route table details. Refer to *Figure A3.22* for a snapshot of the AWS console where this can be viewed:

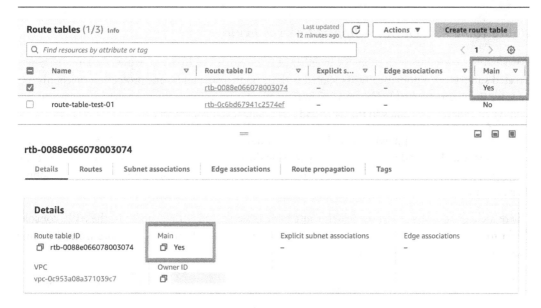

Figure A3.22: VPC Main route table

As shown in the preceding figure, there will be a single route table per VPC that is tagged as the `Main` route table. There can only be a single one of these route tables within every VPC.

Custom Route Tables

It is generally considered best practice to use custom route tables to control routing within and out of your VPC, rather than relying on the main route table. Custom route tables provide more granular control over the routing applied to specific subnets. This is particularly important when determining which subnets need to be public, private, or require specific connectivity settings. For example, you can create a custom route table that directs internet-bound traffic to an internet gateway and associate it with subnets that should be public. Conversely, for subnets that need to remain private, you can create a custom route table without a route to the internet gateway, ensuring they are isolated from the internet. By tailoring route tables to the specific requirements of each subnet, you can enhance security and efficiently manage network traffic within your VPC.

Understanding how to associate these custom route tables with your subnets is crucial. In the next section, we will explore route table associations and the difference between implicit and explicit associations.

Route Table Association – Implicit versus Explicit

Associating route tables with subnets can either be done either **implicitly** or **explicitly**. Since all newly created subnets are automatically associated with the main route table, one protection mechanism is to leave the main route table for the VPC in the default state as when it was created. This means not adding any additional routes to the table and essentially leaving it empty.

Using this workflow, all newly created subnets would have an implicit association with the main route table, which provides no connectivity outside the local VPC. The new subnet will not have the desired routing connectivity until the AWS operator uses an explicit association to attach the subnet to a custom route table.

There are a couple of ways to see which route table is associated with a subnet in the AWS console. Refer to *Figure A3.23* to see how the explicit route table association can be viewed from the route table view within the AWS console:

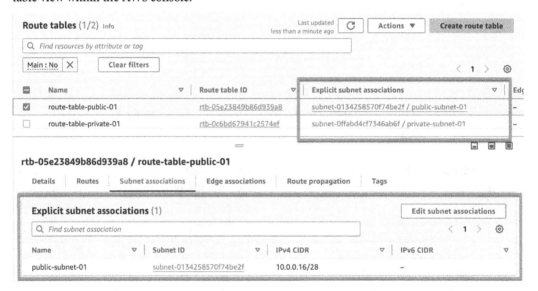

Figure A3.23: Route table subnet association

As shown in the figure, when viewing the list of route tables, you can view all the explicitly associated subnets with that route table. This list refers to any subnets that were manually configured to be associated with this particular route table.

Additionally, *Figure A3.24* shows how the route table association can be viewed from the subnet view within the AWS console:

Figure A3.24: Subnet route table association

As shown in the figure, the subnet view allows you to view the route table that a particular subnet is associated with. This view allows you to view only the route table for this single subnet, so you will not see all the subnets associated with the route table.

Local VPC Route

At the time of creation, all route tables need to be created within a VPC. Inherently, all route tables that are created will have a route entry for the VPC CIDR and the target will be set to `local`. This route ensures that all resources have connectivity within the VPC. A local VPC route will be present for *all* associated VPC CIDRs, including both IPv4 and IPv6.

Resources within the same subnet can communicate directly via their IP addresses without sending the traffic to the VPC router. However, this local VPC route is used by the VPC router for all inter-subnet communication. Refer to *Figure A3.25* to see an example of a local route within a VPC route table:

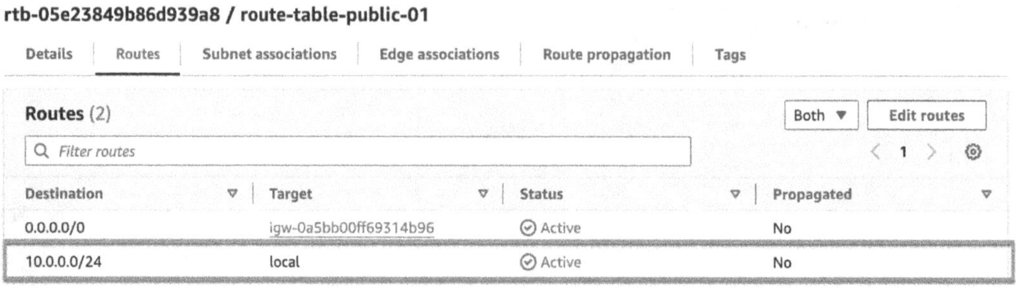

rtb-05e23849b86d939a8 / route-table-public-01

| Details | Routes | Subnet associations | Edge associations | Route propagation | Tags |

Routes (2)   Both ▼   Edit routes

Q *Filter routes*   ⟨ 1 ⟩ ⚙

Destination ▽	Target ▽	Status ▽	Propagated ▽
0.0.0.0/0	igw-0a5bb00ff69314b96	⊘ Active	No
10.0.0.0/24	local	⊘ Active	No

Figure A3.25: Local VPC route

The local route shown in the figure is automatically created in every route table within the VPC with the VPC CIDR as the destination. This ensures that all subnets within the VPC have connectivity from a routing perspective by default.

The local VPC route cannot be deleted. However, the target for these routes can be updated if you want to route traffic destined for the VPC CIDR to either a network interface (eni-xxxxxx) or an EC2 instance (i-xxxxxx). This can be useful in situations where you wish to make sure all inter-subnet communications are sent through an EC2 instance. This could be something like a **network virtual appliance (NVA)** in the form of a third-party firewall.

In addition to overriding the local route for the full VPC CIDR, you can control communications by entering more specific routes than the VPC CIDR, as shown in *Figure A3.26*:

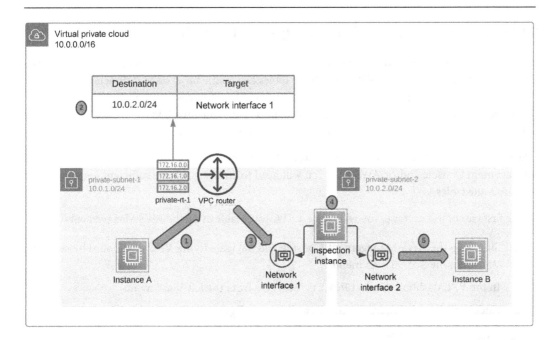

Figure A3.26: Local VPC route override

In *Figure A3.26*, you can see an example where *Instance A* communicates with *Instance B* through an inspection instance. The VPC has a CIDR of 10.0.0.0/16 but a route table is configured with a more specific entry for 10.0.2.0/24 pointed to the network interface of an inspection EC2 instance.

In this section, you successfully learned about VPC route tables. You learned how to associate subnets with a route table and configure route entries to direct traffic within your VPC. This foundational skill ensures that network traffic is routed efficiently and securely in various scenarios, from public internet access to private inter-VPC communication.

While building connectivity is crucial, securing your network is equally important. In the next section, you will explore network security in Amazon VPCs, focusing on two critical elements: security groups and NACLs. These tools are essential for controlling traffic to and from your VPC resources and maintaining a secure environment within AWS.

Exercise: Creating and Managing VPC Route Tables

In this exercise, you will learn how to create and manage VPC route tables using both the AWS Management Console and the AWS CLI. Route tables determine how traffic is directed within your VPC and are crucial for ensuring proper network routing to resources inside and outside your VPC. By the end of this exercise, you will be able to create a route table in a VPC using the AWS console and AWS CLI, associate a subnet with the route table, and add route entries to control traffic flow.

The prerequisites for the exercise are an existing VPC in your AWS account, and access to the AWS Management Console and/or AWS CLI. You will need to ensure you have sufficient permissions to modify route tables.

In the first part of this exercise, you will create a VPC route table using the AWS Management Console:

1. Sign in to the AWS Management Console and navigate to the VPC dashboard by selecting `VPC` from the `Services` menu.

2. In the VPC dashboard, select `Route tables` from the left-hand menu.

3. Click on `Create route table`, as shown in *Figure A3.27*:

Figure A3.27: Route table creation

4. To name the route table, select your VPC and add any desired tags. A snapshot of these fields is shown in *Figure A3.28*:

Figure A3.28: Route table creation details

5. Click Create route table to complete the process.

6. Once the route table is created, select it from the list.

7. Choose Actions and select Edit subnet associations.

8. Choose the subnet to associate with this route table and save your changes.

9. From the `Actions` menu, select `Edit routes`. A snapshot of the `Actions` menu is shown in *Figure A3.29*:

Figure A3.29: Route table actions

10. Add a new route with the desired destination CIDR block (for example, 0.0.0.0/0 for all traffic).

11. Choose the target (for example, an internet gateway) and save the changes.

Once you have completed these steps, you will have created a route table, added a route, and chosen a target.

Alternatively, you can achieve the same outcome by using the CLI, as follows:

1. Open Terminal or Command Prompt and use the following command to create a route table in your specified VPC:

```
aws ec2 create-route-table --vpc-id vpc-12345678
```

Replace vpc-12345678 with your VPC ID. Note the RouteTableId value returned by the command.

2. Use the following command to associate a subnet with the route table:

    ```
    aws ec2 associate-route-table --route-table-id
    rtb-12345678 --subnet-id subnet-12345678
    ```

    Replace `rtb-12345678` and `subnet-12345678` with the appropriate route table and subnet IDs.

3. Run the following command to create a new route entry in the route table:

    ```
    aws ec2 create-route --route-table-id rtb-12345678
    --destination-cidr-block 0.0.0.0/0 --gateway-id
    igw-12345678
    ```

    Replace `rtb-12345678` with your route table ID, and `igw-12345678` with your internet gateway ID.

By following these steps, you will have set up a route table and routing rules for a VPC and its associated subnet.

Summary

This chapter provided a foundational understanding of Amazon VPC networking, an essential element of both the *AWS Certified Advanced Networking - Specialty (ANS-C01)* exam and real-world cloud networking. Through the core components of VPCs—CIDR blocks, subnets, route tables, and gateways—you've learned how to design and deploy isolated network environments on AWS.

The chapter emphasized effective IP addressing management with both IPv4 and IPv6, ensuring the scalability and flexibility of network designs. You also gained insight into subnet configuration best practices, such as sizing, segmentation, and distributing subnets across AZs for enhanced availability and fault tolerance. Understanding how to leverage route tables allows you to optimize traffic flow within and outside your VPC, contributing to both secure and efficient network operations. Lastly, you explored how to manage connectivity to external networks using internet gateways, NAT gateways, and Egress-only internet gateways, learning how to secure outbound access for both public and private subnets.

With these foundational concepts in VPC networking, you are now prepared to delve into security mechanisms that further protect your VPC, setting the stage for securing instances and subnets against unauthorized access.

Appendix 4
VPC Security and External Connectivity

As you build VPC networks within AWS, maintaining security is paramount to protecting your resources and ensuring a secure environment for applications and data. In Amazon VPC, security is achieved through multiple layers, including subnet-level and instance-level controls that regulate traffic flow. This chapter covers the essentials of VPC security, focusing on **network access control lists** (**NACLs**) and security groups. These tools enable granular control over both inbound and outbound traffic, allowing you to enforce strict security policies and maintain the integrity of your network infrastructure.

In this chapter, you will cover the following topics:

- Network security in Amazon VPC

- Internet, NAT, and egress-only gateways

By the end of this chapter, you'll have a solid understanding of the best practices for securing VPCs, ensuring that both subnets and instances are protected against unauthorized access and configured for optimal security.

Network Security in Amazon VPC

As you build connectivity within the AWS cloud, it's crucial to recognize that networking and security are inherently interconnected. **Network security** is essential for protecting your resources from unauthorized access, ensuring data integrity, and maintaining the confidentiality of your data. It involves implementing measures that safeguard your network infrastructure against threats and vulnerabilities, thereby ensuring the smooth and secure operation of your applications.

In Amazon VPC, network security is achieved through a combination of features that control traffic flow at both the subnet and instance levels. By carefully designing your network architecture and applying security controls, you can create a robust environment that meets your organization's security requirements. Two key components that play a vital role in securing your VPC are **security groups** and **NACLs**. Security groups act as virtual firewalls for your instances, regulating inbound and outbound traffic based on defined rules. NACLs provide an additional layer of defense by controlling traffic at the subnet level, allowing you to set granular rules for traffic entering and exiting your subnets.

NACLs

For those of you who have worked in the networking industry, you may already be familiar with the concept of **access control lists (ACLs)**. The AWS implementation of NACLs is very similar to the traditional implementation. NACLs belong to a **single** Amazon VPC and, like route tables, there is the concept of a VPC default NACL. Network ACLs are applied at the subnet level.

A network ACL is a mechanism to either allow or deny IP-based traffic in and out of a VPC subnet. As traffic is processed by the VPC router for a subnet (either ingress or egress), the network ACL for that subnet will be used to process the traffic. Network ACLs are comprised of a set of rules that are processed in **sequential order** from the top down until a match is found. Traditional ACLs are essentially the same as a NACL applied to the physical interface of the VPC router for the particular subnet. However, one difference is that traditional ACLs need to be manually applied in both the inbound and outbound directions. A NACL in Amazon VPC is automatically applied in both the **inbound** and **outbound** directions. This is why network ACLs are comprised of separate sets of inbound rules and outbound rules.

> **Note**
>
> In the context of network ACLs, the term **inbound** would mean inbound *to* the applicable subnet. The term **outbound** means outbound *from* the subnet.

Refer to *Figure A4.1*. You can see that all traffic going in/out of a particular subnet would be processed by a network ACL.

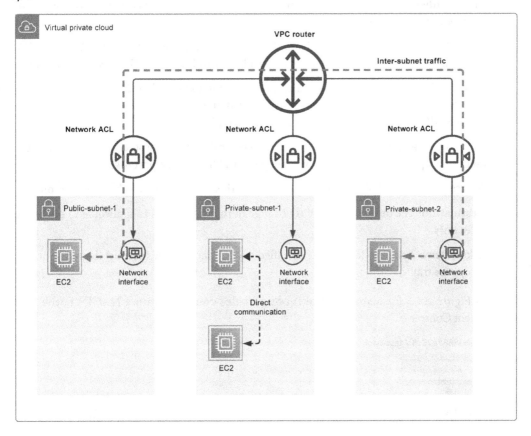

Figure A4.1: Network ACLs

The NACL processes any traffic sent to the local VPC router. This includes traffic that is going between subnets within the same VPC, as seen by the green dotted line between the EC2 instances in public-subnet-1 and private-subnet-2. However, the communication between the two instances within the same subnet, private-subnet-1, does not get processed by a network ACL.

Network ACL Rules

As mentioned, a network ACL is comprised of a set of both inbound and outbound rules. As traffic traverses a network ACL, either the inbound or outbound ruleset is processed, depending on the direction of the traffic. Each set is a sequential list of rules that are processed from the top down until there is a match on one of the rules. After there is a match, the action to either **permit** or **deny** the traffic is taken, and no further rules are processed.

Each network ACL rule is composed of the following:

- **Rule number:** Used for sequential ordering. Lower rule numbers are processed first. This number can range from 1 to 32766.

- **Type:** Defines the type of traffic at Layer 4. This can be as simple as TCP/UDP/ICMP, or AWS also offers some predefined options for protocols such as HTTPS, SSH, DNS, and many more.

- **Protocol:** Layer 4 protocol number. The most common usage will be TCP (6), UDP (17), or ICMP (1). A custom protocol can also be used. If ICMP is selected, options are presented for all ICMP types and codes (echo request, echo reply, etc.).

- **Port range:** The listening range of port numbers for the Layer 4 protocol. This value can be a range or a single number value – for example, HTTPS would be port 443.

- **Source:** The source IP CIDR range of the traffic. This is applicable to **inbound rules only**.

- **Destination:** The destination IP CIDR range of the traffic. This is applicable to **outbound rules only**.

- **Allow/Deny:** Action to take on the specified traffic, that is, allow (permit) the traffic or deny (drop) the traffic.

Refer to *Figure A4.2* for a snapshot of inbound rules configured on a NACL via the AWS Management Console:

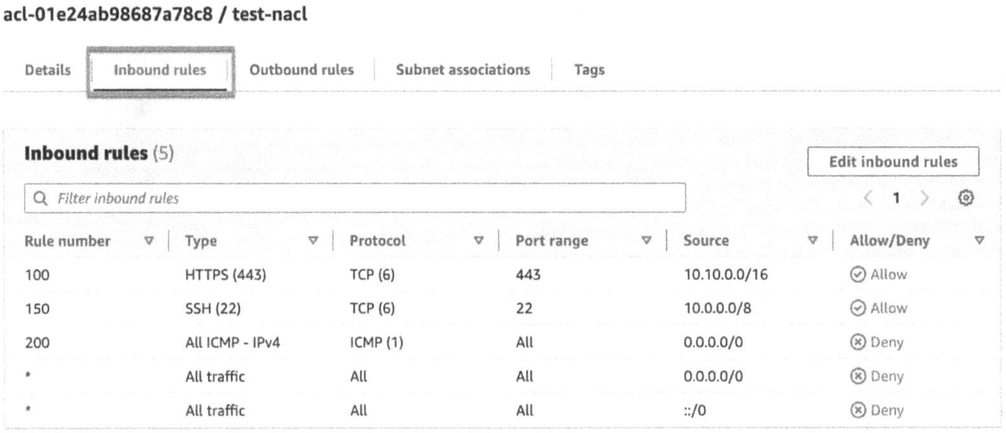

acl-01e24ab98687a78c8 / test-nacl

Details	Inbound rules	Outbound rules	Subnet associations	Tags

Inbound rules (5)   Edit inbound rules

Q Filter inbound rules   < 1 > ⚙

Rule number ▽	Type ▽	Protocol ▽	Port range ▽	Source ▽	Allow/Deny ▽
100	HTTPS (443)	TCP (6)	443	10.10.0.0/16	⊘ Allow
150	SSH (22)	TCP (6)	22	10.0.0.0/8	⊘ Allow
200	All ICMP - IPv4	ICMP (1)	All	0.0.0.0/0	⊗ Deny
*	All traffic	All	All	0.0.0.0/0	⊗ Deny
*	All traffic	All	All	::/0	⊗ Deny

Figure A4.2: Network ACL – Inbound rules

As you can see in the inbound rules, rules `100` and `150` allow inbound communications for HTTPS and SSH traffic from specific source IP addresses. Rule number `200` denies all inbound ICMP traffic. Lastly, there are two implicit deny rules for both IPv4 and IPv6.

Next, refer to *Figure A4.3* for a look at the outbound rules:

acl-01e24ab98687a78c8 / test-nacl

Details	Inbound rules	Outbound rules	Subnet associations	Tags

Outbound rules (4)   Edit outbound rules

Q *Filter outbound rules*   ‹ 1 › ⚙

Rule number ▽	Type ▽	Protocol ▽	Port range ▽	Destination ▽	Allow/Deny ▽
100	All traffic	All	All	0.0.0.0/0	⊘ Allow
200	All traffic	All	All	::/0	⊘ Allow
*	All traffic	All	All	0.0.0.0/0	⊗ Deny
*	All traffic	All	All	::/0	⊗ Deny

Figure A4.3: Network ACL – Outbound rules

The outbound rules in the preceding figure show all outbound IPv4 and IPv6 traffic being permitted by rules `100` and `200`.

Within *Figure A4.2* and *Figure A4.3*, you can see a mix of both user-created rules and implicit deny rules that are system-generated.

One thing you will also notice in *Figure A4.2* and *Figure A4.3* are the two rules at the bottom of each list marked with an asterisk (*) in the `Rule number` column. These are known as the **implicit deny rule**. This rule *cannot* be changed or deleted.

By default, every network ACL will have these implicit deny rules at the end of the sequence. This means that if there is no match for the traffic in any of the previous rules, the traffic will automatically be dropped by the implicit deny rule. There are situations where you may want to change the behavior of the network ACL by configuring a **catch-all** allow rule at the bottom of the sequence. You can see an example of this in *Figure A4.4*:

Inbound rules (7)   Edit inbound rules

Q *Filter inbound rules*   < 1 > ⚙

Rule number	Type	Protocol	Port range	Source	Allow/Deny
100	HTTPS (443)	TCP (6)	443	10.10.0.0/16	⊘ Allow
150	SSH (22)	TCP (6)	22	10.0.0.0/8	⊘ Allow
200	All ICMP - IPv4	ICMP (1)	All	0.0.0.0/0	⊗ Deny
3000	All traffic	All	All	0.0.0.0/0	⊘ Allow
3001	All traffic	All	All	::/0	⊘ Allow
*	All traffic	All	All	0.0.0.0/0	⊗ Deny
*	All traffic	All	All	::/0	⊗ Deny

Figure A4.4: Network ACL – catch-all allow

In this figure, you can see rules 3000 and 3001 are inserted as catch-all permit rules to permit traffic before the implicit deny rules.

Network ACL rules can either define traffic for IPv4 traffic or IPv6 traffic. You cannot combine them into a single rule. This is why you'll notice in the previous figures that there are separate implicit deny rules for both 0.0.0.0/0 and ::/0.

If a VPC is configured with IPv4 only, you will only see a single implicit deny rule on all network ACLs within that VPC for 0.0.0.0/0. If you later add an IPv6 CIDR, another implicit deny rule for ::/0 will be added to all network ACLs.

Default Network ACLs and Custom Network ACLs

Like the behavior with AWS route tables, all VPCs are configured with a **default network ACL**. By default, any newly created subnet within a VPC will be assigned to the default network ACL. The default network ACL is initially configured to always allow all inbound and outbound traffic, using the logic of the **catch-all** allow rule, mentioned previously.

Refer to *Figure A4.5* to see the default network ACL.

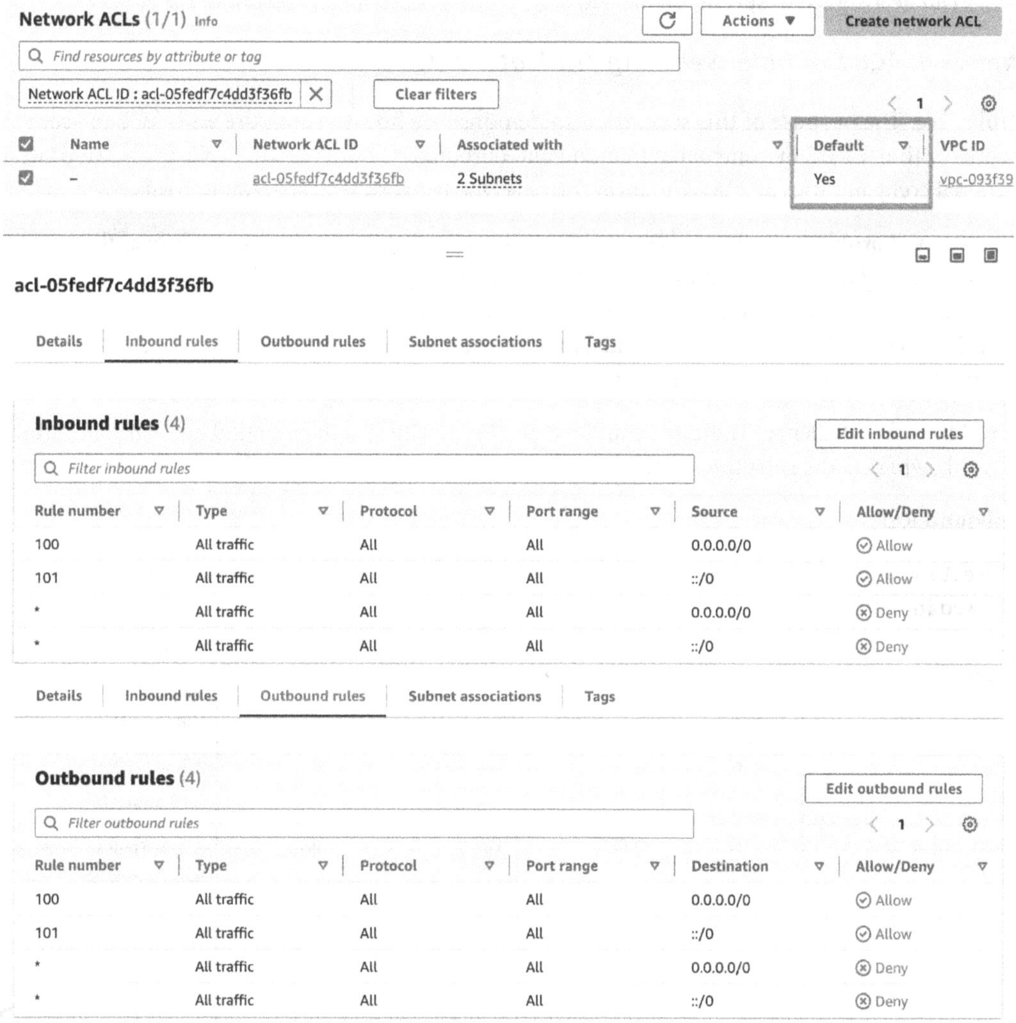

Network ACLs (1/1) Info

Q *Find resources by attribute or tag*

Network ACL ID : acl-05fedf7c4dd3f36fb ☒   Clear filters

⟨ 1 ⟩ ⚙

☑	Name	▽	Network ACL ID	▽	Associated with	▽	Default	▽	VPC ID
☑	–		acl-05fedf7c4dd3f36fb		2 Subnets		Yes		vpc-093f39

acl-05fedf7c4dd3f36fb

| Details | Inbound rules | Outbound rules | Subnet associations | Tags |

Inbound rules (4)

Edit inbound rules

Q *Filter inbound rules*

⟨ 1 ⟩ ⚙

Rule number	▽	Type	▽	Protocol	▽	Port range	▽	Source	▽	Allow/Deny	▽
100		All traffic		All		All		0.0.0.0/0		⊘ Allow	
101		All traffic		All		All		::/0		⊘ Allow	
*		All traffic		All		All		0.0.0.0/0		⊗ Deny	
*		All traffic		All		All		::/0		⊗ Deny	

| Details | Inbound rules | Outbound rules | Subnet associations | Tags |

Outbound rules (4)

Edit outbound rules

Q *Filter outbound rules*

⟨ 1 ⟩ ⚙

Rule number	▽	Type	▽	Protocol	▽	Port range	▽	Destination	▽	Allow/Deny	▽
100		All traffic		All		All		0.0.0.0/0		⊘ Allow	
101		All traffic		All		All		::/0		⊘ Allow	
*		All traffic		All		All		0.0.0.0/0		⊗ Deny	
*		All traffic		All		All		::/0		⊗ Deny	

Figure A4.5: Default network ACL

You will notice that default network ACLs are marked as `Yes` in the `Default` column within the AWS Management Console as well as the default rules that exist.

> **Note**
>
> If a default network ACL for a VPC that is IPv4-only is ever modified by adding or removing rules, the default allow rule will not be added automatically for IPv6 if the VPC is updated to include an IPv6 CIDR. This rule will need to be added manually if desired.

Also, like route tables, **custom network ACLs** can also be created and used. These custom network ACLs can be assigned to specific subnets to control network traffic in and out of the subnet.

Network ACL Example: Securing Trailcats' VPC

Objective: The purpose of this scenario is to demonstrate how to configure an NACL to secure a subnet within a VPC by controlling inbound and outbound traffic. We will explore how Trailcats, a fictional company, uses an NACL to meet their specific network security requirements.

Context and problem: Trailcats operates web applications that need to be accessible over the internet while ensuring that internal management and monitoring are secure. They face the challenge of allowing necessary inbound and outbound traffic for services such as HTTPS, SSH, and RDP while blocking unauthorized access and potential threats. Their goal is to implement an NACL that provides the right balance between accessibility and security.

To address this challenge, Trailcats configured an NACL with specific inbound and outbound rules. The following are the rules they applied.

Inbound Rules

Table A4.1 outlines the inbound rules configured in the NACL, specifying which types of traffic are allowed into the subnet:

Inbound Rules						
Rule #	**Type**	**Protocol**	**Port Range**	**Source**	**Allow/Deny**	**Comments**
100	HTTPS	TCP (6)	443	0.0.0.0/0	Allow	Allow HTTPS from any IPv4 address.
110	RDP=	TCP (6)	3389	10.0.0.0/8	Allow	Allow **Remote Desktop Protocol (RDP)** from any IPv4 address within 10.0.0.0/8.
120	SSH	TCP (6)	22	10.0.0.0/8	Allow	Allow SSH from any IPv4 address within 10.0.0.0/8.
130	Custom ICMP – IPv4	ICMP (1): Echo Reply	N/A	0.0.0.0/0	Allow	Allow ICMP echo replies from any IPv4 address.
140	Custom TCP	TCP (6)	32768 – 65535	0.0.0.0/0	Allow	Allow inbound return IPv4 traffic on ephemeral ports from any IPv4 address.

Inbound Rules						
Rule #	Type	Protocol	Port Range	Source	Allow/Deny	Comments
*	All Traffic	All	All	0.0.0.0/0	Deny	Implicit deny rule. Denies all traffic if not matched by a previous rule.

Table A4.1: Network ACL example – inbound rules

These inbound rules enable the subnet to accept HTTPS traffic from any source, allow remote management via RDP and SSH from internal IP addresses (10.0.0.0/8), permit ICMP echo replies for network diagnostics, and ensure proper handling of return traffic on ephemeral ports. The final rule denies all other inbound traffic not explicitly allowed, enhancing the security posture.

Outbound Rules

Table A4.2 lists the outbound rules set in the NACL, defining what traffic is permitted to leave the subnet:

Outbound Rules						
Rule #	Type	Protocol	Port Range	Destination	Allow/Deny	Comments
100	HTTPS	TCP (6)	443	0.0.0.0/0	Allow	Allow outbound HTTPS traffic to any IPv4 address.
110	SSH	TCP (6)	1024 – 65535	10.0.0.0/8	Allow	Allow SSH return traffic to any IPv4 address within 10.0.0.0/8.
120	Custom ICMP – IPv4	ICMP (1): Echo Request	N/A	0.0.0.0/0	Allow	Allow ICMP echo replies from any IPv4 address.
130	Custom TCP	TCP (6)	32768–65535	0.0.0.0/0	Allow	Allow inbound return IPv4 traffic on ephemeral ports to any IPv4 address.
*	All Traffic	All	All	0.0.0.0/0	Deny	Implicit deny rule. Denies all traffic if not matched by a previous rule.

Table A4.2: Network ACL example – outbound rules

These outbound rules permit the subnet to initiate HTTPS connections to any destination, allow return traffic for SSH sessions to internal addresses, enable ICMP echo requests for diagnostics, and manage return traffic on ephemeral ports. Similar to the inbound rules, any outbound traffic not explicitly allowed is denied, providing an additional layer of security.

By implementing these NACL rules, Trailcats achieved the following:

- **Secure internet access**: Allowed inbound and outbound HTTPS traffic (port 443) from any IPv4 address, enabling secure web communication for their applications accessible over the internet

- **Controlled internal access**: Permitted inbound RDP (port 3389) and SSH (port 22) only from their internal network (10.0.0.0/8), ensuring that remote management is restricted to trusted internal users

- **Network diagnostics**: Enabled outbound ICMP echo requests and inbound echo replies, facilitating ping operations for network troubleshooting without compromising security

- **Managed return traffic**: Allowed return traffic on TCP ephemeral ports (ports 32768–65535) for both inbound and outbound directions, ensuring the proper functioning of established TCP connections

- **Enhanced security**: Applied an implicit deny rule to block all traffic not explicitly allowed, providing a default layer of protection against unauthorized access

Through this carefully configured NACL, Trailcats addressed their security challenges by allowing essential services while preventing unauthorized traffic. This example highlights the importance of using NACLs to enforce subnet-level traffic control, contributing to a robust and secure VPC network architecture.

Network ACL Considerations and Best Practices

Now that you've read over the functionality of network ACLs, this section will cover a few considerations and best practices you should consider when configuring network ACLs within your Amazon VPCs.

Network ACLs Are Stateless

If you refer back to *Table A4.1* and *Table A4.2*, specifically inbound rules 130–140 and outbound rules 120–130, you'll see references to allowing "return" traffic. This is due to the **stateless** nature of network ACLs. This means that no information about previously transmitted traffic is stored or saved. This behavior differs from security groups, as you will in the *Security Groups* section. Since network traffic is commonly bidirectional, network ACLs need to account for both flows of traffic.

For example, if you have an application hosted within your VPC that acts as a public-facing web server, then the configured network ACL on the subnet where the app resides needs to be configured with at least two rules that allow the following:

- Inbound traffic from 0.0.0.0/0 destined for TCP port 443

- Outbound traffic to 0.0.0.0/0 destined for TCP on ports 32768 through 65535

In *Figure A4.6*, you can see the bidirectional flow of traffic and the need for appropriate inbound and outbound rules.

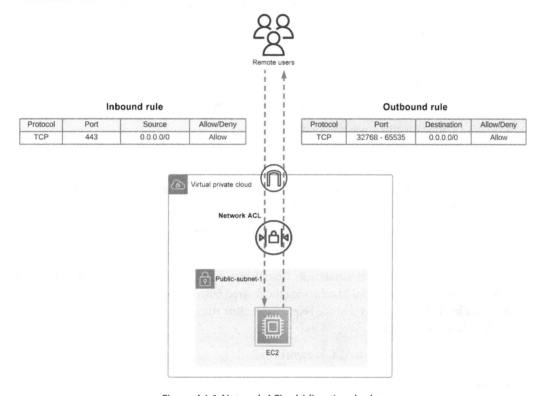

Figure A4.6: Network ACL – bidirectional rules

As you can see in *Figure A4.6*, there are both inbound and outbound rules configured to allow the single flow of traffic to and from the EC2 instance.

The use of ephemeral ports is common across many forms of network connectivity. The client that initiates the network connection is responsible for choosing the ephemeral port range. Typically, the process is initiated by the client with the well-known port as the destination protocol port within the IP header (port 443 for HTTPS in the preceding example). The source port is commonly a random port chosen out of the ephemeral port range so that the client-server can maintain bidirectional communication across those two ports. The server will respond with the well-known port as the source, and the chosen ephemeral port as the destination.

Some common ephemeral port ranges based on the source operation system are as follows:

- Many Linux kernels (including the Amazon Linux kernel) use ports `32768-61000`

- Requests originating from Elastic Load Balancing use ports `1024-65535`

- Windows operating systems through Windows Server 2003 use ports `1025-5000`

- Windows Server 2008 and later versions use ports `49152-65535`

- A **network address translation (NAT)** gateway uses ports `1024-65535`

- AWS Lambda functions use ports `1024-65535`

> **Note**
>
> More details from AWS on ephemeral port usage can be found at this link: `https://docs.aws.amazon.com/vpc/latest/userguide/nacl-ephemeral-ports.html`

Rule Sequencing

Network ACL rules are processed sequentially from the top down. This means that the sequencing of your network ACL rules is crucial to achieving the desired outcome of allowing or denying specific network traffic. This also means that the sequence of your rules should allow for easy insertion of new rules when you need them.

For example, refer to the network ACL in *Figure A4.7*:

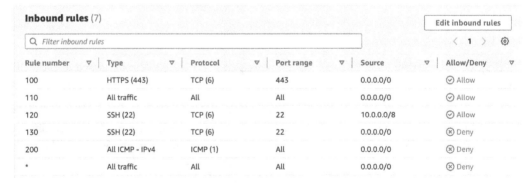

Inbound rules (7)

Rule number	Type	Protocol	Port range	Source	Allow/Deny
100	HTTPS (443)	TCP (6)	443	0.0.0.0/0	⊘ Allow
110	All traffic	All	All	0.0.0.0/0	⊘ Allow
120	SSH (22)	TCP (6)	22	10.0.0.0/8	⊘ Allow
130	SSH (22)	TCP (6)	22	0.0.0.0/0	⊗ Deny
200	All ICMP - IPv4	ICMP (1)	All	0.0.0.0/0	⊗ Deny
*	All traffic	All	All	0.0.0.0/0	⊗ Deny

Figure A4.7: Network ACL – sequence example

The desired outcome of the NACL shown is as follows:

- HTTPS traffic from any IPv4 address is allowed
- SSH traffic is allowed only from `10.0.0.0/8`
- All inbound ICMP is blocked
- Any other IPv4 traffic is permitted

The network ACL does have all the rules to achieve this outcome; however, the sequencing prevents this from being the case. Remember that network ACLs are processed in order from the lowest to the highest rule number, and once a matching rule is found, the specified `Allow` or `Deny` action is applied. *No further rules are evaluated after a match is made*, which means that the first rule that matches the traffic determines the outcome. Using this logic, if an inbound SSH packet with a source IPv4 address of `172.16.10.1` were to pass through the NACL, this traffic would be permitted because of rule number `110` in *Figure A4.7*. In addition, *any* IPv4 traffic would be allowed by this rule, and *none* of the deny rules would be used.

Look at the same NACL in *Figure A4.8* with the same rules but rearranged. Now you will see the desired outcome because rules `120` and `130` take the deny action prior to the catch-all permit rule at rule `200`.

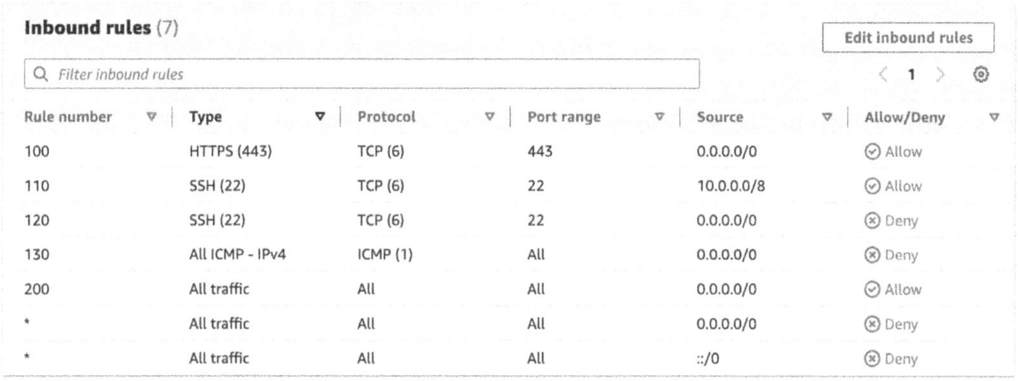

Inbound rules (7)

Rule number	Type	Protocol	Port range	Source	Allow/Deny
100	HTTPS (443)	TCP (6)	443	0.0.0.0/0	⊘ Allow
110	SSH (22)	TCP (6)	22	10.0.0.0/8	⊘ Allow
120	SSH (22)	TCP (6)	22	0.0.0.0/0	⊗ Deny
130	All ICMP - IPv4	ICMP (1)	All	0.0.0.0/0	⊗ Deny
200	All traffic	All	All	0.0.0.0/0	⊘ Allow
*	All traffic	All	All	0.0.0.0/0	⊗ Deny
*	All traffic	All	All	::/0	⊗ Deny

Figure A4.8: Network ACL – sequence example adjusted

In addition, this highlights the importance of following a pattern when defining your rules. Since network ACL rule numbers can range from 1 to 32766, this leaves a lot of room to use them somewhat loosely. A common pattern is to choose a simple number for the first rule, such as `100`. Each new rule after that increases by increments of `10`. Then, if you have any catch-all rules to put last, you can use a very high number such as `1000`. By using a pattern like this, it will become easier to insert rules in between other rules if required.

Path MTU Discovery and Traceroute

ICMP is commonly used for some discovery activities and many network troubleshooting tools. Both path MTU discovery and traceroute fall within these categories.

The **maximum transmission unit** (MTU) is the maximum size packet that can traverse between two systems or across a network link. When a packet size larger than the MTU is sent, IPv4 and IPv6 both have similar ways to handle this, which are both based on ICMP. When the receiving network device cannot send the packet because it is too large, the packet is dropped and an ICMP response is sent back to the source. **Path MTU discovery** is the process of determining the MTU of an end-to-end network path.

Network ACLs can sometimes break the functionality of path MTU discovery if these messages are not permitted. Therefore, in addition to standard echo requests and replies, ICMP messages for `Destination Unreachable` should also be permitted for this functionality. An example of these rules is found in *Figure A4.9*:

Rule number	Type	Protocol	Port range	Source	Allow/Deny
500	Custom ICMP - IPv4	ICMP (1)	Destination Unreachable	0.0.0.0/0	⊘ Allow
501	Custom ICMP - IPv6	IPv6-ICMP (58)	Destination Unreachable	::/0	⊘ Allow

Figure A4.9: ICMP Destination Unreachable rules

In the figure, you can see rules 500 and 501 are allowing all ICMP `Destination Unreachable` messages inbound from any source.

Traceroute is another frequently used network utility tool for troubleshooting network issues. Traceroute provides a hop-by-hop output of each network device along a path between two devices. It does so by sending subsequent ICMP echo request packets with a **time to live** (TTL) value set from 1 up to a maximum of 30. This causes each device along the path to send an ICMP message of `Time To Live Exceeded` back to the source until the true destination is reached.

Traceroute can be blocked by a NACL within an Amazon VPC just like path MTU discovery if these `Time Exceeded` messages are not permitted. Refer to *Figure A4.10* for an example of these rules for both IPv4 and IPv6:

Rule number	Type	Protocol	Port range	Source	Allow/Deny
600	Custom ICMP - IPv4	ICMP (1)	Time Exceeded	0.0.0.0/0	⊘ Allow
601	Custom ICMP - IPv6	IPv6-ICMP (58)	Time Exceeded	::/0	⊘ Allow

Figure A4.10: ICMP Time Exceeded rules

In the figure, you can see rules 600 and 601 allow all ICMP `Time Exceeded` messages inbound from any source.

Default Network ACL Usage

The default NACL associated with each VPC can be simple and easy to use but with caution. It is generally recommended to use only custom NACLs in production environments. This is because you can always be precise and explicit about the network ACL rules applied to specific subnets.

Usage of the default NACL is less predictable. Additionally, it can be difficult to come up with a single set of NACL rules that will apply the appropriate segmentation and security among all your VPC subnets. While the default NACL may be fine to use in development or non-production environments, it is generally recommended to use custom NACLs.

In summary, understanding and applying best practices for network ACLs are essential for securing your Amazon VPC. Key considerations include accounting for the stateless nature of network ACLs by configuring both inbound and outbound rules for bidirectional traffic, carefully sequencing rules to ensure they are processed in the correct order, and allowing necessary ICMP messages to support network utilities such as path MTU discovery and traceroute. Additionally, while default NACLs might suffice in non-production environments, using custom NACLs in production provides greater control and predictability over your network security.

With a thorough understanding of NACLs and their role in securing traffic at the subnet level, it's time to consider additional security measures that operate at a more granular level. Security groups provide this finer control by acting as virtual firewalls for your instances, allowing you to define inbound and outbound traffic rules specific to each resource. In the next section, we'll explore how security groups complement NACLs to enhance the overall security of your Amazon VPC.

Exercise: Creating an NACL

NACLs can be created either via the AWS Management Console or via the AWS CLI.

> **Note**
>
> When following the steps for both the console and the CLI creation of NACLs, replace the VPC ID with one that exists in your account.

AWS Console

An NACL can be deployed from the `Network ACLs` section of the VPC dashboard:

1. Navigate to the Amazon VPC console at `https://console.aws.amazon.com/vpc/`, find the VPC dashboard, and navigate to `Network ACLs`.

2. In the top-right corner, click on the `Create network ACL` button, shown in *Figure A4.11*.

Figure A4.11: Create network ACL

3. On the `Create network ACL` screen, choose which VPC to use from the dropdown. You can also add a name and tags. See *Figure A4.12*.

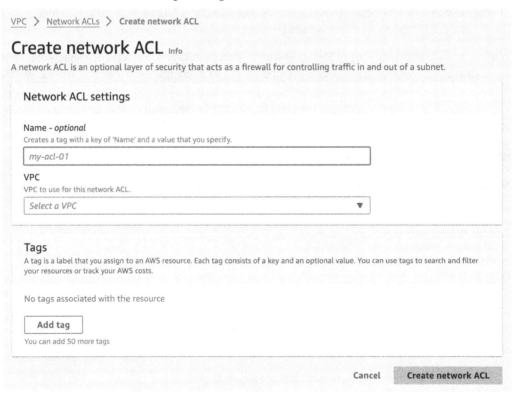

Figure A4.12: Create network ACL details

4. After creation, you can edit the inbound/outbound rules, as shown in *Figure A4.13*.

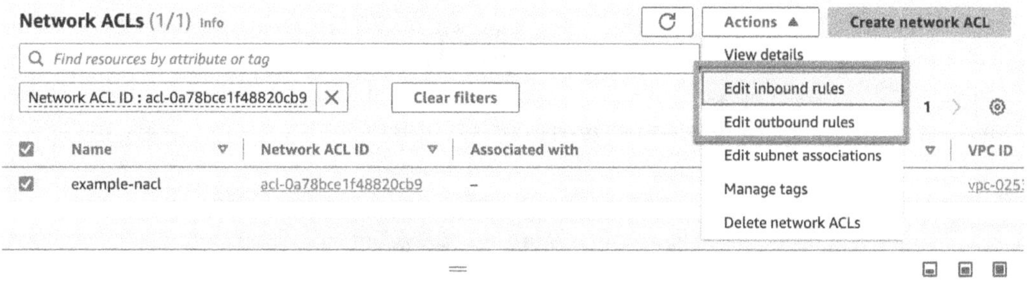

Figure A4.13: Edit NACL rules

Figure A4.13 shows an example NACL, with its ACL ID. From this screen, you can edit inbound rules by clicking on the `Actions` button.

AWS CLI

To create a NACL using the CLI, you can also use the following command followed by the VPC ID: `create-network-acl`.

Here is an example: `aws ec2 create-network-acl --vpc-id vpc-12345678`.

An NACL rule can be configured using the AWS CLI command: `create-network-acl-entry`.

Here is an example:

```
aws ec2 create-network-acl-entry --network-acl-id acl-12345678 --rule-
number 100 --protocol tcp --rule-action allow --ingress --cidr-block
0.0.0.0/0 --port-range From=80,To=80
```

To wrap up our exploration of NACLs, you've now gained a hands-on understanding of how to configure both inbound and outbound rules to secure network traffic at the subnet level. The exercise on creating NACLs provided insight into effectively controlling traffic flow within your Amazon VPC, securing applications and resources with tailored inbound and outbound rules. This approach forms a robust first line of defense for your AWS cloud network.

Now that you have addressed traffic control at the subnet level, it's essential to complement these measures with finer-grained security controls on individual instances. In the upcoming section, you will delve into security groups, which provide a powerful tool for instance-level protection within your VPC. Security groups offer stateful traffic management, allowing you to define access controls that are both dynamic and easily tailored to the specific needs of your applications.

Security Groups

In addition to NACLs, another critical security component of a VPC is **security groups** (**SGs**). While NACLs address applying Layer 4 filtering rules at the subnet level, SGs let you apply security at the EC2 instance level. More specifically, SGs are like firewalls, applied at the **network interface level**. You can read more about **elastic network interfaces** (**ENIs**) in *Chapter 1, Advanced VPC Networking*.

NACLs versus SGs

SGs and NACLs share many similarities. The following are some of the similarities between NACLs and SGs:

- They are both configured within a VPC
- They are both based on IPv4 and IPv6 traffic
- They are both comprised of inbound and outbound rules
- They both have an implicit deny rule
- They both come at no additional cost from a billing perspective

While they share these similarities, there are also some critical differences between NACLs and SGs.

Associated Resources

NACLs can be associated with one or more subnets within a VPC. SGs are associated at the **instance level**, or **network interface level**. While going through the workflow of launching an EC2 instance, you'll notice there is a requirement to associate an SG with the instance. This SG will be associated with the primary network interface of the instance. By default, you can associate up to five security groups with a single network interface.

Figure A4.14 shows how you can assign SGs to instances within a VPC.

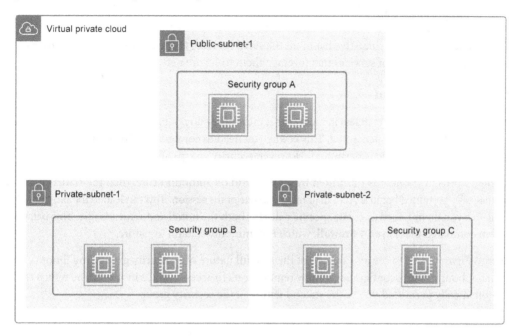

Figure A4.14: SGs in a VPC

As shown in the figure, SGs can be assigned to multiple instances, including across subnets, within a VPC. This allows AWS operators to configure security groups with broader use across applications. For example, if you have a fleet of instances across subnets that are part of an auto-scaling group for an application, they can all use the same security group for ease of defining access for the application.

Since security groups are associated with a network interface, this offers some extensibility in that security groups can be associated with more than just EC2 instances. Essentially, any AWS service that uses (or can use) a network interface within a VPC can have an SG assigned. The following is a list of some of the most common services that utilize security groups:

- EC2 instances
- Elastic load balancer
- Amazon **Elastic Container Service (ECS)**
- AWS Fargate
- Amazon **Elastic File System (EFS)**
- AWS **Database Migration Service (DMS)**
- Amazon RDS instances

- Amazon Redshift

- Amazon ElastiCache

The preceding is not an exhaustive list of all the AWS services that can utilize security groups but speaks to the wide range of services that leverage them to secure network communications.

Stateless versus Stateful

As you'll recall, NACLs are stateless in nature. This means they do not preserve any session information about traffic that has been permitted. This is why you need to configure bidirectional rules to ensure the initial and return traffic is permitted. However, security groups are **stateful**.

In other words, if traffic was permitted by an inbound or outbound rule, then the return traffic is automatically permitted because AWS maintains the state of the session. This can lead to the simplification of rules as you do not need to create specific rules for both the initial and return traffic. This behavior is often compared to a **Layer 4 firewall**, which is commonly stateful in nature.

Refer to *Figure A4.15* to see an example of the stateful nature of a security group. You'll notice that only an inbound rule is configured to allow remote users to access an EC2 instance. The return traffic is automatically permitted due to the stateful nature of the security group.

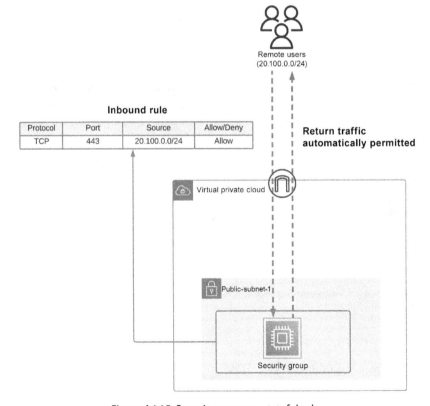

Figure A4.15: Security groups – stateful rules

Rule Processing and Actions

NACLs are composed of a combination of rules that can be configured to either **allow** or **deny** traffic. In contrast, security group rules are **only used to permit specified traffic**. You could consider this a "whitelist" model where traffic is only permitted via a security group if there is an explicit rule to allow it, due to the implicit deny rule. Additionally, **all** security group rules are evaluated in the applicable direction (inbound or outbound) until a match is found. This includes all security groups assigned to the specific network interface. If no match is found, then the traffic is dropped by the implicit deny rule.

This behavior has a couple of implications. Rule sequencing order matters less with security groups as all rules are processed. Since all rules are allow rules, the order of security group rules is less critical. This can, however, limit what security groups can and cannot do from a filtering perspective based on the requirements.

As an example, consider this scenario. You have an EC2 instance that is hosting a public web server for clients residing on the internet. Inbound web traffic comes into the VPC, reaches the instance, and returns via the public internet. Since this is a public-facing resource, both the NACL and the security group have an allow rule configured for HTTPS-based traffic from a source of 0.0.0.0/0. You notice in the instance logs that a specific public IP, 1.2.3.4/32, is making suspicious requests to your instance. You wish to block this traffic to prevent any potential threat.

There is no reasonable way to do this with a security group because of the "whitelist" nature of their rules. Since you cannot add a deny rule, it would not be easy to drop traffic from 1.2.3.4 at the security group level. However, adding a single rule to the NACL could easily solve this problem.

This is not a negative remark targeted at security groups, but instead, a consideration that you need to keep in mind when analyzing what network security tool within your VPC will be able to provide the functionality based on the requirement.

Security Group Rules

Security group rules are composed of many of the same fields that are present within an NACL. However, there is no rule number or allow/deny field because all rules are set to an **allow** action. The following is a brief list of each field within a security group rule:

- **Protocol**: Layer 4 protocol number. The most common usage will be TCP (6), UDP (17), or ICMP (1). Use of a custom protocol can also be used. If ICMP is selected, options are presented for all ICMP types and codes (echo request, echo reply, etc.).

- **Port range**: The listening range of port numbers for the L4 protocol. This value can be a range or a single number value – for example, HTTPS would be port 443.

- **Source**: Applicable to **inbound rules only**.

- **Destination**: Applicable to **outbound rules only**.

- **Description**: A brief description of the security group rule to help with identification. Up to 255 characters are permitted. Allowed characters are a–z, A–Z, 0–9, spaces, and ._-:/()#,@[]+=;{}!$*

Security Group Referencing

A unique and commonly used capability of security group rules is **security group referencing**. This allows AWS operators to further secure their resources within an AWS VPC by ensuring traffic has appropriately traversed a defined path before reaching the instances. This can be done by setting the source of a security group rule to reference **another security group ID**.

In the example shown in *Figure A4.16*, there are separate security groups assigned to an **Application Load Balancer** (**ALB**), web servers, and database servers within a VPC.

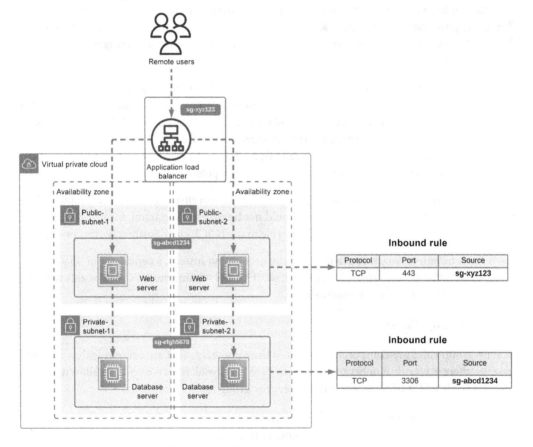

Figure A4.16: Security group referencing

The ALB's security group, with an ID of sg-xyz123, allows ingress traffic from remote users on the internet. The security group assigned to the web servers, sg-abcd1234, allows ingress HTTPS traffic (TCP 443) but only from the security group of the ALB. In the same fashion, the security group of the database servers (sg-efgh5678) only allows TCP 3306 traffic inbound from the security group of the web servers. This creates a series of chaining within the security groups to ensure traffic is only coming from the appropriate sources. This can provide more granular security than using CIDR addresses.

Default Security Groups

A default security group is created within each new VPC that is deployed. The name of this security group is default, and it cannot be deleted. While you can add/remove rules from this default security group, it is generally not recommended to use this security group for production workloads. You should use custom security groups for granularity and defined security for your deployments. Using custom security groups allows you to tailor inbound and outbound traffic rules to the specific requirements of each application or service, providing enhanced security by permitting only necessary network traffic. This granular control reduces the attack surface of your deployments and ensures that the principle of least privilege is enforced within your AWS environment.

All default security groups are configured with a default set of inbound and outbound rules. These rules can be reviewed in *Table A4.3*:

Inbound Rules			
Source	**Protocol**	**Port Range**	**Description**
sg-abcdedf1234567	All	All	Allows all inbound traffic from resources assigned to the default security group. The source of this rule is the ID of this security group.
Outbound Rules			
Source	**Protocol**	**Port Range**	**Description**
0.0.0.0/0	All	All	Allows all outbound IPv4 traffic.
::/0	All	All	Allows all outbound IPv6 traffic. This rule only exists if the VPC that the security group belongs to has an IPv6 CIDR block assigned.

Table A4.3: Default security group rules

You can see there are certain definitions for what is automatically allowed or not allowed. You should be aware of this behavior to know what type of connectivity will be permitted if a resource is using a security group with the default configuration.

> **Note**
>
> Within security groups, a value of `-1` maps to `All`. This value can be used for `Protocol Type`, `Protocol Number`, and `Ports`. When using the AWS Management Console to configure security group rules, you will see the `All` option in the drop-down menus. However, when using something like the AWS CLI or **infrastructure as code (IaC)** to provision them, you may need to define the value as `-1`.

Exercise: Configuring Security Groups

Security groups can either be created from the VPC dashboard or the EC2 dashboard of the AWS Management Console.

AWS Console

1. Navigate to the Amazon VPC console at `https://console.aws.amazon.com/vpc/`.

2. Create a security group by clicking the `Create security group` button at the top right of the VPC dashboard, as shown in *Figure A4.17*:

Figure A4.17: Create security group

On the `Create security group` page, as shown in *Figure A4.18*, add the name and description:

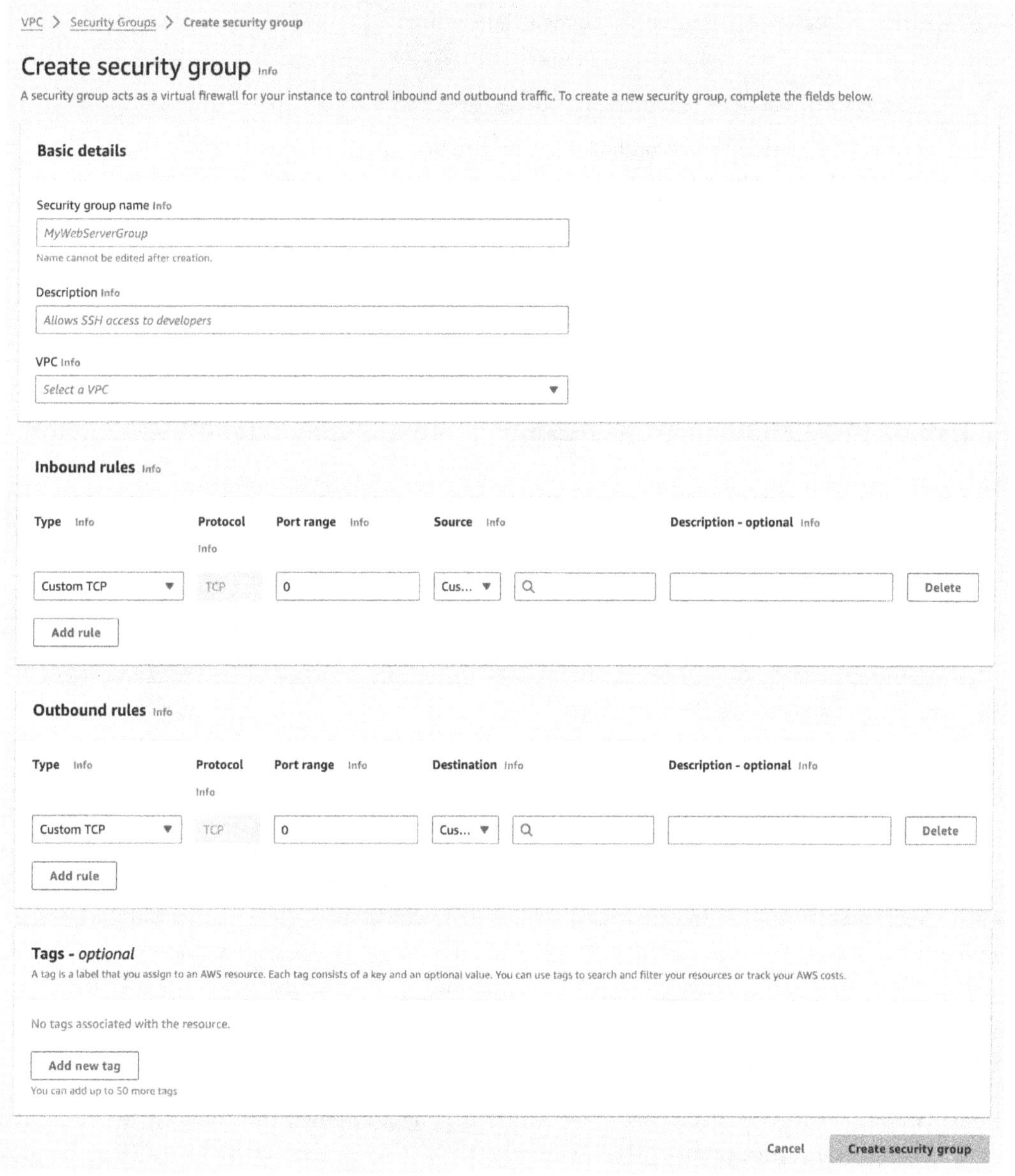

VPC > Security Groups > Create security group

Create security group Info

A security group acts as a virtual firewall for your instance to control inbound and outbound traffic. To create a new security group, complete the fields below.

Basic details

Security group name Info

MyWebServerGroup

Name cannot be edited after creation.

Description Info

Allows SSH access to developers

VPC Info

Select a VPC ▼

Inbound rules Info

Type Info	Protocol Info	Port range Info	Source Info	Description - optional Info	
Custom TCP ▼	TCP	0	Cus... ▼ 🔍		Delete

Add rule

Outbound rules Info

Type Info	Protocol Info	Port range Info	Destination Info	Description - optional Info	
Custom TCP ▼	TCP	0	Cus... ▼ 🔍		Delete

Add rule

Tags - optional

A tag is a label that you assign to an AWS resource. Each tag consists of a key and an optional value. You can use tags to search and filter your resources or track your AWS costs.

No tags associated with the resource.

Add new tag

You can add up to 50 more tags

Cancel   Create security group

Figure A4.18: Create security group details

> **Note**
> The name and description cannot be changed later.

1. Add the VPC that will contain the resources for the security group.

2. You can add inbound rules, outbound rules, and tags at this stage or later.

3. Click on `Create security group`.

AWS CLI

A security group can also be created in one step using this AWS CLI command: `create-security-group`.

Here is an example: `aws ec2 create-security-group --group-name MySecurityGroup --description "My security group" --vpc-id vpc-12345678`

An ingress rule for a security group can be created using this AWS CLI command: `authorize-security-group-ingress`.

Here is an example:

```
aws ec2 authorize-security-group-ingress --group-id sg-12345678
--protocol tcp --port 22 --cidr 0.0.0.0/0
```

The security group is now ready to be used.

Internet, NAT, and Egress-Only Gateways

Connectivity within a VPC is relatively straightforward and easy to configure. But often, you need to build connectivity to resources outside of that VPC. This could mean allowing public access to some web servers you're hosting, allowing your instances to download package updates from the internet, and more. This type of connectivity can be enabled using the gateway services offered by AWS. This section will look at internet gateways, NAT gateways, and egress-only gateways.

Internet Gateways

Resources within a VPC may require connectivity to the public internet. This connectivity is achieved by deploying an **internet gateway** (**IGW**) into the VPC. IGWs are horizontally scaled and highly redundant within a VPC, meaning that they are deployed across all **Availability Zones** (**AZs**) within a particular Region. From a configuration perspective, this is true for all configured IGWs and cannot be altered. They are able to support internet connectivity for both IPv4 and IPv6 traffic. An IGW also does not have bandwidth constraints on network traffic (this limit depends on the size of the EC2 instances using the IGW for connectivity).

Figure A4.19 shows a simple topology where an IGW is attached to a VPC. The public subnets have connectivity to the internet via the IGW, as shown. As mentioned in the *Subnets* section in *Appendix 3, VPC Networking Basics*, a public subnet is any subnet assigned to a route table that has a route entry with the destination set to an IGW.

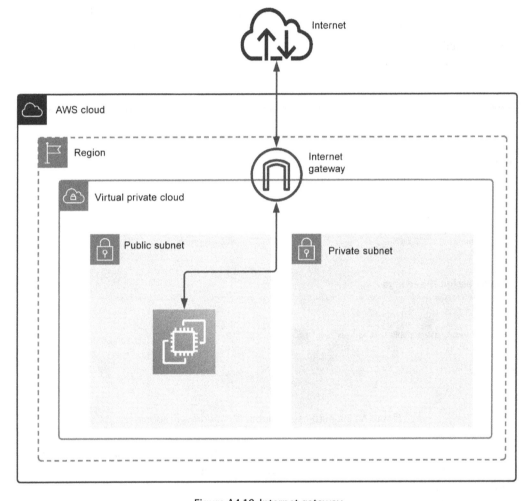

Figure A4.19: Internet gateway

Assume this figure represents a Trailcats VPC where a public-facing web server resides within a public subnet and has direct internet access via an internet gateway. This configuration of the web server within a public subnet associated with the IGW allows for bidirectional communication to and from the internet for the instance.

For an IGW to provide connectivity to the internet, a route table entry pointed toward the IGW is not all that is required. For the traffic to be routable on the internet, the traffic should also use public IP addressing. Amazon EC2 instances can utilize public IP addressing for internet routing in a couple of different ways.

Firstly, for IPv4, EC2 instances could be deployed into a subnet that is configured to auto-assign public IPv4 addresses. This means that all EC2 instances deployed within that subnet will also have a public IPv4 address associated with them. In this scenario with auto-assigned public IPs, the public IPv4 address associated with that instance also shares the lifecycle of that instance, that is, when the instance is terminated, the public IP is also released and cannot be moved to another instance.

Figure A4.20 displays how this setting can be enabled on a per-subnet basis within the subnet settings.

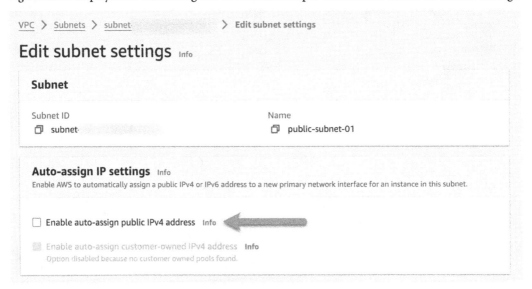

Figure A4.20: Auto-assign public IPv4 address in subnet

As you can see, you can enable auto-assignment of public IPv4 addresses on a subnet basis. This can be used for scenarios where instances can quickly be deployed for testing that need internet access. You would typically avoid this option for long-lived instances as their public IPs are ephemeral and unpredictable since they are tied to the state of the instance.

For consistent public IP addresses and the possibility of reassigning them between EC2 instances, it is better to use Elastic IP addresses. These addresses can be explicitly allocated and associated with EC2 instances and their network interfaces. Elastic IP addresses will be discussed further in *Chapter 1, Advanced VPC Networking*.

With IPv6, the enablement of internet access is quite simple. All assignable IPv6 CIDR blocks from a VPC perspective are globally unique. This means they are all globally routable as well and enabling internet connectivity via an IGW for IPv6 is as simple as having a route toward the IGW. This will commonly be a default route, : : / 0, but may be something more specific if desired.

Use of NAT by IGWs

Within the context of IPv4 addresses, the previously mentioned public IP addresses are never actually assigned directly to the EC2 instances. The only addresses assigned to the network interface of any EC2 instance belong to the private IP address range of the associated subnet.

When a public IP address is assigned to an EC2 instance in AWS, the associated IGW with that VPC configures a **1-to-1 NAT** entry for the public IP to private IP. When a 1-to-1 NAT entry is configured, this means all traffic that traverses the IGW that is **destined to** that public address or **sourced from** that private IP address goes through network translation. This allows the EC2 instance to communicate over the public internet and all traffic is seemingly sourced from the associated public IP address. This NAT entry on the IGW is removed if the public IP address is disassociated from the instance.

Refer to *Figure A4.21* to see an example of an EC2 instance that is associated with a public IPv4 address via an IGW.

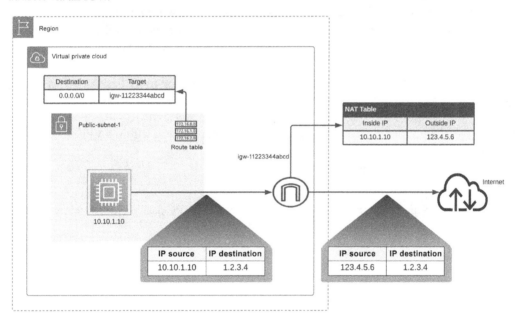

Figure A4.21: 1-to-1 NAT on IGW

In the preceding figure, you can see the EC2 instance is within a subnet that is assigned to a route table with a default route toward the IGW. The IGW is automatically performing a 1-to-1 NAT of the public address 123.4.5.6 to the private address 10.10.1.10. This allows the instance to reach the internet without having any public IP address assigned directly to any of its network interfaces.

Egress-Only Internet Gateways

As noted before, all IPv6 addressing in AWS is globally unique and can therefore be routed publicly. However, that may not always be desired. If you attach an IGW to an IPv6 subnet, that subnet could theoretically be reached by resources out on the public internet. This may not always be desired, as there are security obligations with having EC2 instances completely exposed to the internet. Maybe you only want to allow outbound communication from your IPv6 instances to the internet.

In these situations, you can use an **egress-only IGW**. These gateways can be used in the same way that a standard IGW is used, by attaching them to a VPC and updating the appropriate route tables to route internet-based traffic to them. This is commonly a route entry for ` ::/0`, as these gateways service **only IPv6 traffic**.

As the name suggests, these egress-only internet gateways provide only outbound internet access for IPv6-capable subnets. Typically, an IGW would create a 1-to-1 NAT with a corresponding public IP address. However, an egress-only IGW simply allows outbound connectivity from the VPC but does not allow unsolicited connections from the outside to the resources in the VPC. The egress-only IGW is **stateful**, so the return traffic for permitted outbound connections is automatically allowed back in.

Exercise: How to Configure Internet Gateways and Egress-Only Internet Gateways

You can create an internet gateway or egress-only internet gateway within a VPC via the AWS Management Console or AWS CLI.

AWS Console

1. Navigate to the `Internet gateways` section of the VPC dashboard and select `Create internet gateway` – see *Figure A4.22*:

Figure A4.22: Create internet gateway

2. After creating an IGW, you can attach it to a VPC by clicking the `Actions` menu and choosing `Attach to VPC`, as shown in *Figure A4.23*:

Figure A4.23: Attach IGW to VPC

3. Search for and select a VPC in the `Available VPCs` box, as shown in *Figure A4.24*:

Figure A4.24: IGW attachment page

4. Click `Attach internet gateway`.

Deployment of an egress-only internet gateway is quite similar, as shown in these steps:

1. Navigate to the `Egress-only Internet gateways` section of the VPC dashboard, as shown in *Figure A4.25*.

2. Select the `Create egress only internet gateway` button.

Figure A4.25: Create egress only internet gateway

3. Add a name (optional).

4. *Figure A4.26* shows the details of the egress-only internet gateway settings. To complete it, attach the VPC with the VPC dropdown, by clicking the `Create egress only internet gateway` button in the bottom-right corner.

> **Note**
>
> When creating an egress-only IGW, the VPC is associated with the gateway at the time of creation.

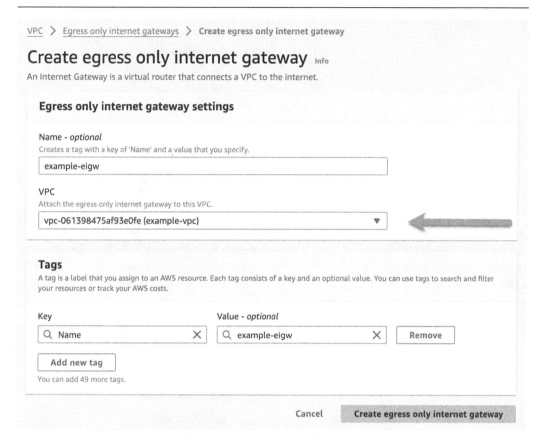

Figure A4.26: Egress only IGW details

AWS CLI

An internet gateway can also be created using this AWS CLI command: `create-internet-gateway`.

Here is an example:

```
aws ec2 create-internet-gateway
```

An egress-only internet gateway can be created using the `create-egress-only-internet-gateway` AWS CLI command followed by the VPC-id.

Here is an example:

```
aws ec2 create-egress-only-internet-gateway --vpc-id vpc-12345678
```

Both IGWs and EIGWs can be attached to a VPC using the `attach-internet-gateway` AWS CLI command followed by the VPC-id and internet gateway ID, or the egress-only internet gateway ID.

Here are some examples:

```
aws ec2 attach-internet-gateway --vpc-id vpc-12345678 --internet-
gateway-id igw-12345678
```

```
aws ec2 attach-internet-gateway --vpc-id vpc-12345678 --internet-
gateway-id eigw-12345678
```

NAT Gateways

A **NAT gateway** is an AWS-managed service that enables instances in a private subnet to connect to the internet, other AWS services, or other private networks while preventing external entities from initiating connections to those instances. NAT gateways perform **source network address translation** (**SNAT**) for outbound traffic, mapping private IP addresses to a single IP address—either public or private—depending on the configuration.

While internet gateways provide external connectivity for public subnets, NAT gateways serve a similar purpose for private subnets that require outbound connectivity without allowing inbound connections from external networks. This includes scenarios where instances in a private subnet need to access the internet or communicate with other private IP addresses in different networks. NAT gateways are created within a specific AZ and can be deployed as either public or private.

In addition to the NAT gateway deployment types, there are several characteristics that you should be aware of when deploying NAT gateways:

- NAT gateways are deployed within a **single AZ**. To ensure NAT gateways are highly available and resilient, you should deploy them across multiple AZs. This is to make sure a single AZ failure doesn't break connectivity for all private instances within the VPC.

- NAT gateways support both **IPv4** and **IPv6** traffic. IPv6 traffic routed to a NAT gateway will use NAT64 to translate the traffic.

- By default, a NAT gateway supports up to **5 Gbps** of traffic but can automatically scale up to **100 Gbps**. This is true for each NAT gateway that is deployed across AZs.

- NAT gateways support **TCP-**, **UDP-**, and **ICMP-based** traffic.

- NAT gateways can support up to **55,000** simultaneous connections per unique destination. A unique destination is classified by the combination of the destination IP address, destination protocol, and destination port. This limit can be increased by assigning additional IP addresses to the NAT gateway. Up to eight IP addresses are supported.

- Public NAT gateways can support **two Elastic IPs** by default. This limit can be increased using a service quota request.

> **Note**
>
> Many of these characteristics and more can be found in the AWS documentation at the following link:
>
> https://docs.aws.amazon.com/vpc/latest/userguide/nat-gateway-basics.html

Refer to *Figure A4.27* for a simplified view of a NAT gateway providing external connectivity for a private subnet within a VPC.

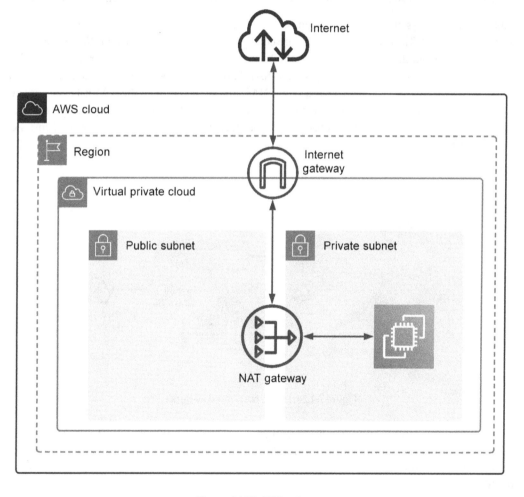

Figure A4.27: NAT gateway

For example, this could be a Linux server within the Trailcats AWS account that resides in a private subnet but requires outbound internet connectivity to do things such as download system patches and access code repositories. The use of the NAT gateway enables the subnet to remain within a private subnet and still have internet access via the public IP of the NAT gateway.

Public NAT Gateways

NAT gateways deployed into public subnets are the most common default configuration. Public NAT gateways allow for instances in private subnets to have outbound connectivity to the internet. They also ensure that no unsolicited inbound connections from the internet can reach these instances.

Public NAT gateways are deployed within public subnets and associated with a public IP address (Elastic IP address). The private route tables are configured with a route toward the NAT gateway. All inbound traffic from the private subnets to the NAT gateway has its source IP address translated to the private IP of the NAT gateway using source NAT. Similar to the public EC2 instances, the traffic is then forwarded to the IGW, which has a 1-to-1 NAT configured between the NAT gateway's private IP and associated public IP. This may seem complicated, but it is all relatively transparent to the user. *Figure A4.28* shows a detailed walk-through of this process from the perspective of an EC2 instance in a private subnet.

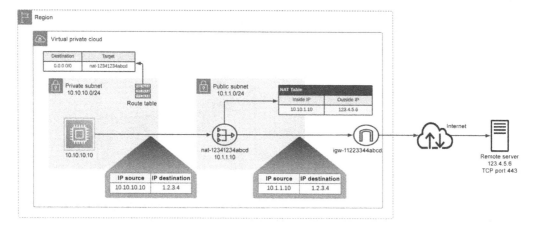

Figure A4.28: Public NAT gateway detail

In this figure, you can see an EC2 instance assigned a private IP address of 10.10.10.10 that uses a route table that sends internet-bound traffic to the NAT gateway. As the NAT gateway resides within a public subnet, this means the NAT gateway also has a public IP associated with its private IP on the IGW. Since all the traffic from the EC2 instance goes through the NAT gateway, which performs SNAT, the instance will have outbound internet connectivity.

Private NAT Gateways

NAT gateways deployed into private subnets function similarly to those in public subnets, with the key difference being that a private NAT gateway resides within a private subnet and does not have a public IP address assigned. This means all traffic sent to the NAT gateway is translated from the source IP address to the private IP address of the NAT gateway. Traffic processed by a private NAT gateway must then be routed to either AWS Transit Gateway or a **virtual private gateway** (**VGW**) to communicate outside the VPC.

While this is a less common deployment, a significant use case for private NAT gateways is when dealing with **overlapping IP address spaces** between your VPC and an on-premises network. In such scenarios, the IP addresses within the VPC might conflict with those in the on-premises network, causing routing and connectivity issues. By using a private NAT gateway, you can perform NAT on the source IP addresses of instances within your VPC, mapping them to a different, non-overlapping IP range. This allows seamless communication between the VPC and the on-premises environment without the need to reconfigure existing IP addresses in either network.

For example, consider a situation where the Trailcats VPC uses the `10.10.0.0/24` IP address range, and the on-premises network also uses the `10.10.0.0/24` range.

Figure A4.29 illustrates this setup, showing how a private NAT gateway enables communication between the Trailcats VPC and the on-premises network by resolving IP address overlap.

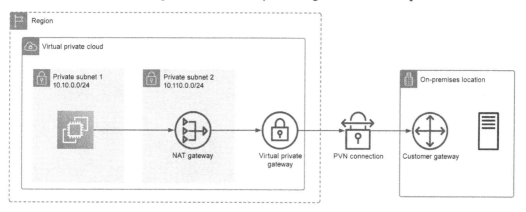

Figure A4.29: Private NAT gateway details

Direct communication between these networks would result in IP conflicts and routing failures. By deploying a private NAT gateway within a private subnet of the VPC, you can translate the source IP addresses of outbound traffic to a different IP range, such as `10.110.0.0/24`, which does not overlap with the on-premises network. This translated traffic can then be routed through a VGW over a VPN connection to the on-premises network without conflict.

Exercise: How to Configure NAT Gateways

You can create a NAT gateway within a VPC via the AWS Console or AWS CLI.

AWS Console

1. Navigate to the `NAT gateways` section of the VPC dashboard.
2. Click on `Create NAT gateway`, as shown in *Figure A4.30*:

Figure A4.30: Create NAT gateway

3. On the `Create NAT gateway` page, specify details about the NAT gateway type, subnet information, and IP configuration, as shown in *Figure A4.31*:

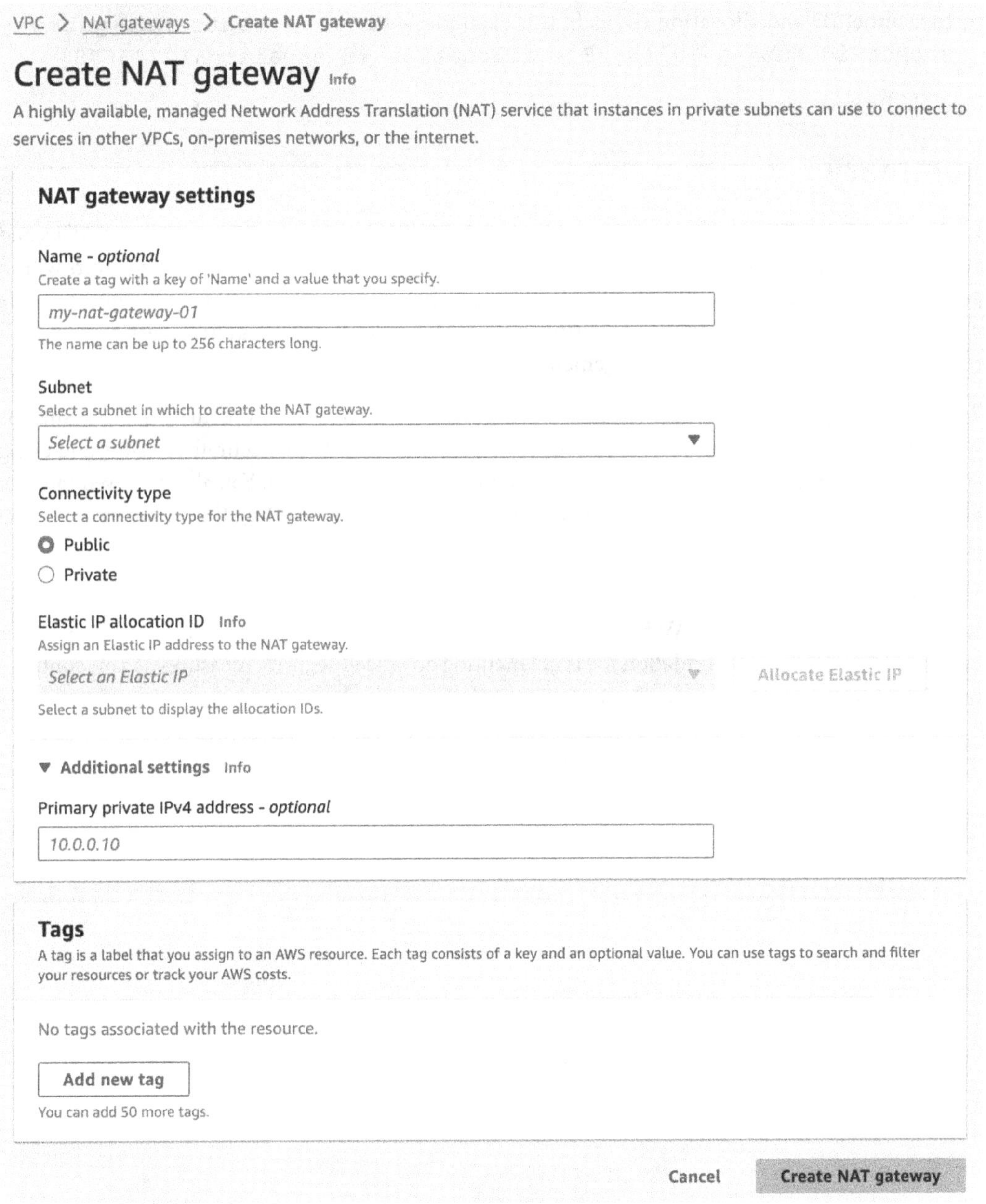

VPC > NAT gateways > Create NAT gateway

Create NAT gateway Info

A highly available, managed Network Address Translation (NAT) service that instances in private subnets can use to connect to services in other VPCs, on-premises networks, or the internet.

NAT gateway settings

Name - *optional*
Create a tag with a key of 'Name' and a value that you specify.

> my-nat-gateway-01

The name can be up to 256 characters long.

Subnet
Select a subnet in which to create the NAT gateway.

> Select a subnet   ▼

Connectivity type
Select a connectivity type for the NAT gateway.

⦿ Public
◯ Private

Elastic IP allocation ID   Info
Assign an Elastic IP address to the NAT gateway.

> Select an Elastic IP   ▼   Allocate Elastic IP

Select a subnet to display the allocation IDs.

▼ **Additional settings**   Info

Primary private IPv4 address - *optional*

> 10.0.0.10

Tags

A tag is a label that you assign to an AWS resource. Each tag consists of a key and an optional value. You can use tags to search and filter your resources or track your AWS costs.

No tags associated with the resource.

> Add new tag

You can add 50 more tags.

Cancel   **Create NAT gateway**

Figure A4.31: NAT gateway creation details

4. Click on `Create NAT gateway`.

AWS CLI

A NAT gateway can be created using the `create-nat-gateway` AWS CLI command followed by the subnet ID and allocation ID, as in this example: `aws ec2 create-nat-gateway --subnet-id subnet-12345678 --allocation-id eipalloc-12345678`

Your NAT gateway is now ready to use.

Summary

In this chapter, you explored the critical components of VPC security in AWS, including NACLs and security groups. Both are essential for managing and securing network access at different layers within your VPC. NACLs provide stateless, subnet-level security by filtering traffic based on sequential rule evaluation, while security groups, operating at the instance level, offer stateful access control with rules that allow dynamic traffic management.

This chapter highlighted best practices for configuring both NACLs and security groups, such as maintaining bidirectional rules for stateless NACLs, rule sequencing for accurate filtering, and using security group references to enable more secure application segmentation. You also reviewed internet gateways, NAT gateways, and egress-only internet gateways, understanding their roles in providing and securing external connectivity.

With this knowledge of VPC security fundamentals, you are well-prepared to design and manage secure cloud environments on AWS, where layered defenses protect both your resources and data. This understanding forms the foundation for implementing advanced security measures as you continue building on your expertise in cloud networking and security.

Index

W

www.packtpub.com

Subscribe to our online digital library for full access to over 7,000 books and videos, as well as industry-leading tools to help you plan your personal development and advance your career. For more information, please visit our website.

Why Subscribe?

- Spend less time learning and more time coding with practical eBooks and videos from over 4,000 industry professionals
- Improve your learning with Skill Plans built especially for you
- Get a free eBook or video every month
- Fully searchable for easy access to vital information
- Copy and paste, print, and bookmark content

At www.packtpub.com, you can also read a collection of free technical articles, sign up for a range of free newsletters, and receive exclusive discounts and offers on Packt books and eBooks.

Other Books You May Enjoy

If you enjoyed this book, you may be interested in these other books by Packt:

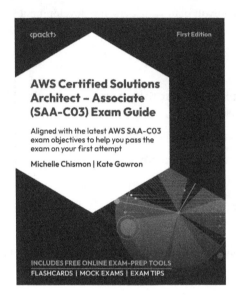

AWS Certified Solutions Architect - Associate (SAA-C03) Exam Guide

Michelle Chismon and Kate Gawron

ISBN: 978-1-83763-000-4

- Identify key AWS services for seamless deployments
- Apply robust security measures for secure AWS solutions
- Utilize efficient data transfer methods to streamline operations
- Compare AWS compute, storage, and database options for best fit
- Design and implement highly resilient architectures on AWS
- Leverage AWS security tools to protect data and applications
- Develop cost-optimized applications that scale dynamically to meet changing demands
- Interpret AWS exam questions strategically to maximize your chances of selecting the correct answers

AWS Certified Developer Associate Certification and Beyond

Rajesh Daswani and Dorian Richard

ISBN: 978-1-80181-929-9

- Host static website content using Amazon S3

- Explore accessibility, segmentation, and security with Amazon VPC

- Implement disaster recovery with EC2 and S3

- Provision and manage relational and non-relational databases on AWS

- Deploy your applications automatically with AWS Elastic Beanstalk

- Use AWS CodeBuild, AWS CodeDeploy, and AWS CodePipeline for DevOps

- Manage containers using Amazon EKS and ECS

- Build serverless applications with AWS Lambda and AWS Cloud9

Share Your Thoughts

Now you've finished *AWS Certified Advanced Networking – Specialty (ANS-C01) Certification Guide*, we'd love to hear your thoughts! Scan the QR code below to go straight to the Amazon review page for this book and share your feedback or leave a review on the site that you purchased it from.

https://packt.link/r/1835080839

Your review is important to us and the tech community and will help us make sure we're delivering excellent-quality content.

Download a Free PDF Copy of This Book

Thanks for purchasing this book!

Do you like to read on the go but are unable to carry your print books everywhere?

Is your eBook purchase not compatible with the device of your choice?

Don't worry – now, with every Packt book, you get a DRM-free PDF version of that book at no cost.

Read anywhere, any place, on any device. Search, copy, and paste code from your favorite technical books directly into your application.

The perks don't stop there: you can get exclusive access to discounts, newsletters, and great free content in your inbox daily.

Follow these simple steps to get the benefits:

1. Scan the QR code or visit the link below:

https://packt.link/free-ebook/9781835080832

2. Submit your proof of purchase.
3. That's it! We'll send your free PDF and other benefits to your inbox directly.